AF572751

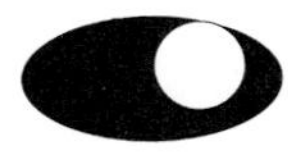

Current Topics in Membranes, Volume 46

Potassium Ion Channels

Molecular Structure, Function, and Diseases

Current Topics in Membranes, Volume 46

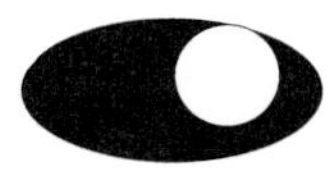

Current Topics in Membranes, Volume 46

Potassium Ion Channels
Molecular Structure, Function, and Diseases

Edited by

Yoshihisa Kurachi
Department of Pharmacology II
Faculty of Medicine
Osaka University
Osaka 565, Japan

Lily Yeh Jan
Howard Hughes Medical Institute
Department of Physiology and Biochemistry
University of California, San Francisco
San Francisco, California

Michel Lazdunski
Institut de Pharmacologie Moléculaire et Cellulaire
CNRS UPR 411
Sophia Antipolis
06560 Valbonne, France

ACADEMIC PRESS
San Diego London Boston New York Sydney Tokyo Toronto

This book is printed on acid-free paper. ∞

Academic Press
a division of Harcourt Brace & Company
525 B Street, Suite 1900, San Diego, California 92101-4495, USA
http://www.apnet.com

Academic Press
24-28 Oval Road, London NW1 7DX, UK
http://www.hbuk.co.uk/ap/

International Standard Book Number: 0-12-153346-8

PRINTED IN THE UNITED STATES OF AMERICA
99 00 01 02 03 04 QW 9 8 7 6 5 4 3 2 1

Contents

PART III G-Protein-Gated Potassium Channels

PART V Other Inwardly Rectifying Potassium Channels

CHAPTER 25 Glial Inwardly Rectifying Potassium Channels

Yoshiyuki Horio and Yoshihisa Kurachi

Contributors

Numbers in parentheses indicate the pages on which the authors' contributions begin.

Makoto Arita (417), Department of Physiology, Oita Medical University, Hasama, Oita 879-5593, Japan

Jacques Barhanin (67), Institut de Pharmacologie Moléculaire et Cellulaire, CNRS UPR 411, Sophia Antipolis, 06560 Valbonne, France

U. Brändle (223), Department of Otolaryngology, University of Tübingen, 72076 Tübingen, Germany

Alice Butler (9), Department of Anatomy and Neurobiology and the Department of Genetics, Washington University School of Medicine, St. Louis, Missouri 63110

David E. Clapham (295), Cardiovascular Division, Children's Hospital and Harvard Medical School, Boston, Massachusetts 02115

Shawn Corey (295), Neuroscience Program, Mayo Foundation, Rochester, Minnesota 55905

B. Fakler (223), Institute of Physiology, University of Tübingen, 72076 Tübingen, Germany

Tohru Gonoi (373), Research Center for Pathogenic Fungi and Microbial Toxicoses, Chiba University, Chiba 260, Japan

Liang Guo (177), Department of Neurophysiology, Tokyo Metropolitan Institute for Neuroscience, Tokyo 183, Japan

Hiroshi Hibino (243), Department of Pharmacology II, Faculty of Medicine, Osaka University, Osaka 565, Japan

Masatsugu Hori (435), The First Department of Medicine, Osaka University Medical School, Suita, Japan

Yoshiyuki Horio (243, 471), Department of Pharmacology II, Faculty of Medicine, Osaka University, Osaka 565, Japan

Yukio Hosoya (355), Department of Nursing, Yamagata School of Health Science, Yamagata, Yamagata 990-22, Japan; Department of

Cell Biology and Signaling, Yamagata University School of Medicine, Yamagata, Yamagata 990-23, Japan

Nobuya Inagaki[1] (373), Division of Molecular Medicine, Center for Biomedical Science, Chiba University School of Medicine, Chiba 260, Japan

Atsushi Inanobe (243), Department of Pharmacology II, Faculty of Medicine, Osaka University, Osaka 565, Japan

Kuniaki Ishii (47), Department of Pharmacology, Yamagata University School of Medicine, Yamagata, Japan

Lily Yeh Jan (1, 143, 321), Howard Hughes Medical Institute, Department of Physiology and Biochemistry, University of California, San Francisco, San Francisco, California 94143

Andreas Karschin (273), Molecular Neurobiology of Signal Transduction, Max-Planck-Institute of Biophysical Chemistry, 37070 Göttingen, Germany

Christine Karschin (273), Molecular Neurobiology of Signal Transduction, Max-Planck-Institute of Biophysical Chemistry, 37070 Göttingen, Germany

Matthew Kennedy (295), Cardiovascular Division, Children's Hospital and Harvard Medical School, Boston, Massachusetts 02115

Masafumi Kitakaze (435), The First Department of Medicine, Osaka University Medical School, Suita, Japan

Kazuhisa Kodama (435), The Cardiovascular Division, Osaka Police Hospital, Osaka, Japan

Chikako Kondo (387), Department of Pharmacology II, Faculty of Medicine, Osaka University, Osaka 565, Japan

Yoshihiro Kubo (177), Department of Neurophysiology, Tokyo Metropolitan Institute for Neuroscience, Tokyo 183, Japan

Maya T. Kunkel (9), Department of Anatomy and Neurobiology and the Department of Genetics, Washington University School of Medicine, St. Louis, Missouri 63110

Yoshihisa Kurachi (243, 355, 387, 471), Department of Cell Biology and Signaling, Yamagata University School of Medicine, Yamagata,

[1] Present address: Department of Physiology, Akita University School of Medicine, Akita 010, Japan.

Yamagata 990-23, Japan; Department of Pharmacology II, Faculty of Medicine, Osaka University, Osaka 565, Japan

Tsunehiko Kuzuya (435), The First Department of Medicine, Osaka University Medical School, Suita, Japan

Michel Lazdunski (67, 199), Institut de Pharmacologie Moléculaire et Cellulaire, CNRS UPR 411, Sophia Antipolis, 06560 Valbonne, France

Florian Lesage (199), Institut de Pharmacologie Moléculaire et Cellulaire, CNRS UPR 411, Sophia Antipolis, 06560 Valbonne, France

Diomedes E. Logothetis (337), Department of Physiology and Biophysics, Mount Sinai School of Medicine, City University of New York, New York, New York 10029-6574

A. N. Lopatin (159), Department of Cell Biology and Physiology, Washington University School of Medicine, St. Louis, Missouri 63110

Eduardo Marbán (449), Section of Molecular and Cellular Cardiology, Department of Medicine, The Johns Hopkins University, Baltimore, Maryland 21205

Pratap Meera (117), Department of Anesthesiology, University of California, Los Angeles, Los Angeles, California 90095

Takashi Miki (373), Division of Molecular Medicine, Center for Biomedical Science, Chiba University School of Medicine, Chiba 260, Japan

Kazuaki Nagashima (373), Division of Molecular Medicine, Center for Biomedical Science, Chiba University School of Medicine, Chiba 260, Japan

Yusuke Nakamura (103), Laboratory of Molecular Medicine, Human Genome Center, Institute of Medical Science, University of Tokyo, Tokyo 108, Japan

Betsy Navarro (295), Neuroscience Program, Mayo Foundation, Rochester, Minnesota 55905

C. G. Nichols (159), Department of Cell Biology and Physiology, Washington University School of Medicine, St. Louis, Missouri 63110

Michael Nonet (9), Department of Anatomy and Neurobiology and the Department of Genetics, Washington University School of Medicine, St. Louis, Missouri 63110

Brian O'Rourke (449), Section of Molecular and Cellular Cardiology, Department of Medicine, The Johns Hopkins University, Baltimore, Maryland 21205

Diane M. Papazian (29), Department of Physiology and Molecular Biology Institute, University of California, Los Angeles, School of Medicine, Los Angeles, California 90095

Jérôme Petit-Jacques (337), Department of Physiology and Biophysics, Mount Sinai School of Medicine, City University of New York, New York, New York 10029

Vez P. Repunte (387), Department of Pharmacology II, Faculty of Medicine, Osaka University, Osaka 565, Japan

Eitan Reuveny (321), Department of Biological Chemistry, Weizmann Institute of Science, Rehovot 76100, Israel

Dmitry N. Romashko (449), Section of Molecular and Cellular Cardiology, Department of Medicine, The Johns Hopkins University, Baltimore, Maryland 21205

Georges Romey (67), Institut de Pharmacologie Moléculaire et Cellulaire, CNRS UPR 411, Sophia Antipolis, 06560 Valbonne, France

J. P. Ruppersberg (223), Institute of Physiology and Department of Otolaryngology, University of Tübingen, 72076 Tübingen, Germany

Yasuhiko Sakata (435), The First Department of Medicine, Osaka University Medical School, Suita, Japan

Lawrence Salkoff (9), Department of Anatomy and Neurobiology and the Department of Genetics, Washington University School of Medicine, St. Louis, Missouri 63110

Michael C. Sanguinetti (85), Department of Medicine, Division of Cardiology, University of Utah, Salt Lake City, Utah 84112

Eisaku Satoh (387), Department of Pharmacology II, Faculty of Medicine, Osaka University, Osaka 565, Japan

U. Schulte (223), Institute of Physiology, University of Tübingen, 72076 Tübingen, Germany

J. Schultz (223), Institute of Physiology, University of Tübingen, 72076 Tübingen, Germany

Susumu Seino (373), Division of Molecular Medicine, Center for Biomedical Science, Chiba University School of Medicine, Chiba 260, Japan

Sakuji Shigematsu (417), Department of Physiology, Oita Medical University, Hasama, Oita 879-5593, Japan

Jin Liang Sui[2] (337), Department of Physiology and Biophysics, Mount Sinai School of Medicine, City University of New York, New York, New York 10029

Toshihiro Tanaka (103), Laboratory of Molecular Medicine, Human Genome Center, Institute of Medical Science, University of Tokyo, Tokyo 108, Japan

Andre Terzic (243), Department of Pharmacology and Medicine, Mayo Clinic, Rochester, Minnesota 55905

Andrew Tinker (143), Centre for Clinical Pharmacology and Toxicology, Department of Medicine/Cruciform Project, Rayne Institute, University College, London WC1E 6JJ, United Kingdom

Ligia Toro (117), Departments of Anesthesiology, Molecular and Medical Pharmacology, and the Brain Research Institute, University of California, Los Angeles, Los Angeles, California 90095

Martin Tristani-Firouzi (85), Department of Pediatrics, University of Utah, Salt Lake City, Utah 84112

Martin Wallner (117), Department of Anesthesiology, University of California, Los Angeles, Los Angeles, California 90095

Zhao-Wen Wang (9), Department of Anatomy and Neurobiology and the Department of Genetics, Washington University School of Medicine, St. Louis, Missouri 63110

Aguan Wei (9), Department of Anatomy and Neurobiology and the Department of Genetics, Washington University School of Medicine, St. Louis, Missouri 63110

Mitsuhiko Yamada (387), Department of Pharmacology II, Faculty of Medicine, Osaka University, Osaka 565, Japan

Alex Yuan (9), Department of Anatomy and Neurobiology and the Department of Genetics, Washington University School of Medicine, St. Louis, Missouri 63110

[2] Present address: Cambridge Neuroscience, Cambridge, Massachusetts 02139.

Preface

Potassium (K^+) ion channels selectively allow K^+ ions to pass through the cell membrane. These channels act to counter the activities of Na^+ and Ca^{2+} channels in controlling cell excitability. Moreover, electrophysiological studies in various tissues have indicated that K^+ channels are essential for various cell functions, including vagal deceleration of the heart beat, glucose-induced insulin secretion, epithelial transport of electrolytes, and suppression of neuronal excitation. Recent progress in the molecular characterization of K^+ ion channels has clearly shown the fundamental importance of this ion channel superfamily in physiology, pharmacology, and diseases.

The field is now rapidly expanding and the number of researchers steadily increasing. It is thus an opportune time to produce an overview of this field. The authors of this volume are active leading researchers in K^+ ion channel research, and many of them participated in the International Symposium on Potassium Ion Channels: Their Molecular Structure, Function, and Diseases, which was held in September 1997 at Yamagata, Japan. This meeting was quite exciting and fruitful, and the editors wish to share the excitement of this meeting with as many people as possible through this volume.

Potassium Ion Channels: Molecular Structure, Function, and Diseases will be useful to electrophysiologists, pharmacologists, molecular biologists, and clinical researchers. It is targeted at graduate and medical students and designed for use as a textbook for graduate and medical courses.

YOSHIHISA KURACHI
LILY YEH JAN
MICHEL LAZDUNSKI

Previous Volumes in Series

Current Topics in Membranes and Transport

Volume 22 The Squid Axon (1984)
Edited by Peter F. Baker

Volume 23 Genes and Membranes: Transport Proteins and Receptors* (1985)
Edited by Edward A. Adelberg and Carolyn W. Slayman

Volume 24 Membrane Protein Biosynthesis and Turnover (1985)
Edited by Philip A. Knauf and John S. Cook

Volume 25 Regulation of Calcium Transport across Muscle Membranes (1985)
Edited by Adil E. Shamoo

Volume 26 Na^+–H^+ Exchange, Intracellular pH, and Cell Function* (1986)
Edited by Peter S. Aronson and Walter F. Boron

Volume 27 The Role of Membranes in Cell Growth and Differentiation (1986)
Edited by Lazaro J. Mandel and Dale J. Benos

Volume 28 Potassium Transport: Physiology and Pathophysiology* (1987)
Edited by Gerhard Giebisch

Volume 29 Membrane Structure and Function (1987)
Edited by Richard D. Klausner, Christoph Kempf, and Jos van Renswoude

Volume 30 Cell Volume Control: Fundamental and Comparative Aspects in Animal Cells (1987)
Edited by R. Gilles, Arnost Kleinzeller, and L. Bolis

Volume 31 Molecular Neurobiology: Endocrine Approaches (1987)
Edited by Jerome F. Strauss III and Donald W. Pfaff

** Part of the series from the Yale Department of Cellular and Molecular Physiology*

Current Topics in Membranes

CHAPTER 1

Studies of Voltage-Dependent and Inwardly Rectifying Potassium Channels

Lily Yeh Jan

Howard Hughes Medical Institute, Department of Physiology and Biochemistry,
University of California, San Francisco, San Francisco, California 94143

I. FAMILIES OF POTASSIUM CHANNELS

Potassium channels of related structure have been found in prokaryotes and eukaryotes (for review, see Jan and Jan, 1997). They are grouped into several families, including one for voltage-gated potassium channels, one for inwardly rectifying potassium channels, and ones for those channels whose α subunits appear to be tandem fusions of α subunits of a more basic design (Salkoff *et al.*, Chapter 2).

Voltage-gated potassium channels belong to a superfamily of channels with four subunits or pseudosubunits, each containing six transmembrane segments (S1–S6) and a P or H5 segment in between S5 and S6 (Noda *et al.*, 1984; Tempel *et al.*, 1987). Inwardly rectifying potassium channels belong to a distantly related superfamily of channels with four subunits or pseudo-subunits, each containing two transmembrane segments (M1 and M2) and a P or H5 segment in between (Ho *et al.*, 1993; Kubo *et al.*, 1993). It thus appears that the latter constitutes the basic pore design whereas additional transmembrane segments could be included, for example, to introduce intrinsic voltage sensors such as the basic residues of the S4 segment into

Current Topics in Membranes, Volume 46

1063-5823/99 $30.00

the membrane-spanning domain (Jan and Jan, 1994). The recent crystallographic study of a prokaryotic potassium channel has revealed the basic pore design: One-third of the pore near the extracellular surface is lined by a loop formed by the signature sequence and by the carboxyl end of the pore helix formed by the remainder of the P or H5 segment that precedes the signature sequence, whereas the rest of the pore is lined by the last transmembrane segment (M2 or S6) (Doyle *et al.*, 1998).

II. VOLTAGE-DEPENDENT POTASSIUM CHANNELS

One distinguishing feature of voltage-gated potassium channels is the presence of intrinsic voltage sensors that can detect changes of voltage difference across the membrane and control channel opening (Papazian, Chapter 3). Characteristic of most potassium channels, the voltage-gated potassium channel has a long pore that accommodates multiple permeant ions (Ishii, Chapter 4). In addition to the α subunits that line the channel pore, various β subunits have been found to be associated with different voltage-gated potassium channels (Barhanin *et al.*, Chapter 5). Mutations of some of these α or β subunits are known to cause episodic cardiac arrhythmia (Sanguinetti and Tristani-Firouzi, Chapter 6; Tanaka and Nakamura, Chapter 7) or idiopathic epilepsy (Singh *et al.*, 1998; Charlier *et al.*, 1998).

The family of voltage-dependent potassium channels also includes calcium-activated potassium channels of large, intermediate, and small conductances. Whereas the calcium-activated potassium channels of large conductance are gated by both voltage and calcium and contribute to action potential repolarization, those of smaller conductances are primarily gated by calcium and are involved in generating the afterhyperpolarization after an action potential (Wallner, *et al.*, Chapter 8).

III. INWARDLY RECTIFYING POTASSIUM CHANNELS

Like voltage-gated potassium channels, inwardly rectifying potassium channels are tetramers, but the regions that determine the compatibility of subunit interactions are different (Tinker and Jan, Chapter 9). The mechanism for inward rectification is accounted for primarily by a block of the channel pore by cytoplasmic cations (Nichols and Lopatin, Chapter 10). These channels exhibit strong interactions with the permeant ions and are well characterized for their ability to accommodate multiple permeant ions in the pore (Kubo and Guo, Chapter 11). Interestingly, a large number of channels with subunits that correspond to a tandem repeat of the inwardly rectifying potassium channel α subunit have been found and may contribute to the leak potassium channels (Lesage and Lazdunski, Chapter 12).

Inwardly rectifying potassium channels are widely distributed in the mammalian brain and other tissues (Karschin and Karschin, Chapter 15; Horio and Kurachi, Chapter 25). Different members of this channel family are regulated by a variety of intracellular signals such as hydrogen ions, redox reactions, kinases, and ATP hydrolysis (Ruppersberg *et al.*, Chapter 13). Association with cytoskeletal proteins is important for channel function and localization (Hibino *et al.*, Chapter 14).

One subfamily of inwardly rectifying potassium channels, *GIRK1-4* (Kir3.1-3.4), can be activated directly by the G-protein $\beta\gamma$ subunits (Reuveny and Jan, Chapter 17; Logothetis, Chapter 18; Hosoya and Kurachi, Chapter 19). A point mutation in the *GIRK2* gene has been found to have gain-of-function effects that lead to neurodegeneration in the *weaver* mutant mice (Navarro *et al.*, Chapter 16).

Another subfamily of inwardly rectifying potassium channels (Kir6.1 and Kir6.2) gives rise to ATP-sensitive potassium channels that also include four β subunits, the sulfonylurea receptor in the ATP-binding cassette superfamily (Miki *et al.*, Chapter 20). Whereas in the pancreatic β cells these channels respond to changes in blood sugar level and control insulin release (Ashcroft, 1988), in the heart the ATP-sensitive potassium channels are sensitive to the cellular metabolic state (O'Rourke *et al.*, Chapter 24) and mediate cardioprotection during ischemia and cardiac arrhythmia (Arita and Shigematsu, Chapter 22; Kitakaze *et al.*, Chapter 23).

IV. CONCLUSION

Potassium channels are present in most cell types and serve a variety of cellular functions. Studies of potassium channel function and regulation are gaining momentum and have benefited from the pooling of resources ranging from biophysics to genomics. More than 40 human genes encode different potassium channel subunits, and mutations of a number of these genes have been found to cause diseases of the heart, kidney, pancreas, and central nervous system (Browne *et al.*, 1994; Curran *et al.*, 1995; Thomas *et al.*, 1995; Simon *et al.*, 1996; Thomas *et al.*, 1996; Wang *et al.*, 1996; Nestorowicz *et al.*, 1997; Splawski *et al.*, 1997; Charlier *et al.*, 1998; Singh *et al.*, 1998). Future studies are likely to further our understanding of how these potassium channels are organized and functionally integrated in a cell, and how these channels may be modulated in the treatment of diseases.

References

Ashcroft, F. M. (1988). Adenosine-5′-tri-phosphate sensitive K^+ channels. *Annu. Rev. Neurosci.* **11,** 97–118.

Browne, D. L., Gancher, S. T., Nutt, J. G., Brunt, E. R. P., Smith, E. A., Kramer, P., and Litt, M. (1994). Episodic ataxia/myokymia syndrome is associated with point mutations in the human potassium channel gene, *KCNA1. Nature Genet.* **8,** 136–140.

Charlier, C., Singh, N. A., Ryan, S. G., Lewis, T. B., Reus, B. E., Leach R. J., and Leppert, M. (1998). A pore mutation in a novel KQT-like potassium channel gene in an idiopathic epilepsy family. *Nature Genet.* **18,** 53–55.

Curran, M. E., Splawski, I., Timoth, K. W., Vincent, G. M., Green, E. D., and Keating, M. T. (1995). A molecular basis for cardiac arrhythmia: HERG mutations cause long QT syndrome. *Cell* **80,** 795–803.

Doyle, D. A., Cabral, J. M., Pfuetzner, R. A., Kuo, A., Gulbis, J. M., Cohen, S. L., Chait, B. T., and MacKinnon, R. (1998). The structure of the potassium channel: Molecular basis of K^+ conduction and selectivity. *Science* **280,** 69–77.

Ho, K., Nichols, C. G., Lederer, W. J., Lytton, J., Vassilev, P. M., Kanazirska, M. V., and Hebert, S. C. (1993). Cloning and expression of an inwardly rectifying ATP-regulated potassium channel. *Nature* **362,** 31–38.

Jan, L. Y., and Jan, Y. N. (1994). Potassium channels and their evolving gates. *Nature* **371,** 119–122.

Jan, L. Y., and Jan, Y. N. (1997). Cloned potassium channels from eukaryotes and prokaryotes. *Annu. Rev. Neurosci.* **20,** 91–123.

Kubo, Y., Baldwin, T. J., Jan, Y. N., and Jan, L. Y. (1993). Primary structure and functional expression of a mouse inward rectifier potassium channel. *Nature* **362,** 127–133.

Nestorowicz, A., Inagaki, N., Gonoi, T., Schoor, K. P., Wilson, B. A., Glaser, B., Landau, H., Stanley C. A., Thorton, P. S., Seino, S., and Permutt, M. A. (1997). A nonsense mutation in the inward rectifier potassium channel gene, *Kir6.2,* is associated with familial hyperinsulinism. *Diabetes* **46,** 1743–1748.

Noda, M., Shimizu, S., Tanabe, T., Takai, T., Kayano, T., Ikeda, T., Takahashi, H., Nakayama, H., Kanaoka, Y., Minamino, N., Kangawa, K., Matsuo, H., Raftery, T., and Numa, S. (1984). Primary structure of *Electrophorus electricus* sodium channel deduced from cDNA sequence. *Nature* **312,** 188–192.

Simon, D. B., Karet, F. E., Rodriguez-Soriano, J., Hamdan, J. H., DiPietro, A., Trachtman, H., Sanjad, S. A., and Lifton, R. (1996). Genetic heterogeneity of Bartter's syndrome revealed by mutations in the K^+ channel, ROMK. *Nature Genet.* **14,** 152–156.

Singh, N. A., Charlier, C., Stauffer, D., DuPont, B. R., Leach, R. J., Melis, R., Ronen, G. M., Bjerre, I., Quattlebaum, T., Murphy, J. V., McHarg, M. L., Gagnon, D., Rosales, T. O., Peiffer, A., Anderson, V. E., and Leppert, M. (1998). A novel potassium channel gene, *KCNQ2,* is mutated in an inherited epilepsy of newborns. *Nature Genet.* **18,** 25–29.

Splawski, I., Tristani-Firouzi, M., Lehmann, M. H., Sanguinetti, M. C., and Keating, M. T. (1997). Mutations in the *hmink* gene cause long QT syndrome and suppress I_{Ks} function. *Nature Genet.* **17,** 338–340.

Tempel, B. L., Papazian, D. M., Schwarz, T. L., Jan, Y. N., and Jan, L. Y. (1987). Sequence of a probable potassium channel component encoded at *Shaker* locus of *Drosophila. Science* **237,** 770–775.

Thomas, P. M., Cote, G. G., Wohllk, N., Haddad, B., Mathew, P. M., Rabl, W., Aguilar-Bryan, L., Gagel, R. F., and Bryan J. (1995). Mutations in the sulfonylurea receptor gene in familial persistent hyperinsulinemic hypoglycemia of infancy. *Science* **268,** 426–429.

Thomas, P., Ye, Y., and Lightner, E. (1996). Mutation of the pancreatic islet inward rectifier Kir6.2 also leads to familial persistent hyperinsulinemic hypoglycemia of infancy. *Hum. Mol. Genet.* **5,** 1809–1812.

Wang, Q., Curran, M. E., Splawski, I., Burn, T. C., Millholland, J. M., VanRaay, T. J., Shen, J., Timonthy, K. W., Vincent, G. M., de Jager, T., Schwartz, P. J., Towbin, J. A., Moss, A. J., Atkinson, D. L., Landes, G. M., Connors, T. D., and Keating, M. T. (1996). Positional cloning of a novel potassium channel gene: KVLQT1 mutations cause cardiac arrhythmias. *Nature Genet.* **12,** 17–23.

PART I

Voltage-Dependent Potassium Channels

CHAPTER 2

The Impact of the *Caenorhabditis elegans* Genome Project on Potassium Channel Biology

Lawrence Salkoff, Maya T. Kunkel, Zhao-Wen Wang, Alice Butler, Alex Yuan, Michael Nonet, and Aguan Wei
Department of Anatomy and Neurobiology and the Department of Genetics, Washington University School of Medicine, St. Louis, Missouri 63110

I. INTRODUCTION

The DNA sequence data produced by the *Caenorhabditis elegans* genome sequencing project has revealed a large extended gene family of potassium channels, the outlines of which are conserved with vertebrates. This initial peek into the nearly complete library of potassium channels from a single "simple" organism has revealed a number and complexity of potassium channel types that surpassed expectations. The vast majority of potassium channel types revealed by expressed sequence tag (EST) sequencing projects from many vertebrate and invertebrate animals fit within the outlines

1063-5823/99 $30.00

of families present in *C. elegans*. Thus, the broad picture of potassium channel families in *C. elegans* may represent family relationships conserved among most vertebrate and invertebrate animals. This conservation implies that the electrical lives of cells from most metazoans have similar requirements and are similarly diverse. Furthermore, the fact that the structures and possible biophysical properties of many distinct potassium channel types are conserved implies that specific functional roles of potassium channels may be conserved, as well. Studies now underway to reveal the tissue types and cellular distribution patterns of the complete library of potassium channels in *C. elegans* could reveal such conserved roles. These studies may be a guide to similar studies in higher animals.

The analysis of potassium channels in *C. elegans* from a "functional genomics" perspective has many unique aspects: *C. elegans* has a genome size of approximately one hundred million base pairs (about 2.5% of the human genome). Currently, greater than 90% of the genome has been sequenced and the project is expected to be completed early in 1999. Data from the *C. elegans* genome sequencing project suggest a total complement of approximately 15,000 to 18,000 genes versus 80,000 to 100,000 genes in the human genome (Wilson *et al.*, 1994). (Thus, the genome of *C. elegans* is more compact than the human genome; the density of genes is higher and the genes themselves contain smaller introns.) The availability of the entire DNA sequence of the *C. elegans* genome provides the first glimpse of the entire set of K^+ channels in one animal. This nearly complete picture of potassium channel diversity is revealed in an unbiased way because no particular cell, tissue type, or abundance class is favored.

A starting point for identifying the primary structures of potassium channels in genomic DNA is the analysis by Genefinder (Wilson *et al.*, 1994) of sequence data produced by the Genome Sequencing Center (GSC) at Washington University and the Sanger Center in England. This program produces a predicted gene structure showing putative intron–exon boundaries and start and stop sites. We have been systematically determining the actual structures by reverse transcriptase polymerase chain reaction (RTPCR) using primers matching the hypothetical start and stop sites. Many genes, however, appear to have complicated structures that require a second level of analysis. This is particularly true of genes that show evidence of alternative splicing. Of particular value in analyzing genomic DNA sequences for alternative exons is the fact that the GSC has sequenced portions of the genome of *C. briggsae,* a species closely related to *C. elegans.* Comparing the genomic sequence of the two organisms has allowed us to identify alternative exons by "reading" the type of DNA by the pattern of conservation: In such a comparative analysis conserved alternative exons become obvious because their open reading frames invariably show third base pair "wobble" (introns are marked by donor and acceptor splice junctions and usually have poor length conservation; regulatory regions appear to be short segments of near identity). RTPCR is then used to verify that predicted alternative forms are transcribed.

In addition to the primary sequence of a gene product and all alternative forms, a complete genomic sequence allows addressing a host of questions that were previously inaccessible. These questions include the genetic determinants of tissue specificity such as enhancers, promoters, and sequences that determine splice-site selection, and other transcribed but nontranslated regions that may control message transport, localization, or lifetime. Eventually, the compilation of all of this information may be used to create a comprehensive picture of potassium channel involvement in all aspects of physiology in both excitable and nonexcitable cells, and at all stages of development. These functional genomics studies might be a model for dealing with the vast data from the human genome project with regard to the usefulness of comparing genome sequence data from related species, the verification of gene structure, and the determination of sequences important in gene regulation.

II. *C. elegans* AS AN ANIMAL MODEL TO STUDY THE GENETIC DETERMINANTS OF MEMBRANE EXCITABILITY

At first glance *C. elegans* may seem to be an unlikely animal model for the study of conserved genes that determine membrane excitability. The evolutionary origin of nematodes dates from the Precambrian era when Protostomes (most invertebrates) separated from Deuterostomes (which includes vertebrates). Both arose from a common ancestral member of Bilateria. Did that common ancestor already posses the extensive set of genes that determine the complex electrical behavior of present day cells, and does the full set remain largely conserved between invertebrates and vertebrates? The answer to both questions is likely to be yes because studies have shown that the evolutionary origins of modern potassium channels extend back to the earliest radiation of multicelled eukaryotes, which produced the Cniderian diploblastic animals (Jegla and Salkoff, 1994, 1995a). These animals, which include jellyfish and sea anemonies, are the most primitive animals to have an organized nervous system. The evolution of these primitive animals preceded the Precambrian era, yet potassium channel subunits cloned from those organisms can be clearly assigned to the recognizable subfamilies of potassium channels shown in Fig. 1. The tree of organized subfamilies shown in Fig. 1 is derived from the *C. elegans* genome project, but appears to be equally valid for classifying potassium

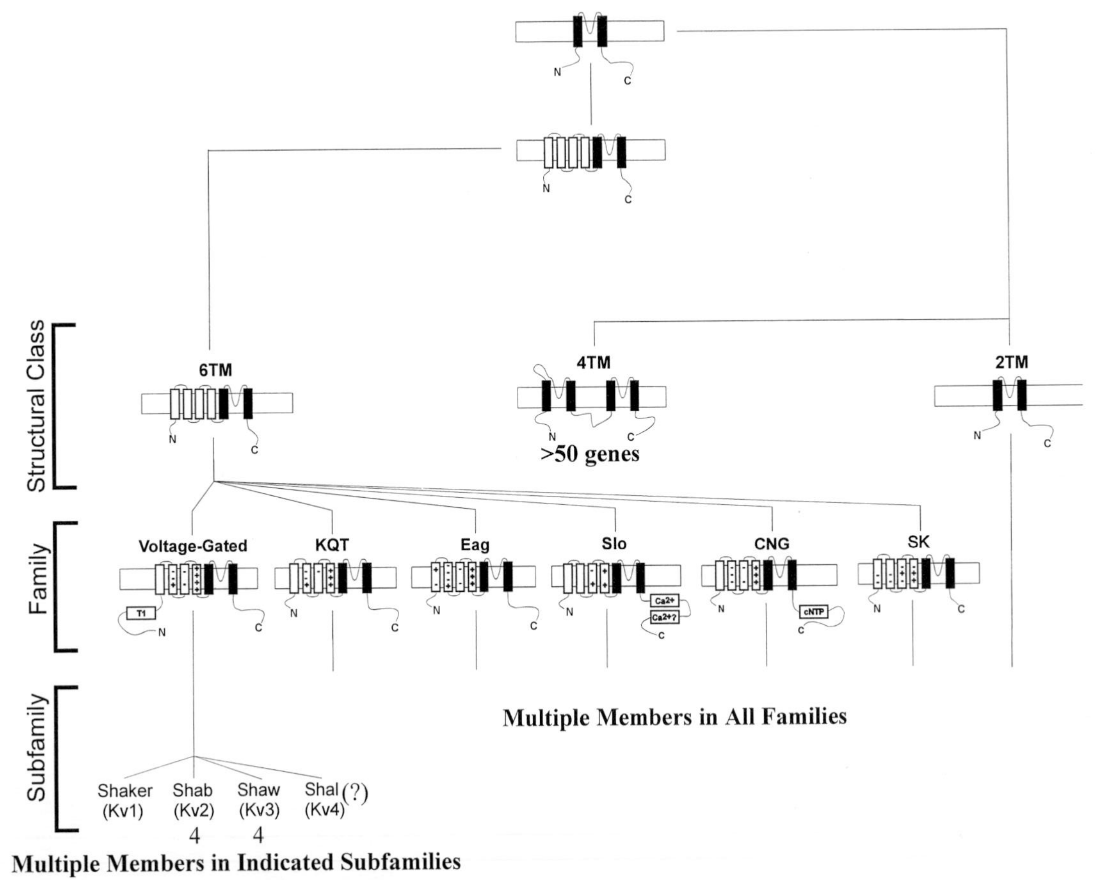

Structural Class
Family
Subfamily
6TM
4TM
2TM
>50 genes
Voltage-Gated
KQT
Eag
Slo
CNG
SK
T1
Ca2+
Ca2+?
cNTP
Shaker (Kv1)
Shab (Kv2) 4
Shaw (Kv3) 4
Shal (Kv4) (?)
Multiple Members in Indicated Subfamilies
Multiple Members in All Families

channels from mammals and *Drosophila,* as well as jellyfish (Jegla and Salkoff, 1994, 1995b). It may be concluded, then, that most genes encoding potassium channels evolved in their nearly present form in the earliest multicelled eukaryotes.

Although the organization of potassium channel subfamilies shown in Fig. 1 appears generally valid for metazoan life, it is apparently not applicable to lower life forms, including eukaryotic protists such as paramecia, yeast, or bacteria. All of these lower life forms have one or more genes encoding transmembrane proteins containing conserved sequence associated with potassium ion selectivity (Jegla and Salkoff, 1994, 1995c; Ketchum *et al.,* 1995; Milkman, 1994), but none of the encoded proteins can be clearly assigned to the scheme of metazoan channels as outlined in Fig. 1 (see also Wei *et al.,* 1996).

Why should such a "simple" animal as *C. elegans* have so many genes encoding potassium channels? *C. elegans* may be simple only in having a small number of cells dedicated to each physiological task. In a classic study that set the groundwork for many future studies, the nervous system of *C. elegans* was reconstructed in its entirety by serial sections of electron micrographs (White *et al.,* 1986). In hermaphrodites, the nervous system consists of 302 neurons and 56 glial and supporting cells. The nervous system of males consists of 381 neurons and 92 glial/supporting cells. Thus, the nervous system accounts for approximately 40% of the somatic cells. As a result of these systematic ultrastructural studies, most of the synaptic connections in the nervous system are known; there are about 5000 chemical synapses, 2000 neuromuscular synapses, and 600 gap junctions. Based on these studies showing the details of morphologies and connections, the 302 neurons of the hermaphrodite can be assigned to 118 classes (White *et al.,* 1986). This heterogeneity of classes in the *C. elegans* nervous system suggests an extremely complex organ. In contrast, based on similar criteria, the 10^{10} neurons of the mammalian cerebellum fall into only five classes.

FIGURE 1 The subunit structures of potassium channels can be assigned to one of three major structural classes (see also Wei *et al.,* 1996). Those possessing two transmembrane domains (2TM) bracketing a highly conserved pore region (P domain) are usually categorized as "inward rectifier" channels. Those possessing four transmembrane domains (4TM) contain two P domains within a single subunit and represent a novel class of channels. Those possessing six transmembrane domains (6TM) with one P domain include the well-known voltage-gated family. Within each structural class, there are conserved families of channels defined by common structural features such as the distinct pattern of charges in S1, S2, and S3. Families, such as voltage-gated channels, can be divided into multiple subfamilies [Shaker, Shab, Shaw, and Shal (Butler *et al.,* 1989)]. Within a single species there are also sometimes multiple members of each subfamily, such as Kv1.1 through Kv1.8, which represent eight mammalian homologs of the Shaker subfamily.

Thus, in *C. elegans,* single neurons, pairs, or quartets of neurons may be accomplishing tasks that whole brain regions undertake in vertebrate species. The difference, then, between *C. elegans* and many "higher" animals may be quantitative rather than qualitative. In *C. elegans* there simply may be fewer cells dedicated to each organ or organ subdivision. Indeed, the repertoire of electrical events in a *C. elegans* neuron may be at least as complex as that in a neuron of any higher animal, since a single cell may serve the functional role of a whole brain subdivision in a higher animal, hence the large number of genes encoding potassium channels in *C. elegans.*

III. POTASSIUM CHANNEL SCOPE AND DIVERSITY

With 80% of the *C. elegans* genome sequenced, about 80 genes that most likely encode potassium channel subunits have been identified. As shown in Fig. 1 this compilation of potassium channel subunits can be divided into three structural classes based on the number of membrane-spanning domains: two (2TM), four (4TM), or six (6TM). The 6TM channels can be further divided into six conserved subfamilies.

A. 4TM Channels

1. An Unusually Large Number of Genes Encode the Structural Class of 4TM Channels in *C. elegans*

The 4TM channels have four transmembrane domains and are unusual among potassium channel subunits in having two "P" regions. The "P" region is responsible in large part for selectivity to potassium ion. The existence of this novel and extensive group of potassium channels has only recently been revealed as a result of DNA sequencing projects (Salkoff and Jegla, 1995). Remarkably, approximately 50 genes that encode 4TM channel subunits have been identified in *C. elegans.* Many questions are raised by these findings with regard to channel structure, function, and tissue distribution in *C. elegans.* (It also has not been determined whether any of the identified loci represent pseudogenes.) The question of whether a similarly large family of 4TM channels exists in vertebrates is also intriguing. The deduced proteins of 4TM subunits are diverse and in general share a low degree of sequence identity. Only a few subunits can be assigned into groups based on overall sequence similarity. Because of this low sequence homology, the techniques of degenerate PCR may not present an efficient means of exploring the scope of this channel family in vertebrates. Thus, although a few ESTs from various EST sequencing projects have appeared in sequence databases, it is likely that the full scope of this family in vertebrates will not be known until the human genome sequencing project progresses.

2. Tissue Distribution of 4TM Channels

The pattern of tissue distribution of 4TM channels may be a key element in understanding their function. Fortunately green fluorescent protein (GFP)–promoter–reporter transformation experiments provide an efficient method for an initial determination of the cellular expression patterns of genes in *C. elegans. C. elegans* are transformed with GFP–promoter constructs via injection of the DNA into the cytoplasm of the hermaphrodite syncytial gonads. A plasmid encoding a selectable marker is also coinjected. Because transformed animals carry multiple copies of the vector in tandem extrachromosomal arrays, GFP–promoter–reporter experiments produce an excellent signal to noise ratio. GFP fluorescence is directly visualized in living animals (Chalfie *et al.,* 1994). Expression patterns for several channels are shown in Fig. 2 (see color plate). Experiments like these have revealed some major (although preliminary) insights.

1. Many 4TM channel genes appear to have their expression limited to small groups of neurons or cells within a single tissue type. In one instance a single interneuron is by far the site of highest expression (Fig. 2, n2Pore16).
2. Apparently many cell types [neurons, muscle, hypoderm (the "epidermis" that secretes the external cuticle)] express 4TM channels.
3. So far there is little or no overlap in expression patterns (genes that are expressed in muscle do not express in hypoderm, and vice versa). Genes that are expressed in small groups of neurons do not have overlapping expression.

Could there be a common thread for the roles of 4TM channels in these different cell types? It is assumed that 4TM channels lack a mechanism of intrinsic voltage dependence. Therefore their native functional roles and properties may depend on a host of factors intrinsic to different cell types, such as various ligands, second messengers, and mechanisms of covalent modification. Thus, it is difficult to extrapolate their native functional properties based on observations of their functional properties from expression in *Xenopus* oocytes. Nevertheless, a few reports describing the functional properties of 4TM channels in *Xenopus* oocytes give some hints to their behavior. Members of this structural class have been characterized in mammals (Lesage *et al.,* 1996, 1997; Duprat *et al.,* 1997; Fink *et al.,* 1996), *Drosophila* (Goldstein *et al.,* 1996), and *Arabidopsis* (Czempinski *et al.,* 1997). As predicted for a K^+ channel lacking a voltage sensing region, the reversal potential of currents from 4TM channels shifts with the change in the equilibrium potential of K^+. Despite their similar topology, these channels differ somewhat in their observed properties. For example, K^+ currents from one subunit are weakly inwardly rectifying (Lesage *et al.*, 1996, 1997), while currents from other members are outwardly rectifying (Goldstein *et al.*, 1996; Fink *et al.*, 1996; Duprat *et al.*, 1997; Czempinski *et al.*, 1997). Furthermore, the activity of one 4TM channel is dependent on intracellular Ca^{2+} (Czempinski *et al.*, 1997) while the activity of another member is sensitive to pH (Duprat *et al.*, 1997). The diverse properties of these many channels may be further increased with the potential formation of heterodimeric channels.

In some respects the high number of these genes encoding novel potassium channels in *C. elegans* is astonishing enough, but apparently this diversity is not sufficient. We have uncovered at least one example of alternative splicing for one gene where at least three alternative exons encoding initiator methionines are combinatorially assembled with several alternative forms at the carboxyl ends of the proteins. The expression of many of these alternative forms has now been verified by RTPCR (Wang *et al.*, 1998). Why are so many different forms of these channels necessary? Although it is premature to make any definitive statements, it appears that almost all cell types will have this voltage-independent class of channels. Those that are inwardly rectifying voltage-independent "leak" channels could set or contribute to the membrane resting potential, while weakly outwardly rectifying channels could tune the excitable properties of the cell. Conceivably such channels could play these roles in all cell types, but why so many different ones? Perhaps because the resting potentials and excitable properties of many diverse cell populations have to be "fine-tuned" and, in some instances, regulated by different mechanisms.

B. 2TM Channels

In contrast to the large number of genes encoding 4TM channels, only three genes in *C. elegans* are known to encode 2TM "inward rectifier" channels. The structural class of 2TM channels is divided into vertebrate subfamilies, identified as Kir1 through Kir6 (Doupnik *et al.*, 1995). These vertebrate subfamily distinctions are apparently not conserved in *C. elegans.*

C. Families of 6TM Channels

1. Voltage-Gated Potassium Channels

These channels are all activated by membrane depolarizations, but display a wide range of gating kinetics and voltage sensitivities. Comprising this family are four gene subfamilies conserved throughout metazoans. The prototypic subfamily members were first cloned in *Drosophila: Shaker, Shab, Shaw,* and *Shal* (Kamb *et al.,* 1987; Papazian *et al.,* 1987; Pongs *et al.,* 1988; Butler *et al.,* 1989; Wei *et al.,* 1990). Multiple members of each subfamily are present in mammals and are widely expressed in diverse excitable and nonexcitable tissues (Chandy and Gutman, 1995). For example, cells of the immune system express several types of voltage-dependent potassium channels (Sutro *et al.,* 1989).

The *C. elegans* genome contains at least one Shaker-type gene, four Shab-type genes, and three Shaw-type genes. Curiously, a Shal-type gene has yet to appear, despite that fact that this subfamily shows the highest sequence conservation among the voltage-gated channels. Shal-type homologs are present in mammals (Pak *et al.,* 1991; Baldwin *et al.,* 1991; Rudy *et al.,* 1991), *Drosophila* (Wei *et al.,* 1990), and significantly, a primitive diploblast (Jegla *et al.,* 1995; Jegla and Salkoff, 1997). In *Drosophila,* all four prototype subfamily genes are expressed in neurons (Baker and Salkoff, 1990; Tsunoda and Salkoff, 1995a,b) and, with the exception of *Shal,* also in muscles. In *C. elegans, Shaw1* is expressed in many neurons and body-wall muscle (Wei and Salkoff, unpublished), while *Shaw2* is expressed in vulva muscles and a more restricted number of neurons (Johnstone *et al.,* 1997). Gain-of-function mutations of the Shaw2 potassium channel are responsible for the defective egg-laying and defecation behaviors observed in *egl-36* mutants (Johnstone *et al.,* 1997).

2. KQT Potassium Channels

These channels are related to the human KvLQT1 channel implicated in one form of long QT syndrome, a heritable risk factor for cardiac arrhythmia (Wang *et al.,* 1996). One common functional hallmark of these channels appears to be unusually slow voltage-dependent activation kinetics, associated with a pronounced Cole-Moore shift (Sanguinetti *et al.,* 1996; Barhanin *et al.,* 1996; Romey *et al.,* 1997; Chouabe *et al.,* 1997; Wei and Salkoff, unpublished). KQT-type potassium channels are encoded by a conserved multigene family. In vertebrates, at least three KQT-type genes are present in public genome databases. *KvLQT1* is expressed in heart and a number of other organs including kidney, pancreas, prostate, and intestines (Wang *et al.,* 1996; Chouabe *et al.,* 1997), but is notably absent in brain. The presence of vertebrate KQT-type genes in neuronal EST databases suggests that other KQT-type channels may be neuronal (Wei *et al.,* 1996; Yokoyama *et al.,* 1996). As do vertebrates, *C. elegans* possesses at least three KQT-type genes (*kqt1-3*). *kqt1* is prominently expressed in the pharyngeal bulb while *kqt2* is expressed in intestinal cells (Fig. 2) (Wei and Salkoff, unpublished). Interestingly, the pharyngeal bulb of *C. elegans* shares many similarities with vertebrate cardiac tissue, including the generation of rhythmic myogenic contractions mediated by action potentials exhibiting a broad plateau phase (Raizen and Avery, 1994; Davis *et al.,* 1995). These results suggest that KQT-type channels may serve similar roles in the control of plateau potentials in analogous tissues across diverse species.

3. *eag*-Like Potassium Channels

The first member of this family was cloned from *Drosophila* (Warmke *et al.,* 1991) as the product of the *ether-a-go-go* (*eag*) locus. Numerous additional homologs cloned from both vertebrate and invertebrate species define a conserved multigene family consisting of at least three subfamilies (*eag, erg,* and *elk*) (Warmke *et al.,* 1991; Ludwig *et al.,* 1994; Titus *et al.,* 1997; Wang *et al.,* 1997). The human *herg* gene encodes a channel with unusual gating properties (Trudeau *et al.,* 1995; Sanguinetti *et al.,* 1995; Smith *et al.,* 1996), and, like *KvLQT1,* is also a long QT locus (Curran *et al.,* 1995). In *C. elegans,* at least two *eag*-like genes are present, one representative each of the *eag*-type and *erg*-type subfamilies.

4. *Slo*-Like Potassium Channels

The first member of this family was cloned from *Drosophila* as the product of the *Slopoke* (*Slo*) locus (Elkins *et al.,* 1986; Atkinson *et al.,* 1991). Functional expression studies of *Drosophila Slo* and its mammalian homolog, *mSlo,* revealed that genes for Slo1 channels encode large conductance "BK-type" potassium channels gated by both voltage and intracellular calcium (Adelman *et al.,* 1992; Butler *et al.,* 1993). Recently, additional members of this gene family have been chacterized (*Slo2* and *Slo3*) (Wei *et al.,* 1996; Schreiber *et al.,* 1998; Yuan, Dourado, and Salkoff, unpublished). In *C. elegans,* two *Slo*-type genes are present, one *Slo1*-like gene and one *Slo2*-like gene. The *Slo1* gene expresses large conductance calcium-activated potassium channels with properties similar to *Drosophila Slo* and mammalian *mSlo* (Jegla and Salkoff, unpublished). In contrast, the *C. elegans Slo2* gene expresses potassium channels that are gated by intracellular chloride in addition to calcium (Yuan, Dourado, and Salkoff, unpublished). Slo3 channels are expressed in mammalian sperm and, despite high sequence conservation with Slo1, are completely insensitive to calcium (however, gating is modulated by pH). Thus, the mechanism of gating common to the Slo family of channels is not calcium, but voltage; voltage-dependent gating can apparently be modified by a host of different factors.

5. Cyclic Nucleotide-Gated (CNG) Cation Channels

These channels are gated by intracellular cyclic nucleotides (Kaupp *et al.,* 1989) and are associated with primary sensory neurons of retinal, olfactory, and gustatory epithelium. However, expression is not exclusive to sensory neurons, since expression has also been observed in hippocampal neurons (Bradley *et al.*, 1997) and sperm (Weyand *et al.*, 1994). In *C. elegans*, at least three distinct genes encode CNG-type channels. Two genes are expressed in cephalic sensory neurons and are disrupted in the chemotaxic and thermotaxic defective mutants *tax-2* and *tax-4* (Coburn and Bargmann, 1996; Komatsu *et al.*, 1996).

6. SK-Like Calcium-Activated Potassium Channels

Members of this gene family encode both small conductance "SK-type" (Kohler *et al.*, 1996) and intermediate conductance "IK-type" (Ishii *et al.*, 1997) calcium-activated potassium channels. These channels are voltage insensitive, inwardly rectifying, and characteristically gated by intracellular calcium in the submicromolar range. SK channels are thought to mediate slow afterhyperpolarization potentials observed in many neurons (Yarom *et al.*, 1985; Hille, 1992). Cells of the immune system also express channels of this type (Grissmer *et al.*, 1992). *C. elegans* possesses at least three SK-type genes.

D. Mutant Analysis

The ability to isolate and characterize mutants in K^+ channel genes complements the characterization of potassium channels and aids in determining the roles of these channels. One excellent example is *egl-36*. Mutations in the *C. elegans egl-36* gene result in defective excitation of egg-laying and enteric muscles. Dominant gain-of-function alleles cause abnormal relaxation of muscles. The *egl-36* gene encodes a Shaw-type (Kv3) voltage-dependent potassium channel (which we have designated nShaw2). As shown in Fig. 3, in *Xenopus* oocytes, wild-type nShaw2 channels express noninactivating channels with slow activation kinetics. One gain-of-function mutation results in a single amino acid substitution in S6 and the other contains a substitution in the cytoplasmic amino-terminal domain. Both mutant alleles produce channels dramatically shifted in their midpoints of activation toward hyperpolarized voltages. An *egl-36::gfp* fusion is expressed in egg-laying muscles and a pair of enteric muscle motorneurons. The mutant *egl-36* phenotypes can thus be explained by expression in these cells of potassium channels that are inappropriately opened at hyperpolarized potentials, causing decreased excitability due to increased potassium conductance. Mutations like these offer insights about the *in vivo* physiological roles of individual types of potassium channels.

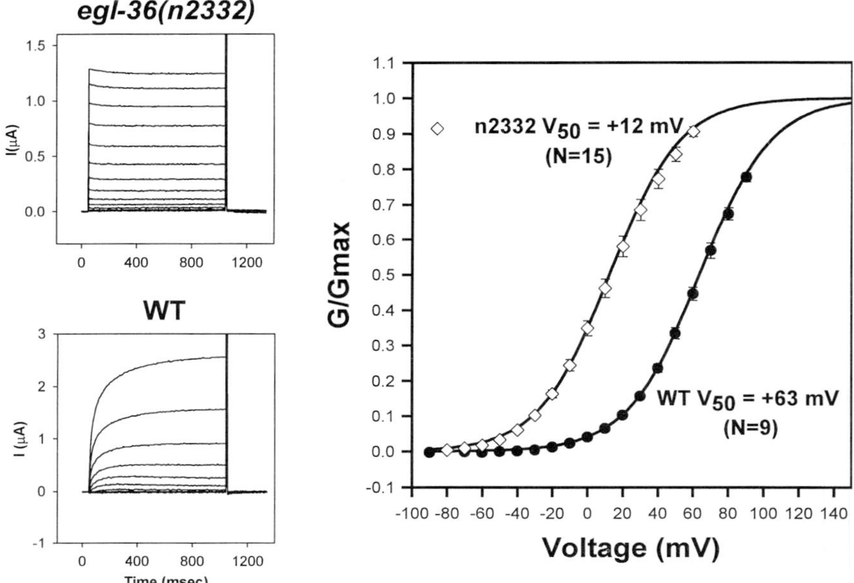

FIGURE 3 Electrophysiological properties of wild-type (WT) and mutant *nShaw2* potassium channels (*egl-36*) expressed in *Xenopus* oocytes. Two electrode voltage-clamp records are shown. Current records are in response to 1-sec voltage steps in 10-mV increments from a holding potential of −90 mV. Normalized GV relationships are shown on the right. The GV relationship for mutant channels is shifted to the left, producing channel openings at more negative voltages. Minor increases of potassium channels near rest can dramatically reduce the electrical responses of cells (see Fig. 5 and Johnstone *et al.*, 1997).

E. Genomic Organization

Many instances of complex gene arrangements exist in *C. elegans* and may represent mechanisms for coordinate gene regulation or the production of alternative protein products. In addition to insights stemming from mutant analysis, interesting aspects of genomic organization that are not discernible without having complete gene sequences are being revealed. For example, we have identified instances of complex splicing patterns, tandem genes, and polycistronic loci. Interesting examples of alternative splicing patterns of potassium channels are present as well. Figure 4 presents one such example with the *nShaw1* gene, which encodes voltage-dependent potassium channel subunits. This splicing pattern has been verified by RT-PCR and cDNA cloning (Wei and Salkoff, unpublished). In this example the first four exons encode a common initiator methionine and T1 domain.

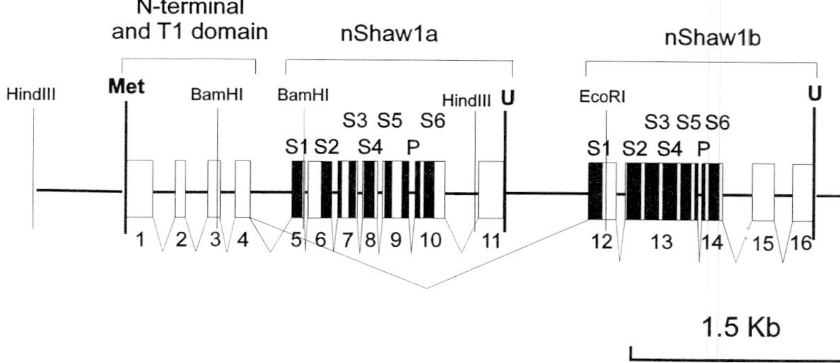

FIGURE 4 Complex organization of the *nShaw1* gene showing alternative splicing patterns. The *nShaw1* gene encodes 6TM voltage-dependent potassium channel subunits. This splicing pattern has been verified by RTPCR and cDNA cloning (Wei and Salkoff, unpublished). The first four exons encode a region common to two alternative forms. In addition to the initiator methionine, the common region encodes the T1 domain, which is important in the assembly of tetrameric channel proteins (Li *et al.*, 1992). Downstream from the common region are two alternative "core" domains, each of which encodes separate sets of membrane-spanning segments S1 through S6. Because both isoforms share an identical T1 domain, they would most likely form heteromeric complexes if expressed in the same cells.

The T1 domain is important in the assembly of tetrameric channel proteins (Li *et al.,* 1992). This common region is spliced to two alternative "core" domains, each of which encodes separate sets of membrane-spanning segments S1 through S6. It remains to be investigated if this gene is organized in this fashion to produce an obligate heteromultimer. This might be the case if a single promoter drives the expression of the two alternative subunits in the same cells. Because both isoforms share an identical T1 domain, they would most likely form heteromeric complexes in individual cells.

IV. DISCUSSION

Why are potassium channel gene families conserved between *C. elegans* and vertebrates? The fact that the entire outline of potassium channels shown in Fig. 1 is equally valid for mammals as well as the simple species, *C. elegans,* might be accounted for by three assumptions: The first assumption is that homologous genes encoding many potassium channels in different species have physiological roles that remain distinct and highly conserved. Examples of such roles could be the control of neurotransmitter release, the shaping of patterned neural output, or the control of membrane resting potential. One indication of such a conserved role from our preliminary results is the observation that *nKqt1* is prominently expressed in the pharyngeal bulb (Fig. 2) (Wei and Salkoff, unpublished). Although *C. elegans* does not have a heart, the pharyngeal bulb has both the structural characteristics (muscle cells electrically coupled) and functional characteristics (rhythmical beating) of the mammalian heart. Since both of these tissues are known to undergo similar electrical events, *nKqt1* may have a conserved role of regulating the height or duration of the plateau potentials.

A second assumption to explain the extent of diversity of channel types is that, even though the overall number of cells in *C. elegans* is lower than in most other animals, the diversity of cell types may be as great. *C. elegans* has a strikingly small number of cells devoted to each of its highly differentiated and complex tissue types. For example, the previously mentioned pharynx, which functions to ingest and process food before pumping it into the intestine, consists of only 20 muscle cells, the main body of the intestine is composed of only 20 cells, and, as mentioned, the entire nervous system is composed of only 302 cells. Thus, *C. elegans* may simply be a "minimalist creature" in having the smallest number of cells devoted to each task, while retaining highly complex and differentiated organs. The third assumption is simply that the required electrical activity of individual cells is as complex in *C. elegans* as that of any metazoan. Assigning complex integrative functions that are normally accomplished by large groups of cells in vertebrates to small groups of cells in *C. elegans* could certainly supply evolutionary pressure to maintain, or even enhance, electrical complexity.

The findings of so many genes encoding voltage-independent (4TM) channels in the database could have important implications if similarly large numbers of these genes are present in mammalian genomes. This is especially true if their tissue distribution is also regionalized. Figure 5 illustrates the dramatic change in a cell's electrical activity that can be elicited by a tiny change in a cell's resting conductance. 4TM channels are excellent candidates to regulate electrical excitability in this way. In some instances the electrical behavior of cells is known to be modulated by changes in resting conductance, and the changes have a behavioral correlate. This has been well studied in the cells of the reticular nucleus of the mammalian thalamus where cells have a bursting pattern when hyperpolarized (during sleep) and switch to a beating pattern when depolarized (during attentiveness) (Bal and McCormick, 1993). Although it is not known how, or even if, 4TM channels can be modulated, the heterogeneity of structure among these channels, along with the very long cytoplasmic structural domains present on some (Wei *et al.,* 1996), certainly presents the possibility of regulation in various ways (after all, the gating of non-voltage-sensitive channels must be regulated in *some* manner). Of potential clinical importance, a regionalized distribution of distinct channel types could present

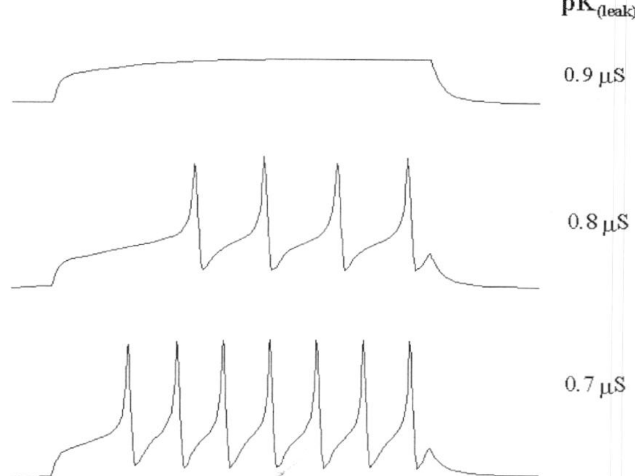

FIGURE 5 A very small change in a cell's resting conductance can elicit a very large change in a cell's electrical activity (hypothetical model). All conductance and ionic factors are equal in the three traces shown, except for the slight changes in resting potassium conductances, as indicated on the right. A slight increase in a cell's resting conductance can result in inhibition of active electrical responses (upper trace). On the other hand, a slight diminution of resting potassium conductance can greatly increase activity (lower trace). Theoretical traces were modeled using a program written by Huguenard and McCormick (1994).

circumscribed targets for pharmacological agents to either enhance or lower the excitability of various organs and brain regions. With regard to human genetic considerations, regionalized expression of a large number of genes in the brain, where mutations could affect the balance of electrical excitability, raises new questions about possible genetic propensities to various syndromes known to have a genetic component, such as essential tremor and epilepsy.

Hence, the *C. elegans* genome sequence data have been an enormous resource in revealing the scope of potassium channel diversity in a simple animal. The extent and diversity of this picture were much greater than anticipated, perhaps because the picture revealed is unbiased with regard to tissue type and level of expression. Prior studies have indicated that some potassium channels seem to be general players present in many different tissues (Tsunoda and Salkoff, 1995a,b). However, it may be the case that many channel genes that have restricted expression will be the last to be characterized molecularly since complete genomic sequence is virtually the only way to detect the full complement of channels with either restricted tissue expression or extremely low level of expression. It is unlikely that this picture will be less elaborate in mammals, and the *C. elegans* genome project has provided a good base for what to anticipate.

Acknowledgment

Supported by a grant from the NIH. We are indebted to Dr. Robert Waterston and the many helpful members of the Washington University Genome Sequencing Center for their kind cooperation.

References

Adelman, J. P., Shen, K.-Z., Kavanaugh, M. P., Warren, R. A., Wu, Y.-N., Lagrutta, A., Bond, C. T., and North, R. A. (1992). Calcium-activated potassium channels expressed from cloned complementary DNAs. *Neuron* **9,** 209–216.

Atkinson, N. S., Robertson, G. A., and Ganetzky, B. (1991). A component of calcium-activated potassium channels encoded by the *Drosophila Slo* locus. *Science* **253,** 551–555.

Baker, K., and Salkoff, L. (1990). The *Drosophila Shaker* gene codes for a distinctive K^+ current in a subset of neurons. *Neuron* **4,** 129–140.

Bal, T., and McCormick, D. A. (1993). Mechanisms of oscillatory activity in guinea-pig nucleus reticularis thalami *in vitro*: A mammalian pacemaker. *J. Physiol.* **468,** 669–691.

Baldwin, T. J., Tsaur, M. L., Lopez, G. A., Jan, Y. N., and Jan, L. Y. (1991). Characterization of a mammalian cDNA for an inactivating voltage-sensitive K^+ channel. *Neuron* **7,** 471–483.

Barhanin, J., Lesage, F., Guillemare, E., Fink, M., Lazdunski, M., and Romey, G. (1996). KvLQT1 and IsK (minK) proteins associate to form the IKS cardiac potassium current. *Nature* **384,** 78–80.

Bradley, J., Zhang, Y., Bakin, R., Lester, H. A., Ronnett, G. V., and Zinn, K. (1997). Functional expression of the heteromeric "olfactory" cyclic nucleotide-gated channel in the hippocampus: A potential effector of synaptic plasticity in brain neurons. *J. Neurosci.* **17,** 1993–2005.

Butler, A., Tsunoda, S., McCobb, D. P., Wei, A., and Salkoff, L. (1993). *mSlo,* a complex mouse gene encoding "maxi" calcium-activated potassium channels. *Science* **261,** 221–224.

Butler, A., Wei, A., Baker, K., and Salkoff, L. (1989). A family of putative potassium channel genes in *Drosophila. Science* **243,** 943–947.

Chalfie, M., Tu, Y., Euskirchen, G., Ward, W. W., and Prasher, D. C. (1994). Green fluorescent protein as a marker for gene expression. *Science* **263,** 802–805.

Chandy, K. G., and Gutman, G. A. (1995). Voltage-gated K^+ channel genes. *In* "CRC Handbook of Receptors and Channels" (N. P. A., Ed.), pp. 1–71. CRC Press, Boca Raton, FL.

Chouabe, C., Neyroud, N., Guicheney, P., Lazdunski, M., Romey, G., and Barhanin, J. (1997). Properties of KvLQT1 K^+ channel mutations in Romano-Ward and Jervell and Lange-Nielsen inherited cardiac arrhythmias. *EMBO J.* **16,** 5472–5479.

Coburn, C. M., and Bargmann, C. I. (1996). A putative cyclic nucleotide-gated channal is required for sensory development and function in *C. elegans. Neuron* **17,** 695–706.

Curran, M., Splawski, I., Timothy, K. W., Vincent, G. M., Green, E. D., and Keating, M. T. (1995). A molecular basis for cardiac arrhythmia: HERG mutations cause long QT syndrome. *Cell* **80,** 299–307.

Czempinski, K., Zimmermann, S., Ehrhardt, T., and Muller-Rober, B. (1997). New structure and function in plant K^+ channels: KCO1, and owtard rectifier with a steep $Ca2^+$ dependency. *EMBO J.* **16,** 2565–2575.

Davis, M. W., Somerville, D., Lee, R. Y. N., Lockery, S., Avery, L., and Fambrough, D. M. (1995). Mutations in the *Caenorhabditis elegans* Na, K^+ -ATPase α-subunit gene, *eat-6,* disrupt excitable cell function. *J. Neurosci.* **15,** 8408–8418.

Doupnik, C. A., Davidson, N., and Lester, H. A. (1995). The inward rectifer potassium channel family. *Curr. Opin. Neurobiol.* **5,** 268–277.

Duprat, F., Lesage, F., Fink, M., Reyes, R., Heurteaux, C., and Lazdunski, M. (1997). TASK, a human background K^+ channel to sense external pH variations near physiological pH. *EMBO J.* **16,** 5464–5471.

Elkins, T., Ganetzky, B., and Wu, C. F. (1986). A Drosophila mutation that eliminates a calcium-dependent potassium current. *Proc. Natl. Acad. Sci. USA* **83,** 8415–8419.

Fink, M., Dupart, F., Lasage, F., Reyes, R., Romey, G., Heurteauz, C., and Lazdunski, M. (1996). Cloning, functional expression and brain localization of a novel uncoventional outward rectifier K^+ channel. *EMBO J.* **15,** 6854–6862.

Goldstein, S. A. N., Price, L. A., Rosenthal, D. N., and Pausch, M. H. (1996). ORK1, a potassium-selective leak channel with two pore domains cloned from *Drosophila melanogaster* by expression in *Saccharomyces cerevisiae. Proc. Natl. Acad. Sci. USA* **93,** 13,256–13,261.

Grissmer, S., Lewis, R. S., and Cahalan, M. D. (1992). Ca(2^+)-activated K^+ channels in human leukemic T cells. *J. Gen. Physiol.* **99,** 63–84.

Hille, B. (1992). "Ionic Channels of Excitiable Membranes." Sinauer Associates, Sunderland, MA.

Huguenard, J., and McCormick, D. A. (1994). "Electrophysiology of the Neuron: A Companion to Shepherd's Neurobiology: An Interactive Tutorial." Oxford Univ. Press, London.

Ishii, T. M., Silvia, C., Hirschberg, B., Bond, C. T., Adelman, A. P., and Maylie, J. (1997). A human intermediate conductance calcium-activated potassium channel. *Proc. Natl. Acad. Sci. USA* **94,** 11,651–11,656.

Jegla, T., and Salkoff, L. (1994). Molecular evolution of K^+ channels in primitive eukaryotes. *Soc. Gen. Physiol. Ser.* **49,** 213–222.

Jegla, T., Grigoriev, N., Gallin, W. J., Salkoff, L. and Spencer, A. (1995). Multiple *Shaker* potassium channels in a primitive metazoan. *J. Neurosci.* **15,** 7989–7999.

Jegla, T., and Salkoff, L. (1995c). A multigene family of novel K^+ channels from *Paramecium tetraurelia. Recept. Chan.* **3,** 51–60.

Jegla, T., and Salkoff, L. (1997). A novel subunit for *shal* K^+ channels radically alters activation and inactivation. *J. Neurosci.* **17,** 32–44.

Johnstone, D. B., Wei, A., Butler, A., Salkoff, L., and Thomas, J. H. (1997). Behavioral defects in *C. elegans* egl-36 mutants result from potassium channels shifted in voltage-dependence of activation. *Neuron* **19,** 151–164.

Kamb, A., Iverson, L. E., and Tanouye, M. A. (1987). Molecular characterization of *Shaker,* a *Drosophila* gene that encodes a potassium channel. *Cell* **50,** 405–413.

Kaupp, B. U., Niidome, T., Tanabe, T., Terada, S., Bonigk, W., Stühmer, W., Cook, N., Kangawa, K., Matsuo, H., Hirose, T., and Numa, S. (1989). Primary structure and functional expression from complementary DNA of the rod photoreceptor cyclic GMP-gated channel. *Nature* **342,** 762–766.

Ketchum, K. A., Joiner, W. J., Sellers, A. J., Kaczmarek, L. K., and Goldstein, S. A. N. (1995). A new family of outwardly rectifying potassium channel proteins with two pore domains in tandem. *Nature* **376,** 690–695.

Kohler, M., Hirschberg, B., Bond, C. T., Kinzie, J. M., Marrion, N., Maylie, J., and Adelman, J. P. (1996). Small conductance, calcium-activated potassium channels from mammalian brain. *Science* **273,** 1709–1714.

Komatsu, H., Mori, I., Rhee, J.-S., Akaike, N., and Ohshima, Y. (1996). Mutations in a cyclic nucleotide-gated channel lead to abnormal thermosensation and chemosensation in *C. elegans. Neuron* **17,** 707–718.

Lesage, F., Guillemare, E., Fink, M., Duprat, F., Lazdunski, M., Romey, G., and Barhanin, J. (1996). TWIK-1, a ubiquitous human weakly inward rectifiying K^+ channel with a novel structure. *EMBO J.* **15,** 1004–1011.

Lesage, F., Lauritzen, I., Dupart, F., Reyes, R., Fink, M., Heurteaux, C., and Lazdunski, M. (1997). The structure, function and distribution of the mouse TWIK-1 K^+ channel. *FEBS Lett.* **402,** 28–32.

Li, M., Jan, Y. N., and Jan, L. Y. (1992). Specification of subunit assembly by the hydrophilic amino-terminal domain of the *Shaker* potassium channel. *Science* **257,** 1225–1230.

Ludwig, J., Terlau, H., Wunder, F., Bruggemann, A., Pardo, L. A., Marquardt, A., Stuhmer, W., and Pongs, O. (1994). Functional expression of a rat homologue of the voltage gated *ether a go-go* potassium channel reveals differences in selectivity and activation kinetics between the *Drosophila* channel and its mammalian counterpart. *EMBO J.* **13,** 4451–4458.

Milkman, R. (1994). An *Escherichia coli* homologue of eukaryotic potassium channel proteins. *Proc. Natl. Acad. Sci. USA* **91,** 3510–3514.

Pak, M. D., Baker, K., Covarrubias, M., Butler, A., Ratcliffe, A., and Salkoff, L. (1991). mShal, a subfamily of A-type K^+ channel cloned from mammalian brain. *Proc. Natl. Acad. Sci. USA* **88,** 4386–4390.

Papazian, D. M., Schwarz, T. L., Tempel, B. L., Jan, Y. N., and Jan, L. Y. (1987). Cloning of genomic and complementary DNA from Shaker, a putative potassium channel gene from *Drosophila. Science* **237,** 749–753.

Pongs, O., Kecskemethy, N., Muller, R., Krah-Jentgens, I., Baumann, A., Kiltz, H. H., Canal, I., Llamazares, S., and Ferrus, A. (1988). *Shaker* encodes a family of putative potassium channel proteins in the nervous system of *Drosophila. EMBO J.* **7,** 1087–1096.

Raizen, D. M., and Avery, L. (1994). Electrical activity and behavior in the pharynx of *Caenorhabditis elegans. Neuron* **12,** 483–495.

Romey, G., Attali, B., Chouabe, C., Abitbol, I., Guillemare, E., Barhanin, J., and Lazdunski, M. (1997). Molecular mechanism and functional significance of the minK control of the KvLQT1 channel activity. *J. Biol. Chem.* **272,** 16,713–16,716.

Rudy, B., Sen, K., Vegas-Saenz de Miera, E., Lau, D., Ried, T., and Ward, D. C. (1991). Cloning of a human cDNA expressing a high voltage-activating, TEA-sensitive, type-A K^+ channel which maps to chromosome 1 band p21. *J. Neurosci. Res.* **29,** 401–412.

Salkoff, L., and Jegla, T. (1995). Surfing the DNA databases for K^+ channels nets yet more diversity. *Neuron* **15,** 489–492.

Sanguinetti, M. C., Curran, M. E., Zou, A., Shen, J., Spector, P. S., Atkinson, D. L., and Keating, M. T. (1996). Coassembly of KvLQT1 and minK(IsK) proteins to form cardiac IKS potassium channel. *Nature* **384,** 80–83.

Sanguinetti, M. C., Jiang, C., Curran, M. E., and Keating, M. T. (1995). A mechanistic link between an inherited and an acquired cardiac arrhythmia: HERG encodes the IKr potassium channel. *Cell* **81,** 299–307.

Schreiber, M., Wei, A., Yuan, A., Gaut, J., M., and Salkoff, L. (1998). Slo3, a novel pH-sensitive K^+ channel from mammalian spermatocytes. *J. Biol. Chem.,* 3509–3516.

Smith, P. L., Baukrowitz, T., and Yellen, G. (1996). The inward rectification mechanism of the HERG cardiac potassium channel. *Nature* **379,** 833–836.

Sutro, J. B., Vayuvegula, B. S., Gupta, S., and Cahalan, M. D. (1989). Voltage-sensitive ion channels in human B lymphocytes. *Adv. Exp. Med. Biol.* **254,** 113–122.

Titus, S. A., Warmke, J. W., and Ganetzky, B. (1997). The *Drosophila erg* K^+ channel polypeptide is encoded by the *seizure* locus. *J. Neurosci.* **17,** 875–881.

Trudeau, M. C., Warmke, J. W., Ganetzky, B., and Robertson, G. A. (1995). HERG, a human inward rectifier in the voltage-gated potassium channel family. *Science* **269,** 92–95.

Tsunoda, S., and Salkoff, L. (1995a). Genetic analysis of *Drosophila* neurons: *Shal, Shaw,* and *Shab* encode most embryonic potassium currents. *J. Neurosci.* **15,** 1741–1754.

Tsunoda, S., and Salkoff, L. (1995b). The major delayed rectifier in both *Drosophila* neurons and muscle is encoded by *Shab. J. Neurosci.* **15,** 5209–5221.

Wang, Z. W., Kunkel, M. T., Wei, A., Butler, A., and Salkoff, L. (1999). Analysis of gene organization of nematode 4TM K+ channels. *Ann. N.Y. Acad. Sci.,* in press.

Wang, Q., Curran, M. E., Splawski, I., Burn, T. C., Millholland, J. M., VanRaay, T. J., Shen, J., Timothy, K. W., Vincent , G. M., de Jager, T., Schwartz, P. J., Towbin, J. A., Moss, A. J., Atkinson, D. L., Landes, G. M., Conners, T. D., and Keating, M. T. (1996). Positional cloning of a novel potassium channel gene: KVLQT1 mutations cause cardiac arrhythmias. *Nature Genet.* **12,** 17–23.

Wang, X. J., Reynolds, E. R., Deak, P., and Hall, L. M. (1997). The *seizure* locus encodes the *Drosophila* homolog of the HERG potassium channel. *J. Neurosci.* **17,** 882–890.

Warmke, J., Drysdale, R., and Ganetzky, B. (1991). A distinct potassium channel polypeptide encoded by the *Drosophila* eag locus. *Science* **252,** 1560–1562.

Wei, A., Covarrubias, M., Butler, A., Baker, K., Pak, M., and Salkoff, L. (1990). K+ current diversity is produced by an extended gene family conserved in *Drosophila* and mouse. *Science* **248,** 599–603.

Wei, A., Jegla, T., and Salkoff, L. (1996). Eight potassium channel families revealed by the *C. elegans* genome project. *Neuropharmacology* **35,** 805–829.

Weyand, I., Godde, M., Frings, S., Welner, J., Muller, F., Altenhofen, W., Hatt, H., and Kaupp, U. B. (1994). Cloning and functional expresion of a cyclic-nucleotide-gated channel from mammalian sperm. *Nature* **368,** 859–863.

White, J. G., Southgate, E., Thomson, J. N., and Brenner, S. (1986). The structure of the nervous system of the nematode *caenorhabditis elegans. Philos. Trans. R. Soc. Lond. Ser. B* **314,** 1–340.

Wilson, R., Ainscough, R., Anderson, K., Baynes, C., Berks, M., Bonfield, J., Burton, J., Connell, M., Copsey, T., Cooper, J., Coulson, A., Craxton, M., Dear, S., Du, Z., Durbin, R., Favello, A., Fulton, L., Gardner, A., Green, P., Hawkins, T., Hillier, L., Jier, M., Johnston, L., Jones, M., Kershaw, J., Kirsten, J., Laister, N., Latreille, P., Lightning, J., Lloyd, C., McMurray, A., Mortimore, B., O'Callaghan, M., Parsons, J., Percy, C., Rifken, L., Roopra, A., Saunders, D., Shownkeen, R., Smaldon, N., Smith, A., Sonnhammer, E., Staden, R., Sulston, J., Thierry-Mieg, J., Thomas, K., Vaudin, M., Vaughan, K., Waterston, R., Watson, A., Weinstock, L., Wilkinson-Sproat, J., and Wohldman, P. (1994). 2.2 Mb of continguous nucleotide sequence from chromosome III of *C. elegans. Nature* **368,** 32–38.

Wood, W. B. (1988). "The Nematode *Caenorhabditis elegans.*" Cold Spring Harbor Press, Cold Spring Harbor, NY.

Yarom, Y., Sugimori, M., and Llinas, R. (1985). Ionic currents and firing patterns of mammalian vagal motoneurons *in vitro. Neuroscience* **16,** 719–737.

Yokoyama, M., Nishi, Y., Yoshii, J., Okubo, K., and Matsubara, K. (1996). Identification and cloning of neuroblastoma-specific and nerve tissue-specific genes through compiled expression profiles. *DNA Res.* **3,** 311–320.

CHAPTER 3

Activation of Voltage-Dependent Potassium Channels

Diane M. Papazian

Department of Physiology and Molecular Biology Institute, University of California, Los Angeles, School of Medicine, Los Angeles, California 90095

I. INTRODUCTION

During the past few years, significant progress has been made toward elucidating the mechanism of voltage-dependent activation in potassium channels. Voltage controls the activity of a large number of potassium channels, including those in the Shaker (Kv1), Shab (Kv2), Shaw (Kv3), and Shal (Kv4) subfamilies (Chandy and Gutman, 1995). These proteins share the structural organization shown in Fig. 1, including six putative transmembrane segments and a re-entrant P loop that contributes to the permeation pathway (Jan and Jan, 1997). Four such subunits surround the pore (MacKinnon, 1991; Liman *et al.*, 1992; Li *et al.*, 1994; Schulteis *et al.*, 1996). The fourth transmembrane segment in each subunit, S4, contains a highly conserved, repeating motif of one positively charged amino acid

Current Topics in Membranes, Volume 46

1063-5823/99 $30.00

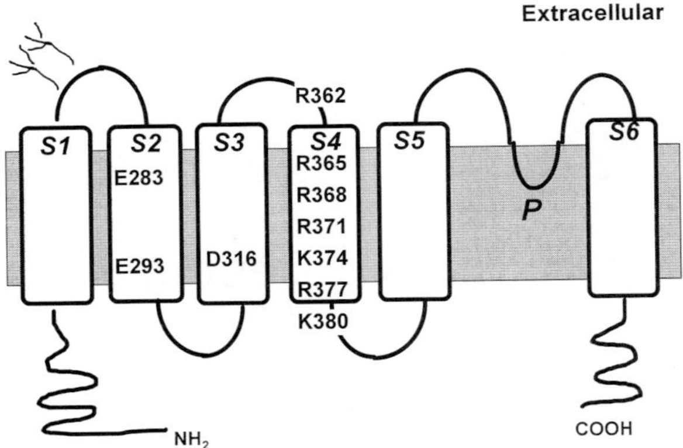

FIGURE 1 Topology of the Shaker potassium channel subunit, showing the approximate position of conserved charged residues in transmembrane segments.

(arginine or lysine) plus two other residues, which are often hydrophobic in character (Tempel *et al.,* 1987). A subset of the charged residues in the S4 segment is now known to make essential contributions to the mechanism of voltage-dependent activation (Aggarwal and MacKinnon, 1996; Seoh *et al.,* 1996). The S4 mechanism of voltage control is used not only by potassium channels but also by voltage-dependent sodium and calcium channels, which resemble four covalently linked potassium channel subunits (Noda *et al.,* 1984; Tanabe *et al.,* 1987; Tempel *et al.,* 1987). In contrast, inwardly rectifying potassium channels of the two transmembrane segment variety lack an intrinsic mechanism of voltage control. Instead, the time and voltage-dependent gating seen in these channels is due to voltage-dependent block of conduction by extrinsic factors (Ficker *et al.,* 1994; Fakler *et al.,* 1995). Among cation channels, therefore, the S4 mechanism appears to account for all known instances of intrinsic voltage control. However, it cannot be assumed that the presence of an S4 segment is sufficient to signify a voltage-dependent mechanism of activation. Cyclic nucleotide-gated channels, which are involved in olfactory and visual sensory transduction, contain recognizable S4 segments (Jan and Jan, 1990). These channels show little voltage sensitivity, however, and are instead opened by the binding of cytoplasmic cAMP or cGMP to the channel protein (Kaupp *et al.,* 1989; Dhallan *et al.,* 1990; Haynes and Yau, 1990; Goulding *et al.,* 1995).

A broad range of technical approaches have contributed to our understanding of the mechanism of voltage-dependent activation. Typically, wild-type and mutant cDNA clones have been expressed in heterologous systems to investigate the functional and structural consequences of the mutations. *Xenopus laevis* oocytes or mammalian cells in culture expressing cloned channel sequences have been used to characterize ionic and gating currents; to probe the orientation, environment, and solvent exposure of specific residues; and to analyze structural interactions.

The studies summarized in this chapter have contributed significantly to our emerging picture of activation gating. Many of the results were obtained using the voltage-dependent Shaker potassium channel of *Drosophila,* in particular the ShakerB or ShakerH4 isoforms (which differ by only two amino acids) containing an amino-terminal deletion to remove N-type inactivation (Schwarz *et al.,* 1988; Kamb *et al.,* 1988; Hoshi *et al.,* 1990). Much of what has been learned is likely to be generally applicable, given the sequence conservation among potassium channels and the fact that similar results have been obtained for voltage-dependent sodium channels (Stühmer *et al.,* 1989; Yang and Horn, 1995; Yang *et al.,* 1996).

II. CONTROL OF CHANNEL ACTIVITY BY VOLTAGE

In voltage-dependent potassium channels, the probability of channel opening (P_o) is steeply dependent on the membrane potential (Fig. 2). In their classic study of the voltage-dependent conductances of the squid giant fiber (Hodgkin and Huxley, 1952), Hodgkin and Huxley estimated that the equivalent of at least four to six elementary charges (e_0) must cross the transmembrane electric field during the activation of a sodium or potassium channel. The control of activity by membrane potential implies that there

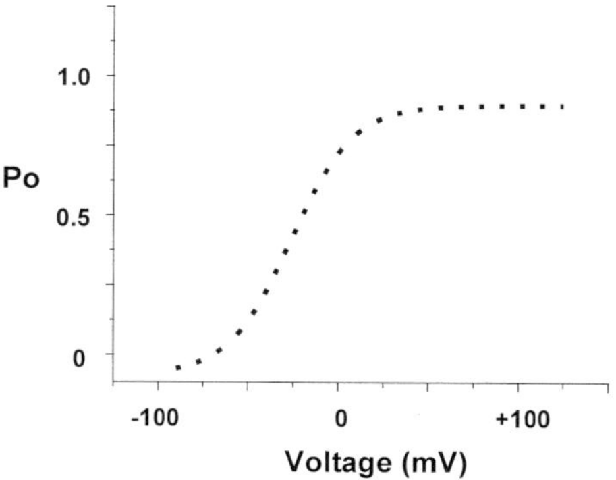

FIGURE 2 The dependence of open probability on voltage is shown for the Shaker-IR (inactivation-removed) channel.

are charged or dipolar groups associated with the channel that sense changes in voltage electrostatically. In fact, the voltage sensor is an intrinsic part of the channel protein that undergoes voltage-dependent conformational changes that alter the chance that the channel will open (Sigworth, 1993).

III. BIOPHYSICAL ANALYSIS OF GATING

Hodgkin and Huxley (1952) predicted that the rearrangements of the voltage sensor upon activation would generate small capacitative currents called gating currents. These nonlinear displacement currents were first recorded from sodium channels in the squid giant fiber after preventing ionic conduction through the pore (Armstrong and Bezanilla, 1973). Gating currents of cloned channels have now been recorded in heterologous expression systems, where the capacitative properties of the membrane can be readily compared in the presence and absence of expressed channels (Bezanilla *et al.,* 1991, 1994). In Shaker channels, a mutation in the re-entrant P loop, W434F, which prevents potassium conduction, has been used to eliminate the large electrical signal of ion permeation, revealing the gating currents (Perozo *et al.,* 1993).

The total gating charge movement is quantified by integrating gating current records. A plot of gating charge versus voltage (q–V) reveals that charge movement occurs at voltages more hyperpolarized than those that activate the ionic conductive (Bezanilla *et al.,* 1991; Stefani *et al.,* 1994). This is one piece of evidence indicating the existence of several charge-moving conformational changes between closed states that prepare the channel for opening.

In Shaker channels, at least two steady-state components of gating charge movement exist. This was apparent from analysis of gating currents in a Shaker S4 mutant, R368Q (Perozo *et al.,* 1994). The q–V curve of R368Q displays two readily distinguished phases differing in their steepness and position along the voltage axis. The R368Q mutation separates these gating charge components, but they are also present in control Shaker channels (Bezanilla *et al.,* 1994; Stefani *et al.,* 1994). In addition, these steady-state components of charge movement appear to correspond to two kinetic components of the gating currents (Bezanilla *et al.,* 1994; Stefani *et al.,* 1994). These results suggest that at least two types of charge-moving conformational changes accompany voltage-dependent gating.

The Shaker channel is steeply voltage-dependent. Schoppa and colleagues (1992) estimated that the equivalent of 12 to 13 e_0 traverses the electric field during the activation of a single channel. This estimate was obtained by macropatch analysis of Shaker channels expressed in *Xenopus* oocytes. The number of channels in the patch was determined by analyzing ionic current fluctuations, after which the pore was blocked and gating currents were recorded from the same patch. The charge per channel was confirmed using the same approach by Seoh and co-workers (1996). A similar value was obtained by Aggarwal and MacKinnon (1996) who counted the number of channels in the patch using a radioactively labeled, high-affinity toxin.

Integration of gating currents reveals the charge that moves upon activation. However, it is possible that some of the charge is not energetically coupled to channel opening, but is essentially carried along for the ride. The charge energetically coupled to activation can be estimated by limiting slope measurements, that is, by determining the slope of the P_o–V curve at extremely low open probabilities (Almers, 1978). In practice, this method usually underestimates the gating charge because of the difficulty of measuring ionic currents at extremely low open probabilities. Recently, a more accurate limiting slope method, combining both gating and ionic current measurements, was applied to Shaker channels (Seoh *et al.,* 1996; Sigg and Bezanilla, 1997). This approach indicates that the entire charge movement, equivalent to 12 to 13 e_0 crossing the transmembrane field, is energetically coupled to pore opening (Seoh *et al.,* 1996).

The elementary transitions of the voltage sensor have been examined by analysis of gating current fluctuations (Sigg *et al.,* 1994). Two phases of gating were distinguished. The first was characterized by multiple small charge movements, whereas the second phase involved a large, shot-like transition that moves a charge of 2.4 e_0 per subunit. The latter phase is likely to represent a transition in which the voltage sensor moves between two well-defined conformational states.

Biophysical analysis of voltage-dependent gating has revealed the complexity of the process. Several kinetic models for the activation of Shaker channels have been proposed (Zagotta and Aldrich, 1990; Sigworth, 1993; Bezanilla *et al.,* 1994; Zagotta *et al.,* 1994). These models include multiple voltage-dependent transitions between closed conformations, followed by a less voltage-dependent transition that opens the channel. These models support a physical mechanism for gating in which each summit contains a voltage sensor that undergoes at least two charge-moving conformational changes between closed states, priming the channel for opening. Opening of the pore, which is likely to involve a concerted transition by all of the subunits, is a distinguishable conformational change that moves little charge.

IV. THE S4 HYPOTHESIS

The conserved S4 sequence was identified by molecular cloning and sequencing of genes encoding sodium, potassium, and calcium channels

(Noda *et al.,* 1984; Tanabe *et al.,* 1987; Papazian *et al.,* 1987; Tempel *et al.,* 1987). The S4 segment became the prime candidate for the voltage sensor because it contains positively charged residues at every third position. These charges were proposed to interact electrostatically with the membrane potential during activation. The S4 sequence, located in the midst of the transmembrane segments, was proposed to adopt a transmembrane orientation. Experimental confirmation was obtained by chemical modification of the flanking S3–S4 and S4–S5 loops on the extracellular and intracellular sides of the membrane, respectively (Larrson *et al.,* 1996). Although much attention has been focused on the S4 segment, it is important to note that S2 and S3 also contain highly conserved negatively charged residues (aspartate and glutamate) that are candidates for voltage-sensing residues (Fig. 1) (Papazian *et al.,* 1995; Planells-Cases *et al.,* 1995).

Several groups tested the hypothesis that the S4 segment contributes to the voltage sensor by making site-directed mutations in the S4 sequence and determining their effects on the steady-state properties of ionic and gating currents (Stühmer *et al.,* 1989; Liman *et al.,* 1991; Papazian *et al.,* 1991; Logothetis *et al.,* 1992: Shao and Papazian, 1993; Perozo *et al.,* 1994). These studies were consistent with the idea that the S4 segment is important in activation. In addition, it was found that the contributions of different S4 residues to the mechanism were not equivalent (Papazian *et al.,* 1991). However, these studies were not conclusive because mutations in some uncharged residues were found to have similar effects on the steady-state activation properties (Schoppa *et al.,* 1992).

V. IDENTIFICATION OF VOLTAGE-SENSING RESIDUES

More conclusive proof that the S4 segment was a fundamental component of the voltage sensor came from studies in which the contribution of individual residues to the gating charge was determined (Aggarwal and MacKinnon, 1996; Seoh *et al.,* 1996). As noted above, the equivalent of 12 to 13 e_0 traverse the transmembrane field during the activation of a single Shaker channel. To identify which residues contribute to this number, two groups measured the charge per channel in mutant channels in which one charged residue per subunit had been neutralized (Fig. 3). Seoh and colleagues (1996) found that neutralization of three residues in the S4 segment, R365, R368, and R371, significantly reduced the charge per channel to approximately 6 to 8 e_0. Aggarwal and MacKinnon (1996) also measured a substantial effect at these positions. Thus, it was concluded that these charged residues are an important part of the voltage sensor. In contrast, the S4 residue K374 does not contribute significantly to the charge per channel.

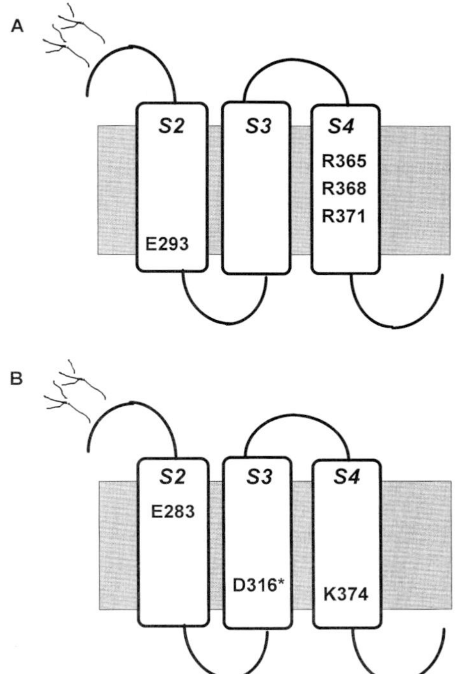

FIGURE 3 The topology models summarize residues that contribute to the gating charge (A) and those that do not move relative to the transmembrane electric field (B) (Seoh *et al.*, 1996).

Seoh and co-workers (1996) also investigated the contribution of three highly conserved, negatively charged residues, E283 and E293 in S2 and D316 in S3. Neutralization of E293 significantly reduced the charge per channel by about 6 to 7 e_0, identifying this residue as an important component of the voltage sensor. Neutralization of D316 led to a much smaller, but consistent decrease in the charge per channel, of approximately 2 e_0. Given the errors inherent in the measurement, this difference from the control channel was not statistically significant. Neutralization of E283 did not alter the charge per channel. The results indicate that E283 and K374 do not move relative to the transmembrane field during activation.

Do E293 in S2 and R365, R368, and R371 in S4 account for all of the charge movement in Shaker channels? To address this question, mutations at these positions were combined so that two, three, or four charges per subunit were neutralized simultaneously (Seoh *et al.*, 1997). The remaining gating charge was estimated by analyzing the limiting slope of the ionic conductance. Neutralization of three of the four residues in various combinations reduced the apparent gating charge to about 5 to 6 e_0. However, after neutralizing all four residues, the apparent gating charge was reduced to 2 to 3 e_0. These results suggest that the four residues E293, R365, R368, and R371 together compose approximately 80% of the gating charge in Shaker channels. To determine whether the S3 residue D316 was responsible for the remaining 20% of the charge, the neutralization mutation D316N was added to E293Q + R365Q + R368N + R371Q. Unfortunately, the additional mutation resulted in a protein that was unable to fold and assemble properly, leaving the issue unresolved.

VI. MOVEMENT OF THE VOLTAGE SENSOR IN THE TRANSMEMBRANE ELECTRIC FIELD

The finding that a residue contributes to the charge per channel could result from two different factors (Perozo *et al.*, 1994; Seoh *et al.*, 1996; Papazian and Bezanilla, 1997). The residue might traverse some fraction of the transmembrane field and thereby be directly involved in the conformational changes of the voltage sensor. Alternatively, the residue might make a significant contribution to the profile of the electric field through which other residues move. These two alternatives, which are not mutually exclusive, arise from the fact that the contribution of any residue to the gating charge is equal to the valence of the residue times the fraction of the electric field that it traverses. Using electrical measurements, these terms cannot be separated. Therefore, neutralization mutations that affect the charge per channel could do so either by reducing the valence of a moving charge or by changing the profile of the local field through which other charged residues move. Because both effects can have dramatic effects on the steepness of a channel's response to voltage, residues in both categories function as essential elements of the voltage sensor. To characterize the physical mechanism of activation, it is therefore important to determine whether a residue moves through the electric field, what fraction of the field it traverses, and what its contribution to the profile of the field is. Progress has been made in addressing the first two issues.

Larsson and colleagues (1996) substituted residues in and near the Shaker S4 segment with the amino acid cysteine, and determined the voltage-dependence of reactivity with hydrophilic sulfhydryl reagents derived from methanethiosulfonate (Akabas *et al.*, 1992). If modification of a specific cysteine residue results in a measurable change in channel function, the reactivity of the position at different voltages and on different sides of the membrane can be assessed. Larsson and co-workers (1996) showed that position 362 in the S4 segment reacts with externally applied, hydrophilic MTS reagents at both hyperpolarized and depolarized potentials, suggesting that this residue is exposed to the extracellular environment in both resting and activated conformations. In contrast, cysteines at positions 365 and 368 were reactive with extracellular reagents only at depolarized potentials. This study, and similar ones conducted with voltage-dependent sodium channels, provided strong evidence that S4 residues do move relative to the transmembrane field during voltage-dependent activation (Yang and Horn, 1995: Yang *et al.,* 1996).

Recently, Starace and co-workers (1997) also addressed this question, using histidine substitutions at positions 365 and 368. The charge on a solvent-accessible histidine residue can be controlled by the pH of the intracellular or extracellular solution, and therefore its exposure can be assessed with an extremely small probe, the proton. Imposition of a pH gradient was sufficient to convert the mutated Shaker channel into a voltage-dependent proton transporter. The results indicate that positions 365 and 368 are solvent accessible on the inside and outside of the membrane under hyperpolarized and depolarized potentials, respectively. Thus, residues 365 and 368 in the Shaker S4 segment traverse the entire electric field during voltage-dependent activation.

As described above, the S2 residue E293 has also been implicated in gating charge movement. Whether E293 contributes to the gating charge by moving through the field or by tuning the field that the S4 residues traverse, or both, is currently unknown.

Conformational changes of the voltage sensor have also been studied by labeling specific, cysteine-substituted positions with fluorescent probes and then measuring changes in fluorescence as a function of voltage (Mannuzzu *et al.,* 1996; Cha and Bezanilla, 1997). Because fluorescent intensity will vary with the environment of the probe, this approach provides a sensitive way to identify conformational changes in the protein. Cha and Bezanilla (1997) have measured fluorescence changes and gating currents in the same preparation, allowing quantitative comparison of the kinetics of fluorescence changes and gating charge movement. Probes inserted near the extracellular end of S4 undergo a voltage-dependent change in fluorescent that parallels the kinetics of gating charge movement. In contrast, probes inserted in the vicinity of the pore undergo changes in fluorescence that are much slower than movement of the gating charge. Instead, the kinetics parallel the onset of ionic conduction and slow inactivation. These results indicate that fluorescent probes in different regions of the protein will be useful in characterizing a variety of functionally important conformational states.

Interestingly, using a probe attached to the extracellular end of S2, Cha and Bezanilla (1997) detected a fluorescent change with very fast kinetics. They propose that this fast rearrangement represents an early step in gating, occurring before movement of the S4 segment. This result is intriguing in light of the contribution of the S2 position E293 to the gating charge (Seoh *et al.,* 1996). However, the fast rearrangement of the extracellular S1–S2 loop may be distinct from the contribution of E293 to the gating process. Between the position of the fluorescent probe and E293 lies E283, a residue that does not move relative to the field during activation (Seoh *et al.,* 1996).

VII. STRUCTURAL INTERACTIONS IN THE VOLTAGE SENSOR

Knowledge of the structure of the voltage sensor is essential for a detailed understanding of the physical mechanism of activation. Biophysical analysis indicates that gating probably involves several functionally important conformations, and that activation involves at least two charge-moving transitions per subunit (Sigworth, 1993; Bezanilla *et al.,* 1994; Zagotta *et al.,* 1994). Thus, the voltage sensor is likely to be a conformationally flexible region of the protein. Ideally, several high-resolution structures of the voltage sensor at various stages of the activation process will be obtained in the future.

In the shorter term, insights into the packing of transmembrane segments in the voltage sensor of Shaker channels have been obtained by biochemical analysis (Papazian *et al.,* 1995; Tiwari-Woodruff *et al.,* 1997). Using a strategy related to intragenic suppression, mutations that disrupt folding of the voltage sensor have been used to identify likely structural interactions among transmembrane segments S2, S3, and S4.

The Shaker protein is made initially as a core-glycosylated precursor in the endoplasmic reticulum (Schulteis *et al.,* 1995; Nagaya and Papazian, 1997). Maturation of the protein involves transfer to the Golgi apparatus where the carbohydrate chains are modified (Nagaya and Papazian, 1997). This process is extremely efficient in diverse expression systems, including *Xenopus* oocytes and transfected mammalian cells in culture. The mature and immature forms of the Shaker protein can be readily distinguished by their differential electrophoretic mobilities (Santacruz-Toloza *et al.,* 1994). Maturation of the Shaker protein is highly correlated with proper folding and assembly into a functional native conformation (Papazian *et al.,* 1995; Tiwari-Woodruff *et al.,* 1997). In fact, Shaker subunits contain structural hallmarks of the native state and fully assemble into tetramers before leaving the endoplasmic reticulum (Nagaya and Papazian, 1997).

Mutations that disrupt folding or assembly trap the protein in an immature form in the endoplasmic reticulum (Papazian *et al.,* 1995; Tiwari-Woodruff *et al.,* 1997). To identify likely structural interactions, these mutations are paired with compensatory second site mutations to identify those that restore maturation and function (Papazian *et al.,* 1995; Tiwari-Woodruff *et al.,* 1997). This rescue is quite specific: Similar second site mutations nearby in the primary structure are unable to restore maturation or functional expression (Fig. 4). These results strongly suggest that the locations of the primary and second site mutations are in close proximity in the native structure. Two charge networks, composed of E283 in S2 with R368 and R371 in S4, and K374 in S4 with E293 in S2 and D316 in S3, stabilize the voltage sensor in Shaker channels (Tiwari-Woodruff *et al.,* 1997). Thus this approach has provided several structural constraints on the packing of transmembrane segments S2, S3, and S4 (Fig. 5).

Interestingly, it has been shown that the second site mutations must be on the same subunit as the primary mutation for rescue to occur (Tiwari-Woodruff *et al.,* 1997). This suggests that the structural interactions stabilize

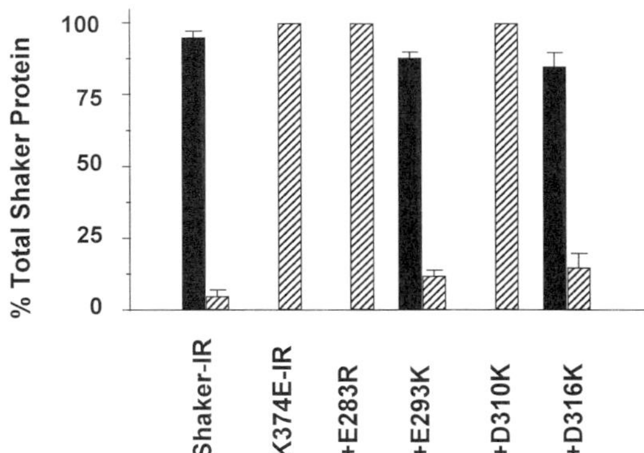

FIGURE 4 Rescue of maturation of the charge reversal mutation K374E. After expression in *Xenopus* oocytes, the amounts of Shaker protein in the mature (solid bars) or immature (striped bars) forms were quantified by densitometry. Data are shown for Shaker-IR, K374E-IR, and double mutant combinations of K374E-IR with E283R, E293K, D310K, or D316K, as indicated. Only combination of K374E with E293K or D316K restores maturation (Tiwari-Woodruff *et al.,* 1997).

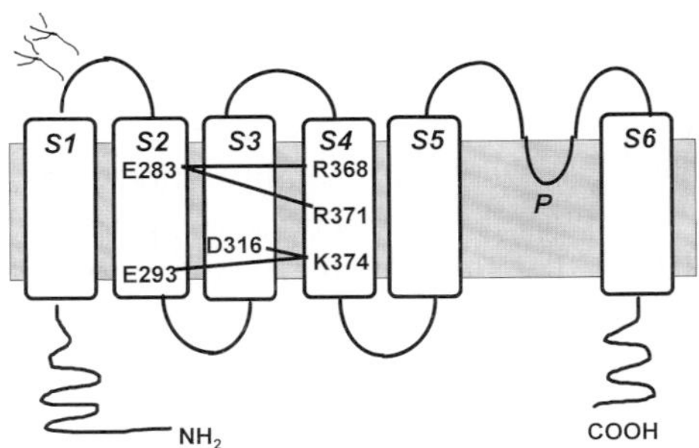

FIGURE 5 The topology model illustrates structural constraints inferred from intragenic suppression strategy (Papazian *et al*, 1995; Tiwari-Woodruff *et al.*, 1997).

the tertiary structure of the subunit, consistent with the expectation that each subunit has its own voltage sensor.

VIII. GENERALITY OF THE S4 MECHANISM

The mechanism of activation observed in Shaker channels is likely to be conserved in other voltage-dependent cation channels. Cysteine residues have been introduced into the S4 segment of domain IV in skeletal sodium channels and probed with the hydrophilic methanethiosulfonate derivatives. As in Shaker channels, several positively charged residues traverse the entire electric field during voltage-dependent activation (Yang and Horn, 1995; Yang *et al.*, 1996).

IX. ROLE OF THE S4 SEGMENT IN VOLTAGE-INSENSITIVE CHANNELS

Given the central role of the S4 segment in the mechanism of voltage-dependent activation, an obvious question concerns the role of the S4 segment in channels that have little or no voltage dependence. This question has been addressed by studying the properties of chimeras made between two closely-related proteins that differ dramatically in their mechanism of activation, the ether-à-go-go potassium channel and a cyclic nucleotide-gated channel from the rat olfactory system (Kaupp *et al.*, 1989; Haynes and Yau, 1990; Brüggemann *et al.*, 1993; Robertson *et al.*, 1996; Tang and Papazian, 1997).

The *Drosophila* ether-à-go-go (eag) potassium channel is the prototype of a subfamily of voltage-gated channels that includes erg (eag-related gene) and elk (eag-like)(Warmke and Ganetzky, 1994). In mammals, this subfamily is extremely important physiologically. In humans, h-erg contributes to the rapid delayed rectifier current that plays a major role in repolarization of the cardiac action potential (Sanguinetti *et al.*, 1995). In fact, mutations in h-erg have been implicated in one form of long QT syndrome, a cardiac arrhythmia associated with sudden death in young people (Sanguinetti *et al.*, 1995).

The membrane topology of eag resembles that of other voltage-dependent potassium channels (Fig. 1). In addition, eag contains a cyclic nucleotide-binding domain following S6. Therefore, the structural organization of eag resembles that of the cyclic nucleotide-gated channels involved in sensory transduction in the visual and olfactory systems (Jan and Jan, 1990; Guy *et al.*, 1991). Whereas eag is a voltage-gated channel regulated by the binding of cyclic nucleotide ligands, the cyclic nucleotide-gated channels are relatively insensitive to voltage despite the presence of an S4 segment (Haynes and Yau, 1990; Jan and Jan, 1990). Instead, the activity of cyclic nucleotide-gated channels is controlled by an allosteric mechanism in which ligand binding stabilizes the open conformation (Goulding *et al.*, 1994; Gordon and Zagotta, 1995).

Chimeras of eag and a rat olfactory channel were made to investigate the basis of their divergent gating properties (Tang and Papazian, 1997). One significant difference between eag and the olfactory channel is the charge in the S4 sequence. Whereas the net S4 charge is +5 in eag, it is only +1 in the olfactory channel. To test whether the olfactory channel S4 segment was capable of supporting a voltage-dependent mechanism of activation despite its reduced charge, it was inserted into eag in place of the eag S4 segment. The resulting chimeric channel was voltage-dependent, although with a somewhat reduced steepness. This result indicates that the olfactory channel S4 can support a voltage-dependent activation mechanism in an appropriate structural context.

Upon replacing the extracellular loop between segments S3 and S4, however, eag was converted into an essentially voltage-independent channel (Tang and Papazian, 1997). Between −200 and +100 mV, the chimeric channel displayed a linear current–voltage relationship. The results indicate that the S3–S4 loop of the olfactory channel increases the relative stability of the open state of eag. Much of this effect is due to a single residue, which is a small hydrophobic residue in eag but a negatively charged residue in the olfactory channel. These results suggest that in the olfactory channel, the S3–S4 loop helps to stabilize the S4 segment in an activated conformation, effectively eliminating the voltage-dependent transitions from the gating mechanism.

X. AN EMERGING PICTURE OF THE PHYSICAL MECHANISM OF VOLTAGE-DEPENDENT ACTIVATION

Recent studies have contributed to an emerging picture of voltage-dependent activation in Shaker channels (Goldstein, 1996; Papazian and Bezanilla, 1997). Four residues are primarily responsible for voltage sensitivity: E293 in S2 and R365, R368, and R371 in S4 (Seoh *et al.,* 1996). Of these, at least two residues, R365 and R368, traverse the entire transmembrane field (Starace *et al.,* 1997). In the four subunits, these residues therefore contribute eight charges to the charge per channel. Several lines of evidence suggest that the voltage-sensing residues are accessible to solvent at hyperpolarized or depolarized potentials, or both. Positions 365 and 368 are accessible to hydrophilic probes on opposite sides of the membrane depending on the voltage (Larsson *et al.,* 1996; Starace *et al.,* 1997). In addition, it has been found that, among the conserved residues in transmembrane segments, the voltage-sensing residues are extremely tolerant of charge reversal mutations (Tiwari-Woodruff *et al.,* 1997). Such mutations do not prevent maturation or functional expression (Table I). In contrast,

TABLE I

Voltage-Sensing Positions Tolerate Charge Reversal Mutations

Mutation	Matures?	Functions?
S2		
E283R	50%	No
E293R	*Yes*	*Yes*
S3		
D316K	No	No
S4		
R365E	*Yes*	*Yes*
R368E	*Yes*	*Yes*
R371E	*Yes*	*Yes*
K374E	No	No

Note. The indicated charge reversal mutations were expressed in *Xenopus* oocytes for analysis of maturation and channel function. Residues that contribute to the gating charge (shown in italics) are more tolerant of mutations than residues that do not contribute to the gating charge (Tiwari-Woodruff *et al.,* 1997).

other positions that do not make large contributions to the charge per channel, including E283 in S2, D316 in S3, and K374 in S4, are much more sensitive to mutations, with neutralization or charge reversal mutations disrupting maturation and function (Tiwari-Woodruff *et al.,* 1997). These results suggest that a distinction can be made among conserved charged residues in transmembrane segments. The conservation of the voltage-sensing residues appears to reflect their important functional rather than structural role. In contrast, E283, D316, and K374 may be conserved because they play important structural roles. It is interesting that D316 corresponds to one of the few residues in the transmembrane segments of potassium channels that have been conserved throughout the three domains of life, including archea, bacteria, and eukaryotes (Jan and Jan, 1997).

Because several voltage-sensing residues are accessible to solvent and traverse the complete electric field, several groups have proposed that the S4 segment may not translocate across the entire width of the membrane during activation (Larsson *et al.,* 1996; Seoh *et al.,* 1996, Yang *et al.,* 1996). Instead, the channel may contain cavities for solvent intrusion in the region of the voltage sensor, serving to focus the transmembrane field on a distance shorter than the width of the bilayer. In this case, large backbone movements may not be needed to transfer the gating charge. Rather, some reorientation of the segments, with a repacking of the side chains into different rotamers, may suffice. In this regard, it is interesting to note that preliminary modeling of transmembrane segments S2, S3, and S4 predicts that S4 may be tilted relative to the two other segments (Tiwari-Woodruff *et al.,* 1997). The tilt of the S4 might provide space for solvent intrusion into the vicinity of the voltage-sensing residues.

Acknowledgments

Work in the author's laboratory was supported by grants from the NIH (GM43459); the American Heart Association, Greater Los Angeles Affiliate; the Muscular Dystrophy Association; the Laubisch Fund for Cardiovascular Research (UCLA); and the Stein/Oppenheimer Endowment (UCLA).

References

Aggarwal, S. K., and MacKinnon, R. (1996). Contribution of the S4 segment to gating charge in the Shaker K^+ channel. *Neuron* **16,** 1169–1177.

Akabas, M. H., Stauffer, D. A., Xu, M., and Karlin, A. (1992). Acetylcholine receptor channel structure probed in cysteine-substitution mutants. *Science* **258,** 307–310.

Almers, W. (1978). Gating currents and charge movements in excitable membranes. *Rev. Physiol. Biochem. Pharmacol.* **82,** 96–190.

Armstrong, C. M., and Bezanilla, F. (1973). Currents related to movement of gating particles of sodium channels. *Nature* **242,** 459–461.

Bezanilla, F., Perozo, E., Papazian, D. M., and Stefani, E. (1991). Molecular basis of gating charge immobilization in Shaker potassium channels. *Science* **254**, 679–683.

Bezanilla, F., Perozo, E., and Stefani, E. (1994). Gating of Shaker K^+ channels. 2. The components of gating currents and a model of channel activation. *Biophys. J.* **66,** 1011–1021.

Brüggemann, A., Pardo, L. A., Stühmer, W., and Pongs, O. (1993). *Ether-à-go-go* encodes a voltage-gated channel permeable to K^+ and Ca^{2+} and modulated by cAMP. *Nature* **365,** 445–448.

Cha, A., and Bezanilla, F. (1997). Characterizing voltage-dependent conformational changes in the Shaker K^+ channel with fluorescence. *Neuron* **19,** 1127–1140.

Chandy, K. G., and Gutman, G. A. (1995). Voltage-gated potassium channel genes. *In* "Ligand- and Voltage-Gated Ion Channels" (R.A. North, Ed.,), pp. 1–71. CRC Press, Ann Arbor, MI.

Dhallan, R. S., Yau, K.-W., Schrader, K. A., and Reed, R. R. (1990). Primary structure and functional expression of a cyclic nucleotide-activated channel from olfactory neurons. *Nature* **347,** 184–187.

Fakler, B., Brändle, U., Glowatzki, E., Weidemann, S., Zenner, H.-P., and Ruppersberg, J. P. (1995). Strong voltage-dependent inward rectification of inward rectifier K^+ channels is caused by intracellular spermine. *Cell* **80,** 149–154.

Ficker, E., Taglialatela, M., Wible, B. A., Henley, C. M., and Brown, A. M. (1994). Spermine and spermidine as gating molecules for inward rectifier K^+ channels. *Science* **266,** 1068–1072.

Goldstein, S. A. N. (1996). A structural vignette common to voltage sensors and conduction pores: Canaliculi. *Neuron* **16,** 717–722.

Gordon, S. E., and Zagotta, W. N. (1995). Localization of regions affecting an allosteric transition in cyclic nucleotide-activated channels. *Neuron* **14,** 857–864.

Goulding, E. H., Ngai, J., Kramer, R. H., Colicos, S., Axel, R., Siegelbaum, S. A., and Chess, A. (1995). Molecular cloning and single-channel properties of the cyclic nucleotide-gated channel from catfish olfactory neurons. *Neuron* **8,** 45–58.

Goulding, E. H., Tibbs, G. R., and Siegelbaum, S. A. (1994). Molecular mechanism of cyclic nucleotide-gated channel activation. *Nature* **372,** 369–374.

Guy, H. R., Durell, S. R., Warmke, J., Drysdale, R., and Ganetzky, B. (1991). Similarities in amino acid sequences of *Drosophila* eag and cyclic nucleotide-gated channels. *Science* **254,** 730.

Haynes, L. W., and Yau, K.-W. (1990). Single channel measurement from the cyclic GMP-activated conductance of catfish retinal cones. *J. Physiol.* (*London*) **429,** 451–481.

Hodgkin, A. L., and Huxley, A. F. (1952). A quantitative description of membrane current and its application to conduction and excitation in nerve. *J. Physiol.* (*London*) **117,** 500–544.

Hoshi, T., Zagotta, W. N., and Aldrich, R. W. (1990). Biophysical and molecular mechanisms of Shaker potassium channel inactivation. *Science* **250,** 533–538.

Jan, L. Y., and Jan, Y. N. (1990). A superfamily of ion channels. *Nature* **345,** 672.

Jan, L. Y., and Jan, Y. N. (1997). Cloned potassium channels from eukaryotes and prokaryotes. *Annu. Rev. Neurosci.* **20,** 91–123.

Kamb, A., Tseng-Crank, J., and Tanouye, M. A. (1988). Multiple products of the *Drosophila Shaker* gene may contribute to potassium channel diversity. *Neuron* **1,** 421–430.

Kaupp, U. B., Niidome, T., Tanabe, T., Terada, S., Bönigk, W., Stühmer, W., Cook, N. J., Kangawa, K., Matsuo, H., Hirose, T., Miyate, T., and Numa, S. (1989). Primary structure and functional expression from complementary DNA of the rod photoreceptor cyclic GMP-gated channel. *Nature* **342,** 762–766.

Larsson, H. P., Baker, O. S., Dhillon, D. S., and Isacoff, E. Y. (1996). Transmembrane movement of the Shaker K^+ channel S4. *Neuron* **16,** 387–397.

Li, M., Unwin, N., Stauffer, K. A., Jan, Y. N., and Jan, L. Y. (1994). Images of purified Shaker potassium channels. *Curr. Biol.* **4,** 110–115.

Liman, E. R., Hess, P., Weaver, F., and Koren, G. (1991). Voltage-sensing residues in the S4 region of a mammalian K^+ channel. *Nature* **353,** 752–756.

Liman, E. R., Tytgat, J., and Hess, P. (1992). Subunit stoichiometry of a mammalian K^+ channel determined by construction of multimeric cDNAs. *Neuron* **9,** 861–871.

Logothetis, D. E., Movahedi, S., Satler, C., Lindpaintner, D., and Nadal-Ginard, B. (1992). Incremental reductions of positive charge within the S4 region of a voltage-gated K^+ channel result in corresponding decreases in gating charge. *Neuron* **8,** 531–540.

MacKinnon, R. (1991). Determination of the subunit stoichiometry of a voltage-activated potassium channel. *Nature* **350,** 232–235.

Mannuzzu, L. M., Maronne, M. M., and Isacoff, E. Y. (1996). Direct physical measure of conformational rearrangement underlying potassium channel gating. *Science* **271,** 213–216.

Nagaya, N., and Papazian, D. M. (1997). Potassium channel α and β subunits assemble in the endoplasmic reticulum. *J. Biol. Chem.* **272,** 3022–3027.

Noda, M., Shimizu, S., Tanabe, T., Takai, T., Kayano, T., Ikeda, T., Takahashi, H., Nakayama, H., Kanaoka, Y., Minamino, N., Kangawa, K., Matsuo, H., Raftery, M. A., Hirose, T., Inayama, S., Hayashida, H., Miyata, Y., and Numa, S. (1984). Primary structure of *Electrophorus electricus* sodium channel deduced from cDNA sequence. *Nature* **312,** 121–127.

Papazian, D. M., and Bezanilla, F. (1997). How does an ion channel sense voltage? *News Physiol. Sci.* **12,** 203–210.

Papazian, D. M., Schwarz, T. L., Tempel, B. L., Jan, Y. N., and Jan, L. Y. (1987). Cloning of genomic and complementary DNA from *Shaker*, a putative potassium channel gene from *Drosophila. Science* **237,** 749–753.

Papazian, D. M., Shao, X. M., Seoh, S.-A., Mock, A. F., Huang, Y., and Wainstock, D. H. (1995). Electrostatic interactions of S4 voltage sensor in Shaker K^+ channel. *Neuron* **14,** 1293–1301.

Papazian, D. M., Timpe, L. C., Jan, Y. N., and Jan, L. Y. (1991). Alteration of voltage-dependence of Shaker potassium channel by mutations in the S4 sequence. *Nature* **349,** 305–310.

Perozo, E., MacKinnon, R., Bezanilla, F., and Stefani, E. (1993). Gating currents from a nonconducting mutant reveal open-closed conformations in Shaker K^+ channels. *Neuron* **11,** 353–358.

Perozo, E., Santacruz-Toloza, L., Stefani, E., Bezanilla, F., and Papazian, D. M. (1994). S4 mutations alter gating currents of Shaker K^+ channels. *Biophys. J.* **66,** 346–355.

Planells-Cases, R., Ferrer-Montiel, A. V., Patten, C. D., and Montal, M. (1995). Mutation in conserved negatively charged residues in the S2 and S3 transmembrane segments of a mammallian K^+ channel selectively modulates channel gating. *Proc. Natl. Acad. Sci. USA* **92,** 9422–9426.

Robertson, G. A., Warmke, J. W., and Ganetzky, B. (1996). Potassium currents expressed from *Drosophila* and mouse *eag* cDNAs in *Xenopus* oocytes. *Neuropharmacology* **35,** 841–850.

Sanguinetti, M. C., Jiang, C., Curran, M. E., and Keating, M. T. (1995). A mechanistic link between an inherited and an acquired cardiac arrhythmia: HERG encodes the I_{Kr} potassium channel. *Cell* **81,** 299–307.

Santacruz-Toloza, L., Huang, Y., John, S. A., and Papazian, D. M. (1994). Glycosylation of Shaker K^+ channel protein expressed in insect cell culture and in *Xenopus* oocytes. *Biochemistry* **33,** 5607–5613.

Schoppa, N. E., McCormack, K., Tanouye, M. A., and Sigworth, F. J. (1992). The size of gating charge in wild-type and mutant Shaker potassium channels. *Science* **255,** 1712–1715.

Schulteis, C. T., John, S. A., Huang, Y., Tang, C.-Y., and Papazian, D. M. (1995). Conserved cysteine residues in the Shaker K^+ channel are not linked by a disulfide bond. *Biochemistry* **34,** 1725–1733.

Schulteis, C. T., Nagaya; N., and Papazian, D. M. (1996). Intersubunit interaction between amino- and carboxyl-terminal cysteine residues in tetrameric Shaker K^+ channels. *Biochemistry* **35,** 12,133–12,140.

Schwarz, T. L., Tempel, B. L., Papazian, D. M., Jan, Y. N., and Jan, L. Y. (1988). Multiple potassium channel components are produced by alternative splicing at the *Shaker* locus in *Drosophila*. *Nature* **331,** 137–142.

Seoh, S.-A., Papazian, D. M., and Bezanilla, F. (1997). Simultaneous neutralization of four voltage-sensing residues in Shaker K^+ channels reduces the apparent gating charge per channel by about 80%. *Biophys. J.* **72,** A28 (abstract).

Seoh, S.-A., Sigg, D., Papazian, D. M., and Bezanilla, F. (1996). Voltage-sensing residues in the S2 and S4 segments of the Shaker K^+ channel. *Neuron* **16,** 1159–1167.

Shao, X. M., and Papazian, D. M. (1993). S4 mutations alter the single channel gating kinetics of Shaker K^+ channels. *Neuron* **11,** 343–352.

Sigg, D., and Bezanilla, F. (1997). Total charge movement per channel: The relation between gating charge displacement and the voltage sensitivity of activation. *J. Gen. Physiol.* **109,** 27–39.

Sigg, D., Stefani, E., and Bezanilla, F. (1994). Gating current noise produced by elementary transitions in Shaker potassium channels. *Science* **264,** 578–582.

Sigworth, F. J. (1993). Voltage-gating of ion channels. *Q. Rev. Biophys.* **27,** 1–40.

Starace, D. M., Stefani, E., and Bezanilla, F. (1997). Voltage-dependent proton transport by the voltage sensor of the Shaker K^+ channel. *Neuron* **19,** 1319–1327.

Stefani, E., Toro, L., Perozo, E., and Bezanilla, F. (1994). Gating of Shaker K^+ channels. 1. Ionic and gating currents. *Biophys. J.* **66,** 996–1010.

Stühmer, W., Conti, F., Suzuki, H., Wang, X., Noda, M., Yahagi, N, Kubo, H., and Numa, S. (1989). Structural parts involved in activation and inactivation of the sodium channel. *Nature* **339,** 597–603.

Tanabe, T., Takeshima, H., Mikami, A., Flockerzi, J., Takahashi, H., Kangawa, K., Kojima, M., Matsuo, H., Hirose, T., and Numa, S. (1987). Primary structure of the receptor for calcium channel blockers from skeletal muscle. *Nature* **328,** 313–318.

Tang, C.-Y., and Papazian, D. M. (1997). Transfer of voltage-independence from a rat olfactory channel to the *Drosophila* ether-à-go-go K^+ channel. *J. Gen. Physiol.* **109,** 301–311.

Tempel, B. L., Papazian, D. M., Schwarz, T. L., Jan, Y. N., and Jan, L. Y. (1987). Sequence of a probable potassium channel component encoded at *Shaker* locus of *Drosophila*. *Science* **237,** 770–775.

Tiwari-Woodruff, S. K., Schulteis, C. T., Mock, A. F., and Papazian, D. M. (1997). Electrostatic interactions between transmembrane segments mediate folding of Shaker potassium channel subunits. *Biophys. J.* **72,** 1489–1500.

Warmke, J. W., and Ganetzky, B. (1994). A family of potassium channel genes related to *eag* in *Drosophila* and mammals. *Proc. Natl. Acad. Sci. USA* **91,** 3438–3442.

Yang, N., and Horn, R. (1995). Evidence for voltage-dependent S4 movement in voltage-gated sodium channels. *Neuron* **15,** 213–218.

Yang, N., George, A. L., and Horn, R. (1996). Molecular basis of charge movement in voltage-gated sodium channels. *Neuron* **16,** 113–122.

Zagotta, W. N., and Aldrich, R. W. (1990). Voltage-dependent gating of Shaker A-type potassium channels in *Drosophila* muscle. *J. Gen. Physiol.* **95,** 29–60.

Zagotta, W. N., Hoshi, T., and Aldrich, R. W. (1994). Shaker potassium channel gating III: Evaluation of kinetic models for activation. *J. Gen. Physiol.* **103,** 321–362.

CHAPTER 4

Permeation of Voltage-Dependent Potassuim Channels

Kuniaki Ishii
Department of Pharmacology, Yamagata University School of Medicine, Yamagata, Japan

I. INTRODUCTION

Permeation of ions across cell membranes efficiently occurs through ion channels. Ion channels are transmembrane proteins forming water-filled pores that permit certain types of ions to pass through. They are considered to be like enzymes that catalyze ion transport across the lipid bilayer membrane. When activated, ion channels select ions and simultaneously allow selected ions to go through their pores at extremely high rates. Ion selectivity (substrate specificity) and rapid ion transport (efficient catalytic reaction) are the two fundamental characteristics of ion channels.

Voltage-dependent K^+ channels need to discriminate between K^+ ions and Na^+ ions; thus they always carry a net outward current to play their physiological roles. The permeability ratio of Na^+ to K^+ is as low as $<1/100$. In spite of this high selectivity, K^+ channels can transport K^+ ions at a rate of $>10^6$ ions per second (Yellen, 1987; Hille, 1992). It is natural to

1063-5823/99 $30.00

think that the channel pores must be narrow to be selective and the narrow region must be short to be highly conductive, which leads to an hourglass shape model. Are voltage-dependent K^+ channels really hourglass-like structures? Since the cloning of the Shaker K^+ channel (Papazian *et al.,* 1987; Timpe *et al.,* 1988), many attempts have been made to reveal the molecular structures that are responsible for the function of voltage-dependent K^+ channels, such as voltage sensing and ion permeation. Recent studies support the idea that voltage-dependent K^+ channels have a very short ion-selective pore (Lü and Miller, 1995; Pascual *et al.,* 1995; Naranjo and Miller 1996; Ranganathan *et al.,* 1996; Gross and MacKinnon, 1996). However, numerous lines of biophysical evidence have indicated that K^+ channels have multiple K^+-binding sites within their pore (Hodgkin and Keynes, 1955; Hille and Schwarz, 1978; Begenisich and De Weer, 1980; Neyton and Miller, 1988). These K^+-binding sites have high affinity for K^+ ions and could work as selectivity filters. This multi-ion nature suggests that the K^+ channels have a long ion-selective pore. In addition, extensive molecular biophysical studies have revealed that several structural elements are involved in K^+ ion permeation, which might be consistent with the multi-ion nature of the K^+ channels.

This chapter focuses especially on molecular structure of the pore of voltage-dependent K^+ channels, which has emerged from molecular biophysical experiments.

II. PROBING THE PORE STRUCTURE

To explore the pore structure, the amino acid residues that influence the pore properties such as conductance, ion selectivity, and susceptibility to pore blockers have been searched. The initial clue to the K^+ channel pore-forming region was obtained by using a peptide toxin from scorpion venom (MacKinnon and Miller, 1989). A collection of K^+ channel toxins has been isolated from the venom of scorpions (Carbone *et al.,* 1982; see Miller, 1995, for review). Leaving the biological role of these toxins aside, they have been extensively used as tools for studying K^+ channels.

A. Outer Vestibule

1. Early Findings

Charybdotoxin (CTX), a 37-amino-acid peptide, is the most widely used scorpion toxin. CTX applied externally inhibits high conductance Ca^{2+}-activated K^+ channels and voltage-dependent K^+ channels as well (Miller *et al.*, 1985; Gimenez-Gallego *et al.*, 1988; Leonard *et al.*, 1992). CTX is too large to enter the narrower part of K^+ channel pores. Blocking by external CTX is relieved by internal application of permeant ions such as K^+ and Rb^+ but not by application of impermeant ions. Permeant ions in the internal solution can enter the CTX-blocked K^+ channels and accelerate CTX dissociation. The CTX blocking mechanism occludes the K^+-conducting pore physically (MacKinnon and Miller, 1988; Goldstein and Miller, 1993). Therefore CTX has been used to probe the structure of the K^+ channel entryway. Because the interaction between K^+ channels and the cationic CTX is highly electrostatic, MacKinnon and Miller (1989) investigated the effects of mutation of negatively charged residues located in putative external loops of the Shaker K^+ channel (S1-S2, S3-S4 and S5-S6 linkers; Fig. 1). They found that residues in the first two external loops did not affect the CTX block but a residue in the third external loop (S5–S6 linker) did affect the block. Further mutagenesis studies of the Shaker K^+ channel have identified several residues in the S5–S6 linker region that affect CTX binding (MacKinnon *et al.*, 1990). The residues whose mutations produce CTX-insensitive channels are located near either end of the H5 sequence (the P region) (at positions D431, T449, and V451 of the Shaker K^+ channel). (Residue numbers are for the Shaker K^+ channel throughout this text, unless otherwise indicated.) The residues whose mutations weakly affect CTX sensitivity reside outside the P region (E422, K427, and G452).

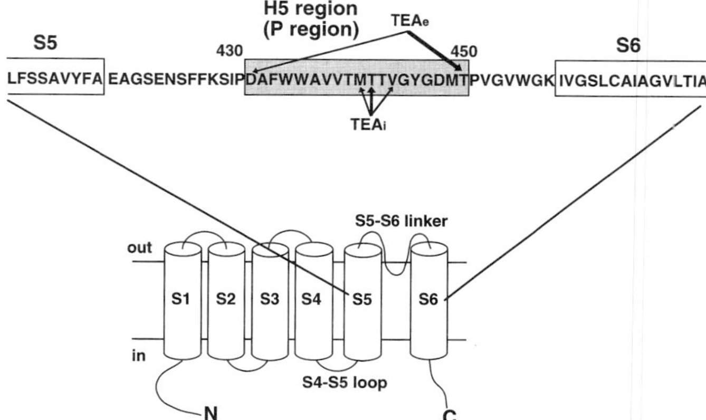

FIGURE 1 Proposed transmembrane topology of voltage-dependent potassium channel. S1 to S6 are hydrophobic transmembrane regions. The amino acid sequence of the pore loop of the Shaker K^+ channel is presented at top in single-letter codes. The H5 (P) region is indicated by a shaded box. Numbers apply to the two prolines at the boundary of the H5 region. The residues that affect external TEA (TEAe) and internal TEA (TEAi) bindings are also shown.

Another K^+ channel blocker commonly used for searching the channel pore is an impermeant open channel blocker, tetraethylammonium (TEA). TEA blocks most of the voltage-dependent K^+ channels from both sides of the membrane (Armstrong and Binstock, 1965; Armstrong, 1971; Hille, 1992). Since TEA is small compared to CTX, it has been used to explore the narrower region of the pore entryway. Mutagenesis studies have shown that the residue at 449 has a profound effect on the sensitivity to external TEA (Fig. 1) (MacKinnon and Yellen, 1990; Heginbotham and MacKinnon, 1992). Mutation of the amino acid residue at this same position also affects CTX binding (MacKinnon *et al.,* 1990). When this residue is tyrosine or phenylalanine, having aromatic side chains, the sensitivity of the channel to external TEA increases. The sensitivity increases with the number of subunits that carry a tyrosine at position 449. It is suggested that the TEA molecule interacts simultaneously with the four tyrosine residues at 449 to block the pore (Heginbotham and MacKinnon, 1992). The amino acid residue at 449 is proposed to be located at the extracellular pore entryway of about 8 Å diameter. Mutations of the residue at 431 weakly affect external TEA binding. The residue at 431 probably interacts with the TEA molecule through an indirect electrostatic reaction (MacKinnon and Yellen, 1990).

From these early data, we gained a vague picture about the outer mouth of the pore: the residues near both ends of the H5 region probably line the perimeter of the external mouth and the residue at 449 might be located somewhat deeper on the external mouth.

2. Recent Findings

The three-dimensional structures of several scorpion toxins have been determined (Bontems *et al.,* 1991; Johnson and Sugg, 1992; Fernandez *et al.,* 1994; Johnson *et al.,* 1994; Krezel *et al.,* 1995). Therefore, taking advantage of their known structure, spatial locations of pore residues have been assessed in detail by investigating specific interactions between the toxins and the K^+ channels. Goldstein and co-workers (1994) mutated all 30 solvent-exposed residues of CTX and examined the effects of each mutation on the kinetics of Shaker K^+ channel block. The residues critical for toxin–channel interaction are mapped on a remarkably flat surface of the CTX molecule; no critical residues are on the sides of the molecule. They also looked for interacting partner residues on CTX and the Shaker K^+ channel. From the complementary mutagenesis of both molecules, Shaker residue 425 (Shaker-425), as a partner of CTX residues 8 and 9, is placed at a ~20 Å radial distance from the central axis and at ~12 Å above the receptor floor. Since high-affinity protein–protein interaction surfaces are generally highly complementary, the CTX interaction surface suggests that the channel pore has a wide and shallow outer vestibule. Several other interacting pairs of residues between toxins and K^+ channels have been identified to present a finer picture of the outer vestibule. Shaker-427 (lysine) and CTX-11 (lysine) have been shown to sense each other through electrostatic forces (Stocker and Miller, 1994). Hidalgo and MacKinnon (1995) using a scorpion toxin, agitoxin 2 (AgTx2), focused on acidic and basic residues because of the highly electrostatic interaction between AgTx2 and K^+ channels. Using thermodynamic mutant cycle analysis, Shaker-431 is placed 12–15 Å away from the central axis. In the linear sequence, Shaker-431 locates at the amino terminus of the P region. Naranjo and Miller (1996) searched for a possible partner for Shaker-449, a major determinant of external TEA binding (Heginbotham and MacKinnon, 1992). Based on the proposed geometry that CTX-27 is close to the central axis and Shaker-449 is located within 5 Å of the central axis, CTX-29 has been identified as an interacting partner for Shaker-449 (Fig. 2). In the linear sequence, Shaker-449 is located at the carboxyl terminus of the P region. Aiyar and colleages (1995) investigated the topology of the P region of Kv1.3 using four structurally related scorpion toxins. They estimate that the outer vestibule of Kv1.3 is 28–34 Å wide and 4–8 Å deep. These and other recent findings on the structure of scorpion toxin receptors all indicate that K^+ channel pores have wide and shallow outer vestibules (Park and Miller, 1992; Stampe *et al.,* 1994; Stocker and Miller, 1994; Aiyar *et al.,* 1995; Hidalgo and MacKinnon, 1995; Naranjo and Miller, 1996; Ranganathan *et al.,* 1996).

B. Ion-Conducting Pore and Inner Vestibule

Several regions have been found to be structural elements of the ion-conducting pore. The H5 region in the S5–S6 linker was the first element

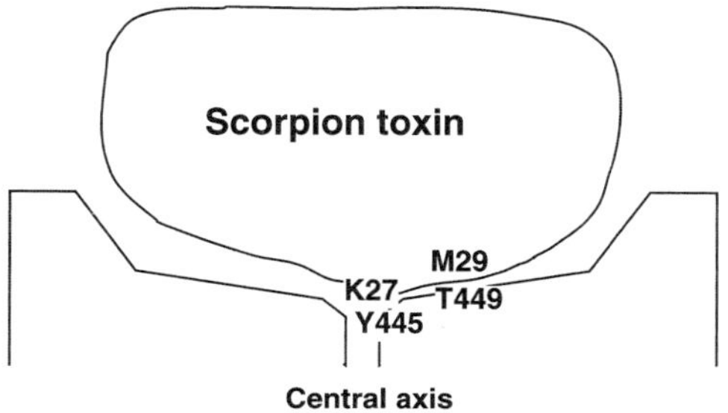

FIGURE 2 Cartoon of a scorpion toxin and outer vestibule of a K^+ channel. Two important pairs of interacting residues are indicated (after Naranjo and Miller, 1996; Ranganathan *et al.,* 1996). The picture is intended to show merely relative locations of the two pairs, and a shallow and wide outer vestibule.

whose contribution to forming the pore was demonstrated. Subsequently, the S4–S5 loop and the S6 transmembrane segment were found to be involved in K^+ ion permeation.

1. H5 Region (P region)

a. Early Findings. Segments have been exchanged between two related K^+ channel clones, Kv2.1 and Kv3.1, which belong to the Shaker superfamily (Frech *et al.,* 1989; Yokoyama *et al.,* 1989; Jan and Jan, 1990; see Chandy, 1991, for nomenclature). These two K^+ channel clones have different pore properties: the single channel conductance of Kv3.1 is three times larger than that of Kv2.1, the sensitivity to external TEA of Kv3.1 is higher than that of Kv2.1, and the sensitivity to internal TEA of Kv3.1 is much lower than that of Kv2.1. Hartmann and colleages (1991) constructed a chimera channel in which a 21-amino-acid stretch containing almost the entire P region of Kv2.1 is replaced with the corresponding region of Kv3.1. The chimera channel showed pore properties similar to the segment donor (Kv3.1), which include not only TEA sensitivity but also channel conductance. The results suggest that the P region contributes to the formation of the ion-conducting pore. Several point mutagenesis studies support this conclusion (Yellen *et al.,* 1991; Yool and Schwarz, 1991; Kirsch *et al.,* 1992). Yellen and colleagues (1991) found that serine substitution for a threonine residue at position 441 in the Shaker K^+ channel specifically reduces the sensitivity to internal TEA. The mutation affected neither the blockade by external TEA nor the channel conductance. The residue at 441 is a determinant of internal TEA binding. Nearby residues (M440 and V443) were later found to affect binding of internally applied TEA (Choi *et al.,* 1993; Aiyar *et al.,* 1994). These residues, which TEA can access from the cytoplasmic side, are located nearly in the middle of the P region (Fig. 1). Because TEA traverses 15% of the transmembrane electric field from the inside and 5% from the outside and the critical residues for internal and external TEA binding are at 441 and 449, respectively, the region flanked by these two residues (8 amino acids long) is thought to traverse 80% of the membrane electric field (Yellen *et al.,* 1991). Although the shape of the transmembrane electric field is unknown, these results lead to a model in which the P region crosses the membrane twice forming a β-strand conformation. Yool and Schwarz (1991) have reported that the relative permeabilities of K^+, Rb^+, and NH_4^+ are altered by the mutations of T441, which is important for the sensitivity to internal TEA, and two other residues in the P region. These data all suggest that the P region definitely constitutes part of the K^+ ion-conducting pore by dipping into the membrane.

b. Recent Findings. The substituted cysteine accessibility method (SCAM; Akabas *et al.*, 1992) is now widely used for analyzing the structures of channel pores. This method takes advantage of the unique chemical properties of the side chain of cysteine residues. In this approach, each residue in a putative pore-forming region is replaced with cysteine, one at a time, and the susceptibility of the mutated channels to sulfhydryl-specific reagents is assessed. If the mutated channel is modified by the sulfhydryl-specific reagents, the substituted residue is assumed to project its side chain into the pore. If the mutated channel is not modified, the substituted residue is assumed either to be inaccessible to the reagents or not to line the ion-conducting pore. Methanethiosulfonate (MTS) derivatives are usually used as the thiol-labeling reagents.

Lü and Miller (1995) probed the P region in the Shaker K^+ channel using a modified cysteine susceptibility analysis. Instead of organic MTS derivatives, they used a K^+ ion analog, Ag^+, as the thiol-labeling reagent to probe a deeper region of the pore. Eleven residues are sensitive to external Ag^+: P0, D1, W4, W5, V8, V13, Y15, D17, M18, T19, and P20. Residues are numbered from the beginning of the P region; P0 is proline at Shaker-430. These residues, some of which are reported to be involved in K^+ ion permeation, are thought to be exposed to external solution. Among the nine residues unaffected by Ag^+, four contiguous residues, T9, M10, T11, and T12, are placed on the cytoplasmic side of the pore, since T11 (Shaker residue T441) is known to be a major determinant of internal TEA binding (Yellen *et al.*, 1991). Surprisingly, the two residues that are thought to locate on the opposite side of the pore, T11 and V13, are separated by only one residue in the linear sequence. The overall pattern of Ag^+ sensitivity is not consistent with a β-strand conformation of the P region that was proposed in earlier models (Yellen *et al.*, 1991; Durell and Guy, 1992; Bogusz *et al.*, 1992). If the P region forms a β strand, a periodic pattern of Ag^+ sensitivity has to be observed. The second half of the pore-forming region in Kv2.1 was similarly studied by Pascual *et al.* (1995). They found that external application of charged MTS reagents modifies substituted cysteine residues at positions 17, 18, 19, and 21, numbered as above. Interestingly, the substituted cysteine residues at positions 9, 11, 12, and 13 in "the signature sequence" can be modified by internal application of charged MTS reagents. The signature sequence is a highly conserved stretch of 8 amino acids within the P region that determines ion selectivity (Heginbotham *et al.*, 1994). Again the SCAM results by Pascual and colleagues contradict a view of a β-stranded P region. The residues accessible from the internal and the external solution are located three residues away each other in the linear sequence. While external Ag^+ modifies the residue at 13 in the Shaker K^+ channel, internal MTS reagents modify the same residue in Kv2.1. Thus the residue at 13 can be accessed from both sides of the channel pore. Similar results showing two-sided access have been obtained in a cGMP-gated channel, as described below (Sun *et al.,* 1996). These findings support the idea that a K^+ channel has a very short ion-conducting pore.

Gross and MacKinnon (1996) used similar site-directed mutagenesis and chemical modification to explore the amino acid residues that fall within "footprint" of AgTx2. They made site-directed mutations in either one or four subunits and interpreted the results in detail, taking into consideration the orientation of the bound toxin. They propose that the amino-terminal side of the P region approaches the central axis forming an α helix and dips below the toxin-binding surface. Then the pore loop reemerges near the central axis and progresses radially in an extended conformation.

2. S4–S5 loop

The "intrinsic blocker" (the inactivation ball) has provided information about the inner mouth of the pore. In the Shaker K^+ channel, the inactivation ball is composed of the initial 20 amino acids, which occludes the channel pore from the cytoplasmic side (Hoshi *et al.,* 1990; Zagotta *et al.,* 1990). Because the positively charged residues in the ball appeared to play an important role in rapid inactivation, Isacoff and co-workers (1991) neutralized four acidic residues located in the cytoplasmic regions of the channel. They found that only the mutation of glutamic acid at 395 in the S4–S5 loop affected rapid inactivation. They further investigated several residues in the S4–S5 loop and suggested that the loop might form part of the receptor for the inactivation gate. Internal application of TEA affected rapid inactivation, which probably reflects the mutual interaction between TEA and the inactivation ball at the binding site (Choi *et al.,* 1991). Therefore, it is likely that the receptor for the inactivation ball forms part of the inner mouth. Increasing the concentration of external K^+ ion accelerates the recovery from rapid inactivation, which is explained by an electrostatic interaction between the inactivation ball and external K^+ ions (Demo and Yellen, 1991). Slesinger and colleagues (1993) have observed that mutations in the S4–S5 loop alter the channel conductance, the Rb^+ selectivity, and the sensitivity to internal TEA, Ba^{2+}, or Mg^{2+}. These results also support the involvement of the S4–S5 loop (the part of the receptor) in forming the inner mouth of the pore. Hybrid K^+ channels were constructed by linking a transient type (Kv1.4) and a delayed rectifier type (Kv1.2) K^+ channel or its S4–S5 loop mutant in tandem array (Nunoki *et al.,* 1994). The difference in inactivation kinetics between wild-type and mutant hybrid K^+ channels suggests that not only the S4–S5 loop of Kv1.4 but also that of Kv1.2 could serve as a receptor site. The results might imply that four

S4–S5 loops, one from each subunit, participate in forming one receptor for the inactivation ball. TEA applied internally blocks the K^+ channels and the blocking mechanism is thought to be occlusion of the pore. But, since TEA enters into the membrane's electric field from the cytoplasmic side to some extent, the shape of the inner vestibule is probably different from that of the outer vestibule. Recently, the NMR structures of two inactivation balls from mammalian transient type K^+ channels (Kv1.4 and Kv3.4) have been determined (Antz *et al.,* 1997). These inactivation ball peptides might serve as useful probes for studying the structure of the inner vestibule, like scorpion toxins do for the outer vestibule.

3. S6 Segment

The S6 segment has been found to be involved in K^+ ion permeation. Substitution of threonine at position 469 within the S6 segment with more hydrophobic residues increases the affinity for a series of alkyl-TEA compounds in a manner consistent with the direct interaction between the compounds and the residues (Choi *et al.,* 1993). The results suggest that the S6 segment is likely to form the part of the pore involved in internal TEA binding. Lopez and co-workers (1994) constructed S6 and H5 chimera channels in which the segment of Kv3.1 was transplanted into the corresponding region of the Shaker K^+ channel. They investigated the possible involvement of the S6 segment in forming the K^+ ion pathway by characterizing the pore properties of the chimera channels. The segment exchanged in the S6 chimera is a part of the S6 segment corresponding to residues 457 to 479 of Shaker and that in the H5 chimera is the 21-amino-acid stretch from the S5–S6 linker of Kv3.1, which is identical to that transplanted into Kv2.1 by Hartmann *et al.* (1991). The single channel conductance, blockade by Ba^{2+}, and sensitivity to internal TEA of the S6 chimera were similar to those of Kv3.1 (the segment donor) but the sensitivity to external TEA was similar to that of Shaker. The results obtained from the S6 chimera support the idea that the S6 segment forms the inner part of the channel pore. In contrast, the H5 chimera showed external TEA sensitivity similar to that of Kv3.1. The other pore properties of the H5 chimera were similar to those of Shaker. The results obtained from the H5 chimera are inconsistent with those obtained by Hartmann *et al.* (1991). They found that the identical H5 segment (P segment) of Kv3.1 conferred most of the pore properties of Kv3.1, not only TEA sensitivity but also single channel conductance, when transplanted into Kv2.1. However, when transplanted into Shaker, the H5 segment conferred only the external TEA sensitivity of Kv3.1. Taken together, these results are interesting because they imply that each structural element of the pore might contribute to the pore properties to a different extent in each K^+ channel. For example, since a large conductance of Kv3.1 was conferred to Kv2.1 but not to Shaker by transplanting the H5 segment, the small conductance of Kv2.1 is probably due to its H5 segment and the small conductance of Shaker K^+ channel is likely to be determined by part of S6 segment not by the H5 segment. Similar segment exchange experiments suggest that the S6 segment and its cytoplasmic extension (post-S6; a short stretch of 9 amino acids) contribute to forming the K^+ channel pore (Taglialatela *et al.,* 1994). Several other studies support that the S6 segment is involved in K^+ ion permeation (Aiyar *et al.,* 1994; Yeola *et al.,* 1996).

Kirsch and colleagues have investigated the structural components of the channel pore using Kv3.1 and Kv2.1. Pore properties examined include single channel conductance, TEA sensitivity, and 4-aminopyridine sensitivity. Based on the changes of these pore properties in the mutant channels, the cytoplasmic halves of the S5 and S6 segments have been proposed to contribute to the inner mouth of the pore (Kirsch *et al.,* 1993; Shieh and Kirsch, 1994).

C. Selectivity Filter

Several mutagenesis experiments have identified the residues that affect ion selectivity of the K^+ channels. Heginbotham and co-workers (1992) took advantage of the similarity of amino acid sequences and the dissimilarity of ion selectivity between voltage-dependent K^+ channels and cyclic nucleotide-gated channels (CNG channels). Whereas voltage-dependent K^+ channels are highly selective to K^+ ion and CNG channels are nonselective, their overall primary structures are very similar, having the P region. An alignment of their P region sequences revealed the lack of two residues in the CNG channel. The two residues correspond to Y445 and G446 of the Shaker K^+ channel, which are located within the signature sequence. Deletion of these two residues from the Shaker K^+ channel resulted in loss of selectivity among monovalent cations. Heginbotham and colleagues (1994) further mutated each amino acid residue in the signature sequence to see changes in ion selectivity. They found that mutations at four residues, T439, V443, G444, and G446, affect ion selectivity. Some mutations at T439 and V443 alter ion selectivity, while some mutations at the same positions do not. However, all the mutations examined at G444 and G446 produced nonselective channels. Thus, the most critical residues for determining ion selectivity seem to be G444 and G446, which is in agreement with their results of deletion mutants. It is now evident that the signature sequence plays fundamental role in determining ion selectivity (Heginbotham *et al.,* 1994; Kirsch *et al.,* 1992; Taglialatela *et al.,* 1993).

It is known that the lysine residue at position 27 (K27) of CTX interacts directly with K^+ ions in the ion conduction pathway located close to the central axis of the channel (Park and Miller, 1992; Goldstein and Miller, 1993). Lysine is present at the homologous position (K27 or K28) in all scorpion toxins called α-K-toxins (Miller, 1995). Ranganathan and colleagues (1996) have identified a partner residue of this important K27 on AgTx2. The partner they identified is Y445, which contributes to the formation of the K^+ ion selectivity filter (Fig. 2). Their results using AgTx2 reveal a shallow outer vestibule formed by the pore loops (the S5–S6 linker), which leads to a central selectivity filter formed by Y445 and nearby residues. The selectivity filter itself at the center of the vestibule is located about 5 Å from the external solution. Aiyar and colleagues (1996) investigated the spatial location of the two residues in the signature sequence. They used Kv1.3 and examined the two residues Y400 and D402, which correspond to Y445 and D447 of Shaker K^+ channel. The spatial location of the two residues relative to H404 (Shaker-449) has been determined. In their model, four alternating D402 and H404 form the external mouth and Y400 (Shaker-445) is located 4–6 Å deeper into the pore. It is now recognized that the selectivity filter is formed by the four pore loops at their meeting point; they approach from the periphery to the central axis and dip below into the pore to a short extent.

A schematic model of the K^+ channel pore structure is shown in Fig. 3. The external portion of the pore is mainly composed of the H5 segment, which contains selectivity filters and binding sites for TEA and scorpion

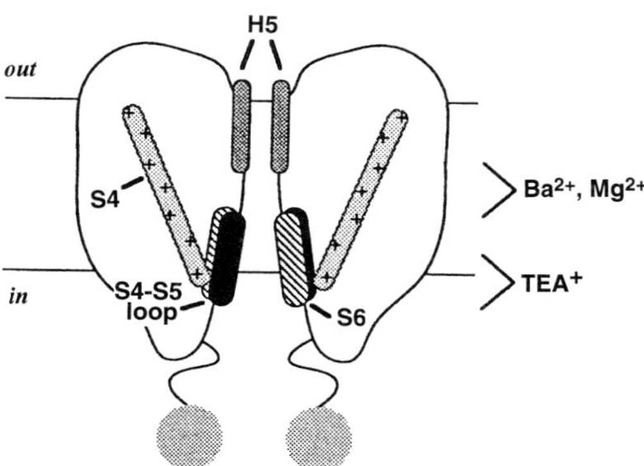

FIGURE 3 A schematic model showing the structural features of the K^+ channel pore. The ion-conducting pore contains several structural elements from the S4–S5 loop, the S6 segment, and the H5 segment. Reproduced with permission from Slesinger *et al.* (1993). Copyright © Cell Press.

toxins. The internal portion of the pore is mainly composed of the S4–S5 loop and the S6 segment. All the available data indicate that these structural elements are involved in K^+ ion permeation. Although complete agreement has not yet been obtained (Durell and Guy, 1996; Goldstein, 1996), it might be better to draw the H5 segment flatter and the narrower region shorter, according to the recent findings on the outer vestibule and the selectivity filter.

III. MULTI-ION NATURE

Numerous lines of biophysical evidence have indicated that voltage-dependent K^+ channels contain more than one K^+ ion at a time in the single-file regions (Hodgkin and Keynes, 1955; Hille and Schwarz, 1978; Begenisich and De Weer, 1980; Neyton and Miller, 1988). When multiple binding sites are occupied, repulsion between bound K^+ ions would greatly lower their binding affinity and facilitate their rapid dissociation. It is also suggested that the binding sites strongly interact with K^+ ions and could work as selectivity filters (Neyton and Miller, 1988; Hille, 1992). Thus, the apparently incompatible combination of the K^+ channel characteristics—high K^+ selectivity and high transport rate—could be explained by the multi-ion nature of the pores. Although the structure–function relationships have been extensively studied since the Shaker K^+ channel was cloned, little effort has been made to test the multi-ion nature of the cloned K^+ channels. It is only in recent years that molecular biophysical studies have established the multi-ion nature of Shaker voltage-dependent K^+ channels. The findings that support the multi-ion nature are as follows:

1. Perez-Cornejo and Begenisich (1994) examined the concentration dependence of the permeability ratios for different ions and voltage-dependent block by external Cs^+. The permeability ratio for K^+ and Rb^+ in the Shaker K^+ channel varied when the concentration of either ion was changed, which is a property of a multi-ion pore. Also in agreement with the multi-ion nature, voltage-dependent Cs^+ block of the Shaker K^+ channel varied with Cs^+ concentration.
2. Single-channel currents were recorded in symmetric solution containing the mixture of K^+ and NH_4^+ with the sum of the two ions being constant (Heginbotham and MacKinnon, 1993). As the mole-fraction of NH_4^+ was gradually increased, current amplitude went through a minimum. This is a feature called the anomalous mole-fraction effect which also suggests that the Shaker K^+ channel has a multi-ion pore.
3. Newland and colleagues (1992) showed in Shaker family K^+ channels that block by TEA applied to one side of the membrane was antagonized by

TEA applied to the opposite side. Their results were not due to competitive antagonism at one TEA binding site but due to mutual antagonism by TEA that occupied both the internal and external binding sites simultaneously. Such a mutual interaction is one of the properties consistent with a multi-ion pore.

4. To evaluate the ratio of ionic fluxes, unidirectional $^{42}K^+$ fluxes through the Shaker K^+ channel were measured by Stampe and Begenisich (1996). The Shaker K^+ channel showed a flux-ratio exponent of ~ 3, which indicates that the channel pore can accommodate at least four K^+ ions at a time in single file.

5. The Ba^{2+} ion, which has a similar unhydrated radius size to K^+ ion, does not permeate K^+ channels but blocks many K^+ channels. It is suggested that the Ba^{2+} ion may interact with the K^+-binding sites but cannot permeate through the pore because of its stronger binding to the sites. In the Shaker K^+ channel, two Ba^{2+}-binding sites have been found (Hurst *et al.,* 1995). This result also indicates the multi-ion nature of the Shaker K^+ channel pore.

If the ion-selective pore is short, as proposed in recent molecular biophysical studies, how can this multi-ion nature be explained?

IV. GATE AND SELECTIVITY FILTER

Compared with the voltage sensors and ion-permeation pore, the molecular nature of the gate is less well understood. As described above, voltage-dependent K^+ channels and CNG channels have homologous structure and their P region is the major determinant of ion selectivity.

In a cGMP-gated channel, in addition to functioning as a selectivity filter, the P region has been reported to function as the activation gate. Sun and co-workers (1996) utilized the SCAM to explore the residues within the P region of a cGMP-gated channel. MTS derivatives were applied to either side of the membrane during the open and closed states. The result was surprising. They found that cysteine at three positions (V4C, T20C, and P22C; numbered from the beginning of the P region) gained access to MTS-etylammonium from both sides of the membrane when the channel was closed. Since one of these mutants (T20C) was modified by negatively charged MTS-etylsulfonate and positively charged Ag^+ as well, the two-sided accessibility to the reagents is not likely to be through a lipophilic pathway. Since the reagents reacted in open and closed states at similar rates, the accessibility does not result from spontaneous openings of the pore. They suppose that the P regions of CNG channels form the thin blade of an iris-like structure, which is only one-residue thick at V4, T20, and P22 so that the side chains of substituted cysteine can rotate and gain access to the reagents from both sides. There must be no gate between external or internal solution and the regions where V4, T20, and P22 reside. Therefore, they further propose that the P region itself must form the gate of the channel, based on the findings of this two-sided accessibility.

In contrast, a different picture has been obtained in the case of the voltage-dependent K^+ channel. Holmgren and colleagues (1997) found that a Shaker K^+ channel S6 mutant in which I470 is substituted for cysteine can trap internally applied quaternary ammonium (QA) compounds, TEA, and decyltrietylammonium. The QA compounds enter the channel pore when the channel opens and are trapped while the channel is closed. The compounds dissociate only when the channel is activated and the trapping occurs with little changes of energetics in gating. They suppose that the activation gate may function as a trap door that prevents access from the intracellular solution to the ion-selective pore. This study has extended to a more detailed analysis of the molecular nature of the gate. Liu and colleagues (1997) introduced cysteine residues into the S6 and nearby region of the Shaker K^+ channel and examined the modification by the internally applied MTS reagents of the substituted cysteine residues during open and closed states. Significant state-dependent accessibility of the MTS reagents was observed when the residues in the deeper region of S6 (470 to 477) were substituted with cysteine. The results imply that the accessibility of the reagents to those residues is regulated by the activation gate, which is consistent with the trapping of QA compounds in the 470C mutant (Holmgren *et al.*, 1997). Modification of 470C and 474C by the MTS reagents was protected by the internal application of an open channel blocker tetrabutylammonium (TBuA), which suggests that these residues are located within the pore, behind or near the TBuA-binding site. Regulation of access to these residues from the intracellular solution by the activation gate could be explained by either of two models (Figs. 4A and 4B). To distinguish between two possibilities, they used Cd^{2+} ion, which binds tightly to the thiol side chains of cysteine. Cd^{2+} strongly blocked the 470C and 474C mutants, which are also state dependent. Especially in the 474C mutant, blocking by Cd^{2+} is essentially irreversible. In spite of this strong interaction between Cd^{2+} and cysteine at 474, voltage dependence for gating in the presence of Cd^{2+} is not altered from that in the absence of Cd^{2+}. Similar results are obtained from the 470C mutant. The results imply that the residues at 470 and 474 lining the pore stay static during channel gating, which favors the trap door model (Fig. 4).

V. CONCLUSION

Although we do not have direct structural information about the voltage-dependent K^+ channel, mutagenesis studies have provided us with basic

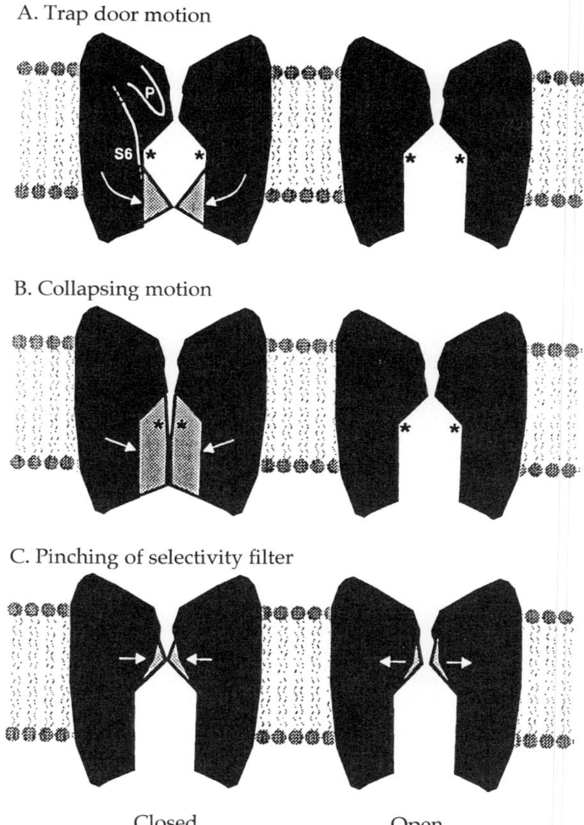

FIGURE 4 Three models of K^+ channel activation gating. (A) Intracellular gate. Between the gate and the selectivity filter is a cavity. In some K^+ channels this cavity can contain and trap organic blockers (Armstrong, 1971; Holmgren *et al.*, 1997). Hypothetical positions are shown for the P loop (forming the selectivity filter) and the S6 segment (part of which lines the pore). Evidence that the P loop forms the selective part of the pore comes from mutagenesis studies (MacKinnon and Yellen, 1990; Yool and Schwarz, 1991; Hartmann *et al.*, 1991; Yellen *et al.*, 1991; Heginbotham *et al.*, 1994). (B) Collapse of the intracellular mouth. In A and B, the asterisks indicate the putative location of 470 and 474. (C) Pinching-shut of the P-region selectivity filter [as proposed by Sun *et al.* (1996) for cyclic nucleotide-gated channels]. Reproduced with permission from Liu *et al.* (1997). Copyright © Cell Press.

information about the structural elements involved in K^+ ion permeation. The pore loops play a crucial role in ion selectivity and permeation. The structure of the outer vestibule is well characterized. The current picture of a voltage-dependent K^+ channel is much like that of an hourglass-like structure. However, there is a big unsolved question concerning the length of the pore where ion selection is established. Whereas recent findings by the SCAM and scorpion toxin bindings support a picture of a short ion-conducting pore, the multi-ion nature established by many biophysical studies in cloned K^+ channels supports a picture of a long pore. How can these two things be reconciled?

Ion channels, like enzymes, catalyze ion transport, and the catalytic site (the ion-selective site) is formed by the pore loops. The validity of the ion-selective site being formed by the pore loops has been recently discussed (MacKinnon, 1995; Marban and Tomaselli, 1997). In Na^+ channel pores, the pore loops are shown to be highly flexible (Benitah *et al.,* 1996; Tsushima *et al.,* 1997). The flexibility of the pore loops is probably a requirement for a catalytic site. Is it possible that the flexibility reconciles the two different pictures about the pore of the voltage-dependent K^+ channel? Although future success in the crystallization of channel proteins may present either of the two pictures, still further molecular biophysical studies will be needed to answer the currently unsolved question.

Acknowledgments

The author thanks Dr. Y. Kurachi for the opportunity to write this chapter and Dr. K. Nunoki for review of the manuscript.

Note added in proof: Recently the crystal structure of the bacterial K^+ channel (KcsA) was reported (Doyle *et al.,* 1998). The structure of the KcsA K^+ channel resembles an inverted teepee with the selectivity filter held at its wide base. The selectivity filter that can accommodate two K^+ ions is only 12 Å long. The poles of the teepee are four inner helices that cross each other near cytoplasmic end, and there is a water-filled cavity between the crossing of the helices and the selectivity filter. Although the KcsA K^+ channel is a two membrane-spanning channel and is gated by proton, its amino acid sequence in the pore region is nearly identical to that found in voltage-dependent K^+ channels. It is likely that the pore structure of voltage-dependent K^+ channels is essentially the same as that of KcsA K^+ channel.

References

Aiyar, J., Nguyen, A. N., Chandy, K. G., and Grissmer, S. (1994). The P-region and S6 of Kv3.1 contribute to the formation of the ion conduction pathway. *Biophys. J.* **67,** 2261–2264.

Aiyar, J., Rizzi, J. P., Gutman, G. A., and Chandy, K. G. (1996). The signature sequence of voltage-gated potassium channels projects into the external vestibule. *J. Biol. Chem.* **271,** 31,013–31,016.

Aiyar, J., Withka, J. M., Rizzi, J. P., Singleton, D. H., Andrews, G. C., Lin, W., Boyd, J., Hanson, D. C., Simon, M., Dethlefs, B., Lee, C-L., Hall, J. E., Gutman, G.A., and Chandy, K.G. (1995). Topology of the pore-region of a K^+ channel revealed by the NMR-derived structures of scorpion toxins. *Neuron* **15,** 1169–1181.

Akabas, M. H., Stauffer, D. A., Xu, M., and Karlin, A. (1992). Acetylcholine receptor channel structure probed in cysteine-substitution mutants. *Science* **258,** 307–310.

Antz, C., Geyer, M., Fakler, B., Schott, M. K., Guy, H. R., Frank, R., Ruppersberg, J. P., and Kalbitzer, H. R. (1997). NMR structure of inactivation gates from mammalian voltage-dependent potassium channels. *Nature* **385,** 272–275.

Armstrong, C. M. (1971). Interaction of tetraethylammonium ion derivatives with the potassium channels of giant axons. *J. Gen. Physiol.* **58,** 413–437.

Armstrong, C. M., and Binstock, L. (1965). Anomalous rectification in the squid giant axon injected with tetraethylammonium. *J. Gen. Physiol.* **48,** 859–872.

Begenisich, T., and De Weer, P. (1980). Potassium flux ratio in voltage-clamped squid giant axons. *J. Gen. Physiol.* **76,** 83–98.

Benitah, J. P., Tomaselli, G. F., and Marban, E. (1996). Adjacent pore-lining residues within sodium channels identified by paired cysteine mutagenesis. *Proc. Natl. Acad. Sci. USA* **93,** 7392–7396.

Bogusz, S., Boxer, A., and Busath, D. D. (1992). An SS1-SS2 beta-barrel structure for the voltage-activated potassium channel. *Prot. Eng.* **5,** 285–293.

Bontems, F., Roumestand, C., Boyot, P., Gilquin, B., Doljansky, Y., Menez, A., and Toma, F. (1991). Three-dimensional structure of natural charybdotoxin in aqueous solution by ^{1}H-NMR. Charybdotoxin possesses a structural motif found in other scorpion toxins. *Eur. J. Biochem.* **196,** 19–28.

Carbone, E., Wanke, E., Prestipino, G., Possani, L. D., and Maelicke, A. (1982). Selective blockage of voltage-dependent K^+ channels by a novel scorpion toxin. *Nature* **296,** 90–91.

Chandy, K. G. (1991). Simplified gene nomenclature [letter]. *Nature* **352,** 26.

Choi, K. L., Aldrich, R. W., and Yellen, G. (1991). Tetraethylammonium blockade distinguishes two inactivation mechanisms in voltage-activated K^+ channels. *Proc. Natl. Acad. Sci. USA* **88,** 5092–5095.

Choi, K. L., Mossman, C., Aube, J., and Yellen, G. (1993). The internal quaternary ammonium receptor site of Shaker potassium channels. *Neuron* **10,** 533–541.

Demo, S. D., and Yellen, G. (1991). The inactivation gate of the Shaker K^+ channel behaves like an open-channel blocker. *Neuron* **7,** 743–753.

Doyle, D. A., Cabral, J. M., Pfuetzner, R. A., Kuo, A., Gulbis, J. M., Cohen, S. L., Chait, B. T., and MacKinnon, R. (1998). The structure of the potassium channel: Molecular basis of K^+ conduction and selectivity. *Science* **280,** 69–77.

Durell, S. R., and Guy, H. R. (1992). Atomic scale structure and functional models of voltage-gated potassium channels. *Biophys. J.* **62,** 238–247; discussion 247–250.

Durell, S. R., and Guy, H. R. (1996). Structural model of the outer vestibule and selectivity filter of the Shaker voltage-gated K^+ channel. *Neuropharmacol.* **35,** 761–773.

Fernandez, I., Romi, R., Szendeffy, S., Martin-Eauclaire, M. F., Rochat, H., Van Rietschoten, J., Pons, M., and Giralt, E. (1994). Kaliotoxin (1-37) shows structural differences with related potassium channel blockers. *Biochemistry* **33,** 14,256–14,263.

Frech, G. C., VanDongen, A. M., Schuster, G., Brown, A. M., and Joho, R. H. (1989). A novel potassium channel with delayed rectifier properties isolated from rat brain by expression cloning. *Nature* **340,** 642–645.

Gimenez-Gallego, G., Navia, M. A., Reuben, J. P., Katz, G. M., Kaczorowski, G. J., and Garcia, M. L. (1988). Purification, sequence, and model structure of charybdotoxin, a potent selective inhibitor of calcium-activated potassium channels. *Proc. Natl. Acad. Sci. USA* **85,** 3329–3333.

Goldstein, S. A. (1996). A structural vignette common to voltage sensors and conduction pores: Canaliculi. *Neuron* **16,** 717–722.

Goldstein, S. A., and Miller, C. (1993). Mechanism of charybdotoxin block of a voltage-gated K^+ channel. *Biophys. J.* **65,** 1613–1619.

Goldstein, S. A., Pheasant, D. J., and Miller, C. (1994). The charybdotoxin receptor of a Shaker K^+ channel: Peptide and channel residues mediating molecular recognition. *Neuron* **12,** 1377–1388.

Gross, A., and MacKinnon, R. (1996). Agitoxin footprinting the shaker potassium channel pore. *Neuron* **16,** 399–406.

Hartmann, H. A., Kirsch, G. E., Drewe, J. A., Taglialatela, M., Joho, R. H., and Brown, A. M. (1991). Exchange of conduction pathways between two related K^+ channels. *Science* **251,** 942–944.

Heginbotham, L., and MacKinnon, R. (1992). The aromatic binding site for tetraethylammonium ion on potassium channels. *Neuron* **8,** 483–491.

Heginbotham, L., and MacKinnon, R. (1993). Conduction properties of the cloned Shaker K^+ channel. *Biophys. J.* **65,** 2089–2096.

Heginbotham, L., Abramson, T., and MacKinnon, R. (1992). A functional connection between the pores of distantly related ion channels as revealed by mutant K^+ channels. *Science* **258,** 1152–1155.

Heginbotham, L., Lu, Z., Abramson, T., and MacKinnon, R. (1994). Mutations in the K^+ channel signature sequence. *Biophys. J.* **66,** 1061–1067.

Hidalgo, P., and MacKinnon, R. (1995). Revealing the architecture of a K^+ channel pore through mutant cycles with a peptide inhibitor. *Science* **268,** 307–310.

Hille, B. (1992). "Ionic Channels of Excitable Membranes." Sinauer Associates, Sunderland, MA.

Hille, B., and Schwarz, W. (1978). Potassium channels as multi-ion single-file pores. *J. Gen. Physiol.* **72,** 409–442.

Hodgkin, A. L., and Keynes, R. D. (1955). The potassium permeability of a giant nerve fibre. *J. Physiol.* **128,** 61–88.

Holmgren, M., Smith, P. L., and Yellen, G. (1997). Trapping of organic blockers by closing of voltage-dependent K^+ channels: Evidence for a trap door mechanism of activation gating. *J. Gen. Physiol.* **109,** 527–535.

Hoshi, T., Zagotta, W. N., and Aldrich, R. W. (1990). Biophysical and molecular mechanisms of Shaker potassium channel inactivation. *Science* **250,** 533–538.

Hurst, R.S., Latorre, R., Toro, L., and Stefani, E. (1995). External barium block of Shaker potassium channels: Evidence for two binding sites. *J. Gen. Physiol.* **106,** 1069–1087.

Isacoff, E. Y., Jan, Y. N., and Jan, L. Y. (1991). Putative receptor for the cytoplasmic inactivation gate in the Shaker K^+ channel. *Nature* **353,** 86–90.

Jan, L. Y., and Jan, Y. N. (1990). A superfamily of ion channels [letter]. *Nature* **345,** 672.

Johnson, B. A., and Sugg, E. E. (1992). Determination of the three-dimensional structure of iberiotoxin in solution by ^1H nuclear magnetic resonance spectroscopy. *Biochemistry* **31,** 8151–8159.

Johnson, B. A., Stevens, S. P., and Williamson, J. M. (1994). Determination of the three-dimensional structure of margatoxin by ^1H, ^{13}C, ^{15}N triple-resonance nuclear magnetic resonance spectroscopy. *Biochemistry* **33,** 15,061–15,070.

Kirsch, G. E., Shieh, C. C., Drewe, J. A., Vener, D. F., and Brown, A. M. (1993). Segmental exchanges define 4-aminopyridine binding and the inner mouth of K^+ pores. *Neuron* **11,** 503–512.

Kirsch, G. E., Drewe, J. A., Taglialatela, M., Joho, R. H., DeBiasi, M., Hartmann, H. A., and Brown, A. M. (1992). A single nonpolar residue in the deep pore of related K^+ channels acts as a K^+:Rb^+ conductance switch. *Biophys. J.* **62,** 136–143; discussion 143–144.

Krezel, A. M., Kasibhatla, C., Hidalgo, P., MacKinnon, R., and Wagner, G. (1995). Solution structure of the potassium channel inhibitor agitoxin 2: Caliper for probing channel geometry. *Prot. Sci.* **4,** 1478–1489.

Leonard, R. J., Garcia, M. L., Slaughter, R. S., and Reuben, J. P. (1992). Selective blockers of voltage-gated K^+ channels depolarize human T lymphocytes: Mechanism of the antiproliferative effect of charybdotoxin. *Proc. Natl. Acad. Sci. USA* **89,** 10,094–10,098.

Liu, Y., Holmgren, M., Jurman, M. E., and Yellen, G. (1997). Gated access to the pore of a voltage-dependent K^+ channel. *Neuron* **19,** 175–184.

Lopez, G. A., Jan, Y. N., and Jan, L. Y. (1994). Evidence that the S6 segment of the Shaker voltage-gated K^+ channel comprises part of the pore. *Nature* **367,** 179–182.

Lü, Q., and Miller, C. (1995). Silver as a probe of pore-forming residues in a potassium channel. *Science* **268,** 304–307.

MacKinnon, R. (1995). Pore loops: An emerging theme in ion channel structure. *Neuron* **14,** 889–892.

MacKinnon, R., and Miller, C. (1988). Mechanism of charybdotoxin block of the high-conductance, Ca^{2+}-activated K^+ channel. *J. Gen. Physiol.* **91,** 335–349.

MacKinnon, R., and Miller, C. (1989). Mutant potassium channels with altered binding of charybdotoxin, a pore-blocking peptide inhibitor. *Science* **245,** 1382–1385.

MacKinnon, R., and Yellen, G. (1990). Mutations affecting TEA blockade and ion permeation in voltage-activated K^+ channels. *Science* **250,** 276–279.

MacKinnon, R., Heginbotham, L., and Abramson, T. (1990). Mapping the receptor site for charybdotoxin, a pore-blocking potassium channel inhibitor. *Neuron* **5,** 757–771.

Marban, E., and Tomaselli, G. F. (1997). Ion channels as enzymes: Analogy or homology? *Trend. Neurosci.* **20,** 144–147.

Miller, C. (1995). The charybdotoxin family of K^+ channel-blocking peptides. *Neuron* **15,** 5–10.

Miller, C., Moczydlowski, E., Latorre, R., and Phillips, M. (1985). Charybdotoxin, a protein inhibitor of single Ca^{2+}-activated K^+ channels from mammalian skeletal muscle. *Nature* **313,** 316–318.

Naranjo, D., and Miller, C. (1996). A strongly interacting pair of residues on the contact surface of charybdotoxin and a Shaker K^+ channel. *Neuron* **16,** 123–130.

Newland, C. F., Adelman, J. P., Tempel, B. L., and Almers, W. (1992). Repulsion between tetraethylammonium ions in cloned voltage-gated potassium channels. *Neuron* **8,** 975–982.

Neyton, J., and Miller, C. (1988). Discrete Ba^{2+} block as a probe of ion occupancy and pore structure in the high-conductance Ca^{2+} -activated K^+ channel. *J. Gen. Physiol.* **92,** 569–586.

Nunoki, K., Ishii, K., Okada, H., Yamagishi, T., Murakoshi, H., and Taira, N. (1994). Hybrid potassium channels by tandem linkage of inactivating and non-inactivating subunits. *J. Biol. Chem.* **269,** 24,138–24,142.

Papazian, D. M., Schwarz, T. L., Tempel, B. L., Jan, Y. N., and Jan, L. Y. (1987). Cloning of genomic and complementary DNA from Shaker, a putative potassium channel gene from Drosophila. *Science* **237,** 749–753.

Park, C. S., and Miller, C. (1992). Mapping function to structure in a channel-blocking peptide: Electrostatic mutants of charybdotoxin. *Biochemistry* **31,** 7749–7755.

Pascual, J. M., Shieh, C. C., Kirsch, G. E., and Brown, A. M. (1995). K^+ pore structure revealed by reporter cysteines at inner and outer surfaces. *Neuron* **14,** 1055–1063.

Perez-Cornejo, P., and Begenisich, T. (1994). The multi-ion nature of the pore in Shaker K^+ channels. *Biophys. J.* **66,** 1929–1938.

Ranganathan, R., Lewis, J. H., and MacKinnon, R. (1996). Spatial localization of the K^+ channel selectivity filter by mutant cycle-based structure analysis. *Neuron* **16,** 131–139.

Shieh, C. C., and Kirsch, G. E. (1994). Mutational analysis of ion conduction and drug binding sites in the inner mouth of voltage-gated K^+ channels. *Biophys. J.* **67,** 2316–2325.

Slesinger, P.A., Jan, Y.N., and Jan, L.Y. (1993). The S4-S5 loop contributes to the ion-selective pore of potassium channels. *Neuron* **11,** 739-749.

Stampe, P., and Begenisich, T. (1996). Unidirectional K^+ fluxes through recombinant Shaker potassium channels expressed in single Xenopus oocytes. *J. Gen. Physiol.* **107,** 449-457.

Stampe, P., Kolmakova-Partensky, L., and Miller, C. (1994). Intimations of K^+ channel structure from a complete functional map of the molecular surface of charybdotoxin. *Biochemistry* **33,** 443–450.

Stocker, M., and Miller, C. (1994). Electrostatic distance geometry in a K^+ channel vestibule. *Proc. Natl. Acad. Sci. USA* **91,** 9509–9513.

Sun, Z. P., Akabas, M. H., Goulding, E. H., Karlin, A., and Siegelbaum, S. A. (1996). Exposure of residues in the cyclic nucleotide-gated channel pore: P region structure and function in gating. *Neuron* **16,** 141–149.

Taglialatela, M., Champagne, M. S., Drewe, J. A., and Brown, A. M. (1994). Comparison of H5, S6, and H5-S6 exchanges on pore properties of voltage-dependent K^+ channels. *J. Biol. Chem.* **269,** 13,867–13,873.

Taglialatela, M., Drewe, J. A., Kirsch, G. E., De Biasi, M., Hartmann, H. A., and Brown, A. M. (1993). Regulation of K^+/Rb^+ selectivity and internal TEA blockade by mutations at a single site in K^+ pores. *Pflu. Arch.—Eur. J. Physiol.* **423,** 104–112.

Timpe, L. C., Schwarz, T. L., Tempel, B. L., Papazian, D. M., Jan, Y. N., and Jan, L. Y. (1988). Expression of functional potassium channels from Shaker cDNA in Xenopus oocytes. *Nature* **331,** 143–145.

Tsushima, R. G., Li, R. A., and Backx, P. H. (1997). P-loop flexibility in Na^+ channel pores revealed by single- and double-cysteine replacements. *J. Gen. Physiol.* **110,** 59–72.

Yellen, G. (1987). Permeation in potassium channels: Implications for channel structure. *Annu. Rev. Biophys. Chem.* **16,** 227–246.

Yellen, G., Jurman, M. E., Abramson, T., and MacKinnon, R. (1991). Mutations affecting internal TEA blockade identify the probable pore-forming region of a K^+ channel. *Science* **251,** 939–942.

Yeola, S. W., Rich, T. C., Uebele, V. N., Tamkun, M. M., and Snyders, D. J. (1996). Molecular analysis of a binding site for quinidine in a human cardiac delayed rectifier K^+ channel. Role of S6 in antiarrhythmic drug binding. *Circ. Res.* **78,** 1105–1114.

Yokoyama, S., Imoto, K., Kawamura, T., Higashida, H., Iwabe, N., Miyata, T., and Numa, S. (1989). Potassium channels from NG108-15 neuroblastoma-glioma hybrid cells. Primary structure and functional expression from cDNAs. *FEBS Lett.* **259,** 37–42.

Yool, A. J., and Schwarz, T. L. (1991). Alteration of ionic selectivity of a K^+ channel by mutation of the H5 region. *Nature* **349,** 700–704.

Zagotta, W. N., Hoshi, T., and Aldrich, R. W. (1990). Restoration of inactivation in mutants of Shaker potassium channels by a peptide derived from ShB. *Science* **250,** 568–571.

CHAPTER 5

IsK: A Novel Type of Potassium Channel Regulatory Subunit

Jacques Barhanin, Georges Romey, and Michel Lazdunski
Institut de Pharmacologie Moléculaire et Cellulaire, CNRS UPR 411, Sophia Antipolis, 06560 Valbonne, France

I. SHORT HISTORY

A. IsK Has an Atypical Structure

The story of IsK (or minK, gene nomenclature *KCNE1*) began 10 years ago, when the group of S. Nakanishi cloned the cDNA of IsK using functional expression in *Xenopus* oocytes (Takumi *et al.*, 1988). This was a very exciting period for K^+ channels since it was the era of cloning the first Shaker-type K^+ channel, immediately followed by the cloning of many related K^+ channel proteins(Miller, 1991; Salkoff and Jegla, 1995). With

1063-5823/99 $30.00

the growing number of available sequences, the picture of what a K^+ channel subunit could be rapidly became elucidated. In particular, the presence of a K^+-selective pore signature sequence, the so-called P domain, was recognized as a hallmark of the whole family (Heginbotham *et al.*, 1994; Mackinnon, 1995; Pascual *et al.*, 1995).

Until very recently, IsK has been regarded as somewhat of an enigma in the ion channel field. The IsK protein is small, 126 to 130 amino acids, depending on the animal species (Fig. 1C), presents only one transmembrane domain (Fig. 1A), and has no sequence homology with any other protein, including all the conventional K^+ channel proteins. In particular, nothing like a consensus P domain exists in its sequence. Nonetheless, when

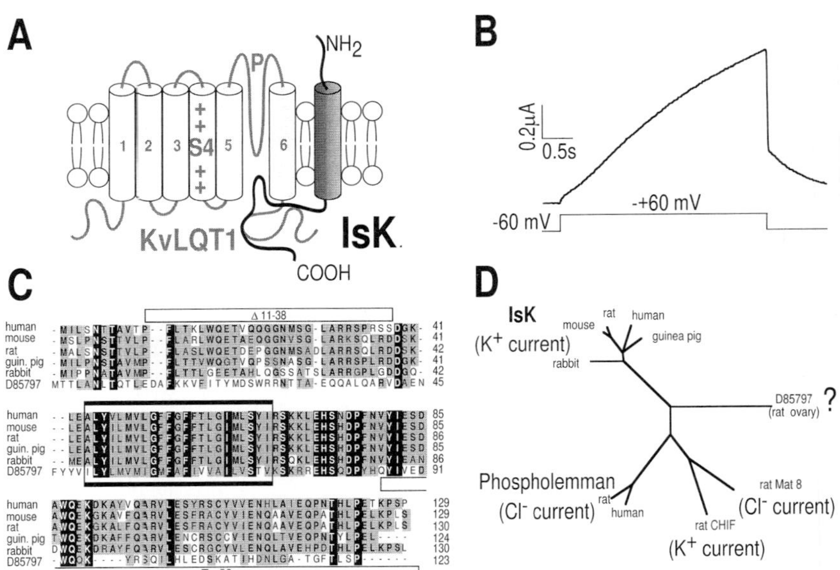

FIGURE 1 General features of IsK protein. (A) Schematic representation of the membrane topology of IsK and of its identified partner, KvLQT1. (B) Typical trace of human IsK expressed in *Xenopus* oocytes. The current activates and deactivates slowly and has not reached a steady state at the end of the 3-sec pulse. (C) Primary sequence alignment of ortholog IsK proteins from different species. The transmembrane segment is boxed and the domains that have been deleted in the Δ11.38 and Tr. 80 mutated proteins are indicated by open rectangles over outlining. Identical residues shared by all sequences are shown by white type on black and conserved residues are in gray areas. D85797 is the GenBank accession number of a rat ovary expressed sequence tag revealed by a BLAST search using the whole human IsK sequence. Note the high conservation in the transmembrane and the immediately following segments. (D) Phylogenetic tree of IsK-like family members generated using Genetic Computer Group software.

expressed in *Xenopus* oocytes, IsK induces a very slow voltage-dependent current with a high K^+ selectivity, very similar to the slow component of the cardiac delayed rectifier current, I_{Ks} (Fig. 1B) (Takumi *et al.,* 1988; Swanson *et al.,* 1993; Honoré *et al.,* 1994; Busch and Suessbrich, 1997; Kaczmarek and Blumenthal, 1997).

B. Controversy over IsK Function

Although a restriction to the channel function of IsK was formulated very early [actually this restriction was already made in the *princeps* cloning paper (Takumi *et al.,* 1988)], many investigators, if not all, were convinced that IsK was acting as a channel by itself despite its unusual structure. Major indications in favor of this hypothesis resulted from site-directed mutagenesis studies showing that mutations in the hydrophobic domain and proximal COOH-terminal segment produced changes in current gating and permeation [reviewed by Kaczmarek and Blumenthal 1997]. Furthermore, it has been observed that alterations in the transmembrane regions also influence the voltage-dependent binding of triethylammonium, a classical pore probe (Wang *et al.,* 1996a). Finally, the channel hypothesis culminated in a model based on complex polymerization analysis, in which it was proposed that the channel was formed of a pentamer of trimers of IsK subunits (Tzounopoulos *et al.,* 1995).

However, in the same period several other lines of evidence were accumulated in favor of a regulatory hypothesis in which IsK acts as an activator of an endogenous silent channel in oocytes. First, the amplitude of IsK currents in oocytes saturates at low levels of cRNA injection, even though the amount of IsK protein in the plasma membrane continues to increase (Attali *et al.,* 1993; Blumenthal and Kaczmarek, 1994). Second, attempts to express IsK currents in numerous eukaryotic cells other than oocytes failed, although the IsK protein was detected in high quantity in cells infected with IsK-expressing vaccinia virus or baculovirus, indicating the lack of an essential cofactor in these cells (Lesage *et al.,* 1993). Third, it was observed that the injection of high quantities of IsK cRNA could induce both a slow outward K^+ current upon depolarization and a hyperpolarization-activated inward Cl^- current (Fig. 2A) (Attali *et al.,* 1993). Such a Cl^- current is occasionally seen in noninjected oocytes, and is identical to the current induced upon injection of another small single transmembrane domain protein, phospholemman (Moorman *et al.,* 1992). This Cl^- current was not recorded upon injection of an equivalent amount of several K^+ channel-specific cRNAs, or of mutated IsK. Expression studies of IsK proteins truncated in either the N- or the C-terminal regions established

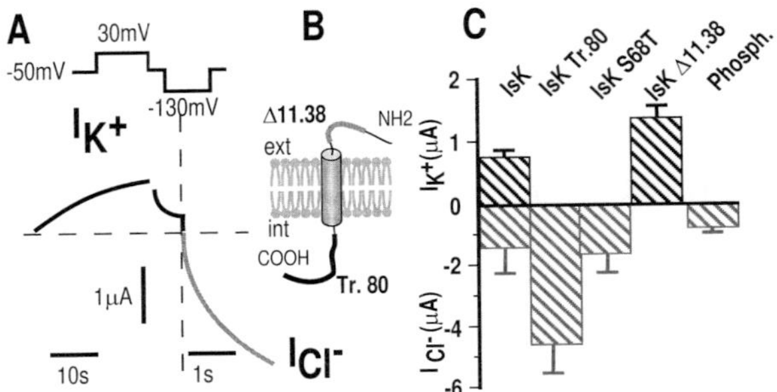

FIGURE 2 IsK is a dual activator of K^+ and Cl^- channels in *Xenopus* occytes. (A) Traces from oocytes injected with 20 ng of IsK cRNA using the pulse-command protocol shown, consisting of a 30-sec depolarization step from −50 to +30 mV followed by a 1-sec repolarization step to −50 mV and a 3-sec hyperpolarizing pulse to −130 mV. (B) Schematic representation of the IsK protein showing the domains deleted in the Δ11.38 and Tr. 80 mutants in gray and black, respectively. (C) Effects of these deletions of the S68T point mutation on IsK-induced K^+ Cl^- Channel induction while increasing K^+ channel activation. The C-terminal truncation (Tr. 80) provokes the reverse situation, enhancing the induced Cl^- currents and eliminating the K^+ currents. The S68T mutation abolishes only the K^+ channel activity and leaves the Cl^- channel activity unmodified. The bar labeled phosph. indicates the K^+ and Cl^- current amplitudes obtained with the protocol shown in A after injection of 50 ng of dog phospholemman cRNA. Reproduced with permission from *Nature* (Attali *et al.*, 1993. The protein IsK a dual activator of K^+ and Cl^- channels. **365,** 850–852) Macmillan Magazines, Limited.

these regions as essential for activation of the Cl^- or K^+ currents, respectively (Figs. 2B, 2C). The simplest explanation for these findings is that IsK (and occasionally phospholemman) activates endogenous oocyte K^+ and Cl^- channels, with a higher affinity for the K^+ channels. The cloning of another K^+-current-inducing factor from rat colon termed CHIF has been regarded as further evidence for the existence of some endogenous silent K^+ channel in oocytes (Attali *et al.*, 1995).

The long-running debate was finally closed with the demonstration of the interaction of IsK with the conventional K^+ channel subunit KvLQT1, identified by positional cloning in patients with the long QT1 syndrome (Barhanin *et al.*, 1996; Sanguinetti *et al.*, 1996; Wang *et al.*, 1996b).

II. ASSOCIATION OF IsK WITH KvLQT1

The *KVLQT1* gene is responsible for the most common form of inherited cardiac arrhythmia known as the chromosome 11-linked long QT syndrome (Wang *et al.*, 1996b). Soon after this discovery, it was shown that the *KVLQT1* gene product was a classic K^+ channel pore-forming subunit that, when associated with IsK, generated an I_{Ks}-like current upon expression in different heterologous systems (Barhanin *et al.,* 1996; Sanguinetti *et al.,* 1996). The physical association of IsK and KvLQT1 has been demonstrated by coimmunoprecipitation of the two proteins (Barhanin *et al.,* 1996) and the endogenous presence of a KvLQT1 subunit in oocytes has been confirmed by the cloning of a *Xenopus* analog (Sanguinetti *et al.,* 1996).

A. *Electrophysiological Findings*

The following results summarize the main conclusions drawn from whole-cell voltage-clamp analysis of K^+ currents on transfected COS cells with IsK, KvLQT1, and KvLQT1+IsK:

1. IsK alone is incapable of expressing any current.
2. The expression of KvLQT1 induces a rapidly activating and a slowly deactivating outward K^+ current (Fig. 3A).
3. Cotransfection of IsK with KvLQT1 leads to the expression of a K^+ current identical to the very slow activating current present in *Xenopus* oocytes injected with cRNA from IsK alone (Fig. 3B). This KvLQT1/IsK current resembles the cardiac delayed rectifier I_{Ks} current not only in the activation and deactivation kinetics but also in the activation threshold potential, which is shifted from −40 mV for KvLQT1 to −20 mV for KvLQT1/IsK, the expected value for I_{Ks}.
4. The amplitude of the KvLQT1/IsK current in COS cells is increased compared to the KvLQT1 current alone (mean current densities of 10.2 ± 1.3 and 43.1 ± 4.1 pA/pF for KvLQT1 and KvLQT1/IsK (at +30 mV, $n = 30$).
5. A striking result is that the deactivation kinetics of the KvLQT1/IsK channel are close to those of KvLQT1 alone (Figs. 3A, 3B). This hallmark of the KvLQT1 channel has been transferred to the KvLQT1/IsK channel, regardless of the difference in the activation kinetics.

At the level of the single-channel analysis, the effects of the association of IsK with KvLQT1 are a dramatic increase in the channel density concomitant with a drastic decrease in the unitary conductance (Romey *et al.,* 1997). These two parameters change from 1–2 channels per patch with a conductance of 7.6 pS for KvLQT1 to 50–100 channels of 0.5–0.6 pS for KvLQT1/IsK (Figs. 3C, 3D). Except in very rare patches where single-channel activity was detectable because the channel number was low and the seal conditions very good, the calculation of the single-channel parameters of the complex KvLQT1/IsK necessitated the use of variance analysis (Romey *et al.,* 1997). Concordant results have been also obtained with cell-

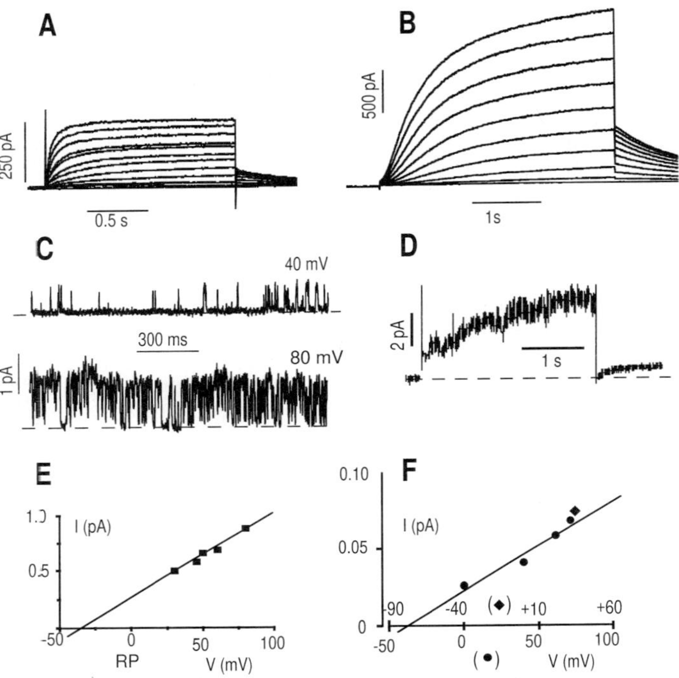

FIGURE 3 Transient expressions of KvLQT1 and KvLQT1/IsK in transfected COS cells. Fast-activating (A) KvLQT1 and slow-activating KvLQT1/IsK (B) channels in COS cells under whole cell recording configuration. Voltage pulses from −80 to +50 mV in 10-mV steps. Tail currents at −40 mV. (C–F) Analysis of detectable (KvLQT1 channel) and nondetectable (KvLQT1/IsK channel) unitary currents in COS membrane patches. (C) KvLQT1 channel: cell-attached patch; activities at two potentials relative to the resting potential (RP). (D) KvLQT1/IsK channel: outside-out patch; representative K^+ current response to step depolarization to +40 mV. (E) KvLQT1 channel: *I–V* relationship for the unitary currents (mean values from three patches). (F) KvLQT1/IsK channel: *I–V* relationship for the unitary currents. (●) Mean values from 30 cell-attached patches and (◆) from the patch shown in D. The unitary currents were estimated from fits of variance–mean current plots with the parabolic function $\sigma^2 = i\, I(t) - I(t)^2\text{–}N$, where σ^2 = variance of the current, i = unitary current, and N-number of functional channels. Adapted from Romey *et al.*, (1997) with permission from The American Society for Biochemistry and Molecular Biology.

attached macropatches in *Xenopus* oocytes with about 10^3 active channels in the patch and a unitary conductance of 0.52 pS.

Hence, the large increase in K^+ current following association of IsK with KvLQT1 results from a mélange of two factors: a large reduction in the unitary conductance overbalanced by a larger increase in the number of functional channels.

B. *Biochemical Findings*

To confirm independently that IsK physically associates with KvLQT1 and to map the interacting protein domains, a yeast two-hybrid assay method was used (Romey *et al.,* 1997). The bait plasmids contained the N terminus (IsKN, aa. 11–38) and C terminus (IsKC, aa. 67–129) of human IsK fused to the GAL4 DNA-binding domain. The prey plasmids were chimeras of various regions of the KvLQT1 channel protein with the GAL4 activation domain. The results essentially showed that the IsK C terminus exhibited a very strong interaction with the region comprising the pore domain of KvLQT1 (KvLQT1P, aa. 218–259) without significant interaction with the entire N terminus (KvLQT1N, aa.1–64) or the entire C terminus of the KvLQT1 channel protein (KvLQT1C, aa. 290–604). Similarly, the IsK N terminus failed to interact with the carboxy- or the amino-terminal domains of KvLQT1. This surprising, but specific interaction of IsK,C-terminal and KvLQT1,P domains has been fully confirmed by affinity chromatography experiments. The IsK protein in its entirety is produced in Sf9 insect cells infected with recombinant baculovirus and assayed for retention on an affinity matrix consisting of GST–KvLQT1 fusion proteins bound to glutathione–Sepharose beads. These experiments effectively demonstrate that the pore region of KvLQT1 is the only region that specifically associates with IsK, and neither the N-terminal nor the C-terminal cytoplasmic domains of KvLQT1 presents any affinity for IsK under these conditions (Romey *et al.,* 1997).

C. *Working Model of the Interaction KvLQT1/IsK*

Both electrophysiological and biochemical data have been used to propose a minimal model that could explain how the KvLQT1/IsK association leads to drastic changes in the properties of the KvLQT1 channel such as activation kinetics and unitary conductance, while preserving K^+ selectivity and deactivation kinetics (Fig. 4). IsK first binds to the outer shell of the KvLQT1 channel, probably via its transmembrane domain. This step provides a closer positioning of the C-terminal domain of IsK to the pore of KvLQT1. Once the KvLQT1 channel reaches the open conformation (O), the C-terminal domain of IsK enters and binds to the pore (OIsK). This leads to a total occlusion, which is later transformed into a partial occlusion resulting in a narrower pore (OIsK*), creating an additional barrier to K^+ mobility and drastically reducing the unitary conductance. The total pore occlusion produced by IsK before relaxation to a partial occlusion is supported by the fact that the normal 7.6-pS KvLQT1 conduc-

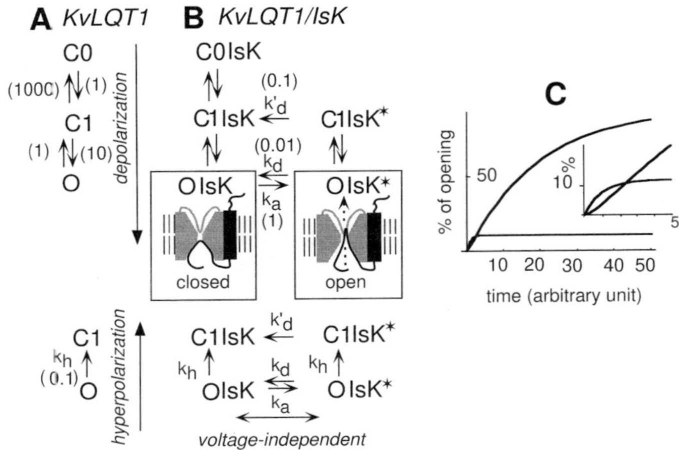

FIGURE 4 Models of KvLQT1 and KvLQT1/IsK channels. (A) KvLQT1 channel with C0, C1, closed states, O, open state, and KvLQT1/IsK channel (B). The C-terminal part of IsK in the inner mouth of the pore (IsK conformation) or inside the pore (IsK* conformation). States are: closed (C0IsK, C1IsK), blocked (OIsK, C1IsK*), partially open (OIsK*). (C) Simulation of the models using the indicated rate constant values. Adapted from Romey *et al.* (1997) with permission from The American Society for Biochemistry and Molecular Biology.

tance is never recorded with the KvLQT1/IsK channel before observing the small conductance behavior. The slow kinetics of activation reflect the conformational change (OIsK $\rightarrow$ OIsK*) leading to the partial opening of the pore. The difficulty for the channel to close when occupied by the C-terminal end of IsK, the "foot-in-the-door" process (Yeh and Armstrong, 1978), leads to the accumulation of open channels (OMK*) and to an increase in the number of functional channels. Actually, long open times (>1 sec) are observed in the few patches with detectable unitary currents (Romey *et al.*, 1997). Assigning an arbitrary value of 1 to the rate constant of the transition between the closed states C0 and C1, one can then set all the other rate constants (Figs. 4A, 4B) for a quantitative treatment of the model. A computer simulation provides a satisfactory fit of the key current properties of KvLQT1/IsK: that is, the slow activation kinetics, and the higher level of K^+ channel expression, and the unchanged rates of deactivation (Fig. 4C).

An extension of the model incorporates basic molecular views on voltage-dependent K^+ channels. The KvLQT1 channel is likely formed by four identical subunits and it is tempting to assume that each of them can bind one IsK subunit. The model in Fig. 5A states that IsK binds independently to each of the KvLQT1 subunits. To simplify, it is hypothesized that IsK

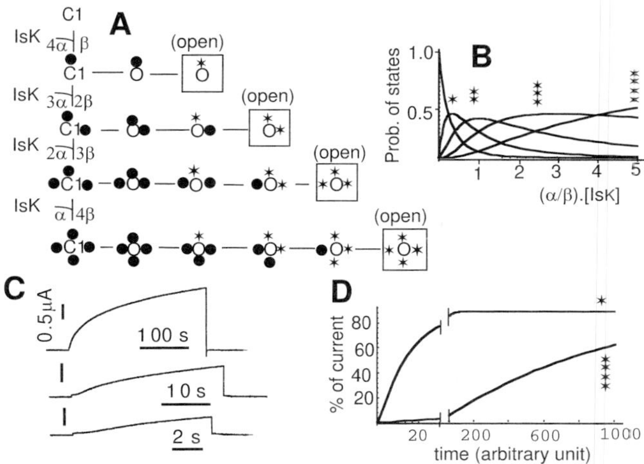

FIGURE 5 Model in which up to four IsK subunits can bind to the KvLQT1 tetramer. (A) States diagrammed with ● represent an IsK subunit in a position that totally plugs the channel, and those with * represent an IsK subunit that only partially occludes the pore. The channel is open if all IsK are in the IsK* conformation and blocked if only one is in the IsK● conformation. (B) Probability of the channel to integrate 0, 1, 2, 3, or 4 IsK as a function of IsK concentration. (C) Time course of activation in the same *Xenopus* oocyte at different time scales of stimulation. Voltage-clamp to +30 from −80 mV. (D) Simulation at two consecutive time scales of the fastest (KvLQT1/IsK) and the slowest (KvLQT1/4IsK) activation kinetics. Adapted from Romey *et al.* (1997) with permission from The American Society for Biochemistry and Molecular Biology.

binds preferentially to KvLQT1 in the C1 state with α and β as kinetic constants for the transition rates. Consequently, up to four IsK could bind to the KvLQT1 channel. The proportion of KvLQT1 channels having incorporated 0, 1, 2, 3, or 4 IsK depends on the effective concentration of IsK in the membrane, as illustrated in Fig. 5B. This model accounts for the complex kinetic behavior of the slow KvLQT1/IsK channel after a long-lasting depolarization. As an example, in the oocyte recordings shown in Fig. 5C at different time scales, it is clear that the current reaches a steady-state level only after several minutes. In fact, the channel can adopt four different open states with distinct kinetics, O^* to O^{4*}, O^* being the fastest and O^{4*} being the slowest states. For short depolarization, the evoked current is shaped only by the fast activation components, while the slowest components become predominant for long-lasting depolarizations. The model is also consistent with an early curious observation that *Xenopus* oocytes injected with large amounts of IsK cRNA have currents that activate more slowly and are delayed, relative to oocytes injected with smaller amounts (Cui *et al.*, 1994; Kaczmarek and Blumenthal, 1997). This behavior is now adequately simulated by our model, which postulates an increase in the number of closed states as the IsK concentration is elevated (Fig. 5D).

III. DIFFERENCES BETWEEN IsK AND ION CHANNEL β SUBUNITS

Auxiliary subunits of voltage-dependent ion channels (Na^+, Ca^{2+}, or K^+ channels) comprise both entirely cytoplasmic intracellular subunits and integral membrane proteins with one, two, or four transmembrane domains (Gurnett and Campbell, 1996). The 12-transmembrane domain SUR sulfonylurea receptor subunits associated with ATP-sensitive K^+ channels (Bryan and Aguilar-Bryan, 1997) are not considered herein as auxiliary subunits. Except for the β subunit associated with the high conductance Ca^{2+}-sensitive K^+ channel purified from smooth muscle, which has two transmembrane domains (Knaus *et al.,* 1994), all the K^+ channel auxiliary subunits are cytosoluble proteins (Isom *et al.,* 1994; Adelman, 1995). With its single transmembrane domain, IsK structurally appears closer to the Na^+ channel β subunits. However, the mode of IsK interaction with the pore of the KvLQT1 subunit and its functional consequences are certainly unprecedented. The main difference between IsK and other auxiliary subunits is that only IsK affects the permeation properties themselves, including the ion selectivity (Wollnik *et al.,* 1997) and the single-channel conductance. The kinetic changes induced by IsK are likely to occur through a steric effect on the closing process rather than through changes in the coupling between gating charges and opening as reported for the Ca^{2+} channel β subunits (Neely *et al.,* 1995). The steric effect is also unrelated to the ball and chain mechanism evoked for the acceleration of K^+ current inactivation by the $\beta 1$ subunit (Rettig *et al.,* 1994). In a way, the interaction of the IsK C-terminal segment with the pore could be regarded as a direct participation of IsK in the pore structure.

In addition, the Na^+, Ca^{2+}, and K^+ channel β subunits have been reported to increase the channel's surface expression by regulating the trafficking process (Scheinman *et al.,* 1989; Varadi *et al.,* 1991; Shi *et al.,* 1996). In this type of regulation, the β subunit assembles with the α subunit in some intracellular pools, before insertion in the plasma membrane(Fink *et al.,* 1996; Shi *et al.,* 1996; Scannevin and Trimmer, 1997). Again, IsK behaves differently from other auxiliary subunits, since its association with KvLQT1 is likely to occur in the plasma membrane, after the KvLQT1 tetramer complex is formed. This conclusion comes from expression experiments in which the rapidly activating KvLQT1 currents are first expressed in *Xenopus* oocytes injected with KvLQT1 cRNA alone and then converted into slowly activating currents following expression of IsK 24 hr later (Fig. 6A). This modification of preexisting channels contrasts with the chaperone-like effect observed with the $\beta 4$ subunit on Kv2.2 expression, that is seen exclusively when both subunits are injected together, not only at the same time (Fig. 6B) but also at the same place (same injection pipet) in the oocyte (Fink *et al.*, 1996). Finally, preliminary immunological experiments with Flag-KvLQT1-transfected COS cells did not reveal enhancement of the surface expression upon co-expression of IsK (unpublished data).

IV. IS IsK UNIQUE?

Clearly, IsK is a new type of ion channel ancillary subunit. It could be a prototypic member of a new family of membrane proteins capable of profound modifications of preformed channels resulting in a large increase in the currents they produce. Other members of this family could include the phospholemman (Moorman *et al.*, 1992) and Mat-8 (Morrison *et al.*, 1995) proteins that induce Cl^- currents in *Xenopus* oocytes, the M2 influenza virus protein responsible for a cationic current activation (Pinto *et al.*, 1992) and CHIF, which induces a K^+ current resembling the IsK current (Attali *et al.*, 1995). All these proteins share the same topology of Type III integral proteins with IsK, but no clear sequence homologies (Fig. 1D). However, it should be noted that, although all of them have been shown

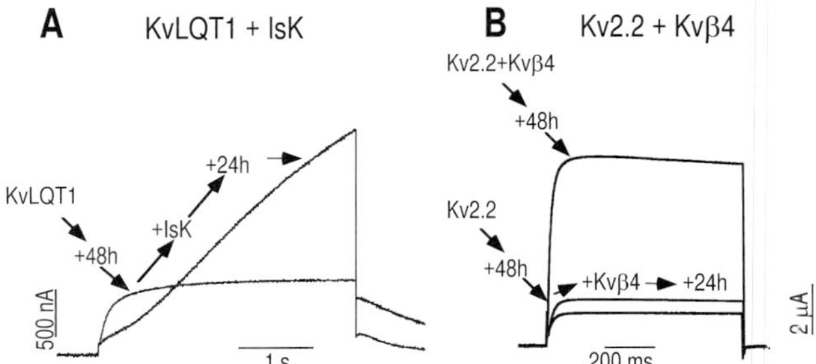

FIGURE 6 Comparison of the time course of the IsK and Kv$\beta 4$ functional expressions. (A) Typical traces evoked at +30 mV in *Xenopus* oocyte injected with KvLQT1 cRNA (KvLQT1 48h) and then reinjected with IsK cRNA for 24 hr (+IsK 24h). Note the disparition of the rapid current transformed in a slow current of higher amplitude. Coinjection of the two cRNAs produces the same currents as the delayed two-step injection (not shown). Adapted from Romey *et al.* (1997) with permission from The American Society for Biochemistry and Molecular Biology. (B) Same protocol as in A using Kv2.2 and Kv$\beta 4$ cRNA. The typical β subunit effect of enhancement of Kv2.2 currents is observed if the two cRNAs are coinjected using the same pipet.

to induce a current in *Xenopus* oocytes, and some of them in artificial bilayers, IsK is the only one for which a pore-forming partner has been identified. Hence, the true channel nature of the other family members cannot be excluded.

A search of the sequence databases for homologous proteins did not yield significant hits as observed with the classic K^+ channel subunits. Only one expressed sequence tag (EST) has been identified (Fig. 1C) that could correspond to an IsK-related protein but the functional expression of this new protein has not yet been reported. It is also intriguing that no IsK homolog has appeared in the genome of the nematode *Caenorhabditis elegans.* With the exception of IsK, genes encoding proteins corresponding to all the classes of K^+ channel proteins, including close homologs of KvLQT1, have been shown to exist in *C. elegans* (Wei *et al.,* 1996). It is possible that the function of subunits like IsK appeared only later in evolution.

V. DIFFERENCES IN TISSUE EXPRESSION OF KvLQT1 AND IsK mRNA: OPEN QUESTIONS

K^+ channel subunits are virtually never exclusively expressed in one single tissue. The KvLQT1 mRNA is quite abundant in many tissues besides the heart where it was first characterized. Northern blot analysis of its distribution in a number of human tissues shows that the 3.2-kb KvLQT1 transcript is present in all the tissues assayed except brain, liver, and skeletal muscle (Fig. 7A). Hybridization of the same blots with an IsK probe did not yield a perfect picture of matching expression. A good correlation between the two transcript levels seems to exist in heart and kidney, but it is clearly not seen in pancreas or testis where either KvLQT1 or IsK transcripts appear in excess compared to the other one (Fig. 7A). These apparent discrepancies suggest several hypotheses. The first one is that in tissues such as pancreas or prostate, KvLQT1 acts on its own, generating a rapidly activating K^+ current. The second one is that other unidentified "IsK-like" proteins (like CHIF, for example) exist in those tissues. The third one is that there are other "KvLQTx-like" proteins that can form heteropolymers with KvLQT1 to generate new types of K^+ currents. The fact that at least two homologs of KvLQT1 are detected in the *C. elegans* genome is probably a good indication that the KvLQT family is not restricted to only one member in mammals (Wei *et al.,* 1996). In fact, a cDNA of a K^+ channel subunit has been cloned (but not functionally expressed) from human neuroblastoma cells that is very similar to KvLQT1 (Yokoyama *et al.,* 1996) and an EST (Accession No. HS32799) also presents close

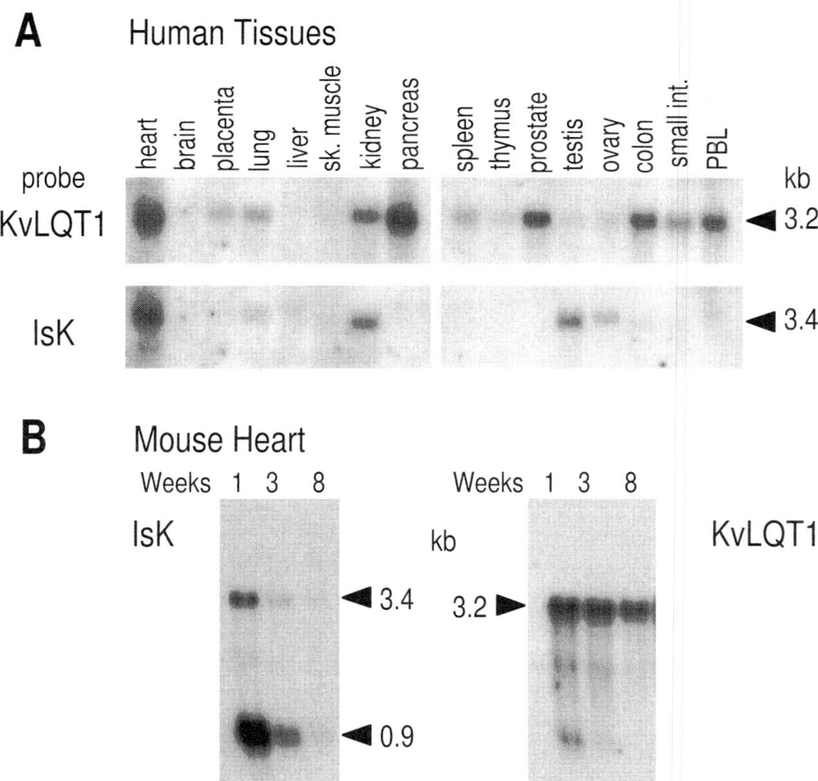

FIGURE 7 Northern blot analysis of the distribution of KvLQT1 and IsK mRNA in human tissues (A) and of expression regulation in the developing mouse heart (B). (A) The same human multiple tissue Northern blots from Clontech were successively probed with human KvLQT1 and IsK ^{32}P-labeled cDNAs. For IsK, only the 3.4-kb band is shown, but a small band of 0.8–0.9 kb is also detected. (B) Time course of diminution of the amount of IsK transcripts in mouse heart from 1-, 3-, and 8-week-old animals. The same blot hybridized with a KvLQT1 probe shows that this mRNA remains at a constant level during this developing period.

similarities, thus indicating the existence of at least two other mammalian family members. A fourth likelihood is that the IsK and KvLQT1 transcriptional activities are differently regulated in the same tissue. Hence, the ratio of both transcripts will not be kept constant but will be variable depending on specific regulating factors, unappreciated in many studies. This situation is found in the developing mouse heart where the KvLQT1 subunit is constitutively present while the IsK expression is submitted to a drastic down regulation (Fig. 7B). Uterus expression is another example. There, the amount of KvLQT1 is independent of the estrogenic impregnation, while the IsK transcript is considerably up-regulated by elevation of estrogen blood concentration [(Folander *et al.* (1990) and results not shown]. Hence, data obtained up to now indicate that, in many tissues, the KvLQT1 subunit is constitutively present while the IsK expression is submitted to regulation. This is precisely the situation that occurs in *Xenopus* oocytes where no current is recorded until IsK cRNA is injected and translated.

VI. IsK NULL MUTANT MICE AND JERVELL AND LANGE-NIELSEN CARDIOAUDITORY DISEASE PROVIDE EVIDENCE FOR THE EXISTENCE OF THE KvLQT1/IsK COMPLEX *in Vivo*

An interaction of IsK and KvLQT1 to form a complex that recapitulates the I_{Ks} current in heterologous expression systems has been clearly demonstrated. It remains important to confirm that this interaction takes place *in vivo.*

Vetter and colleagues (1996) have generated knockout mice for the *isk* gene. Behaviorally, these mice exhibit a so-called shaker/waltzer phenotype indicative of inner ear dysfunctions. Their analysis led to the conclusion that IsK has an essential role in the transport of high concentrations of K^+ ions into the extracellular endolymph that bathes the hair cells. Histological examination of the inner ear of the *isk*-/- mice revealed a pathology closely resembling that seen in human patients who have died from Jervell and Lange-Nielsen syndrome (JLNS) (Vetter *et al.,* 1996). JLNS is a long QT syndrome characterized by profound congenital sensorineural deafness associated with ventricular arrhythmias secondary to abnormal repolarization (long QT). This syndrome is transmitted as an autosomal recessive trait and has been shown to result from mutations in the *KVLQT1* gene (Chouabe *et al.,* 1997; Donger *et al.,* 1997; Neyroud *et al.,* 1997, 1998; Splawski *et al.,* 1997). The presence of KvLQT1 transcripts in the inner ear, co-localized with IsK transcripts, was then demonstrated (Neyroud *et al.,* 1997). Combined, these findings provide strong evidence for the *in vivo* association of IsK with KvLQT1 to form the inner ear K^+ channel that becomes inactive after mutation of one of the two partners. A further confirmation has recently been added by the finding that mutations of the *isk* gene in humans are also found to cause long QT syndromes clinically indistinguishable from cases caused by mutations in *KVLQT1* (Schulze-Bahr *et al.,* 1997; Splawski *et al.,* 1997; Tyson *et al.,* 1997).

In addition to the anatomical study, a functional study of the inner ear in *isk-/-* mice demonstrated the complete absence of the K^+ current in epithelial cells that normally secrete the K^+ in endolymph (Vetter *et al.,* 1996). KvLQT1, which is not affected in mutant mice, is clearly not sufficient to produce active K^+ channels in these cells. This *in vivo* situation contrasts with the *in vitro* functional expression studies in which KvLQT1 is able to form a functional channel by itself. In cardiomyocytes isolated from the heart of *isk-/-* mice it has been impossible to detect any K^+ channel activity that could be attributed to the expression of KvLQT1 alone (unpublished data). It is likely that KvLQT1 channels have a very small intrinsic activity that is undetectable when the KvLQT1 protein is expressed at physiological levels but becomes measurable upon overexpression, as is the case in heterologous expression systems.

VII. CONCLUSIONS

After 10 years of thrilling debates and research, IsK has finally yielded a part of its secret: It is a regulatory subunit and not a channel by itself. In a sense, this finding is a little disappointing as we have to admit that nature is more conformist than we thought. However, this is only partly true since the mode of interaction of IsK and KvLQT1 is really not conformist at all. It is a very sophisticated mechanism dedicated to achieving a drastic slowing down of the activation process in an otherwise classic K^+ channel. The reasons for this slowing down may be understandable in the heart, where it limits the outward current during the plateau phase of the action potential, making the contribution of I_{Ks} to the cardiomyocyte repolarization strongly dependent on the heart rate and on its regulatory state(Barhanin *et al.*, 1998; Romey *et al.*, 1997), but the corresponding reasons in other organs, particularly the inner ear, are not really understood. Nonetheless, because of its unconventional mechanism of interaction in the pore, preventing its closing as a foot in the door, IsK is the first member of a novel type of regulatory channel subunit. We are still far from the complete knowledge of the extent of modifications it produces. They include changes in single-channel conductance, selectivity, kinetics, and even pharmacology (Busch *et al.,* 1997; Busch and Suessbrich, 1997). It is even likely that IsK is absolutely necessary to reveal KvLQT1 activity under physiological conditions. It will be fascinating to see whether structurally related proteins such as phospholemman will turn out to be other members of the same functional family.

Finally, the understanding of the role of IsK has been an important part of a flurry of discoveries leading to the explanation of the mysterious link between congenital long QT syndromes and deafness. It is now hoped that the new molecular knowledge concerning I_{Ks} will be followed by improvements in the management of patients suffering from deafness and from congenital as well as acquired arrhythmias.

References

Adelman, J. P. (1995). Proteins that interact with the pore-forming subunits of voltage-gated ion channels. *Curr. Opin. Neurobiol.* **5,** 286–295.

Attali, B., Guillemare, E., Lesage, F., Honore, E., Romey, G., Lazdunski, M. and Barhanin, J. (1993). The protein IsK is a dual activator of K^+ and Cl^-channels. *Nature* **365,** 850–852.

Attali, B., Latter, H., Rachamim, N., and Garty, H. (1995). A corticosteroid-induced gene expressing an "IsK-like" K^+ channel activity in *Xenopus* oocytes. *Proc. Natl. Acad. Sci. USA* **92,** 6092–6096.

Barhanin, J., Attali, B., and Lazdunski, M. (1998). IKs, a very slow and very intriguing cardiac K^+ channel and its associated long QT diseases. *Trend Cardiovasc. Med.* **8,** 207–214.

Barhanin, J., Lesage, F., Guillemare, E., Fink, M., Lazdunski, M., and Romey, G. (1996). K(v)LQT1 and IsK (minK) proteins associate to form the I-Ks cardiac potassium current. *Nature* **384,** 78–80.

Blumenthal, E. M., and Kaczmarek, L. K. (1994). The MinK potassium channel exists in functional and nonfunctional forms when expressed in the plasma membrane of *Xenopus* oocytes. *J. Neurosci.* **14,** 3097–3105.

Bryan, J., and Aguilar-Bryan, L. (1997). The ABCs of ATP-sensitive potassium channels: More pieces of the puzzle. *Curr. Opin. Cell. Biol.* **9,** 553–559.

Busch, A. E., Busch, G. L., Ford, E., Suessbrich, H., Lang, H. J., Greger, R., Kunzelmann, K., Attali, B., and Stuhmer, W. (1997). The role of the I-sK protein in the specific pharmacological properties of the I-Ks channel complex. *Br. J. Pharmacol.* **122,** 187–189.

Busch, A. E., and Suessbrich, H. (1997). Role of the I-SK protein in the I-minK channel complex. *Trends Pharmacol. Sci.* **18,** 26–29.

Chouabe, C., Neyroud, N., Guicheney, P., Lazdunski, M., Romey, G., and Barhanin, J. (1997). Properties of KvLQT1 K^+ channel mutations in Romano-Ward and Jervell and Lange-Nielsen inherited cardiac arrhythmias. *EMBO J.* **16,** 5472–5479.

Cui, J., Kline, R. P., Pennefather, P., and Cohen, I. S. (1994). Gating of I-sK expressed in *Xenopus* oocytes depends on the amount of mRNA injected. *J. Gen. Physiol.* **104,** 87–105.

Donger, C., Denjoy, I., Berthet, M., Neyroud, N., Cruaud, C., Bennaceur, M., Chivoret, G., Schwartz, K., Coumel, P., and Guicheney, P. (1997). KVLQT1 C-terminal missense mutations causes a forme fruste long QT syndrome. *Circulation* **96,** 2778–2781.

Fink, M., Duprat, F., Lesage, F., Heurteaux, C., Romey, G., Barhanin, J., and Lazdunski, M. (1996). A new K^+ channel beta subunit to specifically enhance Kv2.2 (CDRK) expression. *J. Biol. Chem.* **271,** 26,341–26,348.

Folander, K., Smith, J. S., Antanavage, J., Bennett, C., Stein, R. B., and Swanson, R. (1990). Cloning and expression of the dilayed-rectifier Isk channel from neonatal rat heart and diethylstilbestrol-primed rat uterus. *Proc. Natl. Acad. Sci. USA* **87,** 2975–2979.

Gurnett, C. A., and Campbell, K. P. (1996). Transmembrane auxiliary subunits of voltage-dependent ion channels. *J. Biol. Chem.* **271,** 27,975–27,978.

Heginbotham, L., Lu, Z., Abramson, T., and Mackinnon, R. (1994). Mutations in the K^+ channel signature sequence. *Biophys. J.* **66,** 1061–1067.

Honoré, E., Barhanin, J., Attali, B., Lesage, F., and Lazdunski, M. (1994). External blockade of the major cardiac delayed-rectifier K^+ channel (Kv1.5) by polyunsaturated fatty acids. *Proc. Natl. Acad. Sci. USA* **91,** 1937–1941.

Isom, L. L., Dejongh, K. S., and Catterall, W. A. (1994). Auxiliary subunits of voltage-gated ion channels. *Neuron* **12,** 1183–1194.

Kaczmarek, L. K., and Blumenthal, E. M. (1997). Properties and regulation of the minK potassium channel protein. *Physiol. Rev.* **77,** 627–641.

Knaus, H. G., Folander, K., Garciacalvo, M., Garcia, M. L., Kaczorowski, G. J., Smith, M., and Swanson, R. (1994). Primary sequence and immunological characterization of beta-subunit of high conductance Ca^{2+}-activated K^+ channel from smooth muscle. *J. Biol. Chem.* **269,** 17,274–17,278.

Lesage, F., Attali, B., Lakey, J., Honoré, E., Romey, G., Faurobert, E., Lazdunski, M., and Barhanin, J. (1993). Are *Xenopus* oocytes unique in displaying functional IsK channel heterologous expression? *Recept. Chan.* **1,** 143–152.

Mackinnon, R. (1995). Pore loops: An emerging theme in ion channel structure. *Neuron* **14,** 889–892.

Miller, C. (1991). 1990—Annus-mirabilis of potassium channels. *Science* **252,** 1092–1096.

Moorman, J. R., Palmer, C. J., John, J. E., Durieux, M. E., and Jones, L. R. (1992). Phospholemman expression induces a hyperpolarization-activated chloride current in *Xenopus* oocytes. *J. Biol. Chem.* **267,** 14,551–14,554.

Morrison, B. W., Moorman, J. R., Kowdley, G. C., Kobayashi, Y. M., Jones, L. R., and Leder, P. (1995). Mat-8, a novel phospholemman-like protein expressed in human breast tumors, induces a chloride conductance in *Xenopus* oocytes. *J. Biol. Chem.* **270,** 2176–2182.

Neely, A., Olcese, R., Baldelli, P., Wei, X. Y., Birnbaumer, L., and Stefani, E. (1995). Dual activation of the cardiac Ca^{2+} channel alpha(1C)-subunit and its modulation by the beta-subunit. *Am. J. Physiol.–Cell. Physiol.* **37,** C732–C740.

Neyroud, N., Tesson, F., Denjoy, I., Leibovici, M., Donger, C., Barhanin, J., Faure, S., Gary, F., Coumel, P., Petit, C., Schwartz, K., and Guicheney, P. (1997). A novel mutation in the potassium channel gene KVLQT1 causes the Jervell and Lange-Nielsen cardioauditory syndrome. *Nature Genet.* **15,** 186–189.

Neyroud, N., Denjoy, I., Donger, C., Villain, E., Leenhardt, A., Gary, F., Coumel, P., Schwartz, K., and Guicheney, P. (1998). Heterozygous mutation in the pore of the potassium channel gene KvLQT1 causes an apparently normal phenotype in long QT syndrome. *Eur. Hum. Genet.* **6,** 129–133.

Pascual, J. M., Shieh, C. C., Kirsch, G. E., and Brown, A. M. (1995). K^+ pore structure revealed by receptor cysteines at inner and outer surfaces. *Neuron* **14,** 1055–1063.

Pinto, L. H., Holsinger, L. J., and Lamb, R. A. (1992). Influenza virus M2 protein has ion channel activity. *Cell* **69,** 517–528.

Rettig, J., Heinemann, S. H., Wunder, F., Lorra, C., Parcej, D. N., Dolly, J. O., and Pongs, O. (1994). Inactivation properties of voltage-gated K^+ channels altered by presence of beta-subunit. *Nature* **369,** 289–294.

Romey, G., Attali, B., Chouabe, C., Abitbol, I., Guillemare, E., Barhanin, J., and Lazdunski, M. (1997). Molecular mechanism and functional significance of the MinK control of the KvLQT1 channel activity. *J. Biol. Chem.* **272,** 16,713–16,716.

Salkoff, L., and Jegla, T. (1995). Surfing the DNA databases for K^+ channels nets yet more diversity. *Neuron* **15,** 489–492.

Sanguinetti, M. C., Curran, M. E., Zou, A., Shen, J., Spector, P. S., Atkinson, D. L., and Keating, M. T. (1996). Coassembly of K(v)LQT1 and MinK (IsK) proteins to form cardiac I-Ks potassium channel. *Nature* **384,** 80–83.

Scannevin, R. H., and Trimmer, J. S. (1997). Cytoplasmic domains of voltage-sensitive K^+ channels involved in mediating protein–protein interactions. *Biochem. Biophys. Res. Commun.* **232,** 585–589.

Scheinman, R. I., Auld, V. J., Goldin, A. L., Davidson, N., Dunn, R. J., and Catterall, W. A. (1989). Developmental regulation of sodium channel expression in the rat forebrain. *J. Biol. Chem.* **264,** 10,660–10,666.

Schulze-Bahr, E., Wang, Q., Wedekind, H., Haverkamp, W., Chen, Q., and Sun, Y. (1997). KCNE1 mutations cause jervell and Lange-Nielsen syndrome. *Nature Genet.* **17,** 267–268.

Shi, G. Y., Nakahira, K., Hammond, S., Rhodes, K. J., Schechter, L. E., and Trimmer, J. S. (1996). Beta subunits promote K^+ channel surface expression through effects early in biosynthesis. *Neuron* **16,** 843–852.

Splawski, I., Timothy, K. W., Vincent, G. M., Atkinson, D. L., and Keating, M. T. (1997). Molecular basis of the long-QT syndrome associated with deafness. *N. Engl. J. Med.* **336,** 1562–1567.

Swanson, R., Hice, R. E., Folander, K., and Sanguinetti, M. C. (1993). The IsK protein, a slowly activating voltage-dependent K^+ channel. *Semin. Neurosci.* **5,** 117–124.

Takumi, T., Ohkubo, H., and Nakanishi, S. (1988). Cloning of a membrane protein that induces a slow voltage-gated potassium current. *Science* **242,** 1042–1045.

Tyson, J., Tranebjærg, L., Bellman, S., Wren, C., Taylor, J., Bathen, J., Aslaksen, B., Sørland, S. J., Lund, O., Malcolm, S., Pembrey, M., Bhattacharya, S., and Bitner-Glindzicz, M. (1997). IsK and KvLQT1: Mutation in either of the two subunits of the slow component of the delayed rectifier potassium channel can cause Jervell and Lange-Nielsen syndrome. *Hum. Mol. Genet.* **6,** 2179–2185.

Tzounopoulos, T., Guy, H. R., Durell, S., Adelman, J. P., and Maylie, J. (1995). MinK channels form by assembly of at least 14 subunits. *Proc. Natl. Acad. Sci. USA* **92,** 9593–9597.

Varadi, G., Lory, P., Schultz, D., Varadi, M., and Schwartz, A. (1991). Acceleration of activation and inactivation by the beta subunit of the skeletal muscle calcium channel. *Nature* **352,** 159–162.

Vetter, D. E., Mann, J. R., Wangemann, P., Liu, J. Z., McLaughlin, K. J., Lesage, F., Marcus, D. C., Lazdunski, M., Heinemann, S. F., and Barhanin, J. (1996). Inner ear defects induced by null mutation of the IsK gene. *Neuron* **17,** 1251–1264.

Wang, K. W., Tai, K. K., and Goldstein, S. A. N. (1996a). MinK residues line a potassium channel pore. *Neuron* **16,** 571–577.

Wang, Q., Curran, M. E., Splawski, I., Burn, T. C., Millholland, J. M., Vanraay, T. J., Shen, J., Timothy, K. W., Vincent, G. M., Dejager, T., Schwartz, P. J., Towbin, J. A., Moss, A. J., Atkinson, D. L., Landes, G. M., Connors, T. D., and Keating, M. T. (1996b). Positional cloning of a novel potassium channel gene: KVLQT1 mutations cause cardiac arrhythmias. *Nature Genet.* **12,** 17–23.

Wei, A., Jegla, T., and Salkoff, L. (1996). Eight potassium channel families revealed by the C. elegans genome project. *Neuropharmacology* **35,** 805–829.

Wollnik, B., Schroeder, B. C., Kubisch, C., Esperer, H. D., Wieacker, P., and Jentsch, T. J. (1997). Pathophysiological mechanisms of dominant and recessive KVLQT1 K^+ channel mutations found in inherited cardiac arrhythmias. *Hum. Mol. Genet.* **6,** 1943–1949.

Yeh, J. Z., and Armstrong, C. M. (1978). Immobilisation of gating charges by a substance that simulates inactivation. *Nature* **273,** 387–389.

Yokoyama, M., Nishi, Y., Yoshii, J., Okubo, K., and Matsubara, K. (1996). Identification and cloning of neuroblastoma-specific and nerve tissue-specific genes through compiled expression profiles. *DNA Res.* **3,** 311–320.

CHAPTER 6

Delayed Rectifier Potassium Channels in Normal and Abnormal Cardiac Repolarization

Michael C. Sanguinetti* and Martin Tristani-Firouzi†
*Department of Medicine, Division of Cardiology, and †Department of Pediatrics, University of Utah, Salt Lake City, Utah 84112

I. ACTIVATION OF K^+ CHANNELS MEDIATES CARDIAC REPOLARIZATION

A hallmark property of cardiac action potentials is their long duration. The action potentials of neurons and skeletal myocytes last only a few milliseconds, whereas cardiac action potentials have a prolonged plateau phase that can persist for several hundred milliseconds. Action potentials of heart cells are long because the potassium channels activated by membrane depolarization that mediate repolarization: (i) inactivate rapidly, (ii) activate slowly, (iii) inwardly rectify, or (iv) have a combination of these properties. These gating properties result in a small net outward current during

Current Topics in Membranes, Volume 46

1063-5823/99 $30.00

the plateau phase of the cardiac action potential. Regional heterogeneity in action potential duration is an important determinant of differential refractoriness. For example, the long plateau phase and refractory period of Purkinje fibers prevents reexcitation of this conduction pathway from the adjacent ventricular myocardium. This heterogeneity is caused in part by cell-specific variation in the types and density of K^+ channels that contribute to repolarization. These channels can be classified into three distinct groups based on their relative rate of activation and rectification properties. The transient outward K^+ current (I_{to}) activates and inactivates rapidly upon membrane depolarization and is responsible for the early phase of membrane repolarization that produces a notch in the action potential. The delayed rectifier K^+ current (I_K) has multiple components with variable rates of activation and rectification properties that sum to contribute to repolarization during the plateau phase. The inward rectifier K^+ current (I_{K1}) is negligible at positive potentials but mediates the terminal phase of repolarization. Recent advances in molecular biology and biophysics have provided a molecular basis for these cardiac K^+ channels first identified by electrophysiological techniques (Table I).

Recent genetic findings have linked an inherited arrhythmia, long QT syndrome (LQT), to mutations in genes that encode some of the delayed rectifier K^+ channels. Mutations in *HERG, KVLQT1,* and *minK* genes are associated with a prolonged QT interval, polymorphic ventricular tachycardia, and an increased risk of sudden death. Functional characterization of these gene products led to identification of the molecular basis of two delayed rectifier K^+ currents, I_{Kr} and I_{Ks}. This chapter focuses on the molecular basis of these two currents, which have a prominent role in repolarization of human cardiac action potentials, and how mutations in the genes encoding these channels cause an inherited cardiac arrhythmia.

TABLE I

Current Understanding of the Molecular Basis of Human Cardiac Potassium Channels

Channel type	Abbreviation	Protein subunits
Delayed rectifiers		
Ultrarapid	I_{Kur}	Kv1.5
Rapid	I_{Kr}	HERG
Slow	I_{Ks}	KvLQT1 + hminK
Transient outward	I_{to}	Kv4.3
Inward rectifier	I_{K1}	Kir2.1 (+Kir2.2 and Kir2.3?)
Acetylcholine-activated	I_{KACh}	Kir3.4 + Kir3.1
ATP-inhibited	I_{KATP}	Kir6.2 + SUR2A

II. THE ACTIVITIES OF MULTIPLE POTASSIUM CHANNELS SUM TO FORM THE CARDIAC DELAYED RECTIFIER CURRENT

The delayed rectifier K^+ current (I_K) of ventricular myocytes is composed of three distinct currents that can be distinguished based on their rate of activation and pharmacology. All three currents have been recorded from isolated human cardiac myocytes (Li *et al.*, 1996; Wang *et al.*, 1993b, 1994). I_{Ks} activates very slowly, I_{Kr} activates more rapidly, and I_{Kur} activates ultra-rapidly. I_{Kr} is specifically blocked by Class III antiarrhythmic agents such as *d*-sotalol, dofetilide, almokalant, and E-4031 (Carmeliet, 1992, 1993; Sanguinetti and Jurkiewicz, 1990). I_{Ks} is specifically blocked by indapamide (Turgeon *et al.*, 1994) and the benzodiazepine, L-735,821 (Salata *et al.*, 1996). I_{Kur} is blocked by low concentrations of 4-aminopyridine (Wang *et al.*, 1993b). The current–voltage relationships of I_{Ks} and I_{Kur} are nearly linear. In contrast, the current–voltage relationship of I_{Kr} is bell-shaped, with a maximum near 0 mV when currents are measured using 200-msec voltage pulses.

A. HERG Subunits Coassemble to Form I_{Kr} Channels

HERG was discovered from a high-stringency screen of a human hippocampus cDNA library (Warmke and Ganetzky, 1994) using a mouse *ether-a-go-go* (*eag*) polymerase chain reaction fragment (Warmke *et al.*, 1991). *HERG* encodes a protein of 1159 amino acids with a predicted molecular weight of 127 kDa and is 49% identical to mouse *eag* and 24% identical to the rat cAMP-gated channel at the amino acid level (Warmke and Ganetzky, 1994). Hydropathy plots suggest it has the usual voltage-gated K^+ channel topology with six-transmembrane spanning regions (S1–S6), a K^+ channel pore signature sequence, and cytoplasmic amino- and carboxyl-terminal regions. The biophysical properties of HERG channels expressed in heterologous systems are similar to I_{Kr} measured in cardiac myocytes, including a bell-shaped current–voltage relationship and a half-point for activation near −15 mV (Fig. 1) (Sanguinetti *et al.*, 1995; Trudeau *et al.*, 1995). HERG channels are highly selective for K^+, with a permeability ratio of 140 for K^+/Na^+ (Sanguinetti *et al.*, 1995).

The single-channel conductance of HERG is 12 pS between −50 and −110 mV. At positive test potentials (+40 to +80 mV), the probability of channel opening is very low (0.011 at +40 mV) and the slope conductance is reduced to 5 pS (Zou *et al.*, 1997). The probability of channel opening remains low even after patch excision into a divalent cation-free bathing solution. Thus, rectification of whole-cell HERG results from an intrinsic

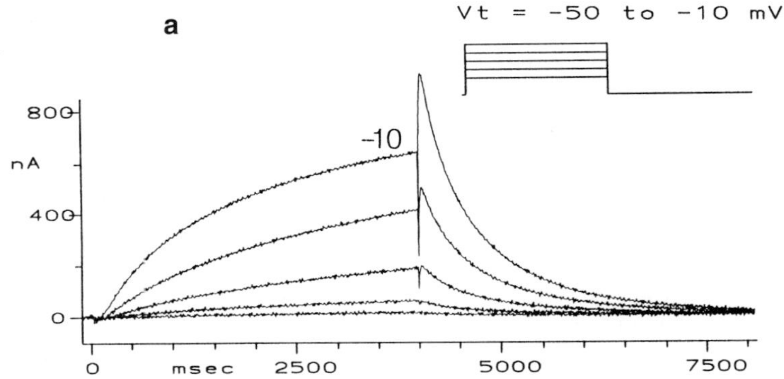
a
Vt = -50 to -10 mV
800
nA
400
0
-10
0
msec
2500
5000
7500

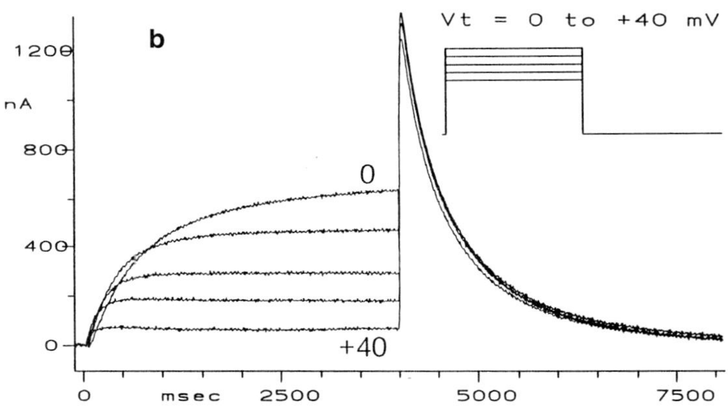
b
Vt = 0 to +40 mV
1200
nA
800
400
0
0
+40
msec
2500
5000
7500

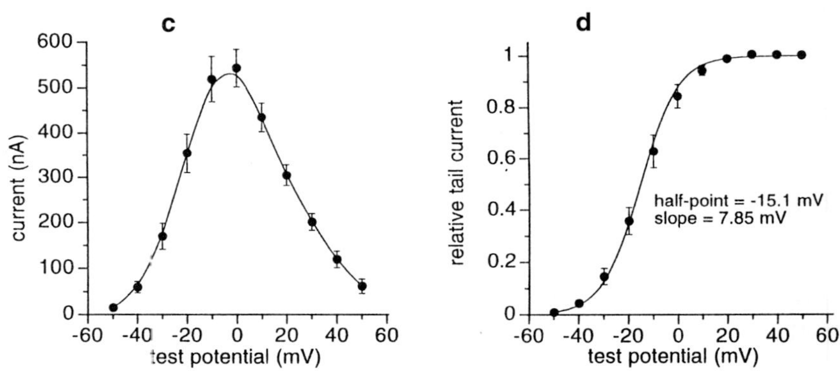
c
600
500
400
300
200
100
0
current (nA)
-60
-40
-20
0
20
40
60
test potential (mV)
d
1
0.8
0.6
0.4
0.2
0
relative tail current
half-point = -15.1 mV
slope = 7.85 mV
test potential (mV)

gating process and is not mediated by block of channels by Mg^{2+}, polyamines, or other intracellular blocking particles (Smith *et al.,* 1996; Spector *et al.,* 1996b) that are known to cause rectification of IRK channels (Fakler *et al.,* 1995, Ficker *et al.,* 1994; Lu and MacKinnon, 1994; Wible *et al.,* 1994).

Rectification of HERG results from a rapid, voltage-dependent inactivation that proceeds at a much faster rate than activation (Shibasaki, 1987; Sanguinetti and Jurkiewicz, 1990; Sanguinetti *et al.,* 1995; Trudeau *et al.,* 1995). For example, at 0 mV the time constant for activation is about 800 msec, whereas the time constant for inactivation is 9 msec (Sanguinetti *et al.,* 1995; Spector *et al.,* 1996b). Outward current during depolarization is reduced because channels inactivate either directly from a closed state ($C \rightarrow I$) or immediately after opening ($C \rightarrow O \rightarrow I$). In either case, channels pass from the I to an O state to reach a closed state ($I \rightarrow O \rightarrow C$) when the membrane is repolarized. Rapid recovery from inactivation precedes deactivation, resulting in a hooked tail current (Sanguinetti *et al.,* 1995; Shibasaki, 1987). The onset of rapid inactivation can be directly measured using a triple-pulse protocol (Liu *et al.,* 1996; Smith *et al.,* 1996; Spector *et al.,* 1996b; S. Wang *et al.,* 1996, 1997; T. Yang *et al.,* 1997). Current is first activated (and inactivated) by a depolarizing pulse followed by repolarization to a negative potential (e.g., −110 mV) for a time (e.g., 25 msec) sufficient to allow channels to recover from inactivation to an open state, but too short for significant channel deactivation. The third pulse is a depolarization that initiates reinactivation of current. The time constants describing the time course for reinactivation and recovery from inactivation are a bell-shaped function of voltage, indicating that rectification is an intrinsic gating process. Inactivation can be removed by mutation of two amino acids (G628C/S631A) in the pore-forming region of the channel (Smith *et al.,* 1996). As predicted, the current–voltage relationship of this mutant HERG channel is linear.

HERG channels are blocked by methanesulfonanilide Class III antiarrhythmic drugs such as E-4031, MK-499, and dofetilide (Snyders and Chaudhary, 1996; Spector *et al.,* 1996a; Trudeau *et al.,* 1995). The onset of block by dofetilide (Kiehn *et al.,* 1996) and MK-499 (Zou *et al.,* 1997) is faster in inside-out patches than in whole-cell recordings of HERG, indicating that

FIGURE 1 I_{Kr}-like currents recorded using two-microelectrode voltage clamp in a *Xenopus* oocyte injected with HERG cRNA. (a and b) Currents elicited by 4-sec depolarizations to test potentials ranging from −50 to +40 mV. (c) Current–voltage relationship for HERG. (d) Voltage dependence of HERG activation. Adapted from Sanguinetti *et al.* (1995).

methanesulfonanilides gain access to their binding site from the intracellular side of the membrane. These drugs preferentially bind either to the open state (Snyders and Chaudhary, 1996; Spector *et al.*, 1996a) or to a closed state that immediately precedes the open state of the channel (Kiehn *et al.*, 1996). The onset and recovery from block of HERG channels by methanesulfonanilides is very slow. In whole-cell recordings of HERG, the onset of block by 10 μM MK-499 was a monoexponential process with a time constant (τ_{on}) of 5.3 s (Spector *et al.*, 1996a). Once bound by drug, channels remain blocked for several minutes (Zou *et al.*, 1997). These slow kinetics explain why block of I_{Kr} by these drugs does not vary as a function of heart rate (Jurkiewicz and Sanguinetti, 1993), but do not explain why these compounds preferentially prolong action potentials and QT interval at slow heart rates.

With the exception of the kinetics of activation and deactivation, most of the properties of HERG are very similar to I_{Kr} measured in cardiac myocytes. The time constants for activation and deactivation of HERG are between 4 and 10 times slower than I_{Kr} characterized from isolated guinea pig ventricular myocytes or cultured atrial tumor (AT-1) cells. This apparent discrepancy between the gating kinetics of cloned HERG channels and I_{Kr} recorded from native myocytes has been explained by the discovery of an alternatively spliced variant of HERG (HERG B) that activates and deactivates faster than HERG, but at a rate similar rate to I_{Kr}. The N terminus of HERG B has 29 amino acids, shorter than the 376 amino acids of HERG A (Lees-Miller *et al.*, 1997). Similar alternatively spliced variants of *merg,* the mouse homolog of *HERG,* have been described (Lees-Miller *et al.*, 1997; London *et al.*, 1997). Unlike full-length *HERG* and *merg,* their alternatively spliced variants are specifically expressed in the heart (Lees-Miller *et al.*, 1997; London *et al.*, 1997). Coexpression of both forms of *merg* (*merg1a and merg1b*) in *Xenopus* oocytes induced a heteromeric channel current with rapid deactivation properties that was blocked by E-4031 with equal potency to that measured for *merg1a* alone (London *et al.*, 1997). It is likely that HERG A and HERG B also coassemble in human cardiac myocytes to form channels with a variable rate of deactivation. The presence of multiple forms of *erg* channels may also explain the variability in the kinetics of I_{Kr} noted between different species.

As described below, minK is a small protein that coassembles with KvLQT1 to form I_{Ks} channels. minK can also interact with HERG to regulate its activity. The first clue that such an interaction might be functionally important was the demonstration that transfection of AT-1 atrial tumor cells with minK antisense decreased the magnitude of I_{Kr} (Yang *et al.*, 1995). When stably HERG-transfected Chinese hamster ovary (CHO) cells were transiently transfected with minK, the amplitude of HERG was doubled. This effect was probably caused by an increase in the number of functional channels (McDonald *et al.*, 1997). Physical interaction of minK and HERG was also demonstrated by coimmunoprecipitation of the two subunits. This interaction did not alter the amount of protein expressed on the cell surface or the single-channel conductance of HERG, but did cause a slight leftward shift in the voltage dependence of activation and inactivation. Thus, minK regulates the activity of both I_{Kr} and I_{Ks} channels.

B. KvLQT1 and minK Subunits Coassemble to Form I_{Ks} Channels

KVLQT1 was discovered by a positional cloning strategy undertaken to find the gene associated with the LQT1 locus on chromosome 11 (Q. Wang *et al.*, 1996). The original clone was incomplete, missing a large region of the 5′ end. Subsequently, different full-length clones were isolated from human pancreas and heart (Sanguinetti *et al.*, 1996b; Wollnik *et al.*, 1997), mouse heart (Barhanin *et al.*, 1996), human heart (W. P. Yang *et al.*, 1997), and human kidney (Chouabe *et al.*, 1997) cDNA libraries. The most commonly found transcript encodes a protein with 676 amino acids having the usual voltage-dependent K^+ channel topology. Northern analyses demonstrated that this transcript is expressed most abundantly in the heart, pancreas, kidney, small intestine, and prostate (Chouabe *et al.*, 1997). When expressed in oocytes or in cultured CHO cells, *KVLQT1* induced a K^+-selective current with activation and inactivation properties unlike any previously described cardiac current (Fig. 2a). Therefore, it was hypothesized that KvLQT1 proteins might coassemble with another protein to form a functional channel with properties that more closely resembled a known cardiac K^+ current. A likely candidate for this accessory protein was minK, a small (129 amino acids) protein that induced an I_{Ks}-like current when cRNA encoding this protein was injected into oocytes. In contrast, transfection of CHO or other mammalian cells with cDNA encoding minK did not induce any current (Fig. 2c). These observations suggested that minK might coassemble with another protein constitutively expressed in oocytes (but not in CHO cells) to form I_{Ks} channels. Based on the fact that mutations in *KVLQT1* caused LQT, it made physiological sense that KvLQT1 subunits might coassemble with minK subunits to form heteromultimeric I_{Ks} channels. This prediction proved to be correct (Barhanin *et al.*, 1996; Sanguinetti *et al.*, 1996b). When CHO cells were cotransfected with *KVLQT1* plus *minK*, a current that had properties indistinguishable from I_{Ks} of cardiac myocytes

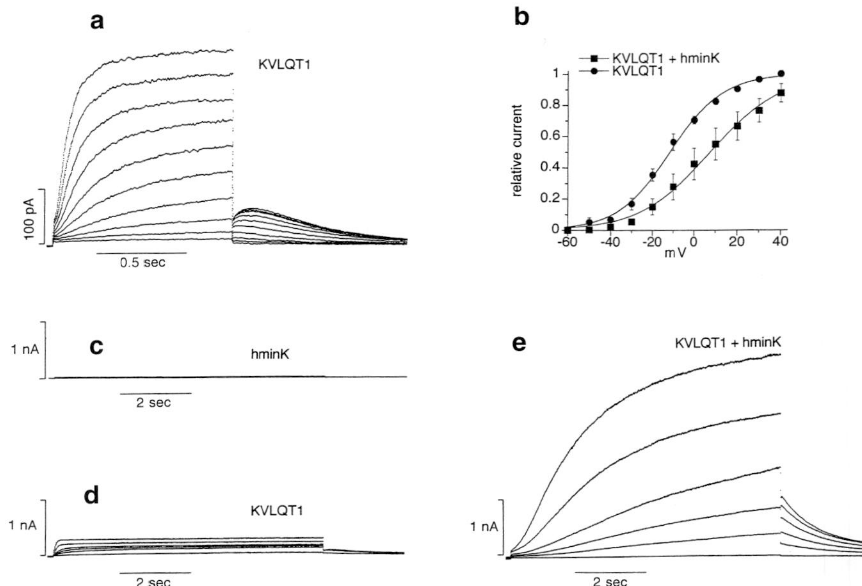

FIGURE 2 KvLQT1 and minK coassemble to form cardiac I_{Ks} channels. (a) Currents recorded during 1-sec pulses from a Chinese hamster ovary (CHO) cell transfected with KVLQT1 cDNA. (b) Relative tail current amplitudes define the voltage dependence of KvLQT1 and I_{Ks} activation. (c) hminK does not induce currents when expressed in CHO cells. (d) KvLQT1 activated in response to 7.5-sec pulses. Note change in current scale compared to panel a. (e) I_{Ks} in a cell cotransfected with both cDNAs. Currents were recorded in response to 7.5-sec pulses to test potentials from −40 to +40 mV. Reprinted with permission from *Nature* (Sanguinetti *et al.,* 1996b. Coassembly of KvLQT1 and minK (IsK) proteins to form cardiac I_{Ks} potassium channel. **384,** 80–83) Macmillan Magazines, Limited.

was induced. Activation of current was delayed upon membrane depolarization, increased in magnitude slowly over several seconds (Fig. 2e), and reached half-maximal activation near +10 mV. These properties closely match I_{Ks} recorded from cardiac myocytes under similar recording conditions. The stoichiometry for assembly of KvLQT1 and minK subunits is not known, but previous findings by Wang and Goldstein (1995) suggested that two minK subunits were likely to coassemble with constitutively expressed subunits (of unknown identity at the time) in oocytes to form a single functional I_{Ks} channel. Alpha subunits of voltage-activated K^+ channels that share some homology with KvLQT1 are believed to coassemble as tetramers to form functional channels. Therefore, a plausible model for a single I_{Ks} channel is two minK plus four KvLQT1 subunits.

III. MUTATIONS IN ION CHANNEL GENES CAUSE LONG QT SYNDROME AND INCREASE THE RISK OF LIFE-THREATENING CARDIAC ARRHYTHMIA

Block of I_{Kr} by Class III antiarrhythmic drugs lengthens action potentials and prolongs membrane refractoriness. This activity can suppress certain types of arrhythmia. Unfortunately, these drugs can sometimes cause excessive action potential prolongation (Hondeghem and Snyders, 1990) and increase the dispersion of refractoriness in the heart, leading to LQT. These unwanted side effects, sometimes but not always associated with overdose, are exacerbated by bradycardia, hypokalemia, or hypomagnesemia (Roden, 1988). The hallmark phenotype for LQT is a prolonged QT interval on the body surface ECG and syncope caused by torsade de pointes arrhythmia (Schwartz *et al.,* 1993). This ventricular tachyarrhythmia is characterized by a sinusoidal twisting of the QRS axis around the isoelectric line, and oftentimes is self-terminating. On rare occasions, torsade de pointes can degenerate into ventricular fibrillation and cause sudden death.

LQT can also be inherited, either as an autosomal dominant (Romano Ward syndrome) or an autosomal recessive (Jervell and Lange-Nielsen syndrome) disorder (Roden *et al.,* 1996; Schwartz *et al.,* 1993). Recently, it was determined that mutations in several different genes can cause Romano Ward syndrome (Curran *et al.,* 1995; Q. Wang *et al.,* 1995, 1996). These genes, *SCN5A, HERG, KVLQT1,* and *hminK,* encode ion channels important for modulation of myocellular repolarization. *SCN5A* encodes the α subunit of the cardiac Na^+ channel. Gain of function mutations in *SCN5A* cause an impaired Na^+ channel inactivation that results in a very small maintained inward current during membrane depolarization (Bennett *et al.,* 1995). As discussed above, *HERG* encodes the I_{Kr} channel (Sanguinetti *et al.,* 1995) and *KVLQT1* and *minK* encode the two subunits that combine to form the I_{Ks} channel (Barhanin *et al.,* 1996; Sanguinetti *et al.,* 1996b). Mutations in *HERG,* or in either *KVLQT1* or *hminK,* cause a decrease, respectively in I_{Kr} or I_{Ks} channel function. The functional consequence of either a persistent inward Na^+ current or a decreased outward K^+ current during the plateau phase of the cardiac action potential is delayed repolarization and an increased QT interval.

A. Mutations in HERG Cause Long QT Syndrome

Mutations in HERG cause LQT2, the dominant form of LQT linked to chromosome 7 (Curran *et al.,* 1995). The functional consequences of several LQT2-associated mutations in *HERG* have been studied in *Xenopus* oocytes (Sanguinetti *et al.*, 1996a). One of these mutations was a single base pair deletion (Δbp1261) that results in a frameshift and a premature stop codon. The encoded protein contains only the first 420 amino acids, including the N-terminal region and a portion of the S1 transmembrane domain. Another mutation was an in-frame intragenic deletion of 27 bp that results in deletion of 9 amino acids within the S3 transmembrane domain. These deletion mutants did not functionally express and did not alter HERG current when coexpressed with wild-type HERG in *Xenopus* oocytes. Thus, these mutations are expected to decrease I_{Kr} in myocytes by haploinsufficiency. The functional effects of Δbp1261 HERG have also been studied in transfected COS cells (Li *et al.*, 1997). In these cells, Δbp1261 HERG had a variable, but significant dominant-negative effect when cotransfected with wild-type HERG. This finding led to study of the domains required for coassembly of HERG. It was concluded that a region in the N-terminal domain that is relatively conserved between eag and HERG and corresponding to the first 135 amino acids of HERG was required for subunit interaction (Li *et al.*, 1997). In oocytes, this region is evidently not important for subunit interaction because functional HERG channels are formed from subunits lacking the majority of the N-terminal region (Schonherr and Heinemann, 1996; Spector *et al.*, 1996b).

Three LQT2-associated missense mutations of HERG (G628S, A561V, N470D) have been characterized in oocytes (Sanguinetti *et al.*, 1996a). N470D HERG was able to form functional homotetramers in oocytes, but the resulting current was activated at more negative potentials and the rate of current deactivation was slower than that of wild-type HERG. When coexpressed with wild-type HERG, N470D had a weak dominant-negative effect (Fig. 3). Neither G628S nor A561V HERG formed functional channels when expressed alone. However, both mutants suppressed the expression of coinjected wild-type HERG by a dominant-negative mechanism. G628S disrupts the pore-forming region of the HERG subunit and had a lethal dominant-negative effect, indicating that a heterotetrameric channel containing even a single mutant subunit was nonfunctional. All *HERG* mutations are predicted to result in a diminished magnitude of I_{Kr}, consistent with the prolonged QT interval observed in affected individuals.

B. Mutations in KVLQT1 or hminK Cause Long QT Syndrome

Mutations in *KVLQT1* cause the dominant form of Romano-Ward syndrome first linked to chromosome 11, and now referred to as LQT1 (Q. Wang *et al.*, 1996). At the time of this writing, 29 different mutations of *KVLQT1* that cause LQT1 have been published (Chouabe *et al.*, 1997, Q.

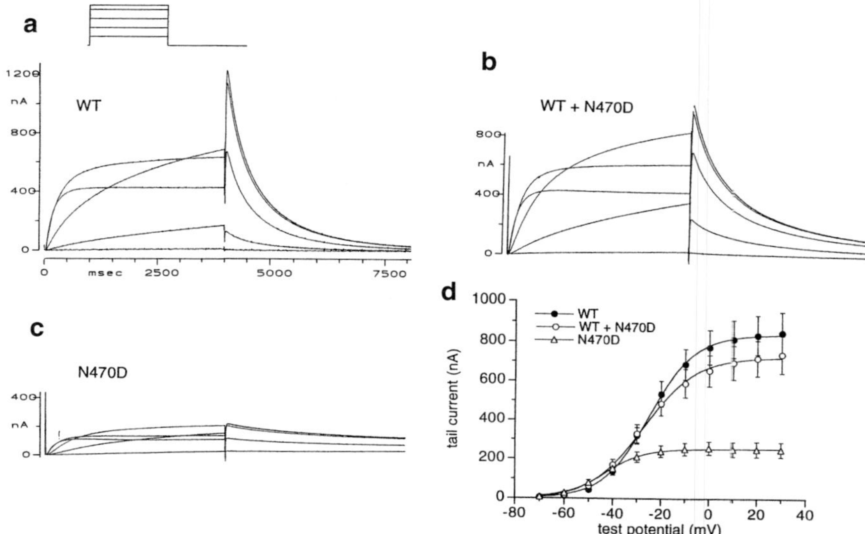

FIGURE 3 Altered and dominant negative suppression of wild-type (WT) HERG function by N470D. (a–c) Currents were recorded in response to 4-sec test pulses (−60, −40, −20, 0, +10 mV) in oocytes injected with cRNA encoding WT-HERG (a), N470D (c), or WT-HERG plus N470D (b). (d) Current–voltage relationship for HERG tail currents. The tail current of N470D was reduced by 70% relative to that of WT-HERG alone. Reprinted from Sanguinetti *et al.* (1996a). Spectrum of HERG K^+ channel dysfunction in an inherited cardiac arrhythmia. *Proc. Natl. Acad. Sci. USA* **93,** 2208–2212 with permission from National Academy of Sciences, U.S.A.

Wang *et al.,* 1996), and it is likely that many more will be described. Most KvLQT1 subunits containing single missense mutations (e.g., R174C, A177P, G269D, T311I, G314S, D317N, Y315S, L342F) did not produce measurable currents when expressed alone or in the presence of hminK (Chouabe *et al.,* 1997; Shalaby *et al.,* 1997; Wollnik *et al.,* 1997). However, these subunits reduced the function of wild-type subunits in coexpression experiments by a dominant-negative effect. Of all the LQT1 missense mutations studied, only one (L272F in S5 transmembrane domain) formed functional channels when expressed alone in *Xenopus* oocytes (Shalaby *et al.,* 1997). Currents induced by coexpression of L272F KvLQT1 and wild-type KvLQT1 were half-activated at a potential 10 mV more negative than that of wild-type KvLQT1 alone. Another KvLQT1 mutant, R555C in the carboxyl terminus, formed functional channels when coexpressed with hminK, but the resulting current density was 75% smaller than current induced by transfection of COS cells with wild-type KVLQT1 alone. In addition, currents deactivated at a faster rate and voltage-dependent activation was shifted by +50 mV (Chouabe *et al.*, 1997).

Dominant missense mutations in *hminK* cause Romano-Ward syndrome, LQT5 (Splawski *et al.*, 1997b). Both of these mutations (S74L and D76N) are located in the cytoplasmic, carboxyl-terminal region of the protein and altered the function of channels formed by coassembly of KvLQT1 and wild-type plus mutant hminK subunits. The voltage dependence of activation was shifted to more positive potentials and the rate of deactivation was enhanced (Fig. 4). These alterations in channel gating would reduce I_{Ks} and lengthen action potential duration of a cardiac myocyte. In addition, D76N has a strong dominant-negative effect.

Homozygous mutations in either *KVLQT1* or *hminK* cause Jervell and Lange-Nielsen syndrome, a severe form of LQT and bilateral deafness (Neyroud *et al.*, 1997; Schulze-Bahr *et al.*, 1997; Splawski *et al.*, 1997a).

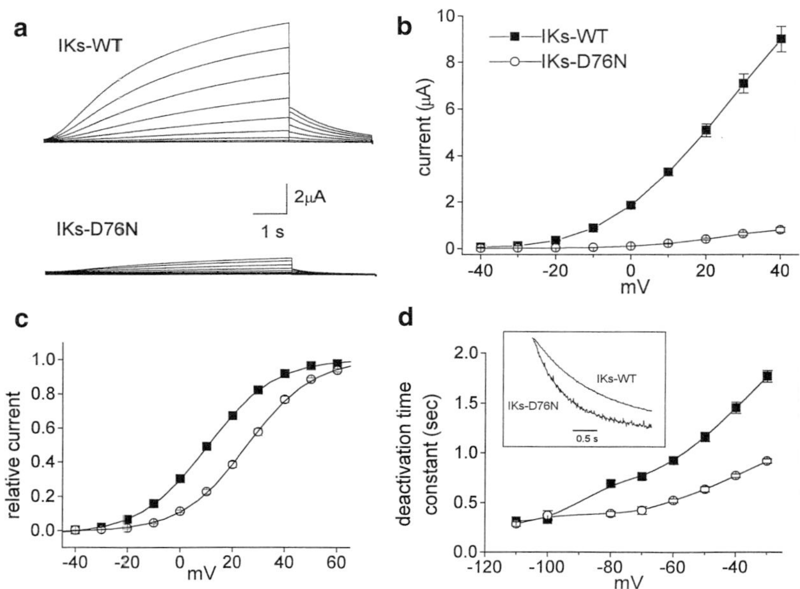

FIGURE 4 Functional effects of D76N hminK on I_{Ks} expressed in *Xenopus* oocytes. (a) I_{Ks} was elicited by depolarizations to test potentials of −20, −10, 0, 10, 20, and 40 mV. Currents were induced by coinjection of oocytes with 1.2 ng wild-type *hminK* + 6 ng *KVLQT1* (IKs-WT), or *KVLQT1* + 0.6 ng wild-type *hminK* + 0.6 ng D76N *hminK* cRNA (IKs-D76N). (b) Isochronal current–voltage relations of wild-type I_{Ks} and IKs-D76N, demonstrating dominant negative suppression of I_{Ks} by D76N mutant subunits. (c) The voltage dependence of IKs-D76N activation is shifted by +17mV compared to that of IKs-WT. (d) IKs-D76N deactivates faster than IKs-WT. From Splawski *et al.* (1997b) with permission from Nature Genetics.

Mutations in *KVLQT1* include a missense mutation (W305S) and a deletion–insertion that leads to a frameshift, a premature stop codon, and truncation of the carboxyl terminus (Neyroud *et al.*, 1997). Both of these mutations have a less severe effect on channel function than the dominant mutations that cause Romano-Ward syndrome (Chouabe *et al.*, 1997; Wollnik *et al.*, 1997). In contrast, the homozygous mutation in *hminK*, D76N, that causes Jervell and Lange-Nielsen syndrome (Schulze-Bahr *et al.*, 1997) is the same as one of the mutations identified for Romano-Ward syndrome that has a strong dominant-negative effect (Splawski *et al.*, 1997b).

IV. MECHANISM-BASED THERAPY FOR LONG QT SYNDROMES

The most common treatment for LQT is administration of β-adrenergic receptor blockers (Roden *et al.*, 1996). It has been estimated that these agents reduce the 10-year mortality of LQT from nearly 50% to 4% in 5 years (Roden *et al.*, 1996; Schwartz, 1985). If feasible, specific gene therapy would be the ideal treatment for these inherited disorders. Another potential gene-specific therapeutic strategy for LQT1, LQT2, and LQT5 would be to augment activity of those functioning K^+ channels not affected by mutant subunits. As discussed above, several drugs are known to block I_{Kr}, but unfortunately, agonists of this channel have not been discovered. However, an interesting and clinically relevant property of HERG (and I_{Kr}) is its paradoxical activation by extracellular K^+. Because the concentration of K^+ inside cells (140 m*M*) is much higher than its concentration outside the cell (4–5 m*M*), the chemical driving force for K^+ movement is outward. An increase in extracellular K^+ concentration would reduce this gradient and therefore, normally reduce the magnitude of outward K^+ current. However, unlike most K^+ currents, HERG is enhanced by elevation of extracellular K^+ (Sanguinetti *et al.*, 1995). The mechanism of this effect is believed to be a K^+-mediated decrease in channel inactivation, a mode of channel gating that is responsible for the bell-shaped relationship between current and voltage of I_{Kr} (Schonherr and Heinemann, 1996; Smith *et al.*, 1996; Spector *et al.*, 1996b). The paradoxical effect of K^+ on HERG may explain why hypokalemia is a risk factor for torsade de pointes. A clinical study supports the idea that control of serum [K^+] is a potential preventative treatment for LQT. An increase in serum K^+ by about 1.3 meq/liter significantly shortened the prolonged QT_c interval and decreased the dispersion of ventricular refractoriness of individuals with LQT caused by mutations in *HERG* (Compton *et al.*, 1996).

A new class of benzodiazepines (e.g., L-735,821) that blocks I_{Ks} and lengthens action potential duration was recently reported (Salata *et al.*,

1996). A structurally similar compound, L-364,373, was found to enhance I_{Ks} in guinea pig ventricular myocytes by causing a leftward shift in the voltage dependence of activation and a slowing of the rate of deactivation (Salata *et al.*, 1998). These effects were stereospecific. The *R*-enantiomer (R-L3) had agonist activity, whereas the *S*-enantiomer was a weak blocker of I_{Ks}. R-L3 at 1 μM also enhanced the magnitude and slowed deactivation of KvLQT1 current expressed in *Xenopus* oocytes, but had no effect on current induced by coexpression of KvLQT1 and minK. This finding indicates either that the binding of R-L3 to KvLQT1 is prevented by association with minK or that the drug still binds but does not affect a change in gating in channels formed by coassembly of KvLQT1 and minK. As expected, when an excess of KvLQT1 compared to minK was expressed in oocytes, R-L3 was able to alter gating of the induced current, but the effect was less than that observed in oocytes expressing only KvLQT1. Drugs such as R-L3 may prove useful in the treatment of inherited or acquired forms of LQT, especially LQT1 and LQT5.

In summary, recent genetic findings linking mutations of K^+ channel genes to LQT have led to rapid advances in our understanding of the molecular basis of two currents, I_{Kr} and I_{Ks}, that modulate cardiac repolarization. These findings add to the growing list of cardiac currents that have a known molecular basis, and should facilitate the design of mechanism-based therapies for the prevention and treatment of LQT and perhaps other life-threatening arrhythmias.

References

Barhanin, J., Lesage, F., Guillemare, E., Fink, M., Lazdunski, M., and Romey, G. (1996). KvLQT1 and IsK (minK) proteins associate to form the I_{Ks} cardiac potassium channel. *Nature* **384**, 78–80.

Bennett, P. B., Yazawa, K., Makita, N., and George, A. L. (1995). Molecular mechanism for an inherited cardiac arrhythmia. *Nature* **376**, 683–685.

Carmeliet, E. (1992). Voltage- and time-dependent block of the delayed K^+ current in cardiac myocytes by dofetilide. *J. Pharm. Exp. Ther.* **262**, 809–817.

Carmeliet, E. (1993). Use-dependent block and use-dependent unblock of the delayed rectifier K^+ current by almokalant in rabbit ventricular myocytes. *Circ. Res.* **73**, 857–868.

Chouabe, C., Neyroud, N., Guichney, P., Lazdunski, M., Romey, G., and Barhanin, J. (1997). Properties of KvLQT1 K+ channel mutations in Romano-Ward and Jervell and Lange-Nielsen inherited cardiac arrhythmias. *EMBO J.* **16**, 5472–5479.

Compton, S., Lux, R., Ramsey, M., Strelich, K., Sanguinetti, M., Green, L., Keating, M., and J. Mason (1996). Genetically defined therapy of inherited long QT syndrome: Correction of abnormal repolarization by potassium. *Circulation* **94**, 1018–1022.

Curran, M. E., Splawski, I., Timothy, K. W., Vincent, G. M., Green, E. D., and Keating, M. T. (1995). A molecular basis for cardiac arrhythmia: *HERG* mutations cause long QT syndrome. *Cell* **80**, 795–804.

Deal, K. K., England, S. K., and Tamkun, M. M. (1996). Molecular physiology of cardiac potassium channels. *Physiol. Rev.* **76**, 49–67.

Fakler, B., Brandle, U., Glowatzki, E., Weidemann, S., Zenner, H.-P., and Ruppersberg, J. P. (1995). Strong voltage-dependent inward rectification of inward rectifier K^+ channels is caused by intracellular spermine. *Cell* **80,** 149–154.

Ficker, E., Taglialatela, M., Wible, B. A., Henley, C. M., and Brown, A. M. (1994). Spermine and spermidine as gating molecules for inward rectifier K^+ channels. *Science* **266,** 1068–1072.

Hondeghem, L. M., and Snyders, D. J. (1990). Class III antiarrhythmic agents have a lot of potential but a long way to go: Reduced effectiveness and dangers of reverse use dependence. *Circulation* **81,** 686–690.

Jurkiewicz, N. K., and Sanguinetti, M. C. (1993). Rate-dependent prolongation of cardiac action potentials by a methanesulfonanilide class III antiarrhythmic agent: Specific block of rapidly activating delayed rectifier K^+ current by dofetilide. *Circ. Res.* **72,** 75–83.

Kiehn, J., Lacerda, A., Wible, B., and Brown, A. M. (1996). Molecular physiology and pharmacology of HERG: Single-channel currents and block by dofetilide. *Circulation* **94,** 2572–2579.

Lees-Miller, J. P., Kondo, C., Wang, L., and Duff, H. J. (1997). Electrophysiological characterization of an alternatively processed ERG K^+ channel in mouse and human hearts. *Circ. Res.* **81,** 719–726.

Li, G.-R., Feng, J., Yue, L., Carrier, M., and Nattel, S. (1996). Evidence for two components of delayed rectifier K^+ current in human ventricular myocytes. *Circ. Res.* **78,** 689–696.

Li, X., Xu, J., and Li, M. (1997). The human Δ1261 mutation of the *HERG* potassium channel results in a truncated protein that contains a subunit interaction domain and decreases the channel expression. *J. Biol. Chem.* **272,** 705–708.

Liu, S., Rasmusson, R. L., Campbell, D. L., Wang, S., and Strauss, H. C. (1996). Activation and inactivation kinetics of an E-4031-sensitive current from single ferret atrial myocytes. *Biophys. J.* **70,** 2704–2715.

London, B., Trudeau, M. C., Newton, K. P., Beyer, A. K., Copeland, N. G., Gilbert, D. J., Jenkins, N. A., Satler, C. A., and Robertson, G. A. (1997). Two isoforms of the mouse *ether-a-go-go*-related gene coassemble to form channels with properties similar to the rapidly activating component of the cardiac delayed rectifier K^+ current. *Circ. Res.* **81,** 870–878.

Lu, Z., and MacKinnon, R. (1994). Electrostatic tuning of Mg^{2+} affinity in an inward-rectifier K^+ channel. *Nature* **371,** 243–245.

McDonald, T. V., Yu, Z., Ming, Z., Palma, E., Meyers, M. B., Wang, K.-W., Goldstein, S. A. N., and Fishman, G. I. (1997). A minK-HERG complex regulates the cardiac potassium current I_{Kr}. *Nature* **388,** 289–292.

Neyroud, N., Tesson, F., Denjoy, I., Leibovici, M., Donger, C., Barhanin, J., Faure, S., Gary, F., Coumel, P., Petit, C., Schwartz, K., and Guicheney, P. (1997). A novel mutation in the potassium channel gene *KVLQT1* causes the Jervell and Lange-Nielsen cardioauditory syndrome. *Nature Genet.* **15,** 186–189.

Roden, D. M. (1988). Arrhythmogenic potential of class III antiarrhythmic agents: Comparison with class I agents. *In* "Control of Cardiac Repolarization by Lengthening Repolarization" (B. N. Singh, ed.), pp. 559–576. Futura Publishing Co., Mt. Kisco.

Roden, D. M., Lazzara, R., Rosen, M., Schwartz, P. J., Towbin, J., and Vincent, G. M. (1996). Multiple mechanisms in the long QT syndrome: Current knowledge, gaps, and future directions. *Circulation* **94,** 1996–2012.

Salata, J. J., Jurkiewicz, N. K., Sanguinetti, M. C., Siegl, P. K., Claremon, D. C., Remy, D. C., Elliott, J. M., and Libby, B. E. (1996). The novel class III antiarrhythmic agent, L-735,821 is a potent and selective blocker of I_{Ks} in guinea pig ventricular myocytes. *Circulation* **94,** I529.

Salata, J. J., Jurkiewicz, N. K., Wang, J., Evans, B. E., Orme, H. T., and Sanguinetti, M. C. (1998). A novel benzodiazepine that activates cardiac slow delayed rectifier K^+ channels. *Mol. Pharmacol.* **53,** 220–230.

Sanguinetti, M. C., Curran, M. E., Spector, P. S., and Keating, M. T. (1996a). Spectrum of HERG K^+ channel dysfunction in an inherited cardiac arrhythmia. *Proc. Natl. Acad. Sci. USA* **93,** 2208–2212.

Sanguinetti, M. C., Curran, M. E., Zou, A., Shen, J., Spector, P. S., Atkinson, D. L., and Keating, M. T. (1996b). Coassembly of KvLQT1 and minK (IsK) proteins to form cardiac I_{Ks} potassium channel. *Nature* **384,** 80–83.

Sanguinetti, M C., Jiang, C., Curran, M. E., and Keating, M. T. (1995). A mechanistic link between an inherited and an acquired cardiac arrhythmia: *HERG* encodes the I_{Kr} potassium channel. *Cell* **81,** 299–307.

Sanguinetti, M. C., and Jurkiewicz, N. K. (1990). Two components of cardiac delayed rectifier K^+ current: Differential sensitivity to block by class III antiarrhythmic agents. *J. Gen. Physiol.* **96,** 195–215.

Schonherr, R., and Heinemann, S. H. (1996). Molecular determinants for activation and inactivation of HERG, a human inward rectifier potassium channel. *J. Physiol.* **493,** 635–642.

Schulze-Bahr, E., Wang, Q., Wedekind, H., Haverkamp, W., Chen, Q., Sun, Y., Rubie, C., Hordt, M., Towbin, J. A., Borggrefe, M., Assmann, G., Qu, X., Somberg, J. C., Breithardt, G., Oberti, C., and Funke, H. (1997). *KCNE1* mutations cause Jervell and Lange-Nielsen syndrome. *Nature Genet.* **17,** 267–268.

Schwartz, P. (1985). Idiopathic long QT syndrome: Progress and questions. *Am. Heart J.* **109,** 399–411.

Schwartz, P. J., Moss, A. J., Vincent, G. M., and Crampton, R. S. (1993). Diagnostic criteria for the long QT syndrome: An update. *Circulation* **88,** 782–784.

Shalaby, F. Y., Levesque, P. C., Yang, W.-P., Little, W. A., Conder, M. L., Jenkins-West, T., and Blanar, M. A. (1997). Dominant-negative *KVLQT1* mutations underlie the LQT1 form of long QT syndrome. *Circulation* **96,** 1733–1736.

Shibasaki, T. (1987). Conductance and kinetics of delayed rectifier potassium channels in nodal cells of the rabbit heart. *J. Physiol.* **387,** 227–250.

Smith, P. L., Baukrowitz, T., and Yellen, G. (1996). The inward rectification mechanism of the HERG cardiac potassium channel. *Nature* **379,** 833–836.

Snyders, D. J., and Chaudhary, A. (1996). High affinity open channel block by dofetilide of *HERG* expressed in a human cell line. *Mol. Pharmacol.* **49,** 949–955.

Spector, P. S., Curran, M. E., Keating, M. T., and Sanguinetti, M. C. (1996a). Class III antiarrhythmic drugs block HERG, a human cardiac delayed rectifier K^+ channel; open channel block by methanesulfonanilides. *Cir. Res.* **78,** 499–503.

Spector, P. S., Curran, M. E., Zou, A., Keating, M. T., and Sanguinetti, M. C. (1996b). Fast inactivation causes rectification of the I_{Kr} channel. *J. Gen. Physiol.* **107,** 611–619.

Splawski, I., Timothy, K. W., Vincent, G. M., Atkinson, D. L., and Keating, M. T. (1997a). Molecular basis of the long-QT syndrome associated with deafness. *N. Engl. J. Med.* **336,** 1562–1567.

Splawski, I., Tristani-Firouzi, M., Lehmann, M. H., Sanguinetti, M. C., and Keating, M. T. (1997b). Mutations in the *hminK* gene cause long QT syndrome and suppress I_{Ks} function. *Nature Genet.* **17,** 338–340.

Trudeau, M., Warmke, J. W., Ganetzky, B., and Robertson, G. A. (1995). HERG, A human inward rectifier in the voltage-gated potassium channel family. *Science* **269,** 92–95.

Turgeon, J., Daleau, P., Bennett, P. B., Wiggins, S. S., Selby, L., and Roden, D. M. (1994). Block of I_{Ks}, the slow component of the delayed rectifier K^+ current, by the diuretic agent indapamide in guinea pig myocytes. *Circ. Res.* **75,** 879–886.

Wang, K.-W., and Goldstein, S. A. N. (1995). Subunit composition of minK potassium channels. *Neuron* **14,** 1303–1309.

Wang, Q., Curran, M. E., Splawski, I., Burn, T. C., Millholland, J. M., VanRaay, T. J., Shen, J., Timothy, K. W., Vincent, G. M., de Jager, T., Schwartz, P. J., Towbin, J. A., Moss, A. J., Atkinson, D. L., Landes, G. M., Connors, T. D., and Keating, M. T. (1996). Positional cloning of a novel potassium channel gene: KVLQT1 mutations cause cardiac arrhythmias. *Nature Genet.* **12,** 17–23.

Wang, Q., Shen, J., Splawski, I., Atkinson, D., Li, Z., Robinson, J. L., Moss, A. J., Towbin, J. A., and Keating, M. T. (1995). *SCN5A* mutations associated with an inherited cardiac arrhythmia, long QT syndrome. *Cell* **80,** 805–811.

Wang, S., Liu, S., Morales, M. J., Strauss, H. C., and Rasmusson, R. L. (1997). A quantitative analysis of the activation and inactivation kinetics of HERG expressed in *Xenopus* oocytes. *J. Physiol.* **502,** 45–60.

Wang, S., Morales, M. J., Liu, S., Strauss, H. C., and Rasmusson, R. L. (1996). Time, voltage and ionic concentration dependence of rectification of h-*erg* expressed in *Xenopus* oocytes. *FEBS Lett.* **389,** 167–173.

Wang, Z., Fermini, B., and Nattel, S. (1993a). Delayed rectifier outward current and repolarization in human atrial myocytes. *Circ. Res.* **73,** 276–285.

Wang, Z., Fermini, B., and Nattel, S. (1993b). Sustained depolarization-induced outward current in human atrial myocytes: Evidence for a novel delayed rectifier K^+ current similar to Kv1.5 cloned channel currents. *Circ. Res.* **73,** 1061–1076.

Wang, Z., Fermini, B., and Nattel, S. (1994). Rapid and slow components of delayed rectifier currents in human atrial myocytes. *Cardiovasc. Res.* **28,** 1540–1546.

Warmke, J., Drysdale, R., and Ganetzky, B. (1991). A distinct potassium channel polypeptide encoded by the *Drosophila eag* locus. *Science* **252,** 1560–1564.

Warmke, J. W., and Ganetzky, B. (1994). A family of potassium channel genes related to *eag* in *Drosophila* and mammals. *Proc. Natl. Acad. Sci. USA* **91,** 3438–3442.

Wible, B. A., Taglialatela, M., Ficker, E., and Brown, A. M. (1994). Gating of inwardly rectifying K^+ channels localized to a single negatively charged residue. *Nature* **371,** 246–249.

Wollnik, B., Schroeder, B. C., Kubisch, C., Esperer, H. D., Wieacker, P., and Jentsch, T. J. (1997). Pathophysiological mechanisms of dominant and recessive *KVLQT1* K^+ channel mutations found in inherited cardiac arrhythmias. *Hum. Mol. Genet.* **6,** 1943–1949.

Yang, T., Kupershmidt, S., and Roden, D. (1995). Anti-minK antisense decreases the amplitude of the rapidly activating cardiac delayed rectifier K^+ current. *Circ. Res.* **77,** 1246–1253.

Yang, T., Snyders, D. J., and Roden, D. M. (1997). Rapid inactivation determines the rectification and $[K^+]_o$ dependence of the rapid component of the delayed rectifier K^+ current in cardiac cells. *Circ. Res.* **80,** 782–789.

Yang, W. P., Levesque, P. C., Little, W. A., Conder, M. L., Shalaby, F. Y., and Blanar, M. A. (1997). KvLQT1, a voltage-gated potassium channel responsible for human cardiac arrhythmias. *Proc. Natl. Acad. Sci. USA* **94,** 4017–4021.

Zou, A., Curran, M. E., Keating, M. T., and Sanguinetti, M. C. (1997). Single HERG delayed rectifier K^+ channels in Xenopus oocytes. *Am. J. Physiol.* **272,** H1309–H1314.

CHAPTER 7

Mutational Analysis of Familial Long QT Syndrome in Japan

Toshihiro Tanaka and Yusuke Nakamura
Laboratory of Molecular Medicine, Human Genome Center, Institute of Medical Science, University of Tokyo, Tokyo 108, Japan

I. INTRODUCTION

Long QT syndrome (LQTS) is characterized by prolongation of the QT interval on electrocardiograms (ECG). Carriers of this disorder may suffer lethal arrhythmias such as ventricular tachycardia, torsade de pointes, or ventricular fibrillation. Although some patients may remain asymptomatic and appear normal except for prolongation of the QT interval, others suffer from recurrent syncope or die suddenly. The variability of QT intervals and clinical signs among carriers makes presymptomatic diagnosis difficult, but early diagnosis is necessary if clinicians are to prevent sudden cardiac death by means of antiarrhythmic drugs and/or implantable defibrillating devices.

1063-5823/99 $30.00

Two subtypes of familial LQTS are defined according to the mode of inheritance and accompanying clinical symptoms. An autosomal recessive type is associated with congenital neural deafness (Jervell and Lange-Nielsen syndrome, JLNS; Jervell and Lange-Nielsen, 1957); another, Romano-Ward syndrome (RWS; Romano, 1965; Ward, 1964), is inherited as an autosomal dominant trait without idiopathic auditory dysfunction.

Earlier genetic linkage analyses of RWS families revealed four LQTS loci. LQT1 lies on chromosome 11p15.5 (Keating *et al.,* 1991); loci for LQT2 on chromosome 7q35 and LQT3 on chromosome 3p21 were reported by Jiang *et al.* (1994); and LQT4 lies on chromosome 4q25-27 (Schott *et al.,* 1995). Three of the responsible genes in these regions have been identified by positional cloning or the positional-candidate approach: affected members of LQT1- and LQT2-linked RWS families have been shown to carry mutations in the *KVLQT1* and *HERG* genes, respectively; both genes encode voltage-gated potassium channels (Curran *et al.,* 1995; Wang *et al.,* 1996a). *SCN5A,* a sodium-channel gene, is responsible for LQT3 (Wang *et al.,* 1995a). Additionally, mutations in *IsK,* a gene located on chromosome 21q22 that encodes a small membrane-spanning protein that forms a potassium channel in association with KvLQT1 (Barhanin *et al.,* 1996; Sanguinetti *et al.,* 1996), have been identified in two RWS families (Splawski *et al.,* 1997b). JLNS also shows genetic heterogeneity, and mutations in both *KVLQT1* and *IsK* have been identified very recently in patients with that disorder (Neyroud *et al.,* 1997; Splawski *et al.,* 1997a; Schluze-Bahr *et al.,* 1997; Tyson *et al.,* 1997).

For patients with clinical symptoms, β-adrenergic blocking agents are the first choice for treatment. Symptoms in about 80% of LQTS patients can be effectively managed with these drugs, but the remaining 20% do not respond (Locati and Schwartz, 1992). Hence, if the different responses reflect mutations at different sites within a given gene or mutations in different genes, genetic diagnosis might provide information to govern the choice of an appropriate therapeutic approach.

So far, 49 distinct LQTS mutations have been reported (Benson *et al.,* 1996; Curran *et al.,* 1995; Dausse *et al.,* 1996; Neyroud *et al.,* 1997; Russell *et al.,* 1996; Satler *et al.,* 1996; Schluze-Bahr *et al.,* 1995, 1997; Splawski *et al.,* 1997a,b; Tanaka *et al.,* 1997; Tyson *et al.,* 1997; van den Berg *et al.,* 1997; Wang *et al.,* 1995a,b, 1996a; Wollnik *et al.,* 1997), but with respect to *KVLQT1* and *HERG,* only parts of the coding regions have been searched. For *SCN5A,* whose genomic structure has been reported (Wang *et al.,* 1996b), the whole coding region can be investigated. In our previous study (Tanaka *et al.,* 1997), we undertook mutational analyses in 32 Japanese LQTS families and identified *HERG* mutations in five of them. This rate seemed low in view of data from linkage analyses (Curran *et al.,* 1995); our inability to examine the entire gene might have accounted for the discrepancy.

It is generally easier to screen genomic DNA for genetic alterations than to screen cDNA reversely transcribed from mRNA when the genes are expressed at a low abundancy and/or in a specific tissue; genomic DNA can be easily obtained from peripheral blood samples, while specific tissue is necessary as a specimen to obtain mRNA. However, for genomic screening DNA sequences around exon/intron boundaries should be determined first, because oligonucleotide primers need to be synthesized to permit polymerase chain reaction (PCR) amplification of the entire coding region.

In this chapter we describe the genomic organization of the *HERG* gene, information that should make it possible to detect most mutations related to *HERG*. We also report results of mutational analyses of the four known LQTS genes in 45 Japanese families.

II. MATERIALS AND METHODS

A. Diagnosis of the Patients

DNA samples were collected from members of 45 LQTS families throughout Japan, along with individual histories and ECG records. To avoid bias, the QT intervals on ECGs were measured by physicians who were not involved in the mutational analyses. Peripheral blood samples were obtained with informed consent of the patients and their family members. DNA was extracted from leukocytes by standard techniques (Sambrook *et al.*, 1989). Phenotypic determinations were made on the basis of the corrected QT interval (Bazett, 1920), as described previously (Fig. 1; Benhorin *et al.*, 1993; Tanaka *et al.*, 1994).

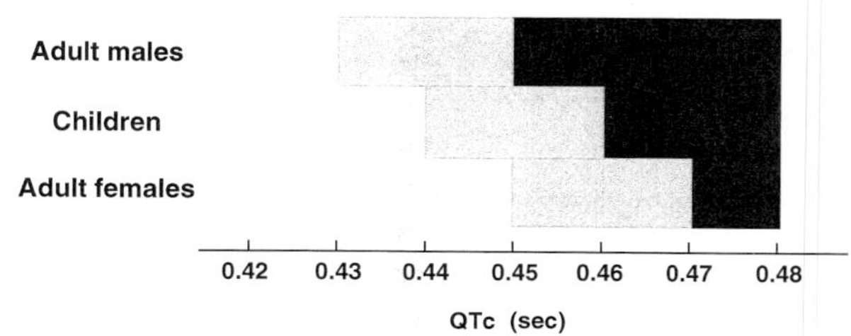

FIGURE 1 Phenotyping of LQTS. Because the QT interval is influenced by age and gender (Hashiba, 1978; Vincent *et al.*, 1992), family members were divided into three subgroups: children under 16, adult males, and adult females. Solid region, affected status; shaded, equivocal.

B. Genomic Library Screening

Cosmid clones containing human DNA inserts were selected by colony hybridization using reverse transcriptase PCR products of the *HERG* gene as probes. Hybridization and washing conditions for cosmid screening were as described by Tokino *et al.* (1991). Partial sequences of clones were determined using a dye-terminator cycle-sequencing kit (Perkin-Elmer) according to the supplier's instructions, with a Perkin-Elmer 377 DNA sequencer.

C. Mutational Analysis

DNA from the proband of each family was analyzed by PCR-SSCP (single strand conformation polymorphism). Some of the primers for genomic amplification and sequencing were described previously (Curran *et al.*, 1995; Wang *et al.*, 1995a, 1996a,b). After amplification, each reaction mixture was diluted with 95% formamide dye, incubated at 80°C for 5 min, and applied to a 6% polyacrylamide gel containing 0.5× TBE and 5% glycerol. Electrophoresis was performed at 4°C. Gels were dried and autoradiographed with intensifying screens.

For DNA sequencing, genomic DNA (20 ng) from each patient showing an aberrant conformer in PCR-SSCP analysis was amplified in a 25-μl PCR reaction as above. After removal of dNTPs and primers by columns, each sample was subjected to cycle sequencing using a dye-terminator cycle-sequencing kit (Perkin-Elmer) according to the supplier's instructions. Electrophoresis was carried out in a Perkin-Elmer 377 DNA sequencer. When a mutation had created a recognition site for an endonuclease, PCR products of DNA from all available family members and from 80 normal individuals were digested by that enzyme and applied to a nondenaturing polyacrylamide gel for confirmation of the mutation. We also confirmed the results by hybridization of mutant allele-specific oligonucleotides (Saiki *et al.*, 1986).

III. RESULTS

A. Genomic Structure of HERG

Figure 2 illustrates the genomic structure of the *HERG* gene, which consists of 15 exons spanning approximately 19 kb on chromosome 7q35. Regions encoding putative transmembrane domains lie within exons 6–8.

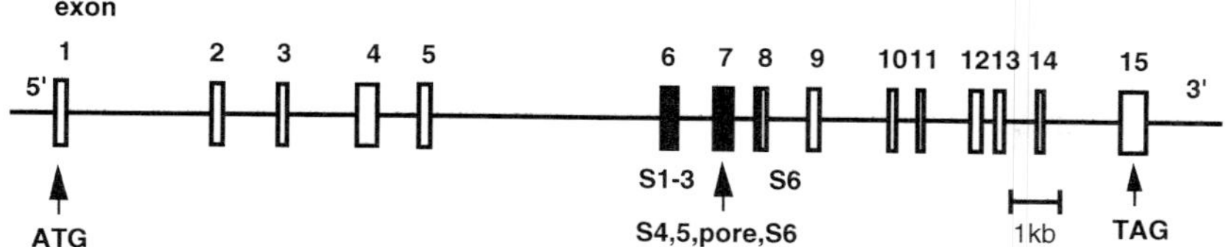

FIGURE 2 Genomic structure of *HERG*. Exons are shown as boxes, numbered 1–15. Locations of the initiation (ATG) and termination (TAG) codons are indicated, and regions encoding putative transmembrane domains (S1–S6, pore) are shown as solid boxes.

Boundaries and splice junctions were identified for all known *HERG* exons, as shown in Table I. The intronic splice sites seemed to be consistent with the GT-AG rule (Shapio and Senapathy, 1987).

To facilitate mutational analysis in *HERG*, we synthesized all the oligonucleotide primers that would be required to cover the whole coding region by means of PCR, except for those whose amplimers had been reported already (Curran *et al.*, 1995). The primer sequences are given in Table II along with the annealing temperature at which each exon could be amplified. Using these primers, we performed PCR-SSCP analysis and DNA sequencing to screen for and characterize disease-causing mutations.

B. Mutational Analyses of the Four LQTS Genes

In total, we detected 14 different mutations in the region distal to S2, either in *KVLQT1* or *HERG* (Table III). As some were missense mutations, we could not completely exclude the possibility of very rare polymorphism. However, since each of the mutations cosegregated among patients showing prolonged QTc intervals on ECGs, and since none of the nucleotide changes was seen among 80 normal individuals, these mutations are likely to have caused the disease. No mutations were found in either *SCN5A* or *IsK* in our panel of 45 families. Figures 3 and 4 illustrate the mutation and cosegregation of *KVLQT1* and *HERG*, respectively, within representative families.

C. Possible Correlation between Mutated Genes and Response to β-Adrenergic Blockade

To find a relationship between genotype and phenotype, we looked retrospectively at the clinical records of patients whose mutations had

TABLE I

Sequences around Exon/Intron Junctions

Acceptor site						Donor site				
Intron	Exon			Exon (bp)	Codon	Exon			Intron	Approximate length of intron
cccggcccgc	CCA	TGG	GCT	1 (89)	1–26	GGC	CAG	A– –	gtgaatgggg	2.9
gcccccctag	– GC	CGT	AAG	2 (231)	27–103	AAA	GAT	G– –	gtaggaacgg	1.0
cactctgcag	– GG	AGC	TGC	3 (165)	104–158	GCC	CCA	G– –	gtaagtgtac	1.3
tctcccgcag	– GC	CGC	GCC	4 (444)	159–306	AGC	ACC	G– –	gtgagggcgc	0.7
ctccacctag	– GG	GCC	ATG	5 (212)	307–376	GTC	ACC	CAG	gtaggcgccc	4.7
ccgggtgcag	GTC	CTG	TCC	6 (429)	377–519	TCT	GAG	GAG	gtggggtcag	0.6
tgtcccccag	CTG	ATC	GGG	7 (388)	520–649	ATT	GGC	T– –	gtgagtgtgc	0.3
acgcccccag	– CC	CTC	ATG	8 (200)	650–715	ATG	AAC	GCG	gtaaggccac	0.7
ctgcccccag	GTG	CTG	AAG	9 (253)	716–799	ATC	CTG	G– –	gtatggggtg	1.3
tggcctccag	– GG	AAG	AAT	10 (194)	800–864	CTG	CGA	GAT	gtgagttggc	0.3
ttggttccag	ACC	AAC	ATG	11 (100)	865–898	GAC	AAG	G– –	gtgaggcggg	0.8
tttcccacag	– AC	ACG	GAG	12 (273)	899–989	CTG	TCA	G– –	gtatcccggg	0.1
ctggctgcag	– GC	GCC	TTC	13 (187)	990–1051	CTC	AAC	AG–	gtgaaggaat	0.5
cctgccccag	– – G	CTG	GAG	14 (178)	1052–1110	CTT	TCT	CAG	gtaagctcca	1.5
cagtgaggag	GTT	TCC	CAG	15 (150)	1111–stop	3′-untranslated region				

TABLE II

PCR Primers Used to Cover the Entire Coding Region of *HERG*

Exon	Forward primer	Reverse primer	Amplified region	Annealing temperature (°C)
1	CCGGCCACCCGAAGCCTAGT	ATCCACACTCGGAAGAAGTC	1–76	60
2	CTGTGTGAGTGGAGAATGTG	CCGTGGTGGTGGCCCCGCGG	77–307	60
3	TGCCCACTGAGTGGGTGC	CCATCACCACCTCGAAATTG	308–400	58
	TGGCTGTCATGTTCATCCT	TGACCTTGGACAGCTCACAG	401–472	58
4	ACGACCACGTGCCTCTCCTCTC	GGACCCACCAGCGCACGCCG	473–710	58
	CCCTGGACGAAGTGACAGCCAT	GGCTGGGGCGGAACGGGTCC	632–916	58
5	GCCTGACCACGCTGCCTCT	CCTCCAAGGTGAGAGGAGA	917–1128	58
6	CAGAGATGTCATCGCTCCTG	GAGCCAGTCCCACACGGCC	1129–1239	58
	CGACGTGCTGCCTGAGTACAA*	CACCTCCTCGTTGGCATTGAC*	1146–1446	58
	TTCCTGCTGAAGGAGACGGAAG*	TACACCACCTGCCTCCTTGCTGA*	1291–1557	60
7	TGCCCCATCAACGGATGTGC*	CAGCCAGCCGATGCGTGAGTCCA*	1558–1758	60
	TAGCCTGCATCTGGTACGC	GCCCGCCCCTGGGCACACTCA*	1691–1945	58
8	CTGTGGGTGGGTGGGGTCC	CAGCATCTGTGTGTGGTAG*	1946–2034	58
	CATCTTCGGCAACGTCGGC	TTTCCCAGCCTGCCACCCACT	1962–2145	58
9	CCAAGGGAGGGTGTGCTGAG	GGTGCATGTGTGGTCTTGAAC	2146–2292	58
	GCACTGCAAACCCTTCCGAG	GGCATTTCCAGTCCAGTGC	2214–2398	58
10	AGAAGGTGCCTGCTGCCTG	TCCAGCTCAGGGCAGCCAA	2399–2592	58
11	AAGGGCCCTGATACTGATTTT	TTCCAGCTCCCAGCCTCA	2593–2692	58
12	CCCCTCTTTGAGGCCCATT	CTTCTCGCAGTCCTCCATCA	2693–2937	58
	TGAGAGCAGTGAGGATGA	TAGACGCACCACCGCTGCC	2814–2965	58
13	TGGCAGCGGTGGTGCGTCT	AGGGAGCTCCTGCTACTGG	2966–3039	58
	TTCTCAGGAGTGTCCAACAT	TGGTCACAGCACTGTAAGC	2976–3152	58
14	ATCCCGGTGGAGGCTGTCA	GAACAAGCGGGTCACGGTAC	3153–3330	58
15	TCCTGTCCTCCCGTCCATC	ACGTGTCCACACTGGGCAG	3331–3480	58

Note. Primers with an asterisk have been reported previously (Wang *et al.*, 1995a). Nucleotide numbering starts with the initiator methionine. Large exons were each divided into several fragments to permit SSCP analysis.

TABLE III
Summary of LQTS Mutations

Gene family	Nucleotide change	Mutation	Region
KVLQT1			
38	GCC to ACC	A49T	Between S2 and S3
56	ATC to ATG	I184M	Pore
29	GGG to AGG	G196R	S6
61	CGG to CCG	R237P	Distal to S6
HERG			
20	ACC to ATC	T474I	Between S2 and S3
76	TAC to TAG	Y493ter	Between S2 and S3
80	CGC to TGC	R534C	S4
27	GCG to GTG	A561V	S5
13	TAT to CAT	Y611H	S5–pore
22	GCG to GTG	A614V	Pore
47	GTC to CTC	V630L	Pore
45	del AT(2134, 2135)	Frameshift	S6
54	CAG to TAG	Q725ter	Distal to S6
68	dup (2539–2569)	Frameshift	Distal to S6

been identified. Among 12 symptomatic patients taking simple β-adrenergic blockade therapy, those who carried mutated *KVLQT1* alleles no longer suffered from syncope. On the other hand, a β-adrenergic blocking agent alone appeared to be effective for only one of five LQTS patients who carried a mutated *HERG* gene ($P = 0.01$, Fisher's exact test; Table IV).

IV. DISCUSSION

As reported here, we determined the genomic structure of *HERG,* a gene responsible for chromosome 7-linked LQTS, synthesized oligonucleotide primers based on flanking intronic sequences for all 15 *HERG* exons, and established conditions for PCR amplification of each exon from genomic DNA. This information will facilitate systematic scanning of all *HERG* exons for mutations in LQTS patients by means of the PCR-SSCP technique.

We identified 14 different mutations in either *KVLQT1* or *HERG* among 45 Japanese LQTS families. Including previous reports, 54 kinds of mutations have been identified so far. Although mutations at amino acid 212 in *KVLQT1* have been reported in 8 families (Russell *et al.,* 1996; Wang *et al.,* 1996a), we have not found this mutation in any of the Japanese families;

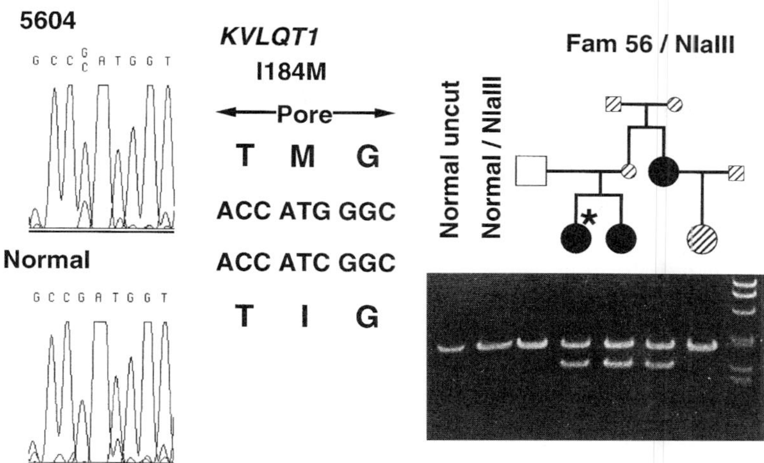

FIGURE 3 *KVLQT1* mutation and its segregation in Family 56. (Left and center) By fluorescent sequencing, Patient 5604 (indicated by an asterisk in the pedigree in the right panel) was shown to carry a C552G mutation, which causes an amino acid change from isoleucine to methionine in the pore region of the *KVLQT1* gene (see also Table III). Sequences of the anti-sense strand are shown. (Right) Endonuclease *Nla*III, which cleaves DNA at CATG sequences, was used for confirmation. Since the mutant allele has a CATG sequence where the normal allele is GATG, the two alleles can be distinguished by *Nla*III digestion and size fractionation using acrylamide gel electrophoresis. In this case, the size of the normal PCR product is 196 bp; *Nla*III digestion of the mutant allele divided it into 148- and 48-bp fragments. Each lane corresponds to the member of family 56 indicated above in the pedigree. Black, affected; white, unaffected; shaded, unknown phenotype. Smaller symbols represent individuals who were not sampled.

the possibility of a founder effect cannot be excluded. The combined data also indicate that detecting high-risk groups by screening for mutations is a difficult strategy.

In our mutational analyses among 45 Japanese LQTS families, 10 separate and novel *HERG* mutations were identified. This proportion seems low in view of results from linkage analyses in Caucasians (Curran *et al.,* 1995) in which 14 of the 23 LQTS families examined appeared to be linked to the *HERG* locus. The low frequency of *HERG* mutations in Japanese LQTS families cannot be attributed mainly to the screening method; PCR-SSCP generally shows greater than 90% sensitivity (Hayashi and Yandell, 1993). The discrepancy could reflect ethnic factors, or it could be due to a significant proportion of mutations lying outside the gene itself, within regulatory regions.

It is not yet clear which regions of *KVLQT1* or *HERG* are critical for their function. Transmembrane domains must be maintained for the channel

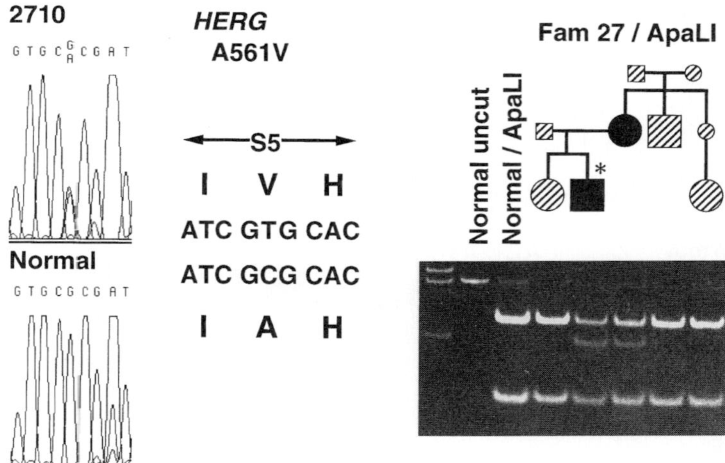

FIGURE 4 *HERG* mutation in Family 27. Patient 2710, indicated by an asterisk in the pedigree, was shown to carry an A561V mutation in the S5 region of *HERG*. Sequences of the sense strand are shown. Endonuclease *Apa*LI, which cleaves DNA at GTGCAC sequences, was used for confirmation. Although the disease status of other family members could not be judged from the ECG, none of them carries the mutant allele.

TABLE IV
Effect of Simple β-Adrenergic Blockade Therapy on Symptomatic LQTS Patients

Gene patient	Sex	Age	QTc (ms)	Symptom after therapy	Symptom-free period (years)
KVLQT1					
3817	F	42	660	—	16
5604	F	17	590	—	5
5605	F	19	460	—	5
2905	M	10	510	—	7
6114	F	20	570	—	9
6115	F	18	550	—	2
6117	F	15	550	—	4
HERG					
2009	M	24	500	Syncope	—
8001	M	18	460	VT	—
2710	M	15	480	Syncope	—
2211	F	22	550	Syncope	—
6801	F	65	520	—	7

structure, but interactions of other regions with different molecules might modify the channel property. Mutational analysis must be done throughout entire coding regions before the pathophysiology of this syndrome can be understood. We believe the information from the study reported in this chapter will be indispensable to that effort.

The relationship between *KVLQT1* mutations and clinical symptoms is not clear. Schwartz and colleagues (1995) suggested that patients with mutated *SCN5A* may be more likely to benefit from Na^+-channel blockers. Our findings may indicate that among symptomatic patients, those who carry mutated *KVLQT1* alleles show a better response to β-blocking agents than patients with mutant *HERG* ($P = 0.01$, Fisher's exact test), but this presumption needs further validation.

In any event, patients carrying the *HERG* mutation are more likely to be at risk for syncope or sudden death under stressful conditions, because the arrhythmogenic effect of catecholamines would be enhanced (Schwartz *et al.,* 1995). Compton and co-workers (1996) have shown that potassium therapy improves repolarization in chromosome 7-linked LQTS (presumed *HERG* mutations), a treatment that needs long-term monitoring. We believe that the information gained from our study will help us identify high-risk groups among LQTS families. Although we investigated the entire coding regions of both *SCN5A* and *IsK,* we identified no mutations in these genes, which therefore appear to have little importance in the pathogenesis of LQTS in Japan.

Some puzzles remain. First, 69% of LQTS mutations are still unidentified. It is possible that additional genes play major roles in the pathogenesis of LQTS; on the other hand, mutations may be lying outside the coding regions of the four identified LQTS genes, in promoter or enhancer elements. It is also a mystery why a single family may contain clinically symptomatic and asymptomatic carriers of a mutant LQTS allele, all of them exhibiting prolongation of the QTc interval. It is likely that factors other than genetic ones, such as hormonal environment that may cause imbalance between activities of sympathetic and parasympathetic nerves, may generate variations in the clinical phenotype. This issue is highly important for understanding the mechanism of ventricular arrhythmia.

Acknowledgments

We are indebted to the family members participating in this study. This work was supported in part by a grant-in-aid from the Ministry of Education, Science, Sports, and Culture of Japan, and by a "Research for the Future" program grant (96L00102) from The Japan Society for the Promotion of Science.

References

Barhanin, J., Lesage, F., Guillemare, E., Fink, M., Lazdunski, M., and Romey, G. (1996). KvLQT1 and IsK (minK) proteins associate to form the IKs cardiac potassium current. *Nature* **384,** 78–80.

Bazett, H. C. (1920). An analysis of the time-relations of electrocardiograms. *Heart* **7,** 353–370.

Benhorin, J., Kalman, Y. M., Medina, A., Towbin, J., Rave-Harel, N., Dyer, T. D., Blangero, J., MacCluer, J. W., and Kerem, B. S. (1993). Evidence of genetic heterogeneity in the long QT syndrome. *Science* **260,** 1960–1962.

Benson, D. W., MacRae, C. A., Vesely, M. R., Walsh, E. P., Seidman, J. G., Seidman, C. E., and Satler, C. A. (1996). Missense mutation in the pore region of *HERG* causes familial Long QT syndrome. *Circulation* **93,** 1791–1795.

Compton, S. J., Lux, R. L., Ramsey, M. R., Strelich, K. R., Sanguinetti, M. C., Green, L. S., Keating, M. T., and Mason, J. W. (1996). Genetically defined therapy of inherited long-QT syndrome. Correction of abnormal repolarization by potassium. *Circulation* **94,** 1018–1022.

Curran, M. E., Splawski, I., Timothy, K. W., Vincent, G. M., Green, E. D., and Keating, M. T. (1995). A molecular basis for cardiac arrhythmia: *HERG* mutations cause long QT syndrome. *Cell* **80,** 795–803.

Dausse, E., Berthet, M., Denjoy, I., Fouet, X. A., Cruaud, C., Bennaceur, M., Faure, S., Coumel, P., Schwartz, K., and Guicheney, P. (1996). A mutation in *HERG* associated with notched T waves in long QT syndrome. *J. Mol. Cell. Cardiol.* **28,** 1609–1615.

Hashiba, K. (1978). Hereditary QT prolongation syndrome in Japan: Genetic analysis and pathological findings of the conducting system. *Jpn. Circ. J.* **42,** 1133–1150.

Hayashi, K., and Yandell, D.W. (1993). How sensitive is PCR-SSCP? *Hum. Mut.* **2,** 338–346.

Jervell, A., and Lange-Nielsen, F. (1957). Congenital deaf mutism, functional heart disease with prolongation of the Q-T interval and sudden death. *Am. Heart J.* **54,** 59–68.

Jiang, C., Atkinson, D., Towbin, J. A., Splawski, I., Lehmann, M. H., Li, H., Timothy, K., Taggart, R. T., Schwartz, P. J., Vincent, G. M., Moss, A. J., and Keating, M. T. (1994). Two long QT syndrome loci map to chromosomes 3 and 7 with evidence for further heterogeneity. *Nature Genet.* **8,** 141–147.

Keating, M., Atkinson, D., Dunn, C., Timothy, K., Vincent, G. M., and Leppert, M. (1991). Linkage of a cardiac arrhythmia, the long QT syndrome, and the Harvey ras-1 gene. *Science* **252,** 704–706.

Locati, E. H., and Schwartz, P. J. (1992). The idiopathic long QT syndrome: Therapeutic management. *PACE* **15,** 1374–1379.

Neyroud, N., Tesson, F., Denjoy, I., Leibovici, M., Donger, C., Barhanin, J., Faure, S., Gary, F., Coumel, P., Petit, C., Schwartz, K., and Guicheney, P. (1997). A novel mutation in the potassium channel gene *KVLQT1* causes the Jervell and Lange-Nielsen cardioauditory syndrome. *Nature Genet.* **15,** 186–189.

Romano, C. (1965). Congenital cardiac arrhythmia. *Lancet* **I,** 658–659.

Russell, M. W., Dick, M., II, Collins, F. S., and Brody, L. C. (1996). *KVLQT1* mutations in three families with familial or sporadic long QT syndrome. *Hum. Mol. Genet.* **5,** 1319–1324.

Saiki, R. K., Bugawan, T. L., Horn, G. T., Mullis, K. B., and Erlich, H. A. (1986). Analysis of enzymatically amplified beta-globin and HLA-DQ alpha DNA with allele-specific oligonucleotide probes. *Nature* **324,** 163–166.

Sambrook, J., Fritsch, E. F., and Maniatis, T. (1989). "Molecular Cloning," 2nd ed. Cold Spring Harbor Laboratory Press, Cold Spring Harbor, NY.

Sanguinetti, M. C., Curran, M. E., Zou, A., Shen, J., Spector, P. S., Atkinson, D. L., and Keating, M. T. (1996). Coassembly of KvLQT1 and minK (IsK) proteins to form cardiac I(Ks) potassium channel. *Nature* **384,** 80–83.

Satler, C. A., Walsh, E. P., Vesely, M. R., Plummer, M. H., Ginsburg, G. S., and Jacob, H. J. (1996). Novel missense mutation in the cyclic nucleotide-binding domain of *HERG* causes long QT syndrome. *Am. J. Med. Genet.* **65,** 27–35.

Schluze-Bahr, E., Haverkamp, W., and Funke, H. (1995). The long-QT syndrome. *N. Engl. J. Med.* **333,** 1783–1784.

Schluze-Bahr, E., Wang, Q., Wedekind, H., Haverkamp, W., Chen, Q., Sun, Y., Rubie, C., Hordt, M., Towbin, J. A., Borggrefe, M., Assmann, G., Qu, X., Somberg, J. C., Breithardt, G., Oberti, C., and Funke, H. (1997). KCNE1 mutations cause Jervell and Lange-Nielsen syndrome. *Nature Genet.* **17,** 267–268.

Schott, J. J., Charpentier, F., Peltier, S., Foley, P., Drouin, E., Bouhour, J. B., Donnelly, P., Vergnaud, G., Bachner, L., Moisan, J. P., Le Marec, H., and Pascal, O. (1995). Mapping of a gene for long QT syndrome to chromosome 4q25-27. *Am. J. Hum. Genet.* **57,** 1114–1122.

Schwartz, P. J., Priori, S. G., Locati, E. H., Napolitano, C., Cantu, F., Towbin, J. A., Keating, M. T., Hammoude, H., Brown, A. M., Chen, L. S. K., and Colatsky, T. J. (1995). Long QT syndrome patients with mutations of the *SCN5A* and *HERG* genes have differential responses to Na^+ channel blockade and to increases in heart rate: Implications for gene-specific therapy. *Circulation* **92,** 3381–3386.

Shapio, M., and Senapathy, P. (1987). RNA splice junction of different class of eukaryotes: Sequence statistics and function implications in gene expression. *Nucleic Acids Res.* **17,** 7155–7175.

Splawski, I., Timothy, K. W., Vincent, G. M., Atkinson, D. L., and Keating, M. T. (1997a). Molecular basis of the long-QT syndrome associated with deafness. *New Engl. J. Med.* **336,** 1562–1567.

Splawski, I., Tristani-Firouzi, M., Lehmann, M. H., Sanguinetti, M. C., and Keating, M. T. (1997b). Mutations in the hminK gene cause long QT syndrome and suppress IKs function. *Nature Genet.* **17,** 338–340.

Tanaka, T., Nakahara, K., Kato, N., Imai, T., Yamazaki, T., Tomita, H., Shimokawa, H., Matsuhashi, H., Sato, N., Matsui, M., Kihira, S., Shimizu, A., Sano, T., Haneda, N., Kino, M., Miyakita, Y., Matsuoka, R., Nagai, R., Yazaki, Y., and Nakamura, Y. (1994). Genetic linkage analyses of Romano-Ward syndrome (RWS) in 13 Japanese families. *Hum. Genet.* **94,** 380–384.

Tanaka, T., Nagai, R., Tomoike, H., Takata, S., Yano, K., Yabuta, K., Haneda, N., Nakano, O., Shibata, A., Sawayama, T., Kasai, H., Yazaki, Y., and Nakamura, Y. (1997). Four novel *KVLQT1* and four novel *HERG* mutations in familial long QT syndrome. *Circulation* **95,** 565–567.

Tokino, T., Takahashi, E., Mori, M., Tanigami, A., Glaser, T., Park, J. W., Jones, C., Hori, T., and Nakamura, Y. (1991). Isolation and mapping of 62 new RFLP markers on human chromosome 11. *Am. J. Hum. Genet.* **48,** 258–268.

Tyson, J., Tranebjaerg, L., Bellman, S., Wren, C., Taylor, J. F. N., Bathen, J., Aslaksen, B., Sorland, S. J., Lund, O., Malcolm, S., Pembrey, M., Bhattacharya, S., and Bitner-Glindzicz, M. (1997). *IsK* and *KVLQT1*: Mutation in either of the two subunits of the slow component of the delayed rectifier potassium channel can cause the Jervell and Lange-Nielsen syndrome. *Hum. Mol. Genet.* **6,** 2179–2185.

van den Berg, M. H., Wilde, A. A., Robles de Medina, E. O., Meyer, H., Geelen, J. L., Jongbloed, R. J., Wellens, H. J., and Geraedts, J. P. (1997). The long QT syndrome: A novel missense mutation in the S6 region of the *KVLQT1* gene. *Hum. Genet.* **100,** 356–361.

Vincent, G. M., Timothy, K. W., Leppert, M., and Keating, M. (1992). The spectrum of symptoms and QT intervals in carriers of the gene for the long-QT syndrome. *N. Engl. J. Med.* **327,** 846–852.

Wang, Q., Shen, J., Splawski, I., Atkinson, D., Li, Z., Robinson, J. L., Moss, A. J., Towbin, J. A., and Keating, M. T. (1995a). *SCN5A* mutations associated with an inherited cardiac arrhythmia, long QT syndrome. *Cell* **80,** 805–811.
Wang, Q., Shen, J., Li, Z., Timothy, K., Vincent, G. M., Priori, S. G., Schwartz, P. J., and Keating, M. T. (1995b). Cardiac sodium channel mutations in patients with long QT syndrome, an inherited cardiac arrhythmia. *Hum. Mol. Genet.* **4,** 1603–1607.
Wang, Q., Curran, M. E., Splawski, I., Burn, T. C., Millholland, J. M., VanRaay, T. J., Shen, J., Timothy, K. W., Vincent, G. M., de Jager, T., Schwartz, P. J., Towbin, J. A., Moss, A. J., Atkinson, D. L., Landes, G. M., Connors, T. D., and Keating, M. T. (1996a). Positional cloning of a novel potassium channel gene: *KVLQT1* mutations cause cardiac arrhythmias. *Nature Genet.* **12,** 17–23.
Wang, Q., Li, Z., Shen, J., and Keating, M. T. (1996b). Genomic organization of the human *SCN5A* gene encoding the cardiac sodium channel. *Genomics* **34,** 9–16.
Ward, O. C. (1964). A new familial cardiac syndrome in children. *J. Ir. Med. Assoc.* **54,** 103–106.
Wollnik, B., Schroeder, B. C., Kubisch, C., Esperer, H. D., Wieacker, P., and Jentsch, T. J. (1997). Pathophysiological mechanisms of dominant and recessive *KVLQT1* K^+ channel mutations found in inherited cardiac arrhythmias. *Hum. Mol. Genet.* **6,** 1943–1949.

CHAPTER 8

Calcium-Activated Potassium Channels in Muscle and Brain

Martin Wallner,* Pratap Meera,* and Ligia Toro*,†,‡
Departments of *Anesthesiology, †Molecular and Medical Pharmacology, and the ‡Brain Research Institute, University of California, Los Angeles, Los Angeles, California 90095

I. INTRODUCTION

Membrane potential and intracellular Ca^{2+} concentration are fundamental physiological regulators. The cytosolic Ca^{2+} concentration is directly linked to the membrane potential by voltage-dependent Ca^{2+} channels and Ca^{2+}-sensitive K^+ channels. The latter can be separated into three classes based on their single channel conductance: large conductance (BK, MaxiK, or slo channels), intermediate conductance (IK), and small conductance (SK) channels. Because of their similarity, in this review, the intermediate conductance (IK) channels are included in the SK channel family. Channels

1063-5823/99 $30.00

of the SK/IK family are voltage independent and open only in response to Ca^{2+} elevations. SK/IK channels form a separate subfamily of K^+ channels with sequence similarity to other classes of K^+ channels restricted to the pore region. In contrast, BK channels are members of the subfamily of voltage-dependent ion channels, but with unique characteristics. BK channels have an intrinsic voltage sensor that is modulated by intracellular Ca^{2+} concentration and additional N- and C-terminal domains, implicated in β-subunit and Ca^{2+} modulation, respectively.

BK and SK channels have different physiological roles. The high Ca^{2+} sensitivity of SK channels allows them to open at relatively low Ca^{2+} concentrations, at hyperpolarized potentials, and evoke long lasting afterhyperpolarizations (AHPs) following action potentials. These AHPs regulate the maximum firing rate and are responsible for the phenomenon of spike frequency adaptation in neurons. In contrast, BK channels require higher Ca^{2+} concentrations (provided by Ca^{2+} sparks or puffs) to efficiently open at hyperpolarized potentials and regulate the smooth muscle resting membrane potential. In addition, their response to depolarization and a concomitant increase in Ca^{2+} concentration through Ca^{2+} channels allow BK channels to contribute to action potential repolarization.

II. LARGE CONDUCTANCE, VOLTAGE-GATED AND Ca^{2+}-SENSITIVE K^+ CHANNELS

BK channels are large conductance, noninactivating, voltage-gated, and Ca^{2+}-modulated potassium channels. Because of their widespread distribution, large single-channel conductance, and stability in artificial and native membranes for extended periods of time, BK channels have served as a model system for physiologists and biophysicists [see reviews by Blatz and Magleby (1987); McManus, (1991); Latorre *et al.* (1989); Latorre, (1994); Conely (1996)]. Their gating kinetics, permeation properties, and blockade by toxins have been extensively studied in native tissues and after reconsititution. Recently, high levels of expression of cloned channels allowed their characterization in macropatches (Cox *et al.*, 1997; Cui *et al.*, 1997) and the measurement of gating currents (Stefani *et al.*, 1997). Cloned and native mammalian BK channels are characterized by their sensitivity to externally applied iberiotoxin (IbTx) ($K_d \sim 1$ nM) and charybdotoxin (ChTx) ($K_d \sim 10$ nM) and by blockade with micromolar concentrations of tetraethylammonium (TEA) ($K_d \sim 250$ μM). For mammalian MaxiK channels, iberiotoxin is more potent than charybdotoxin and is highly specific, whereas, charybdotoxin also blocks other potassium channels including a voltage-independent Ca^{2+} dependent K^+ channel (Ishii *et al.*, 1997b). Characteristics such as conductance and toxin sensitivity are influenced by the measuring conditions. The single-channel conductance of BK channels is ~250 pS in 140 m*M* symmetrical K^+, whereas it is only ~110 pS under physiological conditions (5 m*M* K^+ outside, 140 m*M* K^+ inside). The blocking potencies for iberiotoxin and charybdotoxin, as typical external pore blockers, are reduced by high internal K^+ concentrations (MacKinnon and Miller, 1988).

The probability that the characteristics of BK channels deviate among species is expected to increase with evolutionary distance. For example, the *Drosophila* BK channel is insensitive to nanomolar concentrations of charybdotoxin (Adelman *et al.,* 1992; Perez *et al.,* 1994) and iberiotoxin (Meera *et al.,* 1997), while all studied mammalian BK channels are sensitive to both toxins.

A. Physiological Function

BK channels are found at high density in smooth muscle (for review see Carl *et al.,* 1996), but are less abundant in other tissues. Remarkable is their absence in heart myocytes, which is consistent with the prolonged action potentials in these cells. In many neurons and muscle cells, activation of BK channels by depolarization and Ca^{2+} entry contribute to the fast repolarization phase of the action potential (Fig. 1A), also called fast after-hyperpolarization (fAHP) (Sah, 1996; see Fig. 4). In smooth muscle, blockade of BK channels leads to cell depolarization and increased contractile activity (Fig. 1B; Anwer *et al.,* 1993; Heppner *et al.,* 1997).

BK channels are found colocalized with, and may be selectively associated with, L-type and N-type voltage-dependent Ca^{2+} channels (Gola and Crest, 1993; Robitaille *et al.,* 1993; Davies *et al.,* 1996). These findings go along with the view of restricted diffusion of Ca^{2+} within the cytoplasm (Albritton *et al.,* 1992; Gabso *et al.,* 1997) that would allow Ca^{2+} to rise to micromolar levels in microdomains around Ca^{2+} entry sites (Clapham, 1995; Berridge, 1997). Ca^{2+} entry would then activate BK channels present in the same microdomain with little effect on the spatially averaged Ca^{2+} concentration (Fig. 1C).

In arterial smooth muscle, compelling evidence was provided that BK channels are activated by local Ca^{2+} elevations (so-called Ca^{2+} sparks), which arise from spontaneous openings of Ca^{2+} release channels (ryanodine receptors) of intracellular Ca^{2+} stores (Nelson *et al.,* 1995; Fay, 1995). It is plausible that these local Ca^{2+} elevations close to the plasma membrane activate BK channels resulting in spontaneous transient outward currents (STOCs) and cell hyperpolarization (Fig. 1C). STOCs represent the activation of up to a hundred BK channels and are found in many types of

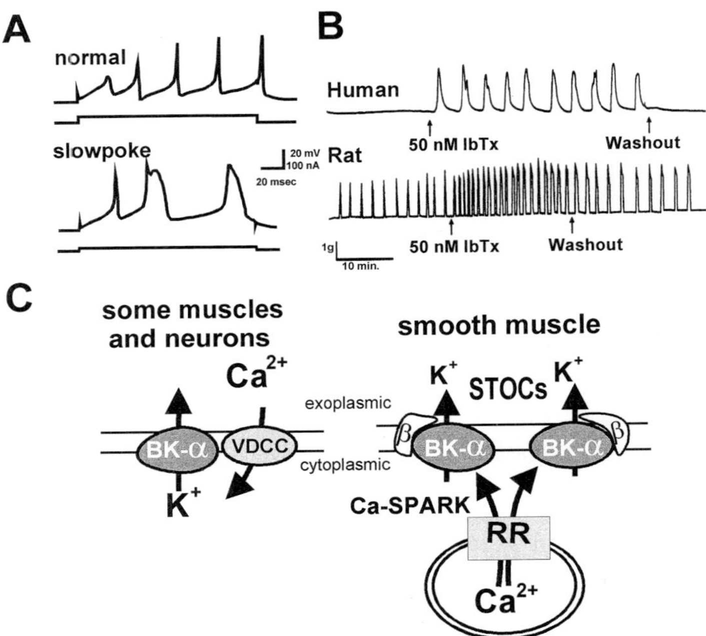

FIGURE 1 Physiological function of BK channels. (A) Action potentials from normal and slowpoke *Drosophila* dorsal longitudinal flight muscle evoked by constant current injection. Redrawn from Elkins *et al.* (1986), with permission from B. Ganetzky. (B) Contraction measurements of uterine smooth muscle strips. BK channel blockade by iberiotoxin induces rhythmic contractions in noncontracting strips, or increases the frequency of contractions in already contracting strips. Modified from Anwer *et al.* (1993), with permission from the American Physiological Society. (C) Models for the activation of BK channels in neurons and muscles. In neurons and some muscles (e.g., *Drosophila* flight muscle) Ca^{2+} influx through voltage-dependent Ca^{2+} channels (VDCCs) during an action potential activates BK channels. The opening of BK channels by voltage and Ca^{2+} then contributes to the repolarization of the action potential. In smooth muscles, the spontaneous opening of ryanodine receptors (RR) leads to local Ca^{2+} elevations resulting in an increase in BK channel activity. These local Ca^{2+} elevations can be observed as Ca^{2+} sparks in Ca^{2+}-imaging experiments (Nelson *et al.*, 1995) and seem to be responsible for spontaneous transient outward currents (STOCs)—a local eruption of BK channel activity observed in many smooth muscle cells (Bolton and Imaizumi, 1996).

smooth muscle cells (Bolton and Imaizumi, 1996; Imaizumi *et al.*, 1996; Bychkov *et al.*, 1997). Such localized Ca^{2+} elevations provide an explanation of how BK channels regulate resting membrane potential and contractile activity in smooth muscle cells (Fig 1B; Anwer *et al.*, 1993; Heppner *et al.*, 1997), despite the fact that at resting membrane potential (~ -50 mV) with a bulk Ca^{2+} concentration of ~100 nM, the activity of the cloned BK channel expressed in *Xenopus* oocytes is apparently too low to accomplish this task (Meera *et al.,* 1996; see Fig. 2C).

B. Structure and Function of the BK Channel α Subunit

The large conductance (BK, MaxiK, or slo) K^+ channel was first cloned utilizing the behavioral *Drosophila* mutant *slowpoke* (slo). Voltage-clamp analysis showed that *slowpoke* flight muscles lack a Ca^{2+}-activated K^+ current. Action potentials from these muscles are significantly longer than those from normal flies, consistent with the role of BK channels in repolarization (see Fig. 1A; Elkins *et al.,* 1986). Cloning of the *slowpoke* locus revealed sequence similarity to voltage-dependent ion channels (Atkinson *et al.,* 1991). Expression of *slowpoke* cDNA produced voltage and Ca^{2+}-activated K^+ currents in *Xenopus* oocytes (Adelman *et al.,* 1993). Homology screening yielded the first mammalian BK channel clone (Butler *et al.,* 1993) and, recently, the sequence of a potential *C. elegans* BK channel ortholog has been reported as well (Wei *et al,* 1996). A schematic view of these channels is given in Fig. 2A. The expressed mammalian clones show all the characteristic features of large conductance Ca^{2+}-activated potassium channels found in native tissues (Butler *et al.,* 1993; Tseng-Crank *et al.,* 1994; Wallner *et al.,* 1995). Differences in the apparent Ca^{2+} sensitivity observed in native tissues may be explained by the differential expression of a modulatory β subunit (Fig. 2B) or splice variants (Fig. 2A).

The BK channel α subunit has a potassium channel pore signature sequence. In addition, charged residues within transmembrane regions S2, S3, and S4, which are highly conserved in all voltage-dependent ion channels, are also present in BK channels, suggesting that BK channels carry an intrinsic voltage sensor. Indeed in the absence of Ca^{2+}, BK channels can still open in response to depolarization (Fig. 2C, inset; Meera *et al.,* 1996). High expression levels of cloned BK channel cDNA allowed the measurement of BK channel gating currents (Fig. 2D), which are a characteristic feature of voltage-gated ion channels. Measurement of gating currents at different Ca^{2+} concentrations (Fig. 2E) indicates that Ca^{2+} acts by facilitating the movement of gating charges that lead to potassium flow through the channel (Stefani *et al.,* 1997). The fact that BK channels have an intrinsic voltage sensor, which can open the channel even in the absence of Ca^{2+} (Fig. 2C), shows that the channel has no absolute dependence on Ca^{2+} to open. Therefore, we prefer the terms "Ca^{2+} acti-

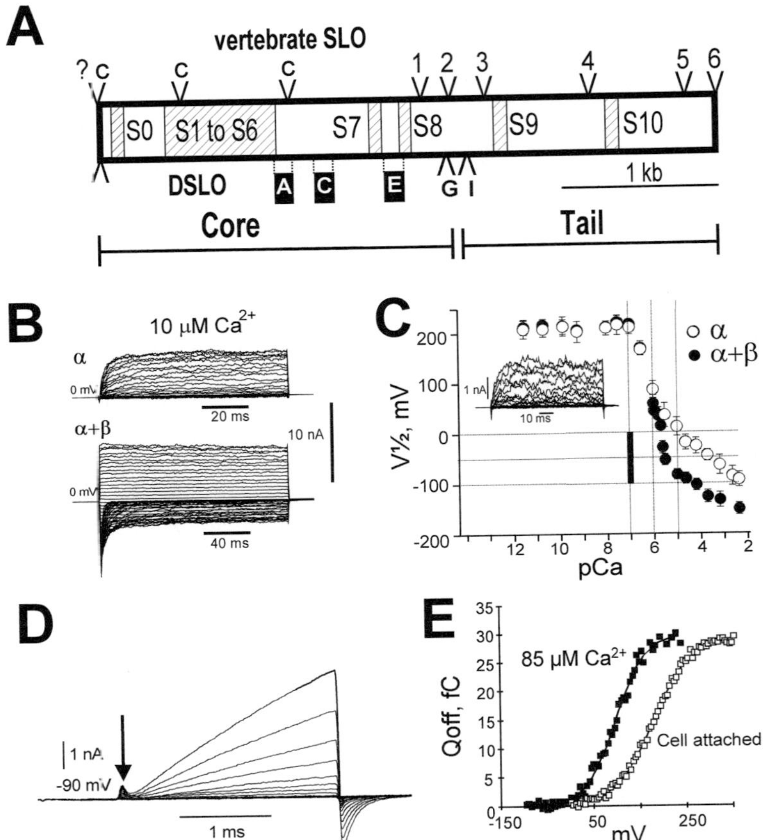

FIGURE 2 Structure and function of BK channel clones. (A) Schematic view of BK channels. Hatched areas mark the location of hydrophobic regions (S0 to S10). The arrows denote the alternative splice sites found in mammalian clones (Butler *et al.*, 1993; Tseng-Crank *et al.*, 1994; Vogalis *et al.*, 1996; Saito *et al.*, 1997). Three additional splice variants have been found in chick cochlea (Rosenblatt *et al.*, 1997; Navaratnam *et al.*, 1997) and are marked with the letter c. Alternative splicing marked with arrows leads to the insertion of short stretches of amino acids into the primary sequence. *Dslo* splice variants are indicated underneath. The black boxes mark regions where alternative splicing leads to an exchange of homologous regions in *Dslo*. In *Dslo* the usage of tissue-specific promoters together with alternative splicing events leads to variants that differ by an extension of 17 amino acids, including an N-linked glycosylation site at the N terminus (Brenner *et al.*, 1996; Wallner *et al.*, 1996). There is uncertainty about the predicted start of the protein and the length of the N-terminal sequence in mammalian BK channel clones (indicated with an arrow and a question mark) (Wallner *et al.*, 1995; McCobb *et al.*, 1995; Saito *et al.*, 1997). (B) Voltage activation measured in excised inside out patches with 110 m*M* symmetrical K^+ from *Xenopus* oocytes injected with human BK channel cRNA (Wallner *et al.*, 1995, Genbank Accession No. U11058) alone and coinjected

vated" or "Ca^{2+} sensitive" over "Ca^{2+} dependent" for referring to this channel.

We have recently provided evidence that in addition to the usual six transmembrane regions of voltage-dependent ion channels, BK channels have an additional unique transmembrane region (S0) at the N terminus (Wallner *et al.*, 1996). Epitope-tagged BK channels expressed in eukaryotic cells confirmed these findings showing that BK channels have an extracellular N terminus (Meera *et al.*, 1997; Jan and Jan 1997; Fig. 3).

Sequence comparison between the *Drosophila* channel and mammalian channels show a C-terminal sequence unique to BK channels, which is the most conserved region. This region is called the "tail" and it is separated from the rest of the protein, the "core", by a nonconserved linker (see Fig. 2A). Wei and colleagues (1994) demonstrated that the covalent linkage between "core" and "tail" is not necessary for channel function. The high

---

with the human BK channel β subunit. Currents are evoked with depolarizing pulses from −199 to +101 mV in steps of 6 mV in 10 μ*M* Ca^{2+}. (C) BK voltage activation curves were measured at different Ca^{2+} concentrations and the voltages required to half activate the channels ($V_{1/2}$) are plotted as a function of Ca^{2+} concentration. Inset: currents in the absence of Ca^{2+} (70 m*M* K_2EDTA: calculated 3 p*M* free Ca^{2+}; −70 to 164 mV every 6 mV, holding potential 0 mV). At each Ca^{2+} concentration it takes about 100 mV to change the open probability from less than 5% to more than 95% (shown as a black bar). The plot shows that: (I) BK channels can open in the practical absence of Ca^{2+} if depolarized to high voltages and show Ca^{2+} independent gating up to a Ca^{2+} concentration of 100 n*M*. (II) As Ca^{2+} concentrations exceed 100 n*M*, there is a sudden and drastic decrease in the voltage required to open the channel. Calcium seems to switch the channel from a purely voltage-gated mode into a Ca^{2+}-modulated mode. (III) The β-subunit sensitization of BK channel activity becomes apparent at Ca^{2+} concentrations above 1 μ*M*. (IV) At resting bulk Ca^{2+} concentration (100 n*M*) the activity of the channel is very low, even at depolarized potentials. (V) Coexpression of β subunits, as in smooth muscle, may bring BK channels into a range where they can be activated by local Ca^{2+} spikes at resting membrane potentials (−50 mV). (VI) Channels without the β subunit (as seems to be the case in neurons) may require both Ca^{2+} elevations and depolarization, as occurs during an action potential, for substantial activation. Experiments in B and C were performed using high extracellular K^+ concentrations and in a heterologous system where additional regulatory subunits may be missing. Modified from Meera *et al.* (1996), with kind permission from Elsevier Science. (D) BK gating currents (indicated with an arrow) precede the ionic current. To allow simultaneous measurement of ionic and gating currents, the ionic current was reduced by replacing K^+ with Cs^+. BK channels are about 400-fold less permeable to Cs^+ over K^+. (E) Measurement of gating currents was performed after blocking Cs^+ currents with TEA. The graph illustrates that an increase in intracellular Ca^{2+} from presumably 100 n*M* in the cell-attached mode to 85 μ*M* in the excised mode leads to a leftward shift of the charge versus voltage curve without modifying the total amount of charge moved. This indicates that micromolar Ca^{2+} facilitates the movement of the gating charge in response to voltage. Figs. D and E are from Stefani *et al.* (1997); Copyright 1997 National Academy of Sciences, U.S.A.

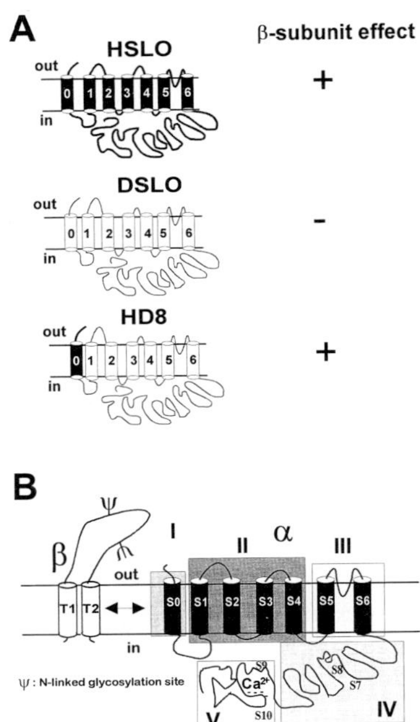

FIGURE 3 Modular structure of MaxiK channel complex. (A) Additional N-terminal transmembrane segment S0 determines β-subunit modulatory effect. By exchanging S0 and the extracellular N-terminus of the *Drosophila* sequence (*Dslo*) with the human sequence (*Hslo*), the previously unresponsive *Drosophila* clone now responds to β-subunit regulation. Modified from Wallner *et al.* (1996); Copyright 1996 National Academy of Sciences, U.S.A. (B) Membrane topology and modular structure of BK channel α and β subunits. BK channel α subunit can be divided into five modules. (I) The unique N-terminal part with transmembrane region S0. (II) The region homologous to voltage-dependent ion channels containing the voltage sensor. (III) The conduction pathway or pore region. (IV) The region starting after S6 containing hydrophobic segments S7 and S8, which are likely intracellular. (V) The "tail" region, which contains a putative Ca^{2+}-binding site, is the highest conserved part among species. Two of these modules (the "tail" and module I, containing S0) are functional when coexpressed as separate parts with the rest of the protein. The β subunit may be considered as an optional sixth module.

sequence conservation and the presence of stretches of negative charges suggests that the tail region might be the region that binds Ca^{2+} (consensus sequences for EF-hand Ca^{2+}-binding motifs seem to be missing in BK channels as well as in cloned SK channels). This view has been supported by functional analysis of mutations in this putative Ca^{2+}-binding site. However, the fact that channels lacking this stretch of negative charges are still regulated by calcium suggests that BK channels may carry more than just one Ca^{2+}-binding site per monomeric subunit (Schreiber and Salkoff, 1997).

The C terminus of BK channels carries four additional hydrophobic regions—S7, S8, S9, and S10. Our results demonstrate that the tail region containing hydrophobic segments S9 and S10 is cytosolic (Meera *et al.*, 1997; Fig. 3) consistent with the view that this region plays a role in Ca^{2+} sensing and, therefore, should be intracellular.

C. Diversity of BK Channels

In contrast to other subfamilies of K^+ channels, which come in sets of closely related members in the same species, the mammalian BK channel α subunit seems to lack close relatives. In *Drosophila* and vertebrates there are, however, a multitude of isoforms produced by alternative splicing (see Fig. 2A). Alternative splicing could be one of the mechanisms explaining the functional diversity of BK channels found in native tissues (Reinhart *et al.*, 1989; Latorre *et al.*, 1989; Toro *et al.*, 1991); their differential distribution may be relevant for specific functions (Perez and Toro, 1994). In this context, splice variants are differently distributed in human brain (Tseng-Crank *et al.*, 1994) and in chick cochlea, where their differential distribution may contribute to the frequency response of hair cells (Navaratnam *et al.*, 1997; Rosenblatt *et al.*, 1997). Alternatively spliced variants of *Drosophila* BK channels show profound differences in apparent Ca^{2+} sensitivities (splice site E) and kinetic behavior (splice site A) (Lagrutta *et al.*, 1994; see Fig. 2A). In vertebrates, the electrophysiological analysis of splice variants has not yielded such dramatic differences. Only two of a multitude of splice variants have been reported to produce functionally different isoforms. A slightly reduced apparent Ca^{2+} sensitivity was reported for a splice variant with a four-amino-acid insertion at splice site 1 (Tseng-Crank *et al.*, 1994). An insertion of a cysteine-rich region into splice site 2 leads to an increase in apparent Ca^{2+} sensitivity ($V_{1/2}$ shifted by ~20 mV to more negative potentials) (Saito *et al.*, 1997). To date, no splice variants have been found in the pore region, which determines the sensitivities to iberiotoxin, charybdotoxin, and TEA (Adelman *et al.*, 1992; Shen *et al.*, 1994; Meera *et al.*, 1997). This is in agreement with the uniform toxin sensitivity of mammalian BK channels found in different tissues.

The molecular identities of ChTx-insensitive large conductance channels in brain (Reinhart *et al.*, 1989), of inactivating types in rat adrenal chromaffin cells (Solaro and Lingle, 1992; Saito *et al.*, 1997), and of a Ca^{2+}- and voltage-insensitive K channel in a smooth muscle cell line (Ehrhardt *et al.,* 1996) remain elusive.

D. BK Channel β Subunit

Purification of BK channels from bovine tracheal smooth muscle and bovine aortic smooth muscle (Garcia-Calvo *et al.,* 1994; Knaus *et al.,* 1994a) revealed an associated protein, which was subsequently cloned. The primary sequence suggests two transmembrane regions separated by a large extracellular loop containing two N-linked glycosylation sites (Knaus *et al.,* 1994b). Experiments using a crosslinking agent showed that charybdotoxin specifically crosslinks to a lysine residue in the β subunit. This shows that the loop between the two hydrophobic segments of the protein is indeed extracellular (Knaus *et al.,* 1994c) and suggests that part of the large extracellular loop in the β subunit is close to the pore. Consistent with this view, β subunit coexpression enhances charybdotoxin binding at low ionic strength (Hanner *et al.,* 1997).

Coexpression of the BK channel β subunit leads to a dramatic increase in the BK channel α-subunit Ca^{2+} sensitivity when measured in symmetrical 110 m*M* K^+ (see Fig. 2B) (McManus *et al.,* 1995; Wallner *et al.,* 1995; Meera *et al.,* 1996). The effect of the β-subunit on Ca^{2+} sensitivity is observed only at Ca^{2+} concentrations above 1 μM Ca^{2+} and not at low Ca^{2+} concentrations (resting levels) (see Fig. 2C).

Drosophila BK channels are not responsive to β-subunit regulation. This difference has been utilized to identify the regions responsible for the β-subunit effect using a chimeric approach. The minimum region to transfer β-subunit responsiveness from the mammalian to the *Drosophila* BK channel was found to be S0 together with the extracellular N terminus (Wallner *et al.,* 1996; Fig. 3A). Therefore, this additional N-terminal transmembrane segment (S0) in BK channels may serve to accommodate the regulatory effects of the β subunit (Fig. 3). The finding that the extracellular loop of the β-subunit is close to the pore, and that S0 may be part of the physical interaction site with the β subunit, suggests that the large extracellular loop of the β subunit folds over the α-subunit protein to come close to the pore, allowing charybdotoxin crosslinking (see Fig. 1C).

The β-subunit mRNA levels are generally high in smooth muscle tissues and low in other tissues, including brain (Tseng-Crank *et al.,* 1996; Jiang *et al.,* in press). This suggests that the β subunit is differentially expressed and that its presence or absence may account for the functional differences observed in native tissues. It is tempting to speculate that in smooth muscle cells, BK channels may need coexpression of the β subunit to enhance their response to Ca^{2+} sparks at resting membrane potentials. In neurons, BK channels may not require the β subunit for their participation in action potential repolarization, because of the concomitant Ca^{2+} influx and membrane depolarization during action potentials (see Fig. 2C). However, the presence of other β subunit homologues cannot be ruled out.

E. Modulation by Other Factors

In addition to the regulation by Ca^{2+} and the β subunit, it has been reported that the activity of BK channels can also be modulated by various other factors like protein kinases (for review see Toro and Stefani, 1993), G proteins (Scornik *et al.,* 1993; Kume *et al.,* 1992), the redox state of the cells (Wang *et al.,* 1997; DiChiara and Reinhart, 1997), nitric oxide (Bolotina *et al.,* 1994), phosphorylation (Esguerra *et al.,* 1994), fatty acids, and membrane stretch (Kirber *et al.,* 1992). However, much more work is required to understand these often complex regulatory events at the molecular level in order to put them into the context of a physiological regulation scheme.

F. Pharmacology and Disease

Cardiovascular diseases are a major public health problem in developed countries. Vascular tone is regulated by the contractile status of smooth muscle cells in blood vessel walls. Inhibition of BK channels causes depolarization and contraction. Conversely, opening of BK channels in smooth muscle cell membranes leads to hyperpolarization and muscle relaxation (Brayden and Nelson, 1992). Therefore, substances that increase the activity of these channels are promising drugs for the treatment of vascular disease.

Three organic compounds present in the extract of the plant *Desmodium adscendens* are activators of BK channels. Extracts of this plant are used in Ghana to treat diseases related to smooth muscle contraction such as asthma and dysmenorrhea. The most potent of these compounds is dehydrosoyasaponin I (DHS-I). Only BK channel complexes containing the BK channel β subunit are activated by DHS-I ($EC_{50} \approx 120\ nM$), showing that modulatory subunits can alter pharmacological properties (McManus *et al.,* 1993; Garcia *et al.,* 1997). In acutely isolated coronary smooth muscle cells, the activation of most BK channels by DHS-I (250 nM) suggests that in this tissue BK channels are predominantly composed of $\alpha + \beta$ complexes (Tanaka *et al.,* 1997). In contrast, DHS-I (500 nM) is without effect in chromaffin cells (Saito *et al.,* 1997), suggesting the absence of this type of

β subunit in these cells. It is interesting that even though DHS-I acts from the cytoplasmic face of the membrane this compound is a modulator of external ^{125}I-ChTx binding (McManus *et al.*, 1993).

Other BK channel openers are fenamates (niflumic and flufenamic acids) (Ottolia and Toro, 1994) and structurally related benzimidazole compounds (NS1619, NS004) (Kaczorowski *et al.*, 1996). Various indole diterpenes (paspalitrem A and C, alfatrem, penitrem, and paspalinine) inhibit ChTx binding and are potent and specific BK channel blockers in electrophysiological experiments (for reviews see Conley, 1996; Garcia *et al.* 1997).

III. SMALL CONDUCTANCE Ca^{2+}-DEPENDENT K^+ CHANNELS

Action potentials are often followed by slow membrane afterhyperpolarizations, AHPs (Fig. 4A). AHPs that last from 50 msec to several seconds are caused by the activation of voltage-independent, Ca^{2+}-dependent potassium channels (for review see Hille, 1992). These channels have been classified according to their conductance into SK and IK conductance channels. They also differ in their toxin (apamin or ChTx) sensitivity.

Native SK channels have recently been classified based on the kinetic properties of the resulting AHPs (Fig. 4A) and the underlying currents (Fig. 4B). Medium afterhyperpolarizations (mAHPs; Fig. 4A, left) have a fast onset (<10 msec) after the action potential and last from 50 to several hundred milliseconds, whereas slow afterhyperpolarizations (sAHPs; Fig. 4A, middle) have a rising phase in the 100 msec range and inactivate within seconds (for review see Sah, 1996). These currents may arise from a direct regulation of SK channels by cytosolic Ca^{2+}. Consistent with this view, a sudden increase in Ca^{2+} concentration by flash photolysis of caged Ca^{2+} almost instantaneously activates sI_{AHP} in hippocampal CA1 neurons and activity can be rapidly terminated by photolytic production of a calcium buffer (Lancaster and Zucker, 1994). Furthermore, the open probability of expressed SK channels in excised inside out patches is directly related to the Ca^{2+} concentration of perfused solutions (Koehler *et al.*, 1996; Ishii *et al.*, 1997b; Joiner *et al.*, 1997). In addition, the Ca^{2+} activation kinetics of a cloned apamin-sensitive SK channel is consistent with the time course of apamin-sensitive mAHPs of native cells (Hirschberg *et al.*, 1998).

A possible explanation for the slow time course of sAHP (for review see Sah, 1996) is that it may result from the activation of slow K channels (SK_s in Fig. 4D) by the release of Ca^{2+} from intracellular stores by Ca^{2+}-induced Ca^{2+} release (CICR). Accordingly, using simultaneous Ca^{2+} and current measurements Garaschuck and colleagues (1997) have shown that

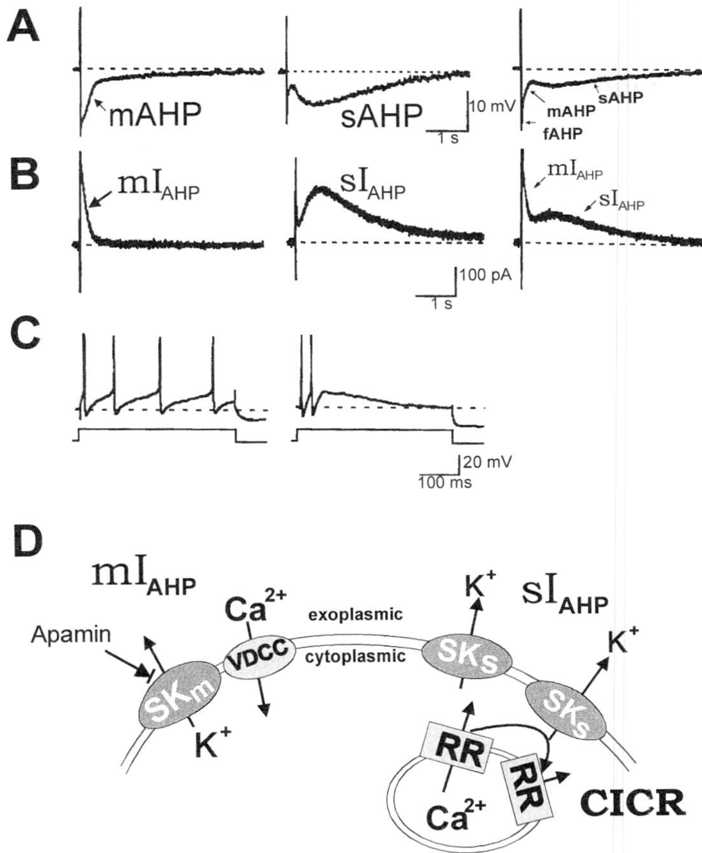

FIGURE 4 Physiological function of SK channels. (A) Time course of fast (fAHP), medium (mAHP), and slow (sAHP) afterhyperpolarizations in neurons showing predominantly mAHP (left), sAHP (middle), and a combination (right). fAHP results from the activation of voltage-dependent K channels including BK channels. (B) Current recordings from neurons expressing mI_{AHP} (left), sI_{AHP} (middle), and a combination of mI_{AHP} and sI_{AHP} (right). (C) Trains of action potentials evoked by constant current injection recorded in neurons expressing mI_{AHP} (left) and in neurons expressing sI_{AHP} (right). Recordings in A, B, and C are from different neurons in the dorsal motor nucleus of the vagus in rats or guinea pigs. Reprinted from Sah (1996) with permission from the author and Elsevier Science. (D) Model for colocalization of SK_m channels (likely formed by cloned channels SK1, SK2, and SK3) with voltage-dependent Ca^{2+} channels (VDCC), and SKs (channels of uncertain molecular identity underlying sAHP) with ryanodine receptors (RR) responsible for Ca^{2+}-induced Ca^{2+} release (CICR).

in CA1 neurons a ryanodine-sensitive Ca^{2+} transient, evoked by caffeine, coincides with a slow K current (possibly sI_{AHP}).

In contrast to the uniform pharmacology of BK channels, there is considerable variability among different types of SK channels. Two of the three cloned small conductance Ca^{2+}-dependent K^+ (SK) channels are blocked by low concentrations of apamin (SK2, SK3), whereas a third channel (SK1) is insensitive to this toxin (Ishii *et al.,* 1997a). An intermediate conductance channel called hIK1 (Ishii *et al.,* 1997b) or SK4 (Joiner *et al.,* 1997), which belongs to the same subfamily, is insensitive to apamin but is highly sensitive to charybdotoxin and clotrimazole. This channel resembles the Gardos channel (Ishii *et al.,* 1997b) first described in erythrocytes (Gardos, 1958). However, for some of the native channels (e.g., sI_{AHP}), there are no specific blockers available, making identification more difficult.

A. Physiological Function

SK channels are found in many neurons where they are responsible for afterhyperpolarizations following action potentials. The activation of apamin-sensitive channels in neurons with medium afterhyperpolarizations limits their firing frequency by slowing the return of the membrane potential to the firing threshold, thereby prolonging the interspike interval (Fig. 4C, left). In agreement, blockade of SK channels by apamin increases firing frequency (Callister *et al.,* 1997; Davies *et al.,* 1996).

Neurons with slow afterhyperpolarization show the phenomenon of spike frequency adaptation (Fig. 4C, right). Spike frequency adaptation limits the duration of bursts of action potentials by steadily increasing the spike intervals until the burst activity ceases. A plausible mechanism is that Ca^{2+} accumulation via CICR during trains of action potentials leads to a steady increase in hyperpolarizing sI_{AHP}, thereby increasing spike intervals until the final termination of burst activity. This process lets neurons respond predominantly to the beginning of a stimulus, with later adaptation. This adaptation has been suggested to be a molecular correlate to paying attention (Hille, 1996) and is modulated by a variety of neurotransmitters (Nicoll, 1988). Application of norephinephrine to CA1 neurons increases the burst durations (Madison and Nicoll, 1982). A similar effect has been reported for a variety of other neurotransmitters (isoproterenol, serotonin, histamine, and carbachol) in hippocampal CA1 and other neurons (Nicoll, 1988; Pedarzani and Storm, 1993; Sah, 1996). This regulation seems to be mediated by protein kinase A (PKA), as it is inhibited by antagonists of PKA (Rp-cAMP and Walsh peptide) and mimicked by forskolin, 8-CPT-cAMP, and by the catalytic subunit of PKA (Pedarzani and Storm, 1993).

Some neurons show both medium and slow afterhyperpolarizations (Fig. 4A, right). In these neurons apamin specifically reduces mI_{AHP} (without effect on sI_{AHP}), whereas ryanodine blocks sI_{AHP} (without effect on mI_{AHP}) (Sah, 1996; Davies *et al.,* 1996; Fig. 4D). The specific block of sI_{AHP} by ryanodine suggests that Ca^{2+} released from internal stores by CICR activates sI_{AHP}, but may not reach apamin-sensitive channels (SK_m in Fig. 4D) responsible for the mI_{AHP}. A likely explanation for these findings is that the distinct Ca^{2+}-entry pathways are colocalized or compartmentalized with specific types of SK channels and that Ca^{2+} elevations remain restricted to these spheres due to the highly restricted mobility of cytosolic Ca^{2+} caused by intracellular Ca^{2+} buffers (Fig. 4D) (Allbritton *et al.,* 1992; Gabso *et al.,* 1997).

Specific coupling of Ca^{2+} sources and Ca^{2+}-sensitive K^+ channels is an emerging concept. Apamin-sensitive currents (I_{AHP}) can be specifically blocked by N-type Ca^{2+} channel blockers without affecting other Ca^{2+} activated K^+ conductances in the same cell (Davies *et al.,* 1996; Callister *et al.,* 1997), whereas L-type Ca^{2+} channels have been found to be selectively coupled to BK channels (Davies *et al.,* 1996); but see Marion and Tavalin (1998). In neurons of superior cervical ganglia, the Ca^{2+} that triggers CICR has been suggested to result from Ca^{2+} entry through R-type Ca^{2+} channels (Davies *et al.,* 1996). The specificity of the coupling between Ca^{2+} entry sites and specific Ca^{2+}-activated K^+ channels may depend on the cytosolic Ca^{2+}-buffering capacity, the amount of Ca^{2+} entering the cytoplasm, and the density and distribution of Ca^{2+} channels and Ca^{2+}-activated K^+ channels.

The differential effects of Ca^{2+}, depending on the route of Ca^{2+} entry, have also been reported for other Ca^{2+}-regulated processes like modulation of synaptic strength and calcium-mediated cell death (for review see Ghosh and Greenberg, 1995).

B. Structure and Function of SK Channels

Small and intermediate conductance Ca^{2+}-activated K^+ channels (Koehler *et al.,* 1996; Ishii *et al.,* 1997b; Joiner *et al.,* 1997) were recently cloned taking advantage of: (I) the fact that all potassium-selective ion channels have a common structural motif—the potassium-selective pore—and (II) cDNA sequences in public expressed sequence tag (EST) databases.

Koehler and colleagues (1997) isolated three closely related cDNA isoforms from human and rat tissue (SK1, SK2, and SK3) with functional properties resembling native SK channels. These proteins constitute a new subfamily of potassium channels with discernible sequence conservation to other K^+ channel families restricted to the pore region. Hydophobicity analysis suggests that similar to voltage-dependent channels, this family may also have six transmembrane segments with a K^+ pore signature sequence located between the fifth and sixth putative transmembrane segment (Fig. 5A). SK1 and SK2 mRNAs were detected in rat brain by *in situ* hybridization. On Northern blots, SK1 mRNA was also found in rat heart and SK2 in rat adrenal gland. Oocytes expressing human SK1 or rat SK2 channels showed a Ca^{2+}-dependent potassium conductance with an EC_{50} of $\sim$0.6 μM Ca^{2+} with a Hill coefficient of $\sim$4; single-channel conductance was $\sim$10 pS for both clones (measured in symmetrical 120 mM K^+ at negative potentials) (Koehler *et al.*, 1996). The high Ca^{2+} sensitivity is conferred by tightly bound calmodulin (Xia *et al.*, 1998). Only SK2 and SK3 channels are sensitive to apamin (IC_{50} of 60 pM for SK2 and 2 nM for SK3). Although highly homologous, the SK1 channel is insensitive to 100 nM apamin. The apamin-sensitive channel SK2 differs from the apamin-insensitive SK1 channel by only three amino acids in the putative pore and surrounding

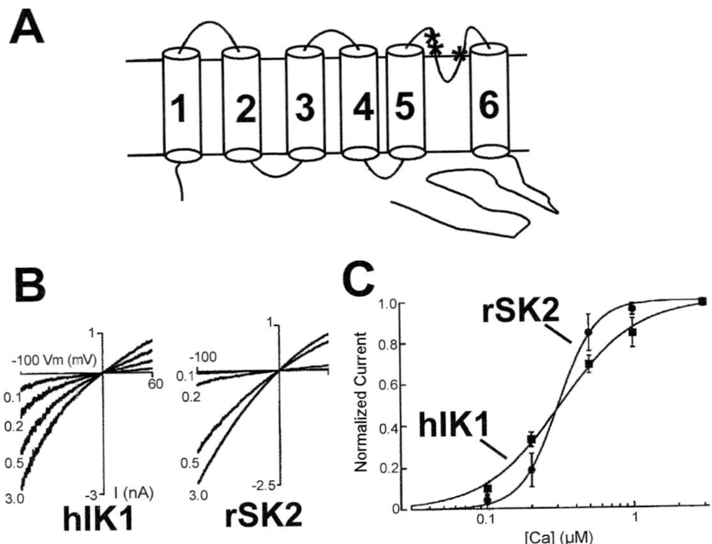

FIGURE 5 Molecular characteristics of SK channels. (A) Proposed membrane topology of SK channels with six transmembrane regions and a pore loop between regions 5 and 6. Stars mark the approximate positions of amino acids responsible for the differences in the apamin sensitivity among cloned SK channels. (B) Functional expression of hIK1 and SK2 channels in *Xenopus* oocytes measured in excised inside out patches using voltage ramps (2.5 sec) at the indicated Ca^{2+} concentrations (in μM). (C) Ca^{2+} concentration response curves for hIK1 and rSK2. Reproduced from Ishii *et al.* (1997b); Copyright 1997, National Academy of Sciences, U.S.A.

vestibular extracellular regions (marked with stars in Fig. 5A). Site-directed mutagenesis showed that each of these three amino acids contributes to the differences in apamin binding (Ishii *et al.*, 1997a). Coexpression of SK1 (apamin insensitive) and SK2 (apamin sensitive) and a SK1–SK2 dimer produced currents with intermediate apamin sensitivity. This suggests that these channels can form heteromultimers (probably heterotetramers) when different subunits are expressed in the same cell. The apamin-sensitive SK2 (Hirschberg *et al.*, 1998) and SK3 channels may be responsible for the apamin-sensitive component (SK_m in Fig. 4D) of the mI_{AHP} current; whereas the apamin-insensitive SK1 channel may be responsible for the small apamin-insensitive component of mI_{AHP} observed in neurons (Davies *et al.*, 1996).

A human cDNA sequence related to the SK channel family has been recently identified by two groups and called hIK1 (IK for intermediate K) by Ishii *et al.* (1997b) and hSK4 by Joiner *et al.* (1997). The single-channel conductance for hIK1 is 42 pS at negative potentials in symmetrical 120 mM K^+ (versus ~10 pS for the SK1 and SK2 isoforms under identical conditions) (Ishii *et al.*, 1997b) and ~12 pS for SK4 at positive potentials in 3 mM K^+ outside/120 mM K^+ inside (Joiner *et al.*, 1997). Because hIK1 and SK4 seem to encode the same protein, the difference in the reported single-channel conductances is likely due to the different voltages used for measurement since these channels show inward rectification properties (Fig. 5B).

The mRNA for hIK1/hSK4 is mainly expressed in nonneuronal tissues (particularly in placenta) and its functional properties (ChTx and clotrimazol sensitivity, insensitivity to 100 nM apamin) resemble the Gardos channel first described in red blood cells (Gárdos, 1958; Ishii *et al.*, 1997b). Similar channels have been described in other cell types (Eder *et al.*, 1997; Jaeger and Grissmer, 1997). Although the Ca^{2+} concentration for half maximum activation of hIK1 is about the same as that of SK2 channel ($K_{0.5} \sim 0.3$ μM), its shallower Ca^{2+} response curve allows hIK1 channels to open at lower Ca^{2+} concentrations (Figs. 5B, 5C; Ishii *et al.*, 1997b). The expression of hIK1 in nonexcitable cells suggests a different physiological function than SK channels that are involved in AHPs.

The molecular identity of native apamin-insensitive SK channels (SK_S in Fig. 4D) underlying sI_{AHP} (see Figs. 4A, 4B, middle), blocked by ryanodine and modulated by a variety of neurotransmitters, remains unclear. It may be formed either by the apamin-insensitive subunits (SK1 or hIK1/SK4) or by a yet unidentified K^+ channel(s).

C. SK Channels in Disease

Because SK channels regulate firing frequency and burst duration in neurons, they may be responsible for neuronal hyperactivity. Indeed the hippocampal CA3 neurons of mutant rats prone to epilepsy show a significant reduction in slow afterhyperpolarization when compared to normal rats, suggesting the involvement of the SK channel current (Verma-Ahuja *et al.*, 1995).

In skeletal muscle, SK channels appear after denervation and may contribute to the symptoms of myotonic dystrophies (Ramirez *et al.*, 1996).

Charybdotoxin-sensitive SK channels are upregulated by growth factor (mitogenic) stimulation and also by the expression of oncogenes (*ras* and *raf*) in fibroblasts and cell lines (Pena and Rane, 1997). Src transformation of NIH3T3 mouse fibroblasts induces the expression of a similar voltage-independent Ca^{2+} dependent K^+ channel, which can be blocked by low concentrations of charybdotoxin (IC_{50} = 20 n*M*) (Draheim *et al.*, 1995). Similar channels have been found in activated T lymphocytes (Jaeger and Grissmer, 1997).

The Gardos channel, likely to be encoded by the hIK1 cDNA, plays an important role in the dehydration of erythrocytes in sickle cell disease. Blockade of this channel by clotrimazole reduces erythrocyte dehydration in patients with this disease (Brugnara *et al.*, 1996).

IV. SUMMARY

There are at least two subfamilies that can account for the variety of Ca^{2+}-sensitive K^+ channels in both excitable and nonexcitable membranes.

BK channels are characterized by their high single-channel conductance, their dual response to voltage and Ca^{2+}, and their blockade by nanomolar concentrations of iberiotoxin and micromolar concentrations of TEA. Heterologously expressed α-subunit cDNA clones mirror these features. Structurally, they are characterized by unique N- and C-terminal sequences, implicated in β-subunit and Ca^{2+} regulation, appended to a six-transmembrane structure typical of voltage-gated ion channels. A regulatory β subunit highly expressed in smooth muscle induces an increase in BK channel Ca^{2+} sensitivity and may allow them to efficiently open in response to local Ca^{2+} elevations.

SK channels are characterized by their small conductance, their voltage independence, and their Ca^{2+}-dependent activation. A family of this class of K^+ channels has been recently cloned. Functional expression shows the expected high Ca^{2+} sensitivity. Two (SK2 and SK3) of three closely related isoforms are highly sensitive to the specific blocker apamin. A fourth, more distant isoform (hIK1 or SK4) has a larger single-channel conductance and is blocked by low concentrations of charybdotoxin and clotrimazole. These characteristics are features of the Gardos channel first described in erythrocytes. Northern blot analysis suggests that hIK1 is expressed in nonneuronal tissues.

BK and different types of SK channels are often coexpressed within the same native cell and seem to be coupled to specific Ca^{2+} sources. Specific Ca^{2+} channel blockers can abolish certain types of calcium-activated K channels without affecting others. The BK channel may be mainly linked to L-type or N-type Ca^{2+} channels in neurons and to ryanodine-sensitive internal stores in smooth muscle. SK channels of the apamin-sensitive type are specifically linked to N-type Ca^{2+} channels in some neurons; whereas the apamin-insensitive slow SK channels are compartmentalized with ryanodine-sensitive Ca^{2+} release channels. In conclusion, the diverse physiological effects of Ca-activated K channels depend on their colocalization and functional coupling with Ca^{2+} sources and the diffusional pattern of Ca^{2+}.

Acknowledgments

Space restrictions made it impossible to cite all original publications. Supported by NIH Grant HL54970 (L.T.) and American Heart Association National Center Grant-in-Aid 9750745N (P.M.). L. Toro is an Established Investigator of the American Heart Association.

References

Adelman, J. P., Shen, K., Kavanaugh, M. P., Warren, R. A., Wu, Y., Lagrutta, A., Bond, C. T., and North, R. A. (1992). Calcium-activated potassium channels expressed from cloned complementary DNAs. *Neuron* **9,** 209–216.

Allbritton, N. L., Meyer, T., and Stryer, L. (1992). Range of messenger action of calcium ion and inositol 1,4,5-triphosphate. *Science* **258,** 1812–1815.

Anwer, K., Oberti, C., Perez, G. J., Perez-Reyes, N., McDougall, J. K., Monga, M., Sanborn, B. M., Stefani, E., and Toro, L. (1993). Calcium-activated K^+ channels as modulators of human myometrial contractile activity. *Am. J. Physiol.* **265,** C976–985.

Atkinson, N. S., Robertson, G. A., and Ganetzky, B. A. (1991). A component of calcium-activated potassium channels encoded by the *Drosophila slo* locus. *Science* **253,** 551–555.

Berridge, M. J., (1997). The AM and FM of calcium signalling. *Nature* **386,** 759–760.

Blatz, A. L., and Magleby, K. L. (1987). Calcium-activated potassium channels. *Trends. Neurosci.* **10,** 463–467.

Bolotina, V. M., Najibi, S., Palacino, J. J., Pagano, P. J., and Cohen, R. A. (1994). Nitric oxide directly activates calcium-dependent potassium channels in vascular smooth muscle. *Nature* **368,** 850–853.

Bolton, T. B., and Imaizumi, Y. (1996). Spontaneous transient outward currents in smooth muscle cells. *Cell Calcium* **20,** 141–152.

Brayden, J. E., and Nelson, M. T. (1992). Regulation of arterial tone by activation of calcium-dependent potassium channels. *Science* **256,** 532–535.

Brenner, R., Thomas, T. O., Becker, M. N., and Atkinson N. S. (1996). Tissue specific expression of a Ca^{2+}-activated K^+ channel is controlled by multiple upstream regulatory elements. *J. Neurosci.* **16,** 1827–1835.

Brugnara, C., Gee, B., Armsby, C. C., Kurth, S., Sakamoto, M., Rifai, N., Alper, S. L., and Platt O. S. (1996). Therapy with oral clotrimazole induces inhibition of the Gardos channel and reduction of erythrocyte dehydration in patients with sickle cell disease. *J. Clin. Invest.* **97,** 1227–12234.

Butler, A., Tsunoda, S., McCobb, D. P., Wei, A., and Salkoff, L. (1993). mSlo, a complex mouse gene encoding "maxi" calcium-activated potassium channels. *Science* **261,** 221–224.

Bychkov, R., Gollasch, M., Ried, C., Luft, F. C., and Haller H. (1997). Regulation of spontaneous transient outward potassium currents in human coronary arteries. *Circulation* **95,** 503–510.

Callister, R. J., Keast, J. R., and Sah, P. (1997). Ca^{2+}-activated K^+ channels in rat otic ganglion cells: Role of Ca^{2+} entry via Ca^{2+} channels and nicotinic receptors. *J. Physiol.* **500,** 571–582.

Carl, A., Lee, H. K., and Sanders, K. M. (1996). Regulation of ion channels in smooth muscles by calcium. *Am. J. Physiol.* **271,** C9–C34.

Clapham, D. (1995). Calcium signaling. *Cell* **80,** 259–268.

Conley, E. C. (1996). Intracellular calcium-activated K^+ channels. *In* "The Ion Channel FactsBook, Intracellular Ligand-Gated Channels," pp. 607–720, Academic Press, San Diego.

Cox, D. H., Cui, J., and Aldrich, R. W. (1997). Allosteric gating of a large conductance Ca-activated K^+ channel. *J. Gen. Physiol.* **110,** 257–281.

Cui, J., Cox, D. H., and Aldrich, R. W. (1997). Intrinsic voltage dependence and Ca^{2+} regulation of mslo large conductance Ca-activated K^+ channels. *J. Gen. Physiol.* **109,** 647–673.

Davies, P. J., Ireland, D. R., and McLachlan, E. M. (1996). Sources of Ca^{2+} for different $Ca(^{2+})$-activated K^+ conductances in neurones of the rat superior cervical ganglion. *J. Physiol.* **495,** 353–366.

DiChiara, T. J. and Reinhart, P. (1997). Redox modulation of *hslo* Ca^{2+}-activated K^+ channels. *J. Neurosci.* **17,** 4942–4955.

Draheim, H. J., Repp, H. and Dreyer, F. (1995). Src-transformation of mouse fibroblasts induces a Ca^{2+} activated K^+ current without changing the T-type Ca^{2+} current. *Biochim. Biophys. Acta* **1269,** 57–63.

Eder, C., Klee, R. and Heinemann U. (1997). Pharmacological properties of Ca^{2+}-activated K^+ currents of ramified murine brain macrophages. *Naunyn-Schmiedebergs Arch. Pharmacol.* **356,** 233–239.

Elkins, T., Ganetzky, B., and Wu, C. (1986). A *Drosophila* mutation that eliminates a calcium dependent potassium current. *Proc. Natl. Acad. Sci. USA* **83,** 8415–8419.

Ehrhardt, A. G., Frankish, N., and Isenberg, G. (1996). A large-conductance K^+ channel that is inhibited by the cytoskeleton in the smooth muscle cell line DDT1 MF-2. *J. Physiol.* **496,** 663–676.

Esguerra, M., Wang, J., Foster, C. D., Adelman, J. P., North, R. A., and Levitan, I. B. (1994). Cloned Ca^{2+}-dependent K^+ channel modulated by a functionally associated protein kinase. *Nature* **369,** 563–565.

Fay, F. S. (1995). Calcium sparks in vascular smooth muscle: Relaxation regulators. *Science* **270,** 588–589.

Gabso, M., Neher, E., and Spira, M. E. (1997). Low mobility of the Ca^{2+} buffers in axons of cultured Aplysia neurons. *Neuron* **18,** 473–481.

Garaschuk, O., Yaari, Y., and Konnerth, A. (1997). Release and sequestration of calcium by ryanodine-sensitive stores in rat hippocampal neurones. *J. Physiol.* **502,** 13–30.

Garcia, M. L., Hanner, M., Knaus, H-G., Koch, R. Schmalhofer, W., Slaughter, R. S., and Kaczorowski, G. J. (1997). Pharmacology of potassium channels. *Adv. Pharmacol.* **39,** 425–471.

Garcia-Calvo, M., Knaus H-G., McManus, O. G., Giangiacomo, K. M., Kaczorowski, G. J., and Garcia, M. L. (1994). Purification and reconstitution of the high-conductance, calcium activated potassium channel from tracheal smooth muscle. *J. Biol. Chem.* **269,** 676–682.

Gárdos, G. (1958). The function of calcium in the potassium permeability of human erythrocytes. *Biochim. Biophys. Acta* **30,** 653–654.

Ghosh, A., and Greenberg, M. E. (1995). Calcium signaling in neurons: Molecular mechanisms and cellular consequences. *Science* **268,** 239–247.

Gola, M., and Crest, M. (1993). Colocalization of active K_{Ca} channels and Ca^{2+} channels within Ca^{2+} domains in helix neurons. *Neuron* **10,** 689–699.

Hanner, M., Schmalhofer, W. A., Munujos, P., Knaus, H.-G., Kaczorowski, G. J., and Garcia, M. (1997). The β-subunit of the high-conductance calcium-activated potassium channel contributes to the high affinity receptor for charybdotoxin. *Proc. Natl. Acad. Sci. USA* **94,** 2853–2858.

Heppner, T. J., Bonev, A. D., and Nelson, M. T. (1997). Ca^{2+}-activated K^+ channels regulate action potential repolarization in urinary bladder smooth muscle. *Am. J. Physiol.* **273,** C110–117.

Hille, B. (1992). "Ionic Channels and Exciteable Membranes," 2nd ed. Sinauer Associates, Sunderland, MA.

Hille, B. (1996). A K^+ channel worthy of attention. *Science* **273,** 1677.

Hirschberg, B., Maylie, J., Adelman, J. P., and Marrion, N. (1998). Gating of recombinant small conductance Ca-activated K^+ channels by calcium. *J. Gen. Physiol.* **111,** 565–581.

Imaizumi, Y., Henmi, S., Nagano, N., Muraki, K., and Watanabe, M. (1996). Regulation of Ca-dependent K current and action potential shape by intracellular Ca storage sites in some types of smooth muscle. *In* "Smooth Muscle Excitation" (T. B. Bolton and T. Tomita, Eds.), pp. 337–354. Academic Press, London.

Ishii, T. M., Maylie, J., and Adelman, J. P. (1997a). Determinants of apamin and d-tubocurarine block in SK potassium channels. *J. Biol. Chem.* **272,** 23,195–23,200.

Ishii, T. M., Silvia, C., Hirschberg, B., Bond, C. T., Adelman, J. P., and Maylie, J. (1997b). A human intermediate conductance calcium-activated potassium channel. *Proc. Natl. Acad. Sci. USA* **94,** 11,651–11,656.

Jaeger, A., and Grissmer, S. (1997). Small Ca^{2+}-activated potassium channels in human leukemic T cells and activated human peripheral blood T lymphocytes. *Cell. Physiol. Biochem.* **7,** 179–187.

Jan, L. Y., and Jan Y. N. (1997). Ways and means for left shifts in the MaxiK channel. *Proc. Natl. Acad. Sci. USA* **94,** 13,383–13,385.

Jiang, Z., Wallner, M., Meera, P., and Toro, L. Human and rodent MaxiK channel β subunit genes: Cloning and characterization. *Genomics.* In press.

Joiner, W. J., Wang, L. Tang, M. D., and Kaczmarek, L. K. (1997). hSK4, a member of a novel subfamily of calcium-activated potassium channels. *Proc. Natl. Acad. Sci. USA* **94,** 11,013–11,018.

Kaczorowski, G. J., Knaus, H.-G., Leonard, R. J., McManus, O. B., and Garcia, M. L. (1996). High-conductance calcium-activated potassium channels; structure, pharmacology, and function. *J. Bioenerg. Biomem.* **28,** 255–267.

Kirber, M. T., Ordway, R. W., Clapp, L. H., Walsh, J. V., and Singer, J. J. (1992). Both membrane stretch and fatty acids directly activate large conductance Ca^{2+}-activated K^+ channels in vascular smooth muscle cells. *FEBS Lett.* **297,** 24–28.

Knaus, H.-G., Garcia-Calvo, M., Kaczorowski, G. J., and Garcia, M. L. (1994a). Subunit composition of the high conductance calcium-activated potassium channel from smooth muscle, a representative of the *mSlo* and *slowpoke* family of potassium channels. *J. Biol. Chem.* **269,** 3921–3924.

Knaus, H.-G., Folander, K., Garcia-Calvo, M., Garcia, M. L., Kaczorowski, G. J., Smith, M., and Swanson, R. (1994b). Primary sequence and immunological characterization of β-

subunit of high conductance Ca^{2+}-activated K^+ channel from smooth muscle. *J. Biol. Chem.* **269,** 17,274–17,278.

Knaus, H.-G., Eberhart, A., Kaczorowski, G., and Garcia, M. L. (1994c). Covalent attachment of charybdotoxin to the β-subunit of the high conductance Ca^{2+} activated K^+ channel. *J. Biol. Chem.* **269,** 23,336–23,341.

Koehler, M., Hirschberg, B., Bond, C. T., Marrion, N. V., Kinzie, J. M., Maylie, J., and Adelman, J. P. (1996). *Science* **273,** 1709–1714.

Kume, H., Graziano, M. P., and Kotlikoff. M. I. (1992). Stimulatory and inhibitory regulation of calcium-activated potassium channels by guanine nucleotide-binding proteins. *Proc. Natl. Acad. Sci. USA* **89,** 11,051–11,055.

Lagrutta, A., Shen, K., North, R. A., and Adelman J. P. (1994). Functional differences among alternatively spliced variants of *slowpoke,* a *Drosophila* calcium-activated potassium channel. *J. Biol. Chem.* **269,** 20,347–20,351.

Lancaster, B., and Zucker, R. S. (1994). Photolytic manipulation of Ca^{2+} and the time course of slow, Ca^{2+}-activated K^+ current in rat hippocampal neurons. *J. Physiol.* **475,** 229–239.

Latorre, R. (1994). Molecular workings of large conductance (maxi) Ca^{2+}-activated K^+ channels. *In* "Handbook of Membrane Channels, Molecular and Cellular Physiology" (C. Peracchia, Ed.), pp. 79–102. Academic Press, San Diego.

Latorre, R., Oberhauser, A., Labarca, P., and Alvarez, O. (1989). Varieties of of calcium-activated potassium channels. *Annu. Rev. Physiol.* **51,** 385–399.

MacKinnon, R., and Miller, C. (1988). Mechanisms of charybdotoxin block of single Ca^{2+} activated K^+ channels. *J. Gen. Physiol.* **91,** 335–349.

Madison, D. V., and Nicoll, R. A. (1982). Noradrenalin blocks accommodation of pyramidal cell discharge in the hippocampus. *Nature* **299,** 636–638.

Marrion, N. V., and Tavalin, S. J. (1998). Selective activation of Ca^{2+}-activated K^+ channels by co-localized Ca^{2+} channels in hippocampal neurons. *Nature* **395,** 900–905.

McCobb, D. P., Fowler, N. L., Featherstone, T., Lingle, C., Saito, M., Krause, J. E., and Salkoff, L. (1995). A human calcium activated potassium channel gene expressed in vascular smooth muscle. *Am. J. Physiol.* **269,** H767–H777.

McManus, O. B. (1991). Calcium-activated potassium channels: Regulation by calcium. *J. Bioenerg. Biomem.* **23,** 537–560.

McManus, O. B., Harris, G. H., Giangiacomo, K. M., Feigenbaum, P., Reuben, J. P., Addy, M. E., Burka, J. F., Kaczorowski, G. J., and Garcia, M. L. (1993). An activator of calcium-dependent potassium channels isolated from a medicinal herb. *Biochemistry* **32,** 6128–6133.

McManus, O. B., Helms, L. M., Pallanck, L., Ganetzky, B., Swanson, R., and Leonard, R. J. (1995). Functional role of the β subunit of high conductance calcium-activated potassium channels. *Neuron* **14,** 645–650.

Meera, P., Wallner, M., Jiang, Z., and Toro, L. (1996). A calcium switch for the functional coupling between α (*Hslo*) and β subunits ($K_{v,Ca}\beta$) of maxi K channels. *FEBS Lett.* **382,** 84–88.

Meera, P., Wallner, M., Song, M., and Toro L. (1997). MaxiK channel, a distinct member of voltage dependent ion channels with seven N-terminal transmembrane segments (S0–S6), an extracellular N-terminus and an intracellular (S9–S10) C-terminus. *Proc. Natl. Acad. Sci. USA,* 14.066–14,071.

Navaratnam, D. S., Bell, T. J., Tu, D. T., Cohen, E. L., and Oberholtzer, J. C. (1997). Differential distribution of Ca^{2+}-activated K^+ channel splice variants among hair cells along the tonotopic axis of chick cochlea. *Neuron* **19,** 1077–1085.

Nelson, M. T., Cheng, H., Rubart, M., Santana, L. F., Bonev, A. D., Knot, H. J., and Lederer, W. J. (1995). Relaxation of arterial smooth muscle by calcium sparks. *Science* **270,** 633–637.

Nicoll, R. A. (1988). The coupling of neurotransmitter receptors to ion channels in the rat brain. *Science* **241,** 545–551.

Ottolia, M., and Toro, L. (1994). Potentiation of large-conductance K_{Ca} channels by niflumic, flufenamic and mefenamic acids. *Biophys. J.* **67,** 2272–2279.

Pedarzani, P., and Storm, J. F. (1993). PKA mediates the effects of monoamine transmitters on the K^+ current underlying the slow spike frequency adaptation in hippocampal neurons. *Neuron* **11,** 1023–1035.

Pena, T. L., and Rane, S. G. (1997). The small conductance calcium-activated potassium-channel regulates ion channel expression in C3H10T1/2 cells ectopically expressing the muscle regulatory factor MRF4. *J. Biol. Chem.* **272,** 21,909–21,916.

Perez, G., Lagrutta, A., Adelman, J. P., and Toro, L. (1994). Reconstitution of expressed KCa channels from *Xenopus* oocytes to lipid bilayers. *Biophys. J.* **66,** 1022–1027.

Perez, G., and Toro, L. (1994). Differential modulation of large conductance K_{Ca} channels by PKA in pregnant and nonpregnant myometrium. *Am. J. Physiol.* **266,** C1459–C1463.

Reinhart, P. H., Cung, S., and Levitan, I. B. (1989). A family of calcium-dependent potassium channels from rat brain. *Neuron* **2,** 1031–1041.

Ramirez, B. U., Behrens, M. I., and Vergara, C. (1996). Neural control of the expression of a Ca(2+)-activated K^+ channel involved in the induction of myotonic-like characteristics. *Cell. Mol. Neurobiol.* **16,** 39–49.

Robitaille, R. P., Garcia, M. L., Kaczorowski, G. J., and Charlton, M. P. (1993). Functional colocalization of calcium and calcium-gated potassium channels in control of transmitter release. *Neuron* **11,** 645–655.

Rosenblatt, K. P., Sun, Z.-P., Heller, S., and Hudspeth, A. J. (1997). Distribution of Ca^{2+}-activated K^+ channel isoforms along the tonotopic gradient of chicken's choclea. *Neuron* **19,** 1061–1075.

Sah, P. (1996). Ca^{2+}-activated K^+ currents in neurons: Types, phsiological roles and modulation. *Trends Neurosci.* **19,** 150–154.

Saito, M., Nelson, C., Salkoff, L., and Lingle C. J. (1997). A cysteine-rich domain defined by a novel exon in a Slo variant in rat adrenel chromaffin cells and PC12 cells. *J. Biol. Chem.* **272,** 11,710–11,717.

Schreiber, M., and Salkoff, L. (1997). A novel calcium-sensing domain in the BK channel. *Biophys. J.* **73,** 1355–1363.

Scornik, F. S., Codina, J., Birnbaumer, L., and Toro, L. (1993). Modulation of coronary smooth muscle KCa channels by Gs alpha independent of phosphorylation by protein kinase A. *Am. J. Physiol.* **265,** H1460–H1465.

Shen, K. Z., Lagrutta, A., Davies, N. W., Standen, N. B., Adelman, J. P., and North, R. A. (1994). Tetraethylammonium block of Slowpoke calcium-activated potassium channels expressed in Xenopus oocytes: Evidence for tetrameric channel formation. *Pflugers Arch.* **426,** 440–445.

Solaro, C. R., and Lingle, C. J. (1992). Trypsin-sensitive, rapid inactivation of a calcium-activated potassium channel. *Science* **257,** 1694–1697.

Stefani, E., Ottolia, M., Noceti, F., Olcese, R., Wallner, M., Latorre, R., and Toro, L. (1997). Voltage-controlled gating in a large conductance Ca^{2+}-sensitive K^+ channel (*hslo*). *Proc. Natl. Acad. Sci. USA* **94,** 5427–5431.

Tanaka, Y., Meera, P., Song, M., Knaus, H.-G., and Toro, L. (1997). Molecular costituents of maxi Kca channels in human coronary smooth muscle: Predominant $\alpha + \beta$ subunit complexes. *J. Physiol.* **502,** 545–557.

Tseng-Crank, J., Foster, C. D., Krause, J. D., Mertz, R., Godinot, N., DiChiara, T. J., and Reinhart, P. H. (1994). Cloning, expression, and distribution of functionally distinct Ca^{2+} activated K^+ channel isoforms from human brain. *Neuron* **13,** 1315–1330.

Tseng-Crank, J., Godinot, N., Johansen, T. E., Ahring, P. K., Strobaek, D., Mertz, R., Foster, C. D., Olesen, S., and Reinhart, P. (1996). Cloning, expression, and distribution of a Ca^{2+} activated K^+ channel β-subunit from human brain. *Proc. Natl. Acad. Sci. USA* **93,** 9200–9205.

Toro, L., and Stefani, E. (1993). Modulation of maxi calcium-activated K channels. Roles of ligands, phosphorylation and G-proteins. *In* "Handbook of Expermental Pharmacology: GTPases in Biology" (B. Dickey and L. Birnbaumer, Eds.), Vol. 108, pp. 561–579. Springer Verlag, New York.

Toro, L., Vaca, L., and Stefani, E. (1991). Calcium-activated potassium channels from coronary smooth muscle reconstituted in lipid bilayers. *Am. J. Physiol.* **260,** H1779–H1789.

Verma-Ahuja, S., Evans, M. S., and Pencek, T. L. (1995). Evidence for decreased calcium dependent potassium conductance in hippocampal CA3 neurons of genetically epilepsy-prone rats. *Epilepsy Res.* **22,** 137–144.

Vogalis, F., Vincent, T., Qureshi, I., Schmalz, F., Ward, M. W., Sanders, K. M., and Horowitz, B. (1996). Cloning and expression of the large-conductance Ca^{2+}-activated K^+ channel from colonic smooth muscle. *Am.J.Physiol.* **271,** G629–G639.

Wallner, M., Meera, P., Ottolia, M., Kaczorowski, G. J., Latorre, R., Garcia, M. L., Stefani, E., and Toro, L. (1995). Characterization of and modulation by a β-subunit of a human Maxi K_{Ca} channel cloned from myometrium. *Recept. Chan.* **3,** 185–199.

Wallner, M., Meera, P., and Toro, L. (1996). Determinant for β-subunit regulation in high conductance voltage-activated and Ca^{2+}-sensitive K^+ channels: An additional transmembrane region at the N-terminus. *Proc. Natl. Acad. Sci. USA* **93,** 14,922–14,927.

Wang, Z. W., Nara, M., Wang, Y. X., and Kotlikoff, M. I. (1997). Redox regulation of large conductance Ca^{2+}-activated K^+ channels in smooth muscle cells. *J. Gen. Physiol.* **110,** 35–44.

Wei, A., Solaro, C., Lingle, C., and Salkoff, L. (1994). Calcium sensitivity of BK-type K_{Ca} channels determined by a separable domain. *Neuron* **13,** 671–681.

Wei, A., Jegla, T., and Salkoff, L. (1996). Eight potassium channel families revealed by the *C. elegans* genome project. *Neuropharmacology* **35,** 805–829.

Xia, X. M., Fakler, B., Rivard, A., Wayman, G., Johnson-Pais, T., Keen, J. E., Ishii, T., Hirschberg, B., Bond, C. T., Lutsenko, S., Maylie, J., and Adelman, J. P. (1998). Mechanism of calcium gating in small-conductance calcium-activated potassium channels. *Nature* **395,** 503–507.

PART II

Inwardly Rectifying Potassium Channels

CHAPTER 9

The Assembly of Inwardly Rectifying Potassium Channels

Andrew Tinker* and Lily Yeh Jan†
*Centre for Clinical Pharmacology and Toxicology, Department of Medicine/Cruciform Project, Rayne Institute, University College, London WC1E 6JJ, United Kingdom; and †Howard Hughes Medical Institute, Departments of Physiology and Biochemistry, University of California, San Francisco, San Francisco, California 94143

I. INTRODUCTION

The primary structures of a large number of potassium channels have been determined and they fall into two major families: the voltage-gated (Kv) and inwardly rectifying (Kir) (Jan and Jan, 1997; Hille, 1991). Kv potassium channels belong to the superfamily of voltage-gated ion channels and, like the cyclic nucleotide-gated family, are tetrameric proteins in which four individual subunits are required to constitute a functional channel. This contrasts with voltage-gated calcium and sodium channels in which four similar but not identical subunits are linked together in the gene and the pore-foming unit is a monomer (Hille, 1991). Kir potassium channels are also tetrameric proteins (Yang *et al.,* 1995b) but have a simpler predicted

1063-5823/99 $30.00

membrane topology than Kv potassium channels. Kv channels have six transmembrane domains (S1–S6), a probable pore-forming hairpin loop (H5), and a cytoplasmic N and C terminus. In contrast, Kir channels have two transmembrane regions (M1 and M2), a similar pore-forming hairpin loop (H5), a cytoplasmic N-terminal domain, and a longer cytoplasmic C-terminal domain (Ho *et al.,* 1993; Kubo *et al.,* 1993; Isomoto *et al.,* 1997; Nichols and Lopatin, 1997; Doupnik *et al.,* 1995). The only region of significant homology between Kv and Kir channels is the H5 region. In addition to the pore-forming subunits there are a number of associating proteins that may be essential or may modulate Kv and Kir channel function.

Kir channels have a large number of family members and are likely to be responsible for a number of important physiological currents. The subfamilies and their properties are summarized in Table I. This diversity and the tetrameric nature of Kir channels leads to the theoretical possibility of one gene product coassembling with another to form heteromultimeric as well as homomultimeric channels. It raises the issue of which regions of the Kir protein are responsible for homotypic interaction (i.e., between one subfamily member and itself) and which are responsible for heterotypic interactions (i.e., regions that influence the ability of two different family members to interact).

II. EXPERIMENTAL STUDIES OF Kir ASSEMBLY

In our studies (Tinker *et al.* 1996) two approaches were used to address the question of how Kir channels assemble. First, a biochemical system involving the transient transfection of a mammalian cell line (HEK293 cells) with tagged channel subunits and subsequent coprecipitation was employed. Second, a more functional assay comprising the injection of *Xenopus laevis* oocytes with *in vitro* transcribed RNA and the measurement of current using a two-electrode voltage clamp was utilized. We then used standard molecular techniques to construct a number of deletion mutants of IRK1/Kir 2.1 with and without an epitope tag and a number of chimeras between IRK1/Kir 2.1 and other family members to determine possible assembly domains.

A. A Biochemical Assay

A biochemical assay system was developed to examine protein–protein interaction between various channel subunits. The approach is similar in principle to coimmunoprecipitation; however, instead of using a precipitat-

TABLE I

The Kir Subfamilies

Subfamily	Members	Key properties	Physiological counterpart/role
1.0/ROMKs	Kir1.1/ROMK1 and a number of N-terminal splice variants	Weakly rectifying, ATP dependent, predominantly renal expression (apical membranes)	Potentially a number of renal tubular K^+ currents
2.0/IRKs	Kir2.1–2.3/IRK1–3	Strongly rectifying; 2.1 and 2.2 widely distributed; 2.3 brain specific	Strongly rectifying currents in a number of tissues including I_{K1} in cardiac cells
3.0/GIRKs	Kir3.1–3.5/GIRK1–5; GIRK2 has a number of splice variants	Strongly rectifying, heteromultimeric assembly, directly gated by G proteins	Physiological activation responsible for slowing of the heart and depression of neuronal excitability
4.0/BIR10	Kir 4.1	Strongly rectifying, ATP dependent, expressed in brain and renal tissue (basolateral membranes)	Responsible for a Muller cell current in the retina
5.0/BIR9	Kir 5.1	Does not express alone, forms heteromultimers with 4.1	Unknown
6.0	Kir 6.1/6.2	Weakly rectifying, interaction with an auxilliary protein (SUR) confers sulfonylurea and potassium channel opening drug sensitivity on current	Responsible for the ATP-sensitive current in pancreatic, cardiac, skeletal, smooth muscle, and neuronal tissues

Note. This table shows the diversity of structure and function of the inward rectifier family of cloned potassium channels (Kir). The field has been reviewed extensively recently (Jan and Jan, 1997; Isomoto *et al.*, 1997; Nichols and Lopatin, 1997; Doupnik *et al.*, 1995).

ing antibody a binding resin (Ni^{2+}-charged iminodiacetic acid coupled to Sepharose beads) recognizing a hexahistidine epitope is used for specific purification. The use of this approach is illustrated in Fig. 1A. A fusion protein of IRK1/Kir 2.1 (IRK1-His_6) in which a six-histidine tag is engineered onto the N terminus, allowing high-affinity specific purification by the binding resin, is made. HEK293 cells are transiently cotransfected with IRK1-His_6 and the test protein of interest, which had been tagged with an eight-amino-acid sequence (Asp–Tyr–Lys–Asp–Asp–Asp–Asp–Lys, known as the FLAG sequence) recognized by a commercially available FLAG antibody. IRK1-His_6-containing complexes are purified under nondenaturing conditions from the detergent-solubilized cell homogenate, washed, and eluted from the resin.The proteins, including those that may copurify, are then subjected to Western blotting. For example, if there is protein–protein interaction between IRK1-His_6 and the FLAG-tagged test protein the two proteins should copurify (Hoffmann and Roeder, 1991) and the interaction will be revealed by probing the blot with the FLAG antibody (Fig. 1A).

Figure 1B shows the practical outcome of such a strategy. The experiment was designed to test the tendency of the two homologous inwardly rectifying potassium channels IRK1/Kir 2.1 and IRK2/Kir 2.2 to coassemble. IRK1/Kir 2.1 and IRK2/Kir 2.2 are tagged with the FLAG epitope (IRK1-FLAG and IRK2-FLAG). A fraction of the cell sample is denatured by sodium dodecyl sulfate (SDS) lysis to examine expression of the channel constructs while the remainder is solubilized under nondenaturing conditions and bound to the resin. Lanes 1–3 (L1–L3) are SDS lysates of a fraction of HEK293 cells transfected with vector as control (L1), IRK1-His_6 and IRK1-FLAG (L2), and IRK1-His_6 and IRK2-FLAG (L3). It is apparent that the FLAG antibody recognizes a specific band not present in control membranes (L1) of approximately 55 kDa for IRK1-FLAG (L2) and 45 kDa for IRK2-FLAG (L3). The mobility of IRK2-FLAG is similar to that expected from its molecular weight while that of IRK1-FLAG is slightly increased from such an estimate. Lanes 4–7 (L4–7) are protein samples eluted from the resin after binding and washing of detergent-solubilized membranes from cells transfected as indicated. The absence of IRK1-FLAG or IRK2-FLAG in L4 and L5 suggests adequate washing conditions and implies that any FLAG activity associated with purification of IRK1-His_6 indicates specific protein–protein interaction. It is apparent in Fig. 1B that while IRK2-FLAG is able to interact with IRK1-His_6 (L7) this is less than the tendency for self-to-self assembly (L6) given equivalent expression of IRK1-FLAG and IRK2-FLAG in L2 and L3 and significant purification of IRK1-His_6 in L6 and L7. It also indicates that IRK1/Kir 2.1 is at least a dimer.

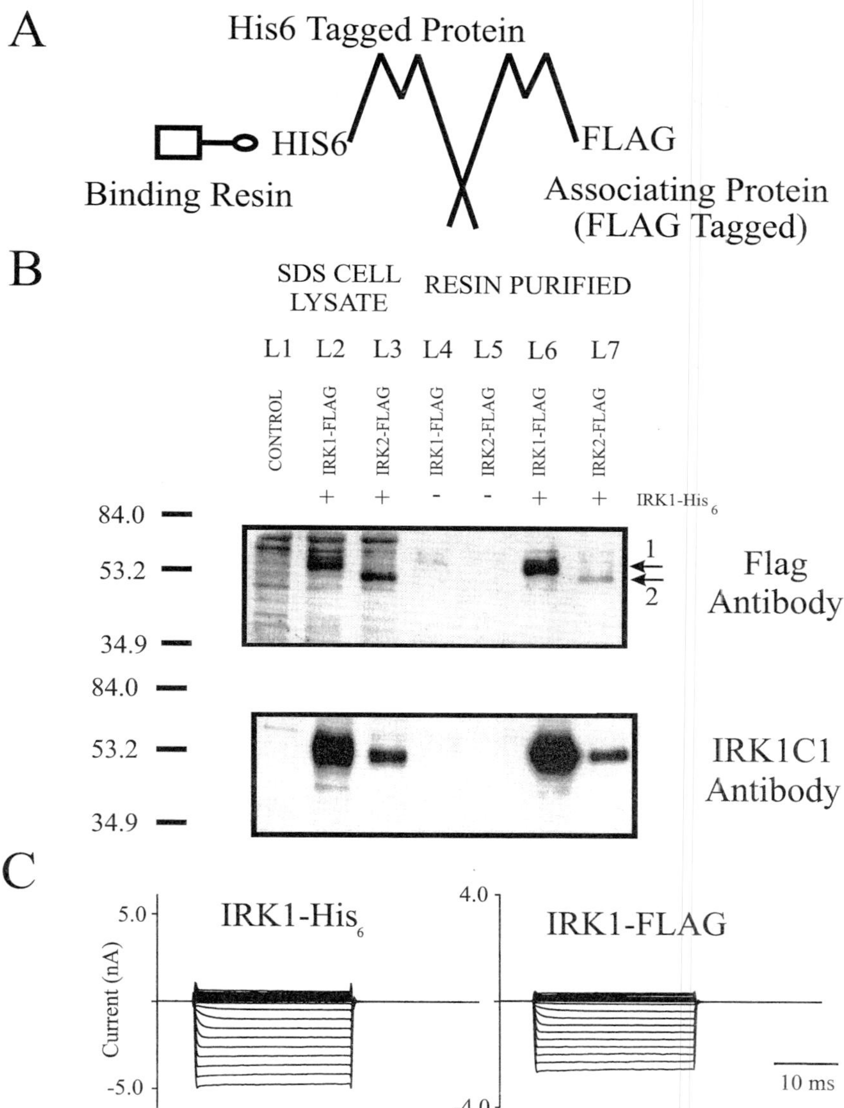

FIGURE 1 (A) An illustration of the biochemical approach. (B) The figures are Western blots of 12% denaturing polyacrylamide gels transferred to nitrocellulose membranes and probed with primary antibodies as indicated. L1–L3, SDS lysates of transfected cells; L4–L7 resin-purified samples (see text for more details). Arrows illustrate the bands corresponding to IRK1-FLAG and IRK2-FLAG in the upper gel. Molecular weight markers in kilodaltons are illustrated beside the gel. (C) Whole-cell configuration of the patch clamp used to record currents in HEK293 cells 48 hr after transient transfection with IRK1-His_6 or IRK1-FLAG.

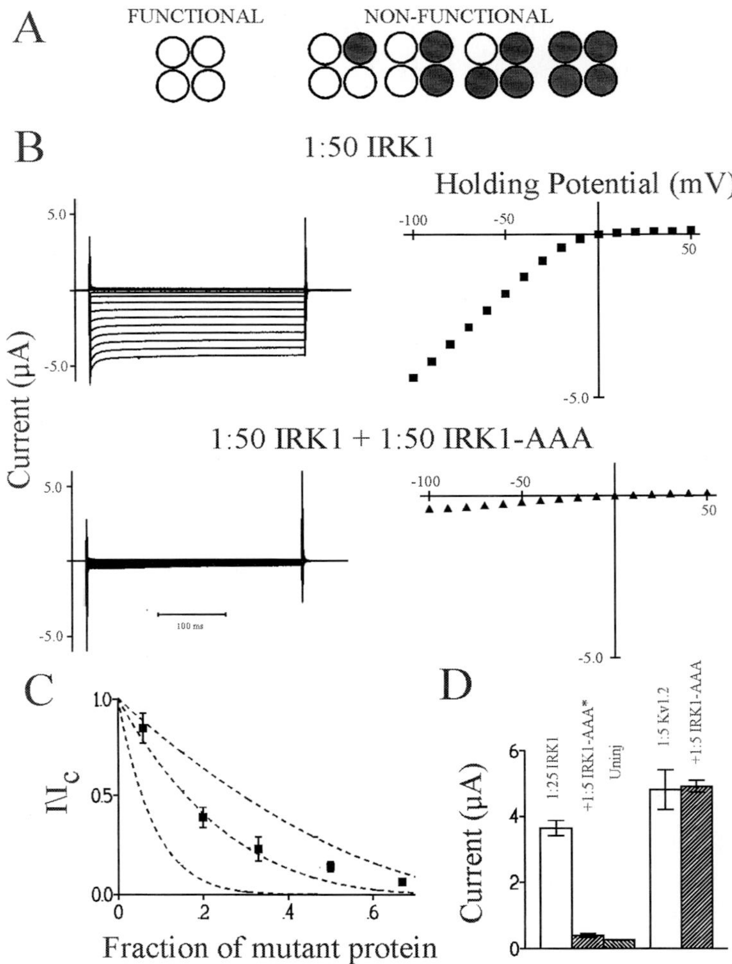

FIGURE 2 (A) In the dominant negative effect the inhibitory subunit (filled circles) associates with wild-type monomers (open circles) and inactivates function. (B) Currents recorded using a two-electrode voltage clamp in oocytes 1 day after injection. The pulse protocol involved stepping to test potentials between 50 and −100 mV in 10 mV steps from a holding potential of 0 mV. Coinjection of the AAA mutant with IRK1/Kir 2.1 leads to suppression of current. (C) A dose–response curve for inhibition of current by IRK1-AAA. $I\backslash I_c$ is the relative current level compared to control. Each data point (■) is an average from five oocytes ±SEM. The dashed lines are theoretical curves based on various stoichiometries: the upper line corresponds to the expectation for a dimer, the middle line for a tetramer, and the bottom line for a dodecamer. The theoretical relationship is $I\backslash I_c = (1-\text{fraction of mutant protein})^N$, where the amount of protein is assumed to be linearly related to the amount of RNA injected. At higher RNA concentrations there is likely to be some saturation of

In order to check that epitope tagging does not lead to any gross alterations in functional channel expression for IRK1-Kir 2.1, we examined whole cell currents under voltage-clamp conditions using the patch-clamp technique. Figure 1C shows recordings from HEK293 cells transiently transfected with IRK1-His_6 and IRK1-FLAG. Approximately 20% of cells contained a strongly rectifying K^+ selective current not present in mock-transfected cells.

B. A Functional Assay

The use of a biochemical approach provides a direct measure of protein-protein interaction but leaves open the question of the functional significance of any interaction. To examine the effect of such interactions on channel function we used a dominant negative approach in *X. laevis* oocytes injected with cRNA and studied with two-electrode voltage clamping (Herskowitz, 1987). The GYG motif in the H5 region of potassium channels plays a key role in selectivity and pore function (Heginbotham *et al.,* 1994; Slesinger *et al.,* 1996). These three amino acids were replaced with three alanines in IRK1/Kir 2.1 (IRK1-AAA). Injected alone this mutant does not lead to current but coinjection with IRK1/Kir 2.1 leads to a strong suppression of current compared to control (Fig. 2B). With the assumption that channels containing one or more mutant subunits leads to a complex that is unable to conduct current (Fig. 2A), titration of the relative reduction of current with increasing concentrations of IRK1-AAA RNA conforms to the behavior expected for a tetramer (Fig. 2C). IRK1-AAA cRNA coinjected with a member of the voltage-gated family (Kv1.2) did not lead to a reduction of current, indicating that the effect was likely to be specific (Fig. 2D).

There are two mechanisms for the dominant negative effect: It can arise either because the mutant subunit coassembles with wild-type subunits and gives rise to nonfunctional channels on the cell membrane or because the multimeric complex containing the mutant protein is recognized and degraded in the endoplasmic reticulum. Our studies do not address which of these possibilities is the likely mechanism.

---

expression and this may explain the deviation of the experimental data from the theoretical curve for $N = 4$. (D) A bar graph showing leak-subtracted currents at -100 mV (IRK1/Kir 2.1) or 50 mV (Kv1.2) with oocytes injected as indicated. Coinjection is indicated by +. $N =$ 9 for IRK1/Kir 2.1±IRK1-AAA injections, $N = 6$ for Kv1.2±IRK1-AAA injections, and $N =$ 4 for uninjected oocytes. The asterisk indicates that there is a statistically significant difference between control and coinjection.

C. Analysis of Possible Assembly Domains

Our aim was to broadly define regions responsible for Kir assembly. To address the question of what determines homotypic interactions we constructed a series of N- and C-terminal deletion mutants of IRK1/Kir 2.1 and used them in the above two assay systems. In other words we tested the ability of IRK1-His_6 to copurify tagged IRK1/Kir 2.1 deletion mutants and the ability of deletion mutants containing the AAA mutation to knockout IRK1/Kir 2.1 current. We also examined the ability of deletion mutants to express current. The data (Tinker *et al.,* 1996) indicate that it is possible to remove the N terminus but only a fraction of the C terminus (at a boundary of approximately amino acid 333 in IRK1/Kir 2.1) before both biochemical and functional evidence of assembly is lost.

The question of regions responsible for heterotypic interactions was then addressed using a chimeric approach. The idea is to replace regions in IRK1/Kir 2.1 with those of a channel that IRK1/Kir 2.1 does not normally coassemble with and then examine biochemically whether this chimera is able to copurify with IRK1-His_6. Construction of chimeras between two channels for which there seems no evidence of coassembly (Kir 6.1 and IRK1/Kir 2.1) shows that replacement of the N terminus and M1 and the distal C terminus of IRK1/Kir 2.1 with Kir 6.1 does not interfere with biochemical association with IRK1/Kir 2.1, but the inclusion of more proximal regions of the C terminus of Kir 6.1 does. In addition to these biochemical data, a double chimera including the N terminus and M1 and distal C terminus of Kir 6.1 transplanted onto H5, M2, and the proximal C terminus of IRK1-AAA was still able to knockout IRK1/Kir 2.1 current. Furthermore chimeras of IRK1/Kir 2.1 with other subfamily and family members (Kir 2.2, 2.3, 1.1/IRK2, IRK3, ROMK1) that included the proximal C terminus led to current knockout in the donor channel and IRK1/Kir 2.1. These data implicate M2 and the proximal C terminus in heterotypic interactions.

We tested the relative importance of these two regions using the functional dominant negative assay. The data indicated that interaction between the two homologous channels IRK1/Kir 2.1 and IRK2/Kir 2.2 seems to be influenced equally by compatibility over the M2C region (amino acids 156–220, with 11 differences) and the proximal C terminus (amino acids 220–300, with 18 differences). However, between channels that belong to different subfamilies and are less homologous, IRK1/Kir 2.1 and Kir 6.1 (29 and 40 differences in M2C and the proximal C terminus, respectively) and IRK1/Kir 2.1 and ROMK1/Kir 1.1 (26 and 45 differences in M2C and the proximal C terminus, respectively), the experiments suggest that compatability between the proximal C-terminal regions is the major contributing factor to whether channels will heteromultimerize.

D. Additional Studies on Assembly

Only a relatively small number of studies have been performed on Kir assembly. In a comprehensive study using immunoprecipiatation on *in vitro* translated GIRK1/Kir 3.1 and GIRK2/Kir 3.2 and deletion mutants, Woodward and colleagues (1997) implicated the N terminus, transmembrane regions, and proximal C-terminus in intersubunit interactions. Another study implicated the transmembrane domains in determining a dominant negative effect between BIR10/Kir 4.1 and GIRK4/Kir 3.4 (Tucker *et al.,* 1996). Fink and co-workers (1996) reached the conclusion that the N terminus was of primary importance. It is obvious that the story may be quite complex and may vary around a central theme depending on the subfamily members studied. More experimental data are needed.

III. A POSSIBLE MODEL FOR Kir ASSEMBLY

The simplest hypothesis is that the M2 and proximal C-terminal domain contacts the adjacent M2 and proximal C-terminal domain in the tetramer. There is little evidence contradicting this possibility. The H5 and M2 segments have been implicated in potassium-selective ion permeation (Ficker *et al.,* 1994; Lopatin *et al.,* 1994; Lu and MacKinnon, 1994; Stanfield *et al.,* 1994; Yang *et al.,* 1995a; Reuveny *et al.,* 1996; Slesinger *et al.,* 1996). The proximal C terminus also contains a residue (E224 in IRK1/Kir 2.1) that seems to play an integral role in pore function and in particular inward rectification (Yang *et al.,* 1995a; Taglialatela *et al.,* 1995). Thus a speculative structural model of the inwardly rectifying potassium channels has this region of the C terminus and the H5/M2 regions making up a cytoplasmic vestibule and core of the pore in which monomer to monomer contact stabilizes the tetramer and determines specificity of association. A cartoon representing a possible model for Kir assembly is shown in Fig. 3. Other possible interactions include an association of the N terminus with the C terminus.

Our studies on IRK1/Kir 2.1 indicate that homomultimerization is favored over heteromultimerization even with the highly homologous IRK2/Kir 2.2. Is it possible to say how important heteromultimerization is for the Kir family? The molecular correlate of the current responsible for slowing of the heart rate in response to vagal nerve stimulation is a heteromultimer of GIRK1/Kir 3.1 and GIRK4/Kir 3.4 (Krapivinsky *et al.,* 1995a). Functional heteromultimers between GIRK1/Kir 3.1 and GIRK2/Kir 3.2 also exist in the central nervous system and constitute an important G-protein-regulated potassium current (Lesage *et al.,* 1994, 1995; Kofuji *et al.,*

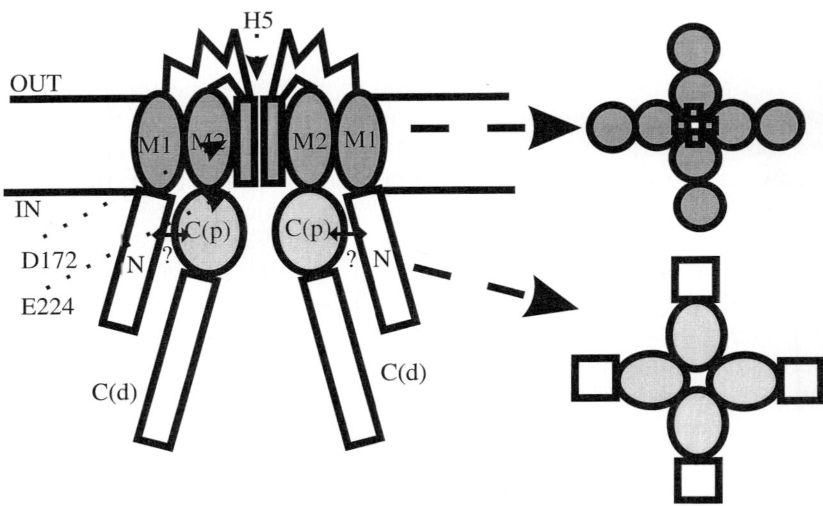

FIGURE 3 A cartoon showing a possible model for Kir assembly.

1995, 1996; Slesinger *et al.*, 1996; Patil *et al.*, 1995; Navarro *et al.*, 1996). There is also evidence for association between members of different subfamilies, namely GIRK4/Kir 3.4 and BIR10/Kir 4.1, though the heteromultimers are degraded in the endoplasmic reticulum (Tucker *et al.*, 1996). A study examining the interaction of proteins from two different subfamilies, BIR10/Kir 4.1 and ROMK1/Kir 1.1, concluded that while heteroligomers did occur they made up a maximum of 30% of the total channel population (Glowatzki *et al.*, 1995). Taken together these observations point to heteromultimerization being case specific, varying in degree depending on the particular combination studied, and not determined rigidly by assignment to a subfamily based on homology or function.

IV. COMPARING THE ASSEMBLY OF Kv AND Kir POTASSIUM CHANNELS

The rules governing the assembly of the voltage-gated potassium channels are better established. The N terminus and the first transmembrane segment determine homotypic interactions. Heteromultimerization occurs generally within a subfamily but not between subfamily members and is determined by the N terminus (Li *et al.*, 1992; Shen *et al.*, 1993; Shen and Pfaffinger, 1995; Pfaffinger and DeRubeis, 1995; Babila *et al.*, 1994). Indeed two assembly boxes have been determined within this domain (Xu *et al.*, 1995). There are some interesting observations in relation to voltage-gated potassium channel assembly. Deletion of the N terminus produces a channel that is able to form functional heteromultimers with other subfamilies (Lee *et al.*, 1994). Fragments of the core and carboxy terminus of Kv1.3 coimmunoprecipitate and suppress an N-terminal deletion of Kv1.3 (Tu *et al.*, 1996). It thus seems that while the N terminus is of central importance, other regions may contribute particularly to homotypic interactions. These differences are summarized in Table II.

The information for more distantly related potassium channels such as the mammalian homologs of the *Drosophila* ether-a-go-go is only starting to be revealed. Although r-eag shows a similar overall predicted topology to Kv channels, assembly seems to be determined by a region in the C terminus (Ludwig *et al.*, 1997). HERG, another member of this family, has an N terminus that self-tetramerizes but deletion of that region does not affect current expression (Spector *et al.*, 1996; Li *et al.*, 1997).

V. OUTSTANDING ISSUES IN RELATION TO KIR ASSEMBLY

The identification of domains involved in homotypic and heterotypic interactions begs the questions of whether the domain can be further narrowed down and whether particular residues are important in the interaction. It may be easier to answer such questions using more *in vitro* biochemical approaches, such as overlay techniques or genetic techniques such as the yeast two hybrid with fusion proteins of short segments of the channel.

The degree to which heteromultimerization occurs with families other than Kir 3.0 is not clearly established. The cell biology of assembly of

TABLE II
A Comparison of Kv and Kir Assembly

Property	Kv	Kir
Homotypic interactions	N terminus/S1 and perhaps other regions contributing	Proximal C terminus/M2 and perhaps other regions contributing
Rules for heteromultimerization	Within a subfamily but only rarely between other subfamilies and families of potassium channel	Case specific; no strict rules apparent yet
Heterotypic interactions	N terminus	Proximal C terminus/M2

GIRK1/Kir 3.1 with GIRK2/Kir 3.2 and GIRK4/Kir 3.4 is an interesting area. It is apparent that coexpression of GIRK2/Kir 3.2 or GIRK4/Kir 3.4 with GIRK1/Kir 3.1 leads to the expression of protein in the plasma membrane. It has been suggested that GIRK1/Kir 3.1 is unable to form a functional homomultimer and in *X. laevis* oocytes couples to a Kir 3.0 subunit, XIR/Kir 3.5 (Hedin *et al.,* 1996). Studies using immunofluorescence of epitope-tagged proteins in heterologous systems suggest that in the absence of GIRK4/Kir 3.4, GIRK1/Kir 3.1 is not translocated to the plasma membrane but is held up in the cytoskeleton (Kennedy *et al.,* 1996). It is interesting to note that the distal C terminus and a single residue (F137) have been implicated in GIRK1/Kir 3.1 in controlling activity and possibly surface membrane expression (Chan *et al.,* 1996, 1997). In particular, the F137 residue may well control activity of the small number of GIRK1/Kir 3.1 homomultimeric channels that insert into the plasma membrane. It seems unlikely that in native heart membranes GIRK4/Kir 3.4 forms a homomultimer as it is possible to remove all GIRK4/Kir 3.4 by immunoprecipitating GIRK1/Kir 3.1 (Krapivinsky *et al.,* 1995b). In addition, a fixed stoichiometry of two of each subunit in the tetramer has been suggested to occur (Silverman *et al.,* 1996). How is this achieved and is there perhaps a preferred sequential assembly pathway, for example, recognition and degradation of GIRK1/Kir 3.1 and GIRK4/Kir 3.4 dimers in the endoplasmic reticulum? Such a process occurs in the assembly of the endplate nicotinic receptor (Gu *et al.,* 1991; Verrall and Hall, 1992).

The ATP-sensitive potassium channel is likely to be a complex of a pore-forming subunit (Kir 6.1/6.2) and a regulatory subunit, the sulfonylurea receptor (SUR1, SUR2A, SUR2B) (Inagaki *et al.,* 1995, 1996; Yamada *et al.,* 1997; Isomoto *et al.,* 1996). Recent studies (Clement *et al.,* 1997) imply a protein–protein interaction and a 1:1 stoichiometry between these two proteins. The domains responsible for the assembly between these two proteins, however, have not been defined. There is a suggestion that the SUR1 subunit may be able to couple to other Kirs such as ROMK1/Kir 1.1 and an endogenous current in HEK293 (Ammala *et al.,* 1996) and that other ABC transporters such as the cystic fibrosis transmembrane regulator may be able to couple to inwardly rectifying potassium channels, specifically ROMK2/Kir 1.2 (McNicholas *et al.,* 1996). This is obviously an area of great biological complexity.

Acknowledgment

Dr. Andrew Tinker is a Wellcome Trust Senior Fellow in Clinical Science.

References

Ammala, C., Moorhouse, A., Gribble, F., Ashfield, R., Proks, P., Smith, P. A., Sakura, H., Coles, B., Ashcroft, S. J., and Ashcroft, F. M. (1996). Promiscuous coupling between the sulphonylurea receptor and inwardly rectifying potassium channels. *Nature* **379,** 545–548.

Babila, T., Moscucci, A., Wang, H., Weaver, F. E., and Koren, G. (1994). Assembly of mammalian voltage-gated potassium channels: Evidence for an important role of the first transmembrane segment. *Neuron* **12,** 615–626.

Chan, K. W., Sui, J. L., Vivaudou, M., and Logothetis, D. E. (1996). Control of channel activity through a unique amino acid residue of a G protein-gated inwardly rectifying K^+ channel subunit. *Proc. Natl. Acad. Sci. USA* **93,** 14,193–14,198.

Chan, K. W., Sui, J. L., Vivaudou, M., and Logothetis, D. E. (1997). Specific regions of heteromeric subunits involved in enhancement of G protein-gated K^+ channel activity. *J. Biol. Chem.* **272,** 6548–6555.

Clement, J. P., 4th, Kunjilwar, K., Gonzalez, G., Schwanstecher, M., Panten, U., Aguilar Bryan, L., and Bryan, J. (1997). Association and stoichiometry of K(ATP) channel subunits. *Neuron* **18,** 827–838.

Doupnik, C. A., Davidson, N., and Lester, H. A. (1995). The inward rectifier potassium channel family. *Curr. Opin. Neurobiol.* **5,** 268–277.

Ficker, E., Taglialatela, M., Wible, B. A., Henley, C. M., and Brown, A. M. (1994). Spermine and spermidine as gating molecules for inward rectifier K^+ channels. *Science* **266,** 1068–1072.

Fink, M., Duprat, F., Heurteaux, C., Lesage, F., Romey, G., Barhanin, J., and Lazdunski, M. (1996). Dominant negative chimeras provide evidence for homo and heteromultimeric assembly of inward rectifier K^+ channel proteins via their N-terminal end. *FEBS Lett.* **378,** 64–68.

Glowatzki, E., Fakler, G., Brandle, U., Rexhausen, U., Zenner, H. P., Ruppersberg, J. P., and Fakler, B. (1995). Subunit-dependent assembly of inward-rectifier K^+ channels. *Proc. R. Soc. Lond. B. Biol. Sci.* **261,** 251–261.

Gu, Y., Forsayeth, J. R., Verrall, S., Yu, X. M., and Hall, Z. W. (1991). Assembly of the mammalian muscle acetylcholine receptor in transfected COS cells. *J. Cell Biol.* **114,** 799–807.

Hedin, K. E., Lim, N. F., and Clapham, D. E. (1996). Cloning of a Xenopus laevis inwardly rectifying K^+ channel subunit that permits GIRK1 expression of IKACh currents in oocytes. *Neuron* **16,** 423–429.

Heginbotham, L., Lu, Z., Abramson, T., and MacKinnon, R. (1994). Mutations in the K^+ channel signature sequence. *Biophys. J.* **66,** 1061–1067.

Herskowitz, I. (1987). Functional inactivation of genes by dominant negative mutations. *Nature* **329,** 219–222.

Hille, B. (1991). "Ionic Channels of Excitable Membranes." Sinauer Associates, Sunderland, MA.

Ho, K., Nichols, C. G., Lederer, W. J., Lytton, J., Vassilev, P. M., Kanazirska, M. V., and Hebert, S. C. (1993). Cloning and expression of an inwardly rectifying ATP-regulated potassium channel. *Nature* **362,** 31–38.

Hoffmann, A., and Roeder, R. G. (1991). Purification of His-tagged proteins in non-denaturing conditions suggests a convenient method for protein interaction studies. *Nucleic Acid Res.* **19,** 6337–6338.

Inagaki, N., Gonoi, T., Clement, J. P., 4th, Namba, N., Inazawa, J., Gonzalez, G., Aguilar Bryan, L., Seino, S., and Bryan, J. (1995). Reconstitution of IKATP: An inward rectifier subunit plus the sulfonylurea receptor [see comments]. *Science* **270,** 1166–1170.

Inagaki, N., Gonoi, T., Clement, J. P., Wang, C. Z., Aguilar Bryan, L., Bryan, J., and Seino, S. (1996). A family of sulfonylurea receptors determines the pharmacological properties of ATP-sensitive K^+ channels. *Neuron* **16,** 1011–1017.

Isomoto, S., Kondo, C., Yamada, M., Matsumoto, S., Higashiguchi, O., Horio, Y., Matsuzawa, Y., and Kurachi, Y. (1996). A novel sulfonylurea receptor forms with BIR (Kir6.2) a smooth muscle type ATP-sensitive K^+ channel. *J. Biol. Chem.* **271,** 24,321–24,324.

Isomoto, S., Kondo, C., and Kurachi, Y. (1997). Inwardly rectifying potassium channels: Their molecular heterogeneity and function. *Jpn. J. Physiol.* **47,** 11–39.
Jan, L. Y., and Jan, Y. N. (1997). Cloned potassium channels from eukaryotes and prokaryotes. *Annu. Rev. Neurosci.* **20,** 91–123.
Kennedy, M. E., Nemec, J., and Clapham, D. E. (1996). Localization and interaction of epitope-tagged GIRK1 and CIR inward rectifier K^+ channel subunits. *Neuropharmacology* **35,** 831–839.
Kofuji, P., Davidson, N., and Lester, H. A. (1995). Evidence that neuronal G-protein-gated inwardly rectifying K^+ channels are activated by G beta gamma subunits and function as heteromultimers. *Proc. Natl. Acad. Sci. USA* **92,** 6542–6546.
Kofuji, P., Hofer, M., Millen, K. J., Millonig, J. H., Davidson, N., Lester, H. A., and Hatten, M. E. (1996). Functional analysis of the weaver mutant GIRK2 K^+ channel and rescue of weaver granule cells. *Neuron* **16,** 941–952.
Krapivinsky, G., Gordon, E. A., Wickman, K., Velimirovic, B., Krapivinsky, L., and Clapham, D. E. (1995a). The G-protein-gated atrial K^+ channel IKACh is a heteromultimer of two inwardly rectifying K(+)-channel proteins. *Nature* **374,** 135–141.
Krapivinsky, G., Krapivinsky, L., Velimirovic, B., Wickman, K., Navarro, B., and Clapham, D. E. (1995b). The cardiac inward rectifier K^+ channel subunit, CIR, does not comprise the ATP-sensitive K^+ channel, IKATP. *J. Biol. Chem.* **270,** 28,777–28,779.
Kubo, Y., Baldwin, T. J., Jan, Y. N., and Jan, L. Y. (1993). Primary structure and functional expression of a mouse inward rectifier potassium channel [see comments]. *Nature* **362,** 127–133.
Lee, T. E., Philipson, L. H., Kuznetsov, A., and Nelson, D. J. (1994). Structural determinant for assembly of mammalian K^+ channels. *Biophys. J.* **66,** 667–673.
Lesage, F., Duprat, F., Fink, M., Guillemare, E., Coppola, T., Lazdunski, M., and Hugnot, J. P. (1994). Cloning provides evidence for a family of inward rectifier and G-protein coupled K^+ channels in the brain. *FEBS Lett.* **353,** 37–42.
Lesage, F., Guillemare, E., Fink, M., Duprat, F., Heurteaux, C., Fosset, M., Romey, G., Barhanin, J., and Lazdunski, M. (1995). Molecular properties of neuronal G-protein-activated inwardly rectifying K^+ channels. *J. Biol. Chem.* **270,** 28,660–28,667.
Li, M., Jan, Y. N., and Jan, L. Y. (1992). Specification of subunit assembly by the hydrophilic amino-terminal domain of the Shaker potassium channel. *Science* **257,** 1225–1230.
Li, X., Xu, J., and Li, M. (1997). The human delta1261 mutation of the HERG potassium channel results in a truncated protein that contains a subunit interaction domain and decreases the channel expression. *J. Biol. Chem.* **272,** 705–708.
Lopatin, A. N., Makhina, E. N., and Nichols, C. G. (1994). Potassium channel block by cytoplasmic polyamines as the mechanism of intrinsic rectification. *Nature* **372,** 366–369.
Lu, Z., and MacKinnon, R. (1994). Electrostatic tuning of Mg^{2+} affinity in an inward-rectifier K^+ channel. *Nature* **371,** 243–246.
Ludwig, J., Owen, D., and Pongs, O. (1997). Carboxy-terminal domain mediates assembly of the voltage-gated rat ether-a-go-go potassium channel. *EMBO J.* **16,** 6337–6345.
McNicholas, C. M., Guggino, W. B., Schwiebert, E. M., Hebert, S. C., Giebisch, G., and Egan, M. E. (1996). Sensitivity of a renal K^+ channel (ROMK2) to the inhibitory sulfonylurea compound glibenclamide is enhanced by coexpression with the ATP-binding cassette transporter cystic fibrosis transmembrane regulator. *Proc. Natl. Acad. Sci. USA* **93,** 8083–8088.

Navarro, B., Kennedy, M. E., Velimirovic, B., Bhat, D., Peterson, A. S., and Clapham, D. E. (1996). Nonselective and G betagamma-insensitive weaver K^+ channels. *Science* **272,** 1950–1953.

Nichols, C. G., and Lopatin, A. N. (1997). Inward rectifier potassium channels. *Annu. Rev. Physiol.* **59,** 171–191.

Patil, N., Cox, D. R., Bhat, D., Faham, M., Myers, R. M., and Peterson, A. S. (1995). A potassium channel mutation in weaver mice implicates membrane excitability in granule cell differentiation [see comments]. *Nature Genet.* **11,** 126–129.

Pfaffinger, P. J., and DeRubeis, D. (1995). Shaker K^+ channel T1 domain self-tetramerizes to a stable structure. *J. Biol. Chem.* **270,** 28,595–28,600.

Reuveny, E., Jan, Y. N., and Jan, L. Y. (1996). Contributions of a negatively charged residue in the hydrophobic domain of the IRK1 inwardly rectifying K^+ channel to K(+)-selective permeation. *Biophys. J.* **70,** 754–761.

Shen, N. V., Chen, X., Boyer, M. M., and Pfaffinger, P. J. (1993). Deletion analysis of K^+ channel assembly. *Neuron* **11,** 67–76.

Shen, N. V., and Pfaffinger, P. J. (1995). Molecular recognition and assembly sequences involved in the subfamily-specific assembly of voltage-gated K^+ channel subunit proteins. *Neuron* **14,** 625–633.

Silverman, S. K., Lester, H. A., and Dougherty, D. A. (1996). Subunit stoichiometry of a heteromultimeric G protein-coupled inward-rectifier K^+ channel. *J. Biol. Chem.* **271,** 30,524–30,528.

Slesinger, P. A., Patil, N., Liao, Y. J., Jan, Y. N., Jan, L. Y., and Cox, D. R. (1996). Functional effects of the mouse weaver mutation on G protein-gated inwardly rectifying K^+ channels. *Neuron* **16,** 321–331.

Spector, P. S., Curran, M. E., Zou, A., Keating, M. T., and Sanguinetti, M. C. (1996). Fast inactivation causes rectification of the IKr channel. *J. Gen. Physiol.* **107,** 611–619.

Stanfield, P. R., Davies, N. W., Shelton, P. A., Sutcliffe, M. J., Khan, I. A., Brammar, W. J., and Conley, E. C. (1994). A single aspartate residue is involved in both intrinsic gating and blockage by Mg^{2+} of the inward rectifier, IRK1. *J. Physiol. London* **478,** 1–6.

Taglialatela, M., Ficker, E., Wible, B. A., and Brown, A. M. (1995). C-terminus determinants for Mg^{2+} and polyamine block of the inward rectifier K^+ channel IRK1. *EMBO J.* **14,** 5532–5541.

Tinker, A., Jan, Y. N., and Jan, L. Y. (1996). Regions responsible for the assembly of inwardly rectifying potassium channels. *Cell* **87,** 857–868.

Tu, L., Santarelli, V., Sheng, Z., Skach, W., Pain, D., and Deutsch, C. (1996). Voltage-gated K^+ channels contain multiple intersubunit association sites. *J. Biol. Chem.* **271,** 18,904–18,911.

Tucker, S. J., Bond, C. T., Herson, P., Pessia, M., and Adelman, J. P. (1996). Inhibitory interactions between two inward rectifier K^+ channel subunits mediated by the transmembrane domains. *J. Biol. Chem.* **271,** 5866–5870.

Verrall, S., and Hall, Z. W. (1992). The N-terminal domains of acetylcholine receptor subunits contain recognition signals for the initial steps of receptor assembly. *Cell* **68,** 23–31.

Woodward, R., Stevens, E. B., and Murrell Lagnado, R. D. (1997). Molecular determinants for assembly of G-protein-activated inwardly rectifying K^+ channels. *J. Biol. Chem.* **272,** 10,823–10,830.

Xu, J., Yu, W., Jan, Y. N., Jan, L. Y., and Li, M. (1995). Assembly of voltage-gated potassium channels. Conserved hydrophilic motifs determine subfamily-specific interactions between the alpha-subunits. *J. Biol. Chem.* **270,** 24,761–24,768.

Yamada, M., Isomoto, S., Matsumoto, S., Kondo, C., Shindo, T., Horio, Y., and Kurachi, Y. (1997). Sulphonylurea receptor 2B and Kir6.1 form a sulphonylurea-sensitive but ATP-insensitive K^+ channel. *J. Physiol. London* **499,** 715–720.
Yang, J., Jan, Y. N., and Jan, L. Y. (1995a). Control of rectification and permeation by residues in two distinct domains in an inward rectifier K^+ channel. *Neuron* **14,** 1047–1054.
Yang, J., Jan, Y. N., and Jan, L. Y. (1995b). Determination of the subunit stoichiometry of an inwardly rectifying potassium channel. *Neuron* **15,** 1441–1447.

CHAPTER 10

Inwardly Rectifying Potassium Channels: Mechanisms of Rectification

C. G. Nichols and A. N. Lopatin
Department of Cell Biology and Physiology, Washington University School of Medicine, St. Louis, Missouri 63110

I. POTASSIUM CHANNELS AND INWARD RECTIFICATION: INTRODUCTION AND DEFINITIONS

Rectification refers to a change of conductance with voltage, and the term has been used to indicate both voltage-dependent channel gating and voltage dependence of the open channel current. Inward rectification of potassium channels means that at any given driving force (voltage), the inward flow of K^+ ions is greater than the outward flow for the opposite driving force. Strong inward rectification was first described in skeletal muscle (Katz, 1949) and is very prominent in cardiac myocytes and in glial cells and neurons in the central nervous system (Nakajima *et al.,* 1988; Hestrin, 1987; Newman, 1993; Brismar and Collins, 1989). Rectification of strong inward rectifiers is sufficient to ensure that conductance declines to 0 about 40 mV positive to the potassium reversal potential (Noble, 1965;

1063-5823/99 $30.00

Vandenberg, 1994). The high conductance at negative voltages allows cells to maintain a stable resting potential, but the reduced conductance at positive voltages avoids short-circuiting the action potential. Weak inwardly rectifier ATP-sensitive K^+ (KATP) channels, on the other hand, allow substantial outward current to flow at positive potentials (Ciani and Ribalet, 1988; Noma, 1983; Nichols and Lederer, 1991). Between these two channels types, K channels showing intermediate rectification properties are found throughout the nervous system, many of them being activated by G proteins or other second messenger systems (Kandel and Tauc, 1966; Constanti and Galvan, 1983; Inoue et al., 1988; Williams *et al.,* 1988; Nakajima *et al.,* 1988; Hestrin, 1987; Newman, 1993; Brismar and Collins, 1989).

Hodgkin and Huxley (1952) developed a common nomenclature to describe the opening (activation) of K^+ and Na^+ channels following depolarization, the subsequent closing of the channels (inactivation), the reversal of the activation process following hyperpolarization (deactivation), and the subsequent recovery of availability of channels at negative voltages (recovery from inactivation). In such channels, there is now much evidence to support the hypothesis that activation and deactivation result from the voltage-dependent movement of the highly charged S4 segment within the membrane (Liman *et al.,* 1991; Papazian *et al.,* 1991; Tytgat and Hess, 1992; Tytgat *et al.,* 1993). Rapid inactivation of many voltage-gated K (Kv) channels results from block of the open channel by a cytoplasmic "ball." In some Kv channels, this ball consists of the amino terminus (Hoshi *et al.,* 1990; Zagotta *et al.,* 1990) of the channel protein, so-called N-type inactivation, or in the case of sodium channels, the linker region between Domains III and IV (Stuhmer *et al.,* 1989). In contrast, the voltage dependence of inwardly rectifier (Kir) channels has been described by various, and rather confusing, terminologies. The increase of current that follows hyperpolarization has been referred to as activation, so that reduction of channel current at positive potentials has been described as deactivation (time-dependent "intrinsic" gating). As discussed below, the reduction of channel current at positive potentials results from block of the open channel (either by Mg^{2+} or by polyamines), essentially an analogous process to inactivation of voltage-gated channels. Some Kv channels also show a mild inward rectification resulting from voltage-dependent block by cytoplasmic Mg^{2+} (Forsythe *et al.,* 1992; Lopatin and Nichols, 1994; Rettig *et al.,* 1992), and most inwardly rectifying channels also show some tendency to close at negative voltages. However, the midpoint voltage for such closure is typically around -80 to -100 mV, and the steepness is much less than for deactivation of Kv channels (Koumi *et al.,* 1994; Lopatin *et al.,* 1995; Nichols *et al.,* 1994). Nevertheless, the parallels between the voltage-dependent behavior of Kv and Kir channels suggest that the voltage-dependent behavior of each, although quantitatively different, might arise from fundamentally similar processes in channels that are actually of fundamentally similar structures; that is, both Kir and Kv channels share the "inner core" of the Kir channel (Nichols, 1993). Hence, closure of both Kv and Kir channels at positive potentials results from pore block by internal cations or inactivating particles. Both Kv and Kir channels deactivate at negative potentials. Although Kir channels do not contain a membrane-spanning S4 region, denoted M0 by Ho *et al.* (1993), with some homology and containing repeated charged residues. Several mutations in S4 were shown to cause negative shifts in the voltage dependence of activation of Kv channels. In confirmation of the above unifying hypothesis, Miller and Aldrich (1996) demonstrated that, by making all of these mutations within one Kv1 subunit, activation was shifted from $V_{1/2} = -42.5$ to -221.9 mV, similar to that observed with real inwardly rectifier channels. Interestingly, these mutations, by shifting activation so far in the negative direction, also uncoupled N-type inactivation from activation and demonstrated the intrinsic voltage dependence of inactivation. Under these conditions, inactivation was steeply voltage dependent, and the overall behavior of this Kv channel was then indistinguishable from that of a strong inward rectifier.

Inward rectification of potassium channels was first recognized by Bernard Katz in 1949. Twenty years later, Clay Armstrong (1969) first suggested that inward rectification might result from voltage-dependent block by an intracellular cation. Another 20 years later, two groups (Matsuda *et al.*, 1987; Vandenberg, 1987) demonstrated that intracellular Mg^{2+} ions were indeed capable of causing inward rectification by just such a mechanism. In the past 5 years, inward rectifier K^+ channel subunits have been cloned and expressed at high levels in recombinant systems. This has led to the realization that intracellular polyamines are in fact major determinants of inward rectification (Lopatin *et al.*, 1994, 1995; Ficker *et al.*, 1994; Fakler *et al.*, 1994, 1995) also acting as cytoplasmic blocking particles. This chapter discusses the mechanism of inward rectification and the structural basis for the phenomenon. The reader is referred to articles by Breitwieser (1991), Carmeliet (1993), Doupnik *et al.* (1995), Kubo (1994), Kurachi *et al.* (1992), Matsuda (1991), Nichols (1993), Nichols and Lopatin (1997), and Vandenberg (1994) for comprehensive reviews of different aspects of inwardly rectifier K channel physiology.

II. THE MECHANISM OF INWARD RECTIFICATION

A. *Pore Block by Internal Cations as a Causal Mechanism of Inward Rectification*

Based on the similarity of pore block by tetraethylammonium ions to inward rectification, Clay Armstrong (1969) suggested that inward rectification might result from a voltage-dependent block of the channel pore by cytoplasmic cations. Mg^{2+} and Na^+ ions were subsequently shown to cause inward rectification of weakly inwardly rectifying K_{ATP} channels (Ciani and Ribalet, 1988; Horie *et al.,* 1987), and internal Mg^{2+} ions were shown to be capable of causing some inward rectification of cardiac ik_1 channels. However, a seemingly intrinsic voltage dependence of the conductance was also clearly a dominant cause of inward rectification in these and other strong inward rectifier channels (Kelly *et al.,* 1992; Kurachi, 1985; Matsuda, 1991; Matsuda *et al.*, 1987, 1989; Oliva *et al.,* 1990; Silver and DeCoursey, 1990; Vandenberg, 1987). For both Mg^{2+}-induced, and intrinsic rectification, a strong dependence on external $[K^+]$ (K_0) was demonstrated; increasing K_0 relieves the rectification. For Mg^{2+}-induced rectification, this effect is explained by K^+ ion binding at external sites and knocking-off Mg^{2+} from sites deeper inside a multi-ion pore (Armstrong, 1971; Hille and Schwartz, 1978; Horie *et al.,* 1987; Yellen, 1984). Intrinsic rectification was then phenomenologically modeled as the opening and closing of an activation gate with transition rates that are also dependent on K_0 (Ciani *et al.,* 1988; Cleemann and Morad, 1979; Ishihara *et al.,* 1989; Oliva *et al.,* 1990; Stanfield *et al.,* 1994a).

B. Polyamines Induce Intrinsic Rectification

An intriguing observation made by Matsuda in 1988 was that intrinsic rectification of cardiac inward rectifier K^+ channels gradually disappears with time when a membrane patch is excised from cells and exposed to bath solution in the inside out configuration. Following the cloning of strong inward rectifier K^+ channel genes (Kir2.x gene family members; Kubo *et al,* 1993), it was possible to observe high levels of expressed inward rectifier currents. In macropatch experiments on Kir2.3 channel expressed in *Xenopus* oocytes, we observed that rectification disappeared when patches were isolated (Lopatin *et al.,* 1994), but was restored when we moved the patch back toward the oocyte (Fig. 1a)! This led us to the realization that rectification disappeared because some factor, or factors, were being lost from the oocyte interior, and that these intrinsic rectifying factors were actually being released from intact oocytes. Solutions could actually be conditioned by exposure to intact oocytes (Fig. 1b). We made some rudimentary biochemical characterization of the intrinsic rectifying factors in solution conditioned by oocytes (Figs. 1c,1d), sufficient to indicate that these factors are polyamines (spermine, spermidine, putrescine), metabolites of amino acids found in almost all cells (Tabor and Tabor, 1984). Application of these polyamines to inside out patches containing Kir2.x channels restores all the essential features of intrinsic rectification (Fig. 2). Less potent than

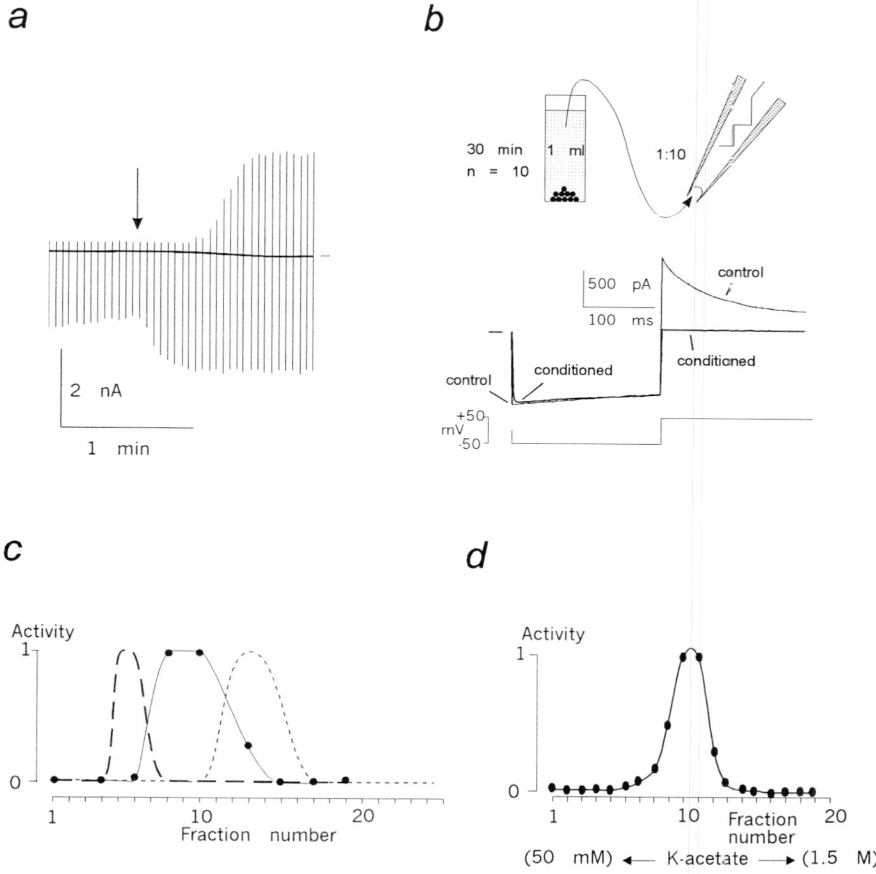

FIGURE 1 (a) Inward rectification disappears in inside out membrane patches. A slow time base current record in response to voltage steps from 0 to −50 and then to +50 mV before and following excision (arrow) of a patch expressing cloned strong inwardly rectifier Kir2.3 channels. (b) Oocyte conditioned solutions restore rectification. Strong inwardly rectifier Kir2.3 currents (inside out patch) in response to voltage steps from −50 to +50 mV after patch isolation (large outward currents, loss of rectification, control) and after subsequent exposure to diluted, oocyte-conditioned solution (rectification restored, conditioned). IRFs are small and positively charged. (c) Elution profile of dextran blue (dashed), potassium (dotted), and biological IRF activity (solid) from a Sephadex G-10 column, after application of conditioned solution. Partial retention of IRF activity indicates a MW in the range 100–1000. (d) Elution profile of IRF bioactivity from S-Sepharose, using a linear K-acetate gradient. Biological activity eluted at about 500 m*M* [K^+], consistent with a strongly positively charged compound. IRF activity did not bind to the column above pH 10, indicating that the charges are titratable amines.

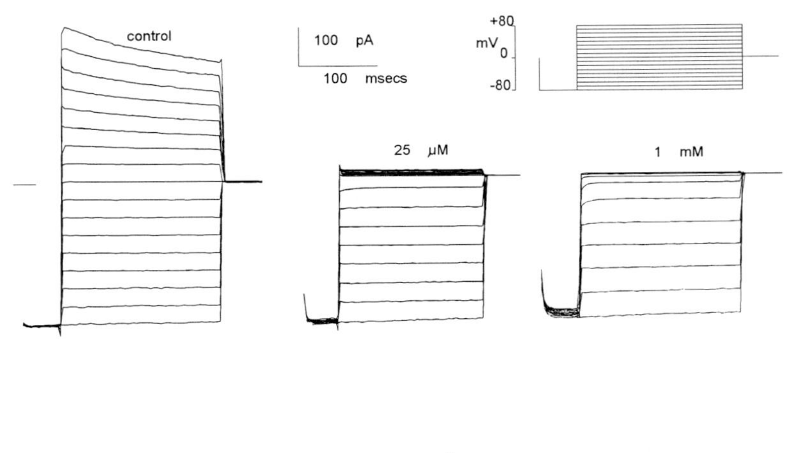

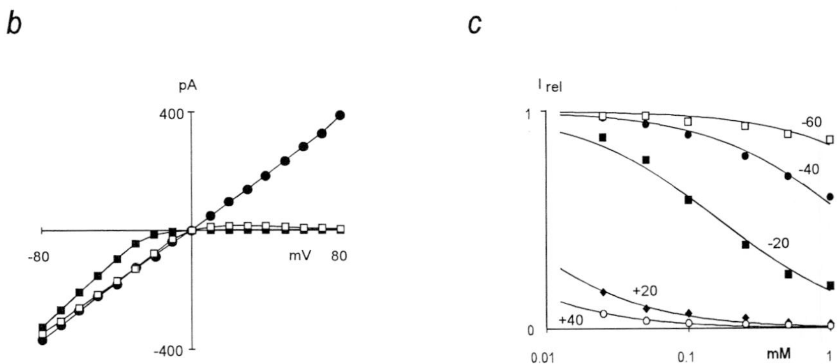

FIGURE 2 Spermidine causes strong inward rectification. (a) Kir2.3 currents in response to voltage steps between −80 and +80 mV, 3 min after patch isolation (left) and in the presence of 25 μM (center and 1 m*M* spermidine (right). (b) Currents at the end of a 20-msec test pulse voltage after patch isolation (circles) and in the presence of 25 μM (open squares) and 1 m*M* (closed squares) spermidine. (c) Current in spermidine relative to the current in control at different voltages (relative to E_K = 0 mV). At E_K, half maximal block would occur at about 30 μM spermidine.

spermine and spermidine, putrescine and cadaverine also cause rectification with similar efficacy to Mg^{2+}, but closely related molecules that contain negative charges or bulky substituents are without effects (Fig. 3a). The voltage dependence of spermine and spermidine block are steeper than Mg^{2+} block (Fakler *et al.,* 1994; Ficker *et al.,* 1994; Lopatin *et al.,* 1994, 1995), which explains why inward rectification in endogenous cells is steeper

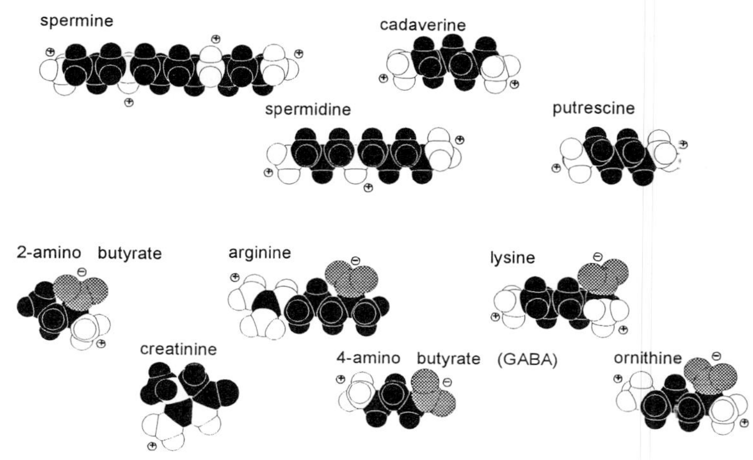

b

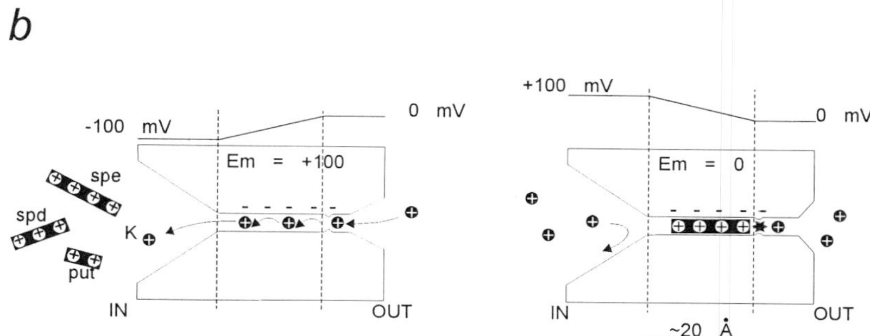

FIGURE 3 (a) Structures of natural polyamines and related metabolites. The linear polyamines (spermine, spermidine, putrescine, cadaverine) all cause inward rectification. The other compounds, which all contain carboxyl side chains, or a ring structure (creatinine) are inactive. Amines are white, oxygen atoms are shaded, methyl groups are black (b) Long-pore plugging as a model for steep inward rectification. As the membrane is depolarized, polyamines are hypothesized to enter deeply into the membrane voltage field and to associate with multiple potassium-binding sites. External K^+ ions relieve rectification by electrostatic repulsion of polyamines.

than that produced by Mg^{2+} ions (Hille, 1992). The voltage dependence of spermine and spermidine unblock rates match the rate constants of channel activation in cell-attached patches (Lopatin *et al.,* 1995). Kir1.1 (ROMK

1) channels, Kir4.1 channels, Kir6.2 (KATP) channels, and delayed rectifier Kv2.1 (DRK1) channels all show only weak inward rectification. In contrast to Kir2.x channels, they are blocked by millimolar concentrations of Mg^{2+} and polyamines (Fakler *et al.,* 1994; Lopatin *et al.,* 1994; Lu and Mackinnon, 1994; Nichols *et al.,* 1994; Shyng *et al.,* 1997), and the block is only weakly voltage dependent. Mutational analyses (see below) further suggest that Mg^{2+} and polyamines may in fact share the same binding sites within Kir channels. This can explain the apparently anomalous effect that Mg^{2+} transiently increases currents in the presence of intrinsic gating induced by polyamines (Ishihara, 1997; Yamashita *et al.,* 1996).

Since the initial discovery of polyamine-induced rectification, polyamines have also been demonstrated to cause strong inward rectification in cardiac iK,ACh channels (Yamada and Kurachi, 1995), and in AMPA/kainate receptors (Bowie and Mayer, 1995; Donevan and Rogawski, 1995; Isa *et al.,* 1995; Kamboj*et al,* 1995; Koh *et al.,* 1995). Block of AMPA/kainate receptor channels is not as potent as the block of Kir channels and is relieved under strong depolarization (Bowie and Mayer, 1995), consistent with polyamines actually permeating the channel. However, there is little or no relief of polyamine block of Kir channels with depolarization (Lopatin *et al.,* 1994), and spermidine does not permeate Kir2.1 channels (Sha *et al.*, 1996).

C. "Long-Pore Plugging" Mechanism of Polyamine Action

The steepness of the voltage dependence of channel block by polyamines increases as the charge on the polyamine increases (Lopatin *et al.,* 1994, 1995; Ficker *et al.,* 1994), and mutations that alter Mg^{2+} block sensitivity also alter polyamine blocking affinity (Fakler *et al.,* 1994; Yang *et al.,* 1995). External potassium ions substantially relieve rectification, by increasing the polyamine off rate (Lopatin and Nichols, 1996), as expected for a channel blocker that interacts with permeant ions within the pore. Much evidence has accumulated to suggest that potassium channels are actually long pores, with binding sites for several K^+ ions within them (Hille and Schwartz, 1978; Hodgkin and Keynes, 1955). Mg^{2+} ions are spherical charges, with diameters similar to K^+ ions, and it is reasonable to suggest that these ions occupy K^+ ion-binding sites within the pore in order to block the channel. On the other hand, spermine is a very long (almost 20 Å long) and thin molecule (diameter ~3 Å), with spatially distributed positive charges. It is a tantalizing possibility that in blocking Kir channels, spermine lies in the long pore, each charge associating with a different site that would otherwise be occupied by K^+ ions (Lopatin *et al.,* 1995). Yang *et al.* (1995) examined steady-state polyamine block of Kir2.1 channels over a wide concentration range, and their data suggest that at least two polyamines bind within the channel, with different affinities. We initially proposed two concentration-dependent binding reactions (i.e., two polyamines independently entering the channel pore) and a voltage-dependent transition deep within the voltage field (that may reflect repulsion between the two polyamines) to account for the voltage and concentration dependence of spermine-induced rectification that we measured (Lopatin *et al.,* 1995). One reason for proposing that two polyamine molecules could reside in the pore is that the apparent charge movement in steady-state spermine block of Kir2.3 (~5.3; Lopatin *et al.,* 1995) is greater than that which would result from movement of one spermine molecule through the whole voltage field. However, this model ignores possible interactions of the blocking particle with permeant ions and hence may underestimate the voltage dependence of block (Ruppersberg *et al.,* 1994). If the channel was blocked by only one spermine molecule, but the entering spermine molecule had to sweep out permeant ions to reach its binding site, excess charge movement could result (Ruppersberg *et al.,* 1994). In examining channel block by a series of diaminoalkanes (putrescine analogs with varying alkyl chain length) we find that the apparent charge associated with block increases as the alkyl chain length increases (Pearson and Nichols, 1998). This is not likely to result from more blocking particles entering the pore; rather this is consistent with more K^+ ions being displaced to the outside as the length of the blocking molecule increases (Fig. 3).

It should be pointed out that other mechanisms of intrinsic rectification have been proposed, either in addition to polyamine-induced rectification or questioning the concept of a simple pore block by polyamines and instead suggesting that polyamines are modifiers of a truly intrinsic gate (Aleksandrov *et al.,* 1996; Shieh *et al.,* 1996; Lee *et al.,* 1997). It has been suggested that there may be an important intrinsic gate, since residual rectification remaining after substantial washout of polyamines from the membrane patch is increased by changing the bath pH from 7 to 9 (Shieh *et al.,* 1996; Lee *et al.,* 1997). Although this residual weak rectification might be attributed to block by traces of polyamines (particularly spermine, which washes out very slowly, τ approximately many seconds to minutes), the pH dependence reported by Shieh *et al.* (1996) is opposite that expected from simple consideration of the effects of pH on polyamine charge (i.e., alkaline pH will tend to deprotonate the amines and hence reduce block). This pH dependence is also opposite that which we reported with Kir2.3 channels (Lopatin *et al.,* 1994). At pH 9, polyamines will remain almost fully protonated in free solution. We cannot state, however, what such a pH change will do to the charge distribution on unknown sites within the pore, and an alternative possibility is that elevated pH titrates pore residues involved with polyamine binding and reduces binding affinity. Indeed, Lee and coworkers (1997) report that mutation of the two cysteine residues in M1 (C90S and C102S) removes the pH-dependent effect on residual rectification. In free solution, the pK of cysteine is 8.5, so that significant deprotonation of cysteine will occur at pH 9, leaving a negatively charged $-CH_2-S(-)$ side chain. If the M1 segment contributes to the lining of the channel inner vestibule, then this increase in electronegativity resulting from cysteine titration might lead to enhanced polyamine binding.

III. THE STRUCTURE OF THE KIR CHANNEL PORE: BINDING SITES FOR POLYAMINES

Both voltage-gated and inwardly rectifying K^+ channels contain an extracellular loop (H5 or P loop) with a –GYG– or –GFG– sequence that forms the K^- selectivity filter (Hartmann *et al.*, 1991). In voltage-gated Kv channels, the S6 transmembrane region that follows the P loop has also been shown to line the channel pore (Aiyar *et al.*, 1994), and Stanfield and colleagues (1994b) demonstrated that a glutamate residue in the second transmembrane domain M2 (which corresponds to S6 in Kv channels) of Kir2.1 (IRK1) is at least partially responsible for Mg^{2+} block. Neutralization of this residue reduced both Mg^{2+} block and intrinsic rectification (Wible *et al.*, 1994), as well as reducing external Cs^+ and Ba^{2+} blocking affinity (Henry *et al.*, 1995). Subsequent experiments showed that this residue also determines polyamine blocking affinity (Fakler *et al.*, 1994; Lopatin *et al.*, 1994; Yang *et al.*, 1995), and when neutralized in the otherwise strong inwardly rectifier Kir4.1 (BIR10) it causes a reduction in Mg^{2+} and spermine affinity of ~5 orders of magnitude (Fakler *et al.*, 1994). In Kir1.1, the equivalent residue (asparagine) is neutral, and expressed Kir1.1 currents exhibit much shallower rectification and Mg^{2+} blocking affinity (Lu and Mackinnon, 1994; Nichols *et al.*, 1994). Lu and Mackinnon (1995) showed that by replacing this residue with a positively charged lysine, channels are permanently rectified and rectification is shallow and polyamine insensitive. A histidine residue at this site also leads to permanent rectification at low internal pH (Lu and Mackinnon, 1995). The rectification is titrated at higher pH, as the histidine residue is neutralized, but is insensitive to external pH. These results indicate that internal, but not external, protons have free access to this site, consistent with the idea that a tight selectivity filter, formed by the H5 region, exists at the outer mouth of Kir channels and blocks access of ions other than K^+ to the long inner vestibule. Studies with chimeras between weakly rectifying Kir1.1 (ROMK1) and strongly rectifying Kir2.1 (IRK1) indicated that the C-terminal region, beyond M2, might contain necessary structural elements for strong inward rectification and high-affinity Mg^{2+} block (Pessia *et al.,* 1995; Taglialatela *et al.,* 1994). Yang and co-workers (1995) further demonstrated that E224 (in the C terminus of Kir2.1) is a major determinant of both Mg^{2+} and polyamine sensitivity, and that dual neutralization of the negative charges in M2 (D172N) and in the C terminus (E224G) reduced polyamine and Mg^{2+} sensitivity almost to that of Kir1.1 (ROMK1).

In the original reports of polyamine-induced rectification, we (Lopatin *et al.,* 1994, 1995) and Ficker and colleagues (1994) reported much slower unbinding of spermine and spermidine than Mg^{2+} from Kir2.1 and Kir2.3 channels. Using a nearly identical Kir2.1 clone, Fakler and co-workers (1994) reported the opposite result: unbinding of Mg^{2+} ions was much slower than spermidine or spermine. The clone used by Fakler *et al.* (1994) contained a threonine residue at position 84 immediately preceding the M1 transmembrane domain. Ruppersberg and colleagues (1996) subsequently demonstrated that both absolute and relative off rates of different polyamines and Mg^{2+} from the channel depend critically on the residue at this position, suggesting that this region contributes to forming the internal entrance to the pore, in analogy to the demonstration of a role for the corresponding S4–S5 linker region in internal Mg^{2+} block of *Shaker* channels (Slesinger *et al.,* 1993).

Thus the Kir channel structural features are being delineated. Rather than H5 looping deeply into the membrane, the weight of evidence, now supported by physical data (Schwalbe *et al.,* 1996), suggests that the H5 regions of the channel form more of a shallow disk at the outer surface. The inner vestibule is formed by at least residues at the cytoplasmic end of M1, residues in M2, and residues in the C-terminal region and must be physically large enough to contain at least one and possibly two, molecules of spermine simultaneously.

IV. THE PHYSIOLOGICAL SIGNIFICANCE OF POLYAMINE-INDUCED RECTIFICATION

Polyamines have been a subject of interest as cellular metabolites since they were first discovered by van Leeuwenhoek (1678). They are essential for normal and neoplastic cell growth and may have a role as stabilizing moieties for DNA (Tabor and Tabor, 1984), but other cellular functions remain undefined. Nanomolar to micromolar concentrations of free polyamines would be required to reproduce the degree of rectification seen in intact cells (Bowie and Mayer, 1995; Fackler *et al.,* 1995; Ficker *et al.,*

1994), and induction of inward rectification may well be the most potent physiological property of polyamines. Although cellular levels are strongly buffered, total cellular polyamine concentrations (10–10,000 μM; Seiler, 1994) are clearly in excess of those required to cause very strong rectification of Kir channels. Given the steep voltage dependence of polyamine block, it is likely that cytoplasmic concentrations will always be within the range necessary to dynamically affect rectification over the physiological voltage range, such that alterations of polyamine levels will change rectification. Fakler and colleagues (1995) demonstrated that inclusion of ATP in the whole-cell patch-clamp pipet could relieve inward rectification of Kir2.1 channels, and this is consistent with a partial chelation of free polyamines. Bianchi and co-workers (1996) demonstrated relief of inward rectification of endogenous currents in RBL-1 cells after treatment with an inhibitor of the polyamine synthetic enzyme *S*-adenosylmethionine decarboxylase. The treatment resulted in an increase in cellular putrescine and a decrease in spermidine and spermine levels, with a shallowing of the *I–V* relationship and a significant increase in outward currents. We reported similar results in *Xenopus* oocytes expressing Kir2.1 channels (Shyng *et al.,* 1996), and also utilized a Chinese hamster ovary cell line that is deficient in ornithine decarboxylase activity (Steglich and Scheffler, 1982) and requires putrescine in the medium for normal cell growth, in order to demonstrate the effects of polyamine depletion on the rectification of expressed Kir2.3 channels. In these cells, removal of putrescine leads to gradual decline in intracellular levels of putrescine, then spermidine, and finally spermine (Steglich and Scheffler, 1982), and these changes correlate with alterations in Kir2.3 kinetics predicted by excised-patch experiments (Lopatin *et al.,* 1995). The effects of altered polyamine levels on inward rectification and excitability in intact tissues remain largely unexplored, but it is an exciting possibility that physiological regulation of excitability might occur through changes in polyamine levels (Nichols *et al.,* 1996).

V. CONCLUSIONS

Exogenous expression of cloned Kir channels has permitted determination of the molecular basis of inward rectification. The realization that intracellular polyamines are primarily responsible now begs multiple questions. From a biophysical perspective, the availability of high-affinity ligands (polyamines) for the inner vestibule, with a very large possible number of structural variants, should permit a detailed probing of the inner vestibule of the Kir channel and analysis of the residues involved in lining the channel pore. Preliminary experiments clearly demonstrate that modulation of polyamine levels, and consequent alteration of rectification, is possible. However, from a physiological and clinical perspective, the possibility that physiological or pathophysiological changes of polyamine levels might affect cellular excitability by altering Kir channel rectification remains essentially unexplored.

Acknowledgments

Our own experimental work has been supported by the NIH (Grants HL45742 and HL54171 to C.G.N.), the American Heart Association (Missouri Affiliate Fellowship to A.N.L; Established Investigatorship to CGN), and Juvenile Diabetes Foundation (Grant-In-Aid to C.G.N.).

References

Aiyar, J., Nguyen, A. N., Chandy, K. G., and Grissmer, S. (1994). The P-region and S6 of Kv3.1 contribute to the formation of the ion conduction pathway. *Biophys. J.* **67,** 2261–2264.

Aleksandrov, A., Velimirovic, B., and Clapham, D. E. (1996). Inward rectification of the IRK1 K^+ channel reconstituted in lipid bilayers. *Biophys. J.* **70,** 2680–2687.

Armstrong, C. M. (1969). Inactivation of the potassium conductance and related phenomena caused by quaternary ammonium ion injected in squid axons. *J. Gen. Physiol.* **54,** 553–575.

Armstrong, C. M. (1971). Interaction of tetraethylammonium ion derivatives with the potassium channel of giant axons. *J. Gen. Physiol.* **58,** 413–437.

Bianchi, L., Roy, M. L., Taglialatela, M., Lundgren, D. W., Brown, A. M., and Ficker, E. (1996). Regulation by spermine of native inward rectifier K^+ channels in RBL-1 cells. *J. Biol. Chem.* **271,** 6114–6121.

Bowie, D., and Mayer, M. L. (1995). Inward rectification of both AMPA and kainate subtype glutamate receptors generated by polyamine-mediated ion channel block. *Neuron* **15,** 453–462.

Breitwieser, G. E. (1991). G protein-mediated ion channel activation. *Hypertension* **17,** 684–692.

Brismar, T., and Collins, V. P. (1989). Inwardly rectifying potassium channels in human malignant glioma cells. *Brain Res.* **480,** 249–258.

Caldarera, C. M., Orlandini, G. Casti, A., and Moruzzi, G. J. (1974). Polyamine and nucleic acid metabolism in myocardial hypertrophy of the overloaded heart. *Mol. Cell. Cardiol.* **6,** 95–104.

Carmeliet, E. (1993). K^+ channels and control of ventricular repolarization in the heart. *Fund. Clin. Pharmacol.* **7,** 19–28.

Ciani, S., Krasne, S., Miyazaki, S., and Hagiwara, S. (1988). A model for anomalous rectification: Electrochemical-potential-dependent gating of membrane channels. *J. Membr. Biol.* **44,** 103–134.

Ciani, S., and Ribalet, B. (1988). Ion permeation and rectification in ATP-sensitive channels from insulin-secreting cells (RINm5F): Effects of K^+, Na^+ and Mg^{2+}. *J. Membr. Biol.* **103,** 171–180.

Cleemann, L., and Morad, M. (1979). Potassium currents in frog ventricular muscle: Evidence from voltage clamp currents and extracellular K accumulation. *J. Physiol.* **286,** 113–143.

Constanti, A., and Galvan, M. (1983). Fast inward-rectifying current accounts for anomalous rectification in olfactory cortex neurones. *J. Physiol.* **335,** 153–178.

Donevan, S. D., and Rogawski, M. A. (1995). Intracellular polyamines mediate inward rectification of Ca(2+)-permeable alpha-amino-3-hydroxy-5-methyl-4-isoxazolepropionic acid receptors. *Proc. Nat. Acad. Sci. USA* **92,** 9298–9302.

Doupnik, C. A., Davidson, N., and Lester, H. A. (1995). The inward rectifier potassium channel family. *Curr. Opin. Neurobiol.* **5,** 268–277.

Fakler, B., Brandle, U., Bond, C., Glowatzki, E., Konig, C., Adelman, J. P., *et al.* (1994). A structural determinant of differential sensitivity of cloned inward rectifier K^+ channels to intracellular spermine. *FEBS Lett.* **356,** 199–203.

Fakler, B., Brandle, U., Glowatzki, E., Weidemann, S., Zenner, H. P., and Ruppersberg, J. P. (1995). Strong voltage-dependent inward rectification of inward rectifier K^+ channels is caused by intracellular spermine. *Cell* **80,** 149–154.

Feuerstein, B. G., Szollozi, J., Basu, H. S., and Marton, L. J. (1992). alpha-Difluoromethylornithine alters calcium signaling in platelet-derived growth factor-stimulated A172 brain tumor cells in culture. *Cancer Res.* **52,** 6782–6789.

Ficker, E., Taglialatela, M., Wible, B. A. *et al.* (1994). Spermine and spermidine as gating molecules for inward rectifier K channels. *Science* **266,** 1068–1072.

Forsythe, I. D., Linsdell, P., and Stanfield, P. R. (1992). Unitary A-currents of rat locus coeruleus neurones grown in cell culture: Rectification caused by internal Mg^{2+} and Na^+. *J. Physiol.* **451,** 553–583.

Hartmann, H. A., Kirsch, G. E., Drewe, J. A., Taglialatela, M., Joho, R. H., and Brown, A. M. (1991). Exchange of conduction pathways between two related K^+ channels. *Science* **251,** 942–944.

Hayashi, Y., Hattori, Y., Moriwaki, A., Lu, Y. F., and Hori, Y. (1993). Increases in brain polyamine concentrations in chemical kindling and single convulsion induced by pentylenetetrazol in rats. *Neurosci. Lett.* **149,** 63–66.

Henry, P., Lopatin, A. N., Makhina, E. N., and Nichols, C. G. (1995). A negative charge in M2 resulates sensitivity of inward rectifier K channel to external cation block. *Biophys. J.* **68,** A264.

Hestrin, S. (1987). The properties and function of inward rectification in rod photoreceptors of the tiger salamander. *J. Physiol.* **390,** 319–333.

Hille, B. (1992). "Ionic Channels of Excitable Membranes." Sinauer Associates, Sunderland, MA.

Hille, B. (1978). Potassium channels as multi-ion single-file pores. *J. Gene. Physiol.* **72,** 409–442.

Ho, K., Nichols, C. G., Lederer, W. J., Lytton, J., Vassilev, P. M., Kanazirska, M. V., *et al.* (1993). Cloning and expression of an inwardly rectifying ATP-regulated potassium channel. *Nature* **362,** 31–38.

Hodgkin, A. L., and Huxley, A. M. (1952). Currents carried by sodium and potassium ions through the membrane of the giant axon of *Loligo*. *J. Physiol.* **116,** 449–472.

Hodgkin, A. L., and Keynes, R. D. (1955). The potassium permeability of a giant nerve fibre. *J. Physiol.* **128,** 28–60.

Horie, M., Irisawa, H., and Noma, A. (1987). Voltage-dependent magnesium block of adenosine-triphosphate-sensitive potassium channel in guinea-pig ventricular cells. *J. Physiol.* **387,** 251–272.

Hoshi, T., Zagotta, W. N., and Aldrich, R. W. (1990). Biophysical and molecular mechanisms of Shaker potassium channel inactivation. *Science* **250,** 533–538.

Inoue, M., Nakajima, S., and Nakajima, Y. (1988). Somatostatin induces an inward rectification in rat locus coeruleus neurones through a pertussis toxin-sensitive mechanism. *J. Physiol.* **407,** 177–198.

Isa, T., Iino, M., Itazawa, S., and Ozawa, S. (1995). Spermine mediates inward rectification of Ca(2+)-permeable AMPA receptor channels. *Neuroreport* **6,** 2045–2048.

Ishihara, K. (1997). Time-dependent outward currents through the inward rectifier potassium channel IRK1. The role of weak blocking molecules. *J. Gen. Physiol.* **109,** 229–243.

Ishihara, K., Mitsuiye, A., Noma, A., and Takano, M. (1989). The Mg^{2+} block and intrinsic gating underlying inward rectification of the K^+ current in guinea-pig cardiac myocytes. *J. Physiol.* **419,** 297–320.

Kamboj, S. K., Swanson, G. T., and Cull-Candy, S. G. (1995). Intracellular spermine confers rectification on rat calcium-permeable AMPA and kainate receptors. *J. Physiol.* **486,** 297–303.

Kandel, E., and Tauc, L. (1966). Anomalous rectification in the metacerebral giant cells and its consequences for synaptic transmission. *J. Physiol.* **183,** 287–304.

Katz, B. (1949). Les constantes electriques de la membrane du muscle. *Arch. Sci. Physiol.* **2,** 285–299.

Kelly, M. E., Dixon, S. J., and Sims, S. M. (1992). Inwardly rectifying potassium current in rabbit osteoclasts: A whole-cell and single-channel study. *J. Membr. Biol.* **126,** 171–181.

Koh, D. S., Burnashev, N., and Jonas, P. (1995). Block of native Ca(2+)-permeable AMPA receptors in rat brain by intracellular polyamines generates double rectification. *J. Physiol.* **486,** 305–312.

Koumi, S., Sato, R., and Hayakawa, H. (1994). Modulation of voltage-dependent inactivation of the inwardly rectifying K^+ channel by chloramine-T. *Eur. J. Pharmacol.* **258,** 281–284.

Kubo, Y. (1994). Towards the elucidation of the structural-functional relationship of the inward rectifying K^+ channel family. *Neurosci. Res.* **21,** 109–117.

Kubo, Y., Baldwin, T. J., Jan, Y. N., and Jan, L. Y. (1993). Primary structure and functional expression of a mouse inward rectifier potassium channel. *Nature* **362,** 127–133.

Kurachi, Y. (1985). Voltage-dependent activation of the inward rectifier potassium channel in the ventricular cell membrane of guinea-pig heart. *J. Physiol.* **366,** 365–385.

Kurachi, Y., Tung, R. T., Ito, H., and Nakajima, T. (1992). G protein activation of cardiac muscarinic K^+ channels. *Prog. Neurobiol.* **39,** 229–246.

Laschet, J., Trottier, S., Grisar, T., and Leviel, V. (1992). Polyamine metabolism in epileptic cortex. *Epilep. Res.* **12,** 151–156.

Lee, J. K., John, S. A., Lu, Y., Shieh, R. C., and Weiss, J. N. (1997). The intrinsic gating mechanism is fundamental to inward rectification in the IRK1 channel. *Biophys. J.* **72,** A253.

Liman, E. R., Hess, P., Weaver, F., and Koren, G. (1991). Voltage-sensing residues in the S4 region of a mammalian K channel. *Nature* **353,** 752–756.

Lopatin, A. N., Makhina, E. N., and Nichols, C. G. (1994). Potassium channel block by cytoplasmic polyamines as the mechanism of intrinsic rectification. *Nature* **372,** 366–369.

Lopatin, A. N., Makhina, E. N., and Nichols, C. G. (1995). The mechanism of inward rectification of potassium channels. *J. Gen. Physiol.* **106,** 923–955.

Lopatin, A. N., and Nichols, C. G. (1994). Inward rectification of outward rectifying DRK1 (Kv2.1) potassium channels. *J. Gen. Physiol.* **103,** 203–216.

Lopatin, A. N., and Nichols, C. G. (1996). $[K^+]$-dependence of polyamine induced rectification in inward rectifier potassium channels (IRK1, Kir2.1). *J. Gen. Physiol.* **108,** 105–113.

Lu, Z., and Mackinnon, R. (1994). Electrostatic tuning of Mg^{2+} affinity in an inward rectifier K^+ channel. *Nature* **371,** 243–246.

Lu, Z., and Mackinnon, R. (1995). Probing a potassium channel pore with an engineered protonatable site. *Biochemistry* **34,** 13,133–13,138.

Matsuda, H. (1988). Open-state substructure of inwardly rectifying potassium channels revealed by magnesium block in guinea-pig heart cells. *J. Physiol.* **397,** 237–258.

Matsuda, H. (1991). Magnesium gating of the inwardly rectifying K^+ channel. *Annu. Rev. Physiol.* **53,** 289–298.

Matsuda, H., Matsuura, H., and Noma, A. (1989). Triple-barrel structure of inwardly rectifying K^+ channels revealed by Cs^+ and Rb^+ block in guinea-pig heart cells. *J. Physiol.* **413,** 139–157.

Matsuda, H., Saigusa, A., and Irisawa, H. (1987). Ohmic conductance through the inwardly rectifying K^+ channel and blocking by internal Mg^{2+}. *Nature* **325,** 156–159.
Miller, A. G., and Aldrich, R. W. (1996). Conversion of a delayed rectifier K^+ channel to a voltage-gated inward rectifier K^+ channel by three amino acid substitutions. *Neuron* **16,** 853–858.
Nakajima, Y., Nakajima, S., and Inoue, M. (1988). Pertussis toxin-insensitive G protein mediates substance P-induced inhibition of potassium channels in brain neurons. *Proc. Natl. Acad. Sci. USA* **85,** 3643–3647.
Newman, E. A. (1993). Inward-rectifying potassium channels in retinal glial (Muller) cells. *J. Neurosci.* **13,** 3333–3345.
Nichols, C. G. (1993). The 'inner core' of inward rectifier potassium channels. *Trends Pharmacol. Sci.* **14,** 320–323.
Nichols, C. G., Ho., K., and Hebert, S. (1994). Mg^{2+} dependent inward rectification of ROMK1 potassium channels expressed in Xenopus oocytes. *J. Physiol.* **476,** 399–409.
Nichols, C. G., and Lederer, W. J. (1991). ATP-sensitive potassium channels in the cardiovascular system. *Am. J. Physiol.* **261,** H1675–H1686.
Nichols, C. G., and Lopatin, A. N. (1997). Inward rectifier potassium channels. *Annu. Rev. Physiol.* **59,** 171–191.
Nichols, C. G., Lopatin, A. N., Makhina, E. N., Pearson, W. L., and Sha, Q. (1996). Inward rectification and implications for cardiac excitability. *Circ. Res.* **78,** 1–7.
Noble, D. (1965). Electrical properties of cardiac muscle attributable to inward going (anomalous) rectification. *J. Cell. Comp. Physiol.* **66,** 127–136.
Noma, A. (1983). ATP-regulated K^+ channels in cardiac muscle. *Nature* **305,** 147–148.
Oliva, C., Cohen, I. S., and Pennefather, P. (1990). The mechanism of rectification of I_{K1} in canine Purkinje myocytes. *J. Gen. Physiol.* **96,** 299–318.
Papazian, D. M., Timpe, L. C., Jan, Y. N., and Jan, L. (1991). Alteration of voltage-dependence of Shaker potassium channel by mutations in the S4 sequence. *Nature* **349,** 305–310.
Pearson, E. L., and Nichols, C. G. (1998). Block of Kir 2.1 channels by alkylamine analogues of endogenous polyamines. *J. Gen. Physiol.* In press.
Pessia, M., Bond, C. T., Kavanaugh, M. P., and Adelman, J. P. (1995). Contributions of the C-terminal domain to gating properties of inward rectifier potassium channels. *Neuron* **14,** 1039–1045.
Rettig, J., Wunder, F., Stocker, M., Lichtinghagen, R., Mastiaux, F., Beckh, S., *et al.* (1992). Characterization of a Shaw-related potassium channel family in rat brain. *EMBO J.* **11,** 2473–2486.
Ruppersberg, J. P., Fakler, B., Brandle, U., Zenner, H.-P., and Schultz, J. H. (1996). An N-terminal site controls blocker-release in Kir2.1 channels. *Biophys. J.* **70,** A361.
Ruppersberg, J. P., vanKitzing, E., and Schoepfer, R. (1994). The mechanism of magnesium block of NMDA receptors. *Sem. Neurosci.* **6,** 87–96.
Schwalbe, R. A., Wang, Z., and Brown, A. M. (1996). N-glycosylation studies of ROMK1 reveal unexpected extracellular regions in the pore-forming segment. *Biophys. J.* **70,** A309.
Seiler, N. (1994). Formation, catabolism and properties of the natural polyamines. *In* "The Neuropharmacology of Polyamines." (C. Carter, Ed.), Academic Press, London/New York.
Sha, Q., Romano, C., Lopatin, A. N., and Nichols, C. G. (1996). Spermidine release from Xenopus oocytes: Electrodiffusion through a membrane channel. *J. Biol. Chem.* **271,** 3392–3397.
Shieh, R. C. John, S. A., Lee, J. K., and Weiss, J. N. (1996). Inward rectification of the IRK1 channel expressed in Xenopus oocytes: Effects of intracellular pH reveal an intrinsic gating mechanism. *J. Physiol.* **494,** 363–376.

Shyng, S. L., Ferrigni, T., and Nichols, C. G. (1997). Control of rectification and gating of cloned K_{ATP} channels by the Kir6.2 subunit. *J. Gen. Physiol.* **110,** 141–153.

Shyng, S. L., Sha, Q., Ferrigni, T., Lopatin, A. N.,, and Nichols, C. G. (1996). Depletion of intracellular polyamines relieves inward rectification of potassium channels. *Proc. Natl. Acad. Sci. USA* **93,** 12,014–12,019.

Silver, M. R., and DeCoursey, T. E. (1990). Intrinsic gating of inward rectifier in bovine pulmonary artery endothelial cells in the presence or absence of internal magnesium. *J. Gen. Physiol.* **96,** 109–133.

Slesinger, P. A., Jan, Y. N., and Jan, L. Y. (1993). The S4-S5 loop contributes to the ion-selective pore of potassium channels. *Neuron* **11,** 739–749.

Stanfield, P. R., Davies, N. W., Shelton, P. A., Khan, I. A., Brammar, W. J., Standen, N. B., and Conley, E. C. (1994a). The intrinsic gating of inward rectifier K channels expressed from the murine IRK1 gene depends on voltage, K^+ and Mg^{2+}. *J. Physiol.* **475,** 1–4.

Stanfield, P. R., Davies, N. W., Shelton, P. A., Sutcliffe, M. J., Khan, I. A., Brammar, W. J., *et al.* (1994b). A single aspartate residue is involved in both intrinsic gating and blockage by Mg^{2+} of the inward rectifier, IRK1. *J. Physiol.* **478,** 1–6.

Steglich, C., and Scheffler, I. E. (1982). An ornithine decarboxylase-deficient mutant of Chinese hamster ovary cells. *J. Biol. Chem.* **257,** 4603–4609.

Stuhmer, W., Conti, F., Suzuki, H., Wang, X. D., Noda, M., Yahagi, N., *et al.* (1989). Structural parts involved in activation and inactivation of the sodium channel. *Nature* **339,** 597–603.

Tabor, C. W., and Tabor, H. (1984). Polyamines. *Annu. Rev. Biochem.* **53,** 749–790.

Taglialatela, M., Wible, B. A., Caporoso, R., and Brown, A. M. (1994). Specification of the pore properties by the carboxyl terminus of inwardly rectifying K^+ channels. *Science* **264,** 844–847.

Tytgat, J., and Hess, P. (1992). Evidence for cooperative interactions in potassium channel gating. *Nature* **359,** 420–423.

Tytgat, J., Nakazawa, K., Gross, A., and Hess, P. (1993). Pursuing the voltage sensor of a voltage-gated mammalian potassium channel. *J. Biol. Chem.* **268,** 23,777–23,779.

Vandenberg, C. A. (1987). Inward rectification of a potassium channel in cardiac ventricular cells depends on internal magnesium ions. *Proc. Nat. Acad. USA* **84,** 2560–2566.

Vandenberg, C. A. (1994). Cardiac inward rectifier potassium channel. *In* "Ion Channels in the Cardiovascular System." (P. M. Spooner and A. M. Brown, Eds.) Futura Publishing, New York.

van Leeuwenhoek, A. (1978). Observationes D. Anthonii Leeuwenhoek, de natis e semine genitali animalculis. *Philos. Trans. R. Soc.* **12,** 1040–1043.

Wible, B. A., Taglialatela, M., Ficker, E., and Brown, A. M. (1994). Gating of inwardly rectifying K^+ channels localized to a single negatively charged residue. *Nature* **371,** 246–249.

Williams, J. T., Colmers, W. F., and Pan, Z. Z. (1988). Voltage- and ligand-activated inwardly rectifying currents in dorsal raphe neurons *in vitro*. *J. Neurosci.* **8,** 3499–3506.

Yamada, M., and Kurachi, Y. (1995). Spermine gates inward-rectifying muscarinic but not ATP-sensitive K^+ channels in rabbit atrial myocytes. Intracellular substance-mediated mechanism of inward rectification. *J. Biol. Chem.* **270,** 9289–9294.

Yamashita, T., Horio, Y., Yamada, M., Takahashi, N., Kondo, C., and Kurachi, Y. (1996). Competition between Mg^{2+} and spermine for a cloned IRK2 channel expressed in a human cell line. *J. Physiol.* **493,** 143–156.

Yang, J., Jan, Y. N., and Jan, L. Y. (1995). Control of rectification and permeation by residues in two distinct domains in an inward rectifier K^+ channel. *Neuron* **14,** 1047–1054.

Yallen, G. (1984). Relief of Na^+ block of Ca^{2+} channels by external cations. *J. Gen. Physiol.* **84,** 187–199.

Zagotta, W. N., Hoshi, T., and Aldrich, R. W. (1990). Restoration of inactivation in mutants of Shaker potassium channels by a peptide derived from ShB. *Science* **250,** 568–571.

CHAPTER 11

Structure–Function Relationship of the Inward Rectifier Potassium Channel

Yoshihiro Kubo and Liang Guo
Department of Neurophysiology, Tokyo Metropolitan Institute for Neuroscience, Tokyo 183, Japan

I. INTRODUCTION

The existence of inward rectifier K^+ channels has been known since the first report by Katz in 1949. The primary structure was revealed by cDNA cloning to be the two-transmembrane type (Ho *et al.,* 1993; Kubo *et al.,* 1993a). Since then, various members of the same family have been isolated, and structure–function studies have also made a remarkable progress. In this chapter, we review two studies that we performed on the structure–

Current Topics in Membranes, Volume 46

1063-5823/99 $30.00

function of the H5 pore region of the cloned inward rectifier K^+ channel, IRK1 (Kubo, 1996; Guo and Kubo, 1998).

In Section II, we focus on the dependence on external K^+, which is important for consistent inward rectification at various external K^+ concentrations. Because a positively charged amino acid, arginine, is conserved at the external mouth of the pore of all inward rectifying K^+ channels cloned so far, we speculated that this residue is critically important to the interaction between the channel and external K^+. We introduced a point mutation at this site and analyzed the properties of the mutant (Kubo, 1996).

In Section III, we discuss the open–close behaviors at a single-channel recording level. The members of the inwardly rectifying K^+ channel family can be roughly classified into two groups by their open–close kinetics: a fast group and a slow group. As the classification goes along with the presence or absence of a negatively charged amino acid, glutamate, at the center of the H5 region, we postulated that the site determines the speed of the open–close kinetics. We made a point mutant of IRK1 whose glutamine at 140 is mutated to glutamate, and analyzed the single-channel behavior in detail (Guo and Kubo, 1998).

In Section IV, we discuss the similarities and differences of the pore structure of the inward rectifier and voltage-gated K^+ channels.

II. DEPENDENCE ON EXTERNAL K^+: R148Y MUTANT

A. Background: Brief Sketch of Inward Rectification Mechanism

The inward rectifier K^+ channel mainly conducts inward current below the equilibrium potential of K^+ (E_K) and allows little permeation of outward current above E_K. This inward rectification has been reported to be caused both by the block of the outward current by cytoplasmic Mg^{2+} (Matsuda *et al.,* 1987; Vandenberg, 1987; Matsuda, 1988) and by intrinsic channel gating (Matsuda, 1988; Ishihara *et al.,* 1989; Silver and DeCoursey, 1990). Recently it was uncovered that the apparent intrinsic gating is mostly due to a block by cytoplasmic polyamines that is actually extrinsic to the channel (Lopatin *et al.,* 1994; Fakler *et al.,* 1995; Ishihara *et al.,* 1996). Furthermore, negatively charged amino acid residues in the center of the M2 region (Stanfield *et al.,* 1994; Lu and MacKinnon, 1994; Wible *et al.,* 1994) and in the C-terminal hydrophilic domain (Yang *et al.,* 1995) were identified to be involved in the binding of these blockers to the channels. The inward rectifier shows consistent inward rectification at various extracellular K^+ concentrations ($[K^+]_o$), as if it senses the shift from E_K ($E - E_K$).

B. Extracellular K^+ as an Activator of IRK1

Although two important cytoplasmic blockers and their putative binding sites have been identified, several questions regarding the mechanism of inward rectification remain to be elucidated. Why are the voltage dependencies of the channel activity, namely the steady-state conductance–voltage (g–V) relationship and the activation time constant–voltage (τ_{act}–V) relationship, shifted in accordance with E_K when $[K^+]_o$ is changed (Ishihara *et al.*, 1989; Stanfield *et al.*, 1994; Ishihara and Hiraoka, 1994) ? Does this channel really sense $E_m - E_K$? If yes, how does this channel sense it?

One simple explanation for the sensing mechanism of $E_m - E_K$ could be that the cytoplasmic blockers sense the direction of the net flow of K^+; they could be dragged into the channel by the outward flux of K^+ and pushed out by the inward flux (blocking particle model) (Pennefather *et al.*, 1992).

However, the results of experiments on native inward rectifier channels where $[K^+]_i$ was changed suggest that these channels sense a combination of $[K^+]_o$ and E_m rather than $E_m - E_K$. Hagiwara and Yoshii (1979) reported that the conductance curve of the inward rectifier K^+ channel shifts when $[K^+]_o$ is changed. Matsuda (1991) showed that Mg^{2+} blocking depends on $[K^+]_o$ and E_m but not on $[K^+]_i$ or $E_m - E_K$. The cloned inward rectifier K^+ channel IRK1 was also shown to sense a combination of K^+_o and E_m because the τ_{act}–V relationship shifted with a change in $[K^+]_o$ but not in $[K^+]_i$ (Figs. 1A,1B) (Kubo, 1996). Considering this dependence of the channel activity (g–V and τ_{act}–V relationship) on E_m and K^+_o and the loss of the outward current of the inward rectifier K^+ channel in the complete absence of K^+_o, the blocking particle model has to be combined with the K^+-activated K^+ channel model (Ciani *et al.*, 1978; Cohen *et al.*, 1989; Pennefather *et al.*, 1992). The fact that K^+_o and Rb^+_o act as activators of this channel with different potencies and results showing different activating effects of K^+_o and Tl^+_o on IRK1 (Kubo, 1996) support the K^+-activated K^+ channel model.

C. Molecular Basis of the Interaction of Extracellular K^+ with IRK1

Given the importance of channel regulation by extracellular cations, as shown above, the molecular identification of the site where K^+_o interacts with the channel is thought to be critical for further biophysical studies. It is natural to expect that the site is located at the extracellular side of the channel, that it is conserved strongly among the inward rectifying K^+ channel family (Figs. 2A,2B), and that mutation of the site causes changes of both the g–V and the τ_{act}–V relationship that reflect the voltage dependencies of

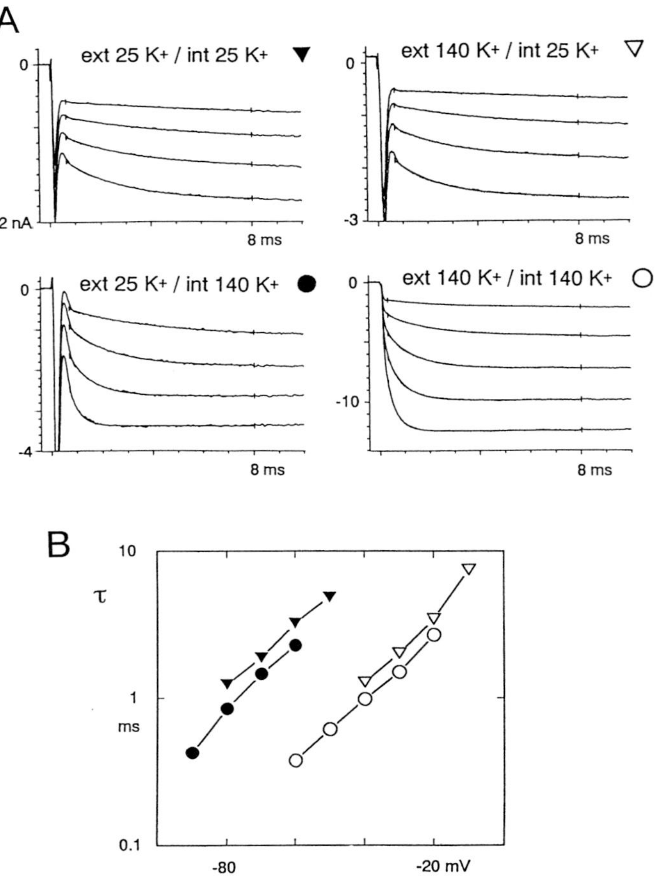

FIGURE 1 Dependence of the IRK1 properties on external K^+. (A) Whole-cell currents recorded from HEK 293 cells stably transfected with the IRK1 channel cDNA. Current traces elicited by applying step pulses from the holding potential of 0 mV to −50, −60, −70, and −80 mV (25/25: $[K^+]_o/[K^+]_i$ in mM); −60, −70, −80, and −90 mV (25/140); −10, −20, −30, and −40 mV (140/25); and −20, −30, −40, −50, and −60 mV (140/140) are shown. The region within the small vertical bars of each trace was fitted by a single exponential, and the fitted traces were overwritten. (B) The activation time constant–voltage relationship of IRK1 obtained from the fittings in A. The $[K^+]_o/[K^+]_i$ (in mM) for each symbol are as shown in A. The voltage to cause *e*-fold change of τ_{act} was 22 mV. When E_K was shifted −43 mV by lowering $[K^+]_o$ from 140 to 25 mM, the plot shifted by −40 mV in experiments with both 25 and 140 mM $[K^+]_i$. Reproduced from Kubo (1996). Effects of extracellular cations and mutations in the pore region of the inward rectifier K^+ channel IRK1. *Recept. Chan.* **4,** 73–83 with permission from Harwood Academic Publishers.

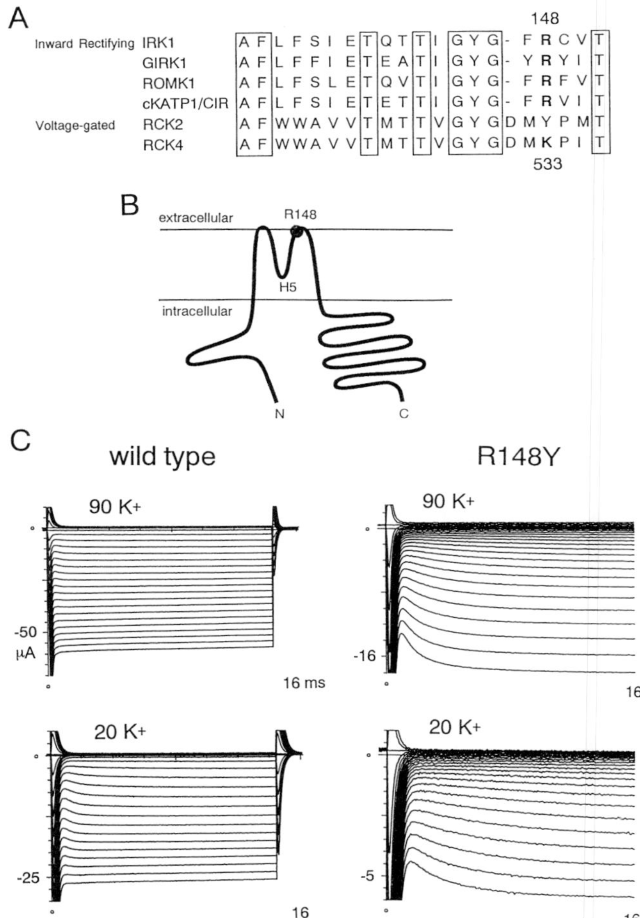

FIGURE 2 Structure and expression of the wild-type IRK1 and the R148Y mutant expressed in *Xenopus* oocytes. (A) Alignment of the H5 regions of several members of the inward rectifier K^+ channel family and of the voltage-gated K^+ channel family. The numbers indicate the residues R148 of IRK1 and K533 of RCK4. (B) A schematic drawing of the structure of the inward rectifier K^+ channels with two transmembrane regions and the location of R148 of the IRK1 channel. (C) Comparison of current traces of wild-type IRK1 and the mutant R148Y in *Xenopus* oocytes recorded under two-electrode voltage clamp. Reproduced from Kubo (1996). Effects of extracellular cations and mutations in the pore region of the inward rectifier K^+ channel IRK1. *Recept. Chan.* **4,** 73–83 with permission from Harwood Academic Publishers.

channel activity. A candidate site that satisfies these criteria was found at the extracellular end of the H5 region (R148).

The only functional mutation at that site, R148Y, caused a left-shift of the g–V and the τ_{act}–V relationships (slower activation upon hyperpolarization) and a reduction of the steepness of the voltage dependencies (Figs. 2C,3A,3B). This resulted in an apparent lack of saturation of the conductance (in K^+_o) and a diminished current amplitude in K^+_o compared with the wild type (Figs. 3A,3B). In Tl^+ solutions, the channel activation changes were qualitatively similar, but quantitatively less prominent. Given these

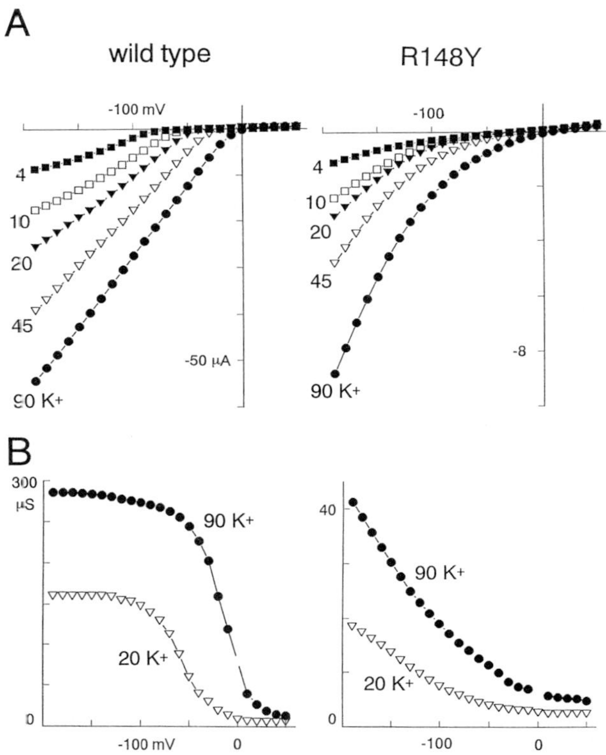

FIGURE 3 Current–voltage relationships and chord conductance–voltage relationships of the wild type and the R148Y mutant expressed in *Xenopus* oocytes. (A) Current–voltage plots; (B) chord conductance–voltage plots for the indicated $[K^+]_o$. The normalized g–V plot of the wild type was shifted by −43 mV, when E_K was shifted by −38 mV, by lowering $[K^+]_o$ from 90 to 20 mM. In the case of the R148Y mutant, the shift was not measurable, as it could not be normalized due to the lack of saturation of the conductance. Reproduced from Kubo (1996). Effects of extracellular cations and mutations in the pore region of the inward rectifier K^+ channel IRK1. *Recept. Chan.* **4,** 73–83 with permission from Harwood Academic Publishers.

results, it can be speculated that R148 is a site where K^+ and Tl^+ exert their effects to activate the channel; in the mutant R148Y the ability of K^+ and Tl^+ to activate the channel seems to be attenuated differentially. How is the R148 residue involved in the activation? As it is difficult to imagine a positively charged amino acid residue forming a binding site for a cation, R148 is not likely to be a binding site. One possibility is that the positive charge of R148 functions as an energy barrier to increase the occupancy of K^+ at a binding site located deeper in the channel. The strong conservation of R148 among cloned inward rectifier channels (Fig. 2A), together with the fact that except for R148Y all other amino acid residue substitutions tested were nonfunctional, also indicates that the site is critical for channel function.

D. Mechanistic Link between Extracellular K^+ *Interaction and Channel Opening*

Based on experiments at various $[K^+]_i$ and a similar theoretical approach as Hille and Schwarz (1978), using the three-barrier two-site model, Matsuda (1991) concluded that the assumption of a multiple-ion single-file pore and the existence of an intracellular blocking ion is not sufficient to explain the results. She interpreted the results by including the assumption that the energy for Mg^{2+} binding in the pore is affected by K^+_o. It is conceivable that K^+_o could then not only affect channel block by cytoplasmic Mg^{2+} but also by polyamines. This is supported by a study of IRK1 (Fig. 1), as the activation kinetics, which were reported to reflect a recovery from polyamine block, were also shown to be regulated by K^+_o.

Attempting to accommodate the experimental results, a scheme that illustrates a possible mechanism of how K^+_o influences channel activation was presented (Fig. 4) (Kubo, 1996). In low $[K^+]_o$ (or $[Tl^+]_o$), K^+ (or Tl^+) is not bound to the channel and the channel is blocked by cytoplasmic blockers (Mg^{2+} and polyamines). When $[K^+]_o$ (or $[Tl^+]_o$) is increased, K^+ (or Tl^+) binds to the outer part of the pore in a $[K^+]_o$ (or $[Tl^+]_o$)-dependent manner. The bound K^+ then expels the blockers at the inner part of the pore and leads to the activation of the channel. At more hyperpolarized potentials, the channel is likely in the state of the lower part of the scheme, due to the voltage dependencies of both the block (less intense at hyperpolarized potential) and the binding of K^+ (more intense at hyperpolarized potential). For the R148Y mutant, the observations were interpreted as follows. As the residue R148, which presumably contributes to an energy barrier to stabilize the occupancy of K^+ in the pore, is mutated, the occupancy of K^+ is not stable enough to expel the blockers, and instead, K^+ is

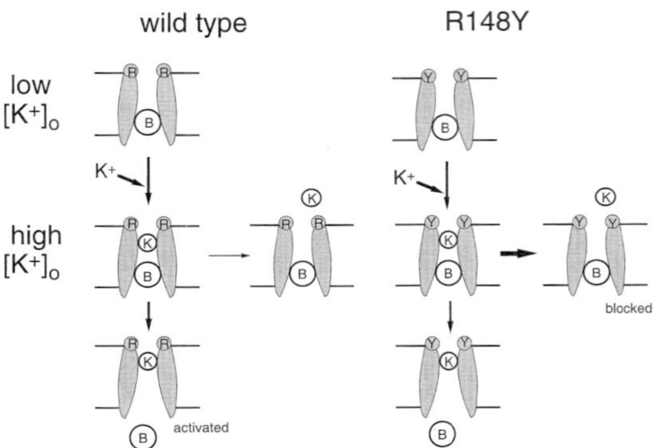

FIGURE 4 Schematic representation of the effect of extracellular K^+ on the activation of the IRK1 channel. The circled K and B represent K^+ ion and the cytoplasmic blockers such as Mg^{2+} or polyamines, respectively. The R and Y in the shaded parts represent the arginine or tyrosine residue at 148 of the IRK1 channel. Reproduced from Kubo (1996). Effects of extracellular cations and mutations in the pore region of the inward rectifier K^+ channel IRK1. *Recept. Chan.* **4,** 73–83 with permission from Harwood Academic Publishers.

expelled by the electrostatic repulsion of the blockers. Thus either a stronger hyperpolarization or higher $[K^+]_o$ is required to achieve the same extent of activation. The results are a negative shift and a less steep voltage dependence of the g–V and the τ_{act}–V plots. In the case of Tl^+, because the stability of the occupancy is not as severely affected by the mutation as that of K^+, the changes caused by the mutation are less prominent.

Moreover, the fact that some rectification remains in the complete absence of Mg^{2+} and polyamines (Fakler *et al.,* 1995) suggests the possibility that the interaction between the channel and K^+_o can also regulate channel activity regardless of the cytoplasmic blockers by an additional gating mechanism (Kubo, 1994). To test this point, it will be informative to compare the rectification properties and the channel activities of the wild type and the R148Y mutant in the complete absence of cytoplasmic Mg^{2+} and polyamines.

III. OPEN–CLOSE KINETICS AT A STEADY STATE: Q140E MUTANT

A. Background

Although all members of the inwardly rectifying K^+ channel family cloned so far share a very high similarity in the H5 region, as illustrated by the boxes in the alignment in Fig. 5A, they display distinct differences in their steady-state single-channel behaviors. Under similar recording conditions, some cloned channels such as ROMK1, IRK1, KAB-2, and sWIRK show a relatively slow open–close kinetics with long open times (Ho *et al.*, 1993; Kubo *et al.*, 1993a, 1996; Takumi *et al.*, 1995), while others such as GIRK1/KGA and CIR show a quite fast open–close kinetics with much shorter open times (Kubo *et al.*, 1993b; Dascal *et al.*, 1993; Krapivinsky *et al.*, 1995). It is likely that the structural difference in the H5 region is responsible for the differences in the open–close kinetics of inwardly rectifying K^+ channels. A comparison of amino acid sequences in the H5 region, showed that, at the center of H5 region (site 140 in IRK1), there is a noncharged glutamine (Q) in channels that show long openings, while there is a negatively charged glutamate (E) in channels that show short openings (Guo and Kubo, 1998). To test the hypothesis that the site (Q/E) at the center of the H5 region is a structural determinant for the gating (open–close kinetics in a steady state at hyperpolarized potentials) of inwardly rectifying K^+ channels, a point mutant of IRK1, Q140E, whose glutamine at 140 was

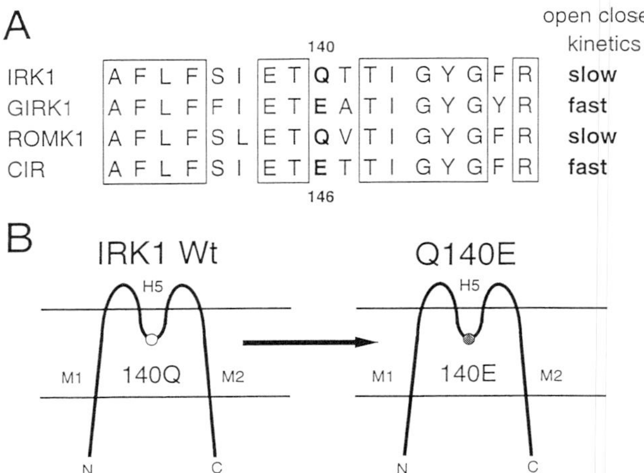

FIGURE 5 Alignment and schematic drawing of the H5 pore region of inwardly rectifying K^+ channels. (A) Alignment of deduced amino acid sequences of the H5 region of some members of the inwardly rectifying K^+ channel family. Bold letters represent the residues at the site equivalent to 140 of IRK1; the numbers indicate the residue Q140 of IRK1 and E146 of CIR. The property of open–close kinetics for each member is shown on the right. (B) The proposed structure with the pore-forming H5 region and two membrane-spanning domains M1 and M2. The location of Q140 and the mutation Q140E is depicted. Reproduced from Guo and Kubo (1998). Comparison of the open–close kinetics of the cloned inward rectifier K^+ channel IRK1 and its point mutant (Q140E) in the pore region. *Recept. Chan.* **5,** 273–289.

mutated to glutamate, was prepared, and the electrophysiological properties were compared with those of wild type (Wt) at both macroscopic and single-channel levels. Our results, which supported the hypothesis (Guo and Kubo, 1998) are reviewed in this chapter.

B. Comparison of Gating of the Single-Channel Current

The macroscopic current and the single-channel conductance of the mutant did not differ from those of the wild type significantly. As illustrated in Fig. 6A, the gating (open–close kinetics at a steady state) of IRK1 Wt was relatively slow. At more hyperpolarized membrane potentials, the open time became shorter and the open probability became lower. This feature was consistent with previous studies on both native inward rectifier K^+ channels of cardiac ventricular cells (Kameyama *et al.,* 1983; Sakmann and Trube, 1984; Kurachi, 1985) and cloned channels expressed in *Xenopus* oocytes (Kubo *et al.,* 1993a). Q140E displayed a faster open–close kinetics than Wt at each membrane potential tested. The most striking feature of Q140E was the frequent appearance of very short closing events, which subsequently caused the shortening of open times.

1. Analysis of the Open–Close Kinetics

Examples of single-channel current recordings and the open (Fig. 7A), and closed (Fig. 7B) time distributions of Wt and Q140E at −80 mV are shown. In both Wt and Q140E, the open time distribution could be well fitted by a single exponential function. The open time constant (τ_o) was 436 ± 37 msec (n = 6) for Wt and 36 ± 5 msec (n = 6) for Q140E, which was over 10-fold shorter than the former. In Wt, the distribution of closed times shorter than 500 msec could be well fitted by a sum of two exponentials with time constants of a slow component ($\tau_{c\text{-sec}}$), 122 msec, and a fast component ($\tau_{c\text{-f}}$), 11.6 msec. In Q140E, the closed time shorter than 500 msec could also be fitted by a sum of two exponentials with a slow component ($\tau_{c\text{-s}}$) of 117 msec and another component ($\tau_{c\text{-complex}}$) of 2.7 msec. This faster time constant was termed $\tau_{c\text{-c}}$, because we found that it actually consists of two separate components by analyzing the initial part of the distribution on an expanded time scale (Fig. 7C). The two components were termed $\tau_{c\text{-ef}}$ of 0.9 msec and $\tau_{c\text{-f}}$ of 12.4 msec. The same diagram of Wt revealed that the incidence was too low to be analyzed (Fig. 7C). In summary, the distribution of total closing events of Q140E actually consisted of three time constants of 1.2 ± 0.2 msec ($\tau_{c\text{-ef}}$), 11.0 ± 1.5 msec ($\tau_{c\text{-f}}$) and 116 ± 14 msec ($\tau_{c\text{-s}}$) (n = 6), while the closed time distribution of Wt had only two time constants of 12.2 ± 2.6 msec ($\tau_{c\text{-f}}$) and 132 ± 24 msec ($\tau_{c\text{-s}}$)

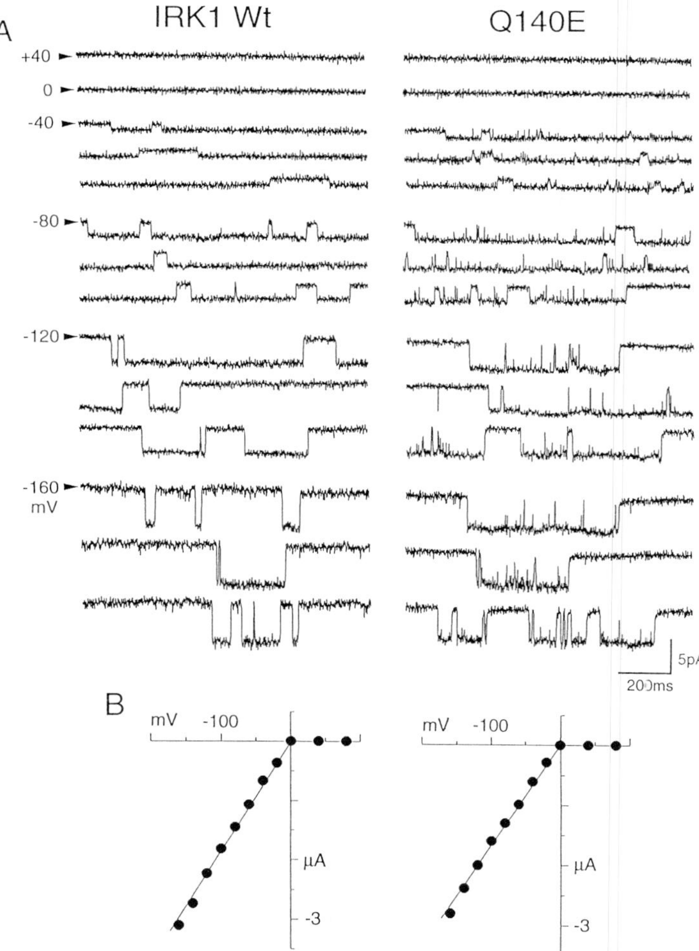

FIGURE 6 Single channel recordings and *I–V* plots of IRK1 Wt and Q140E. (A) Single-channel currents at various membrane potentials from +40 to −160 mV as indicated on the left of traces. Arrowheads on the left indicate the zero current level. (B) Plots of single-channel current–voltage relationship. The estimated single-channel conductance was 22.6 pS for Wt and 21.8 pS for Q140E. Reproduced from Guo and Kubo (1998). Comparison of the open–close kinetics of the cloned inward rectifier K^+ channel IRK1 and its point mutant (Q140E) in the pore region. *Recept. Chan.* **5,** 273–289.

(n = 6). Obviously, the values of $\tau_{c\text{-}s}$ and $\tau_{c\text{-}f}$ of Q140E were in the same range as those values of Wt. Therefore, the most distinct difference of closed times between Wt and Q140E was the occurrence of the frequent extra-fast closure in Q140E.

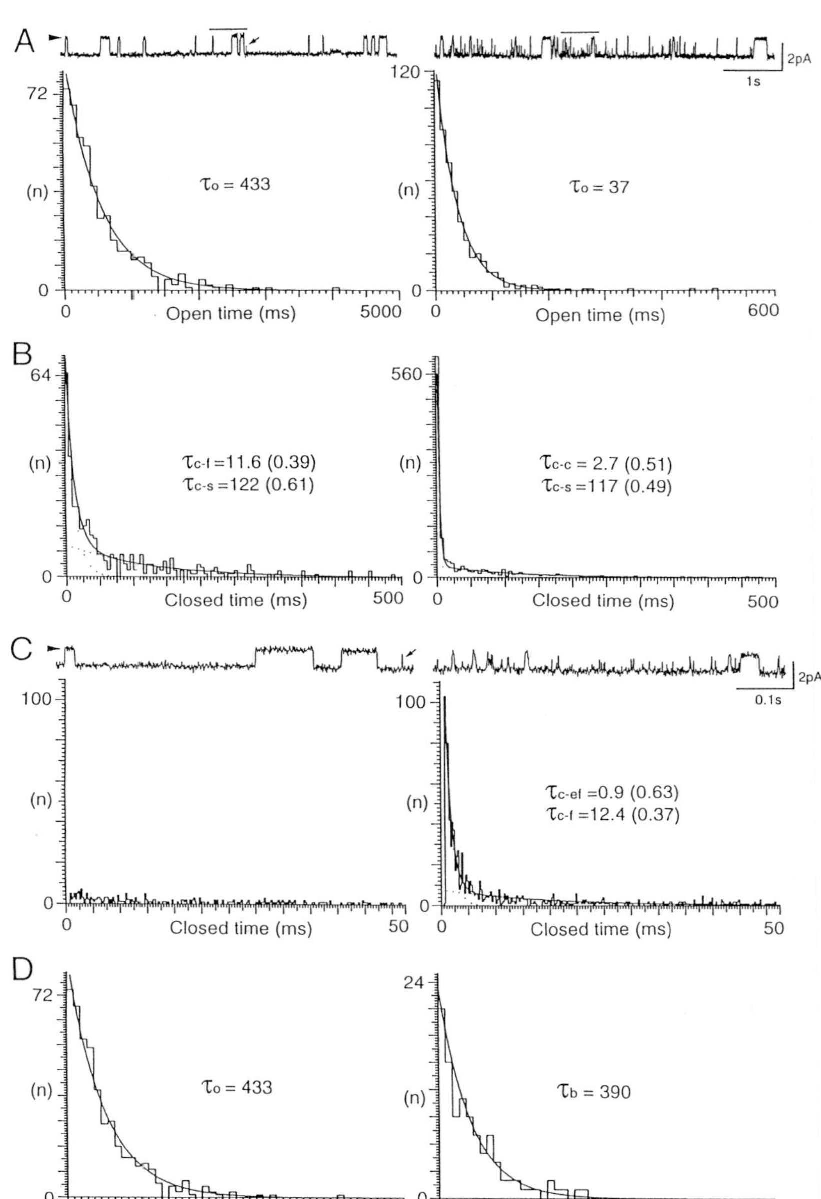
IRK1 Wt
Q140E
A
B
C
D
2pA
1s
0.1s
τo = 433
τo = 37
τc-f =11.6 (0.39)
τc-s =122 (0.61)
τc-c = 2.7 (0.51)
τc-s =117 (0.49)
τc-ef =0.9 (0.63)
τc-f =12.4 (0.37)
τo = 433
τb = 390
(n)
Open time (ms)
Closed time (ms)
Burst duration (ms)
72
120
64
560
100
24
0
5000
600
500
50

2. Burst Analysis

In contrast with IRK1 Wt, whose open–close kinetics did not show a clear burst behavior, the openings of Q140E were grouped as bursts of openings that included the extra-fast closures. The criterion for definition of a burst (τ_{crit}) was obtained by solving the equation proposed by Jackson *et al.* (1983), based on the analysis of the closed time distribution. A burst was defined as a group of openings separated by closures no shorter than τ_{crit} (2.7–4.9 msec in this study).

Figure. 7D shows the example of the burst analysis. A histogram of a burst time, obtained by neglecting the closed events shorter than τ_{crit} (3.2 msec), showed a distribution similar to that of τ_o (Wt) and could be fitted with a single exponential. The time constant of the exponential (τ_b) of 390 msec, was in the same range as τ_o of Wt (433 msec). The mean and SD of τ_b (Q140E) from six patches was 369 $\pm$ 17 msec. In combination with the results of the closed time analysis that, in addition to the extra-fast component, Q140E also had fast and slow components corresponding to those of Wt, this result indicates that gating similar to that of Wt exists in Q140E if the extra-fast closures are ignored. In other words, Q to E mutation introduced the frequent extra-fast closures, which shortened the open time, on top of the gating of Wt.

3. Voltage Dependence of the Open–Close Kinetics

Consistent with the results from previous studies (Kameyama *et al.,* 1983; Sakmann and Trube, 1984), we also observed a clear voltage dependence of open times of IRK1 Wt, which decreased with increasing hyperpolarization (Fig. 6A). The voltage dependencies of both IRK1 Wt and Q140E were analyzed here in more detail. The distributions of open, closed, and

---

FIGURE 7 Analysis of the kinetics of the single-channel current of IRK1 Wt and Q140E. (A) Examples of the single-channel current traces and its open time distribution at −80 mV for IRK1 Wt (left) and Q140E (right). The short bars above the current traces indicate the periods, which are shown expanded in C. Arrowheads on the left indicate the zero current level. The arrow on the trace indicates the fast closing event that was infrequently observed in Wt. (B) The closed time distribution of Wt and Q140E from the same current recordings as A. (C) Expanded current traces from A and the distribution of closed times shorter than 50 msec of Wt and Q140E. Although the distribution of Wt in this range could not be fitted by exponential functions, that of Q140E was well fitted by double exponentials, with an extra-fast time constant ($\tau_{c\text{-}ef}$) and a fast time constant (τ_f). (D) Burst analysis of Q140E. By neglecting the extra-fast closing events shorter than τ_{crit} (3.2 msec), the burst duration of Q140E was calculated and fitted by a single exponential with a time constant (τ_b) that is in the same range as τ_o of Wt. The open time distribution of Wt is the same as in A. Reproduced from Guo and Kubo (1998). Comparison of the open–close kinetics of the cloned inward rectifier K^+ channel IRK1 and its point mutant (Q140E) in the pore region. *Recept. Chan.* **5,** 273–289.

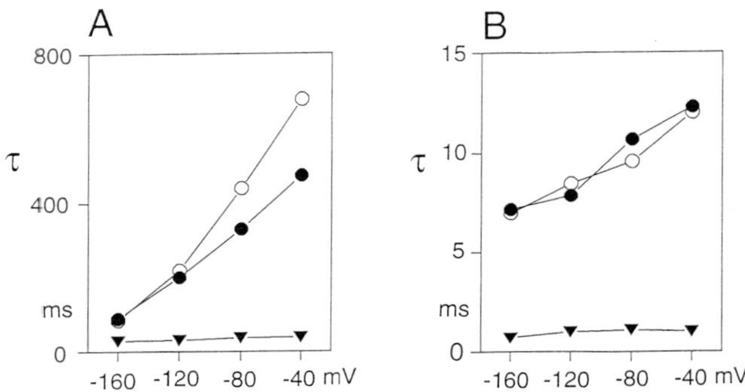

FIGURE 8 Voltage dependence of the parameters of the single-channel kinetics. (A) Voltage dependence of Wt τ_o (open circles), Q140E τ_b (filled circles), and Q140E τ_o (filled triangles). (B) Voltage dependence of Wt $\tau_{c\text{-}f}$ (open circles), Q140E $\tau_{c\text{-}f}$ (filled circles), and Q140E $\tau_{c\text{-}ef}$ (filled triangles). Reproduced from Guo and Kubo (1998). Comparison of the open–close kinetics of the cloned inward rectifier K^+ channel IRK1 and its point mutant (Q140E) in the pore region. *Recept. Chan.* **5,** 273–289.

burst times at four different potentials were analyzed, and the time constant values were plotted in Fig. 8. In Wt, τ_o was strongly voltage dependent (Fig. 8A). $\tau_{c\text{-}f}$ showed a weak voltage dependency (Fig. 8B). Both of them decreased at more hyperpolarized membrane potentials. However, $\tau_{c\text{-}s}$ was almost insensitive to voltage changes. In Q140E, τ_o (Fig. 8A) showed only a weak voltage dependency in contrast with τ_o of Wt. τ_b (Fig. 8A) and $\tau_{c\text{-}f}$ (Fig. 8B) of Q140E displayed a voltage dependence similar to τ_o and $\tau_{c\text{-}f}$ of Wt. $\tau_{c\text{-}ef}$ (Fig. 8B) of Q140E lacked voltage dependence.

C. The Open–Close Kinetics Model for IRK1 Wt

The open time distribution of IRK1 Wt was fitted by a single exponential and displayed a strong dependence on membrane potentials, while its closed time distribution was fitted by a sum of two exponentials; a fast component ($\tau_{c\text{-}f}$, 12 msec) and a slow component ($\tau_{c\text{-}s}$, 132 msec).

The open–close kinetics of IRK1 Wt in our study could be described by a three-state model, a single open state and two closed states. The state diagram is either the C–C–O model (Kameyama *et al.,* 1983; Sakmann and Trube, 1984) or the C–O–C model (Sakmann and Trube, 1984). In both schemes, the main point of our discussion—the presence of an additional closed state in Q140E—is the same. To simplify our discussion, here we describe the open–close kinetics using the C–C–O model, in which there is only one exit from the open state. The term "closed" state in the scheme means the same as "nonconducting" state:

$$\text{Closed 2} \underset{\alpha 2}{\overset{\beta 2}{\rightleftarrows}} \text{Closed 1} \underset{\alpha 1}{\overset{\beta 1}{\rightleftarrows}} \text{Open.}$$

The α and β represent the rate constants of each transition. The relationships of τ and the rate constants are known as follows (Kameyama *et al.*, 1983):

$$\tau_o = 1/\alpha 1, \qquad \tau_{c\text{-}f} = 1/x1, \qquad \tau_{c\text{-}s} = 1/x2.$$

$x1$ and $x2$ are the roots of

$$X^2 - (\alpha 2 + \beta 1 + \beta 2)X + \beta 1 \beta 2 = 0 \qquad (x1 > x2).$$

When $\beta 1 \gg \alpha 2, \beta 2$, $\tau_{c\text{-}f}$ and $\tau_{c\text{-}s}$ are approximately

$$\tau_{c\text{-}f} = 1/(\beta 1 + \alpha 2), \qquad \tau_{c\text{-}s} = 1/\beta 2.$$

The properties of τ_o (Wt) indicate that $\alpha 1$ is dependent on membrane potentials and $[K^+]_o$. The greater hyperpolarization of V_m and the higher $[K^+]_o$ caused the shortening of τ_o (Wt), in other words, an increase in $\alpha 1$.

D. Q140E Has an Extra-fast Gating

The open time distribution of the Q140E mutant was fitted with a single exponential, similar to Wt. On the other hand, the closed time distribution was fitted by a sum of three exponentials; an extra-fast component ($\tau_{c\text{-}ef}$, 1.2 msec), a fast one ($\tau_{c\text{-}f}$, 11 msec), and a slow one ($\tau_{c\text{-}s}$, 116 msec). The values of $\tau_{c\text{-}f}$ and $\tau_{c\text{-}s}$ of Q140E corresponded to those of Wt quite well, and the burst time distribution, obtained by neglecting the extra-fast closings, were well fitted by a single exponential (τ_b), which is equivalent to τ_o of Wt. Thus, we concluded that Q140E acquired extra-fast closings in addition to the gating of Wt. The state diagram of Q140E is as follows:

$$\begin{array}{ccccc} & & & & \text{Closed 3} \\ & & & & \alpha 3 \uparrow\downarrow \beta 3 \\ \text{Closed 2} & \underset{\beta 1}{\overset{\beta 2}{\rightleftarrows}} & \text{Closed 1} & \underset{\alpha 1}{\overset{\beta 1}{\rightleftarrows}} & \text{Open} \end{array}$$

Closed 3 is the extra-fast close state acquired by the Q140E mutation. The relationships between τ and the rate constants are

$$\tau_o = 1/(\alpha 1 + \alpha 3), \qquad \tau_{c\text{-}ef} = 1/\beta 3, \qquad \tau_b = 1/x1.$$

$x1$ is one of the roots ($x1 < x2$) of

$$X^2 - (\alpha1+\beta3+\alpha3)X + \beta3^*\alpha1 = 0.$$

When $\beta3 \gg \alpha1,\alpha3$, τ_b is approximately

$$\tau_b = 1/\alpha1.$$

Similar to $\alpha1$, $\alpha3$ exhibited a dependence on $[K^+]_o$. $\alpha3$ was, however, quite different from $\alpha1$ in that it lacks voltage dependence.

E. Effects of External and Internal Cations on Single-Channel Properties

The results presented above confirmed our speculation that a fast gating was acquired by the glutamate (E) introduced at site 140. Therefore, the next focus of this study was to elucidate how E140 exerts its effect on the gating. As the replacement of Q with E introduced a negative charge at the center of the proposed H5 pore region, it is natural to speculate that this fast gating is due to the block by positively charged molecule(s) from either side of the membrane. Because the solutions contained cations H^+, Mg^{2+}, and K^+, we tested them individually. The effect of cytoplasmic polyamines was also tested. Because we found no evidence that the transition to the extra-fast closed state is due to the block by H^+, Mg^{2+}, or polyamines from either side of the membrane, we postulated that the K^+ ion itself might be causing it. To test this hypothesis, the effect of $[K^+]_o$ on the single-channel properties of IRK1 Wt and Q140E was investigated (Fig. 9).

In Wt, the open time histogram in 20 m*M* K^+_o at -160 mV was well fitted by a single exponential with τ_o of 408 msec, about fourfold longer than 106 msec in 140 m*M* K^+_o. The closed time histogram of Wt in 20 m*M* K^+_o could be fitted by a sum of two exponentials with $\tau_{c\text{-}f}$ and $\tau_{c\text{-}s}$ of 13.2 and 130 msec; the corresponding values in 140 m*M* K^+_o were 8.3 and 140 msec, respectively. Analysis of the closed times shorter than 50 msec did not give a good fit in either 20 or 140 m*M* K^+_o. In Q140E, τ_o in 20 m*M* K^+_o at -160 mV was 93 msec, about threefold longer than 27 msec in 140 m*M* K^+_o. The total closing events of Q140E were still fitted by three exponentials with $\tau_{c\text{-}ef}$ of 1.4 msec, $\tau_{c\text{-}f}$ of 19 msec and $\tau_{c\text{-}s}$ of 138 msec, which were similar to the corresponding values of 1.0, 20, and 90 msec in 140 m*M* K^+_o. τ_b of Q140E was 398 msec in 20 m*M* K^+_o, about fourfold longer than 113 msec in 140 m*M* K^+_o. These τ_b values were in the same range as τ_o (Wt). All these results indicate that the decrease in $[K^+]_o$ slows down the transition to the extra-fast closed state of Q140E as well as to the closed state in common with Wt and Q140E.

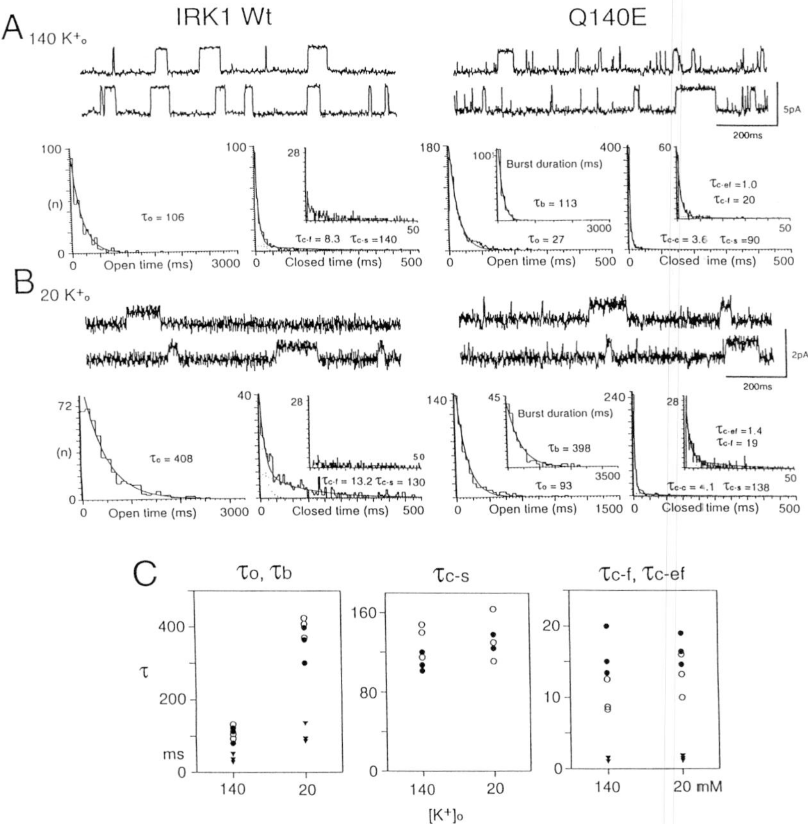

FIGURE 9 Effects of $[K^+]_o$ on the open–close kinetics. (A) Current traces of Wt (left) and Q140E (right) at −160 mV in 140 mM K^+_o. Distributions of the open time and the closed time are also shown below the traces. In the case of Q140E, distributions of burst duration (inset of the open time analysis) and of closed time shorter than 50 msec (inset of closed time analysis) are also shown. (B) Current traces and distributions in 20 mM K^+_o. (C) Plots of τ_o and τ_b (left), $\tau_{c\text{-}s}$ (middle), and $\tau_{c\text{-}f}$ and $\tau_{c\text{-}ef}$ (right) of IRK1 Wt and Q140E in various $[K^+]_o$. The symbols represent (left): τ_o of Wt (open circles), τ_b of Q140E (filled circles), τ_o of Q140E (filled triangles); (middle): $\tau_{c\text{-}s}$ of Wt (open circles), $\tau_{c\text{-}s}$ of Q140E (filled circles); (right): $\tau_{c\text{-}f}$ of Wt (open circles), $\tau_{c\text{-}f}$ of Q140E (filled circles), $\tau_{c\text{-}e}$f of Q140E (filled triangles). Reproduced from Guo and Kubo (1998). Comparison of the open–close kinetics of the cloned inward rectifier K⁺ channel IRK1 and its point mutant (Q140E) in the pore region. *Recept. Chan.* **5,** 273–289.

F. A Possible Mechanism of the Transition to the Extra-fast Closed State of Q140E

We observed that increases in $[Mg^{2+}]_o$ caused increases but not decreases in τ_o, and that changes of pH_o, pH_i, $[Mg^{2+}]_i$, or cytoplasmic polyamines did not affect τ_o. Thus, the possibility that Mg^{2+}, H^+, or polyamines act as a gating molecule at highly hyperpolarized potentials was denied. Since the decrease in $[K^+]_o$ caused increases in both τ_o (Wt) and τ_o (Q140E), in other words, the reduction of $\alpha 1$ and $\alpha 3$, we postulated that K^+ itself causes the transition to the closed state. If the binding of the permeating ion to the channel is too strong, conduction could be stopped. The block by the permeating ion itself is acceptable, since it could actually be observed when Tl^+ was used as a charge carrier even in IRK1 Wt. Thus, it is fair to think that these two steps involve the binding of K^+o and that they are closely related to each other. In terms of the voltage dependency, however, these two steps are clearly different. $\alpha 1$ is highly dependent on membrane potential, whereas $\alpha 3$ is almost independent. Judging from the voltage dependency, we speculate that the site responsible for $\alpha 3$ (Q140) is very close to the external mouth of the pore, and the site that determines $\alpha 1$ is located deeper in the electric field.

IV. DISCUSSION: COMPARISON WITH VOLTAGE-GATED K^+ CHANNEL

In the following we discuss relevant or different aspects of the pore structure of the inward rectifier and voltage-gated K^+ channels.

A. Dependence on External K^+

The voltage-gated K^+ channel RCK4 also shows a marked dependence on K^+_o. Pardo and colleagues (1992) identified a site that is involved in the regulation of this channel by K^+_o. By mutating K533 to Y (Fig. 2A), the RCK4 channel lost its dependence on K^+_o. The same site was also found to affect the sensitivity to blocking by external triethylammonium as well as by cytoplasmic Mg^{2+} (Ludwig *et al.,* 1993). They discussed that the mutation of this site modulates state occupancy of the outer part of the pore by K^+, and thereby the extent of the Mg^{2+} block at the inner part of the pore by electrostatic repulsion (Ludwig *et al.,* 1993). The equivalent site of the *Shaker* B channel, T449, is also involved in the interaction with K^+_o as well as in C-type inactivation. If IRK1 has a blocker-independent gating mechanism, it might be related to the mechanism of C-type inactivation, in that the vacancy of K^+_o in the outer part of the pore leads to nonconducting states. The electrophysiological properties of the inward rectifier K^+ channel seem to be strikingly different from those of the voltage-gated K^+ channel at first glance due to the absence of the voltage sensor and the extremely high sensitivity to cytoplasmic blockers. These two types of channels, however, might be quite related to each other, in that they have a gating mechanism that depends on the occupancy of K^+ at the outer part of the pore involving the equivalent sites R148 (IRK1), K533 (RCK4), and T449 (*Shaker* B): (1) the occupancy links to the unblock of the cytoplasmic blockers and (2) apart from the blockers, the vacancy of K^+ at the site leads to C-type inactivation or related nonconducting states. The charged residue R is conserved without exception in the inwardly rectifying K^+ channel family members cloned so far. Thus, the link of K^+ interaction at this site and the activation of the channel is thought not to be specific for IRK1 but to be a general mechanism that links permeation, gating, and block of all members of the inwardly rectifying K^+ channel family, as well as some members of the voltage-gated K^+ channel family.

B. Pore Structure

There is a possibility that the structure of H5 of the inward rectifier K^+ channel is quite different from that of the voltage-gated K^+ channel: (1) Yang and co-workers (1997) reported that two highly conserved charged amino acids of the H5 region, E138 and R148, form a salt bridge defining the structure of the pore region. (2) Schwalbe and co-workers (1996) reported that most of the H5 region of the inwardly rectifying K^+ channel ROMK1 can be glycosylated from outside, suggesting that most of the H5 region is accessible from the extracellular side. (3) It was reported that the amino acid residues in the C-terminal (Taglialatela *et al.,* 1994; Yang *et al.,* 1995) and the N-terminal (Fakler *et al.,* 1996) cytoplasmic regions affect the extent of inward rectification, suggesting that these regions might be involved in the permeation pathway. Our mutant Q140E acquired an additional extra-fast closed state, which is K^+_o dependent, but almost voltage independent. This result suggests that Q140 is involved in the permeation pathway, but is located at a very shallow position in the electric field. It is possible that the H5 region of IRK1 does not go too far into the membrane, in contrast with voltage-gated K^+ channels (Yellen *et al.,* 1991; Pascual *et al.,* 1995), or that the Q140 is not located in the deepest part of H5. To clarify these points, a systematic cysteine scan mutagenesis study, as was done on the voltage-gated K^+ channel (Pascual *et al.,* 1995), should be effective.

Acknowledgments

These studies were supported by research grants from the Human Frontier Science Program Organization (to Y.K.) and the Ministry of Education, Science, Sports and Culture of Japan (to Y.K.). L.G. was supported by a fellowship from the Sasagawa Health Science Foundation.

References

Ciani, S., Krasne, S., Miyazaki, S., and Hagiwara, S. (1978). A model of anomalous rectification: Electrochemical-potential-dependent gating of membrane channels. *J. Membr. Biol.* **44,** 103–134.

Cohen, I. S., DiFrancesco, D., Pennefather, P., and Mulrine, N. K. (1989). Internal and external K help gate the inward rectifier. *Biophys. J.* **55,** 197–202.

Dascal, N., Schreibmayer, W., Lim, N. F., Wang, W., Chavkin, C., Dimagno, L., Labarca, C., Kieffer, B. L., Gaveriaux-Ruff, C., Trollinger, D., Lester, H., and Davidson, N. (1993). Atrial G protein-activated K^+ channel: Expression cloning and molecular properties. *Proc. Natl. Acad. Sci. USA* **90,** 10,235–10,239.

Fakler, B., Bond, C. T., Adelman, J. P., and Ruppersberg, J. P. (1996). Heterooligomeric assembly of inward-rectifier K^+ channels from subunits of different subfamilies: Kir2.1 (IRK1) and Kir4.1(BIR10). *Pflugers. Arch.* **433,** 77–83.

Fakler, B., Brandle, U., Glowatzki, E., Weidemann, S., Zenner, H-P., and Ruppesberg, J. P. (1995). Strong voltage-dependent inward rectification of inward rectifier K^+ channels is caused by intracellular spermine. *Cell* **80,** 149–154.

Guo, L., and Kubo, Y. (1998). Comparison of the open–close kinetics of the cloned inward rectifier K^+ channel IRK1 and its point mutant (Q140E) in the pore region. *Recept. Chan.* **5,** 273–289.

Hagiwara, S., and Yoshii, M. (1979). Effects of internal potassium and sodium on the anomalous rectification of the starfish egg as examined by internal perfusion. *J. Physiol.* **292,** 251–265.

Hille, B., and Schwarz, W. (1978). Potassium channels as multi ion single file pores. *J. Gen. Physiol.* **72,** 409–442.

Ho, K., Nichols, C. G., Lederer, W. J., Lytton, J., Vassilev, P. M., Kanazirska, M. V., and Hebert, S. C. (1993). Cloning and expression of an inwardly rectifying ATP-regulated potassium channel. *Nature* **362,** 31–38.

Ishihara, K., and Hiraoka, M. (1994). Gating mechanism of the cloned inward rectifier K^+ channel from the mouse heart. *J. Membr. Biol.* **142,** 55–64.

Ishihara, K., Hiraoka, M., and Oochi, R. (1996). The tetravalent organic cation spermine causes the gating of the IRK1 channel expressed in murine fibroblast cells. *J. Physiol.* **491,** 367–381.

Ishihara, K., Mitsuiye, T., Noma, A., and Takano, M. (1989). The Mg^{2+} block and intrinsic gating underlying inward rectification of the K^+ current in guinea pig cardiac myocytes. *J. Physiol.* **419,** 297–320.

Jackson, M. B., Wong, B. S., Morris, C. E., Lecar, H., and Christian, C. N. (1983). Successive openings of the same acetylcholine receptor channel are correlated in open time. *Biophys. J.* **42,** 109–114.

Kameyama, M., Kiyosue, T., and Soejima, M. (1983). Single channel analysis of the inward rectifier K^+ current in the rabbit ventricular cells. *Jpn. J. Physiol.* **33,** 1039–1056.

Katz, B. (1949). Les constantes electriques de la membrane du muscle. *Arch. Sci. Physiol.* **2,** 285–299.

Krapivinsky, G., Gordon, E. A., Wickman, K., Velimirovic, B., Krapivinsky, L., and Clapham, D. E. (1995). The G-protein-gated atrial K^+ channel I_{KACH} is a heteromultimer of two inwardly rectifying K^+-channel proteins. *Nature* **374,** 135–141.

Kubo, Y. (1994). Towards the elucidation of the structural-functional relationship of inward rectifying K^+ channel family. *Neurosci. Res.* **21,** 109–117.

Kubo, Y. (1996). Effects of extracellular cations and mutations in the pore region on the inward rectifier K^+ channel IRK1. *Recept. Chan.* **4,** 73–83.

Kubo, Y., Baldwin, T. J., Jan, Y. N., and Jan, L. Y. (1993a). Primary structure and functional expression of a mouse inward rectifier potassium channel. *Nature* **362,** 127–133.

Kubo, Y., Miyashita, T., and Kubokawa, K. (1996). A weakly inward rectifying potassium channel of the salmon brain. *J. Biol. Chem.* **271,** 15,729–15,735.

Kubo, Y., Reuveny, E., Slesinger, P. A., Jan, Y. N., and Jan, L. Y. (1993b). Primary structure and functional expression of a rat G protein coupled muscarinic potassium channel. *Nature* **364,** 802–806.

Kurachi, Y. (1985). Votage-dependent activation of the inward-rectifier potassium channel in the ventricular cell membrane of guinea-pig heart. *J. Physiol.* **366,** 365–385.

Lopatin, A. N., Makhina, E. N., and Nichols, C. G. (1994). Potassium channel block by cytoplasmic polyamines as the mechanism of intrinsic rectification. *Nature* **372,** 366–369.

Lu, Z., and MacKinnon, R. (1994). Electrostatic tuning of Mg^{2+} affinity in an inward-rectifier K^+ channel. *Nature* **371,** 243–246.

Ludwig, U., Lorra, C., Pongs, O., and Heinemann, S. H. (1993). A site accessible to external TEA^+ and K^+ influences intracellular Mg^{2+} block of cloned potassium channels. *Eur. Biophys. J.* **22,** 237–247.

Matsuda, H. (1988). Open-state substructure of inwardly rectifying potassium channels revealed by magnesium block in guinea-pig heart cells. *J. Physiol.* **397,** 237–258.

Matsuda, H. (1991). Effects of external and internal K^+ ions on magnesium block of inwardly rectifying K^+ channels in guinea pig heart cells. *J. Physiol.* **435,** 83–99.

Matsuda, H., Saigusa, A., and Irisawa, H. (1987). Ohmic conductance through the inwardly rectifying K^+ channel and blocking by internal Mg^{2+}. *Nature* **325,** 156–159.

Pardo, L. A., Heinemann, S. H., Terlau, H., Ludewig, U., Lorra, C., Pongs, O., and Stuhmer, W. (1992). Extracellular K^+ specifically modulates a rat brain K^+ channel. *Proc. Natl. Acad. Sci. USA* **89,** 2466–2470.

Pascual, J. M., Shieh, C-C., Kirsch, G. E., and Brown, A. M. (1995). K^+ pore structure revealed by reporter cysteines at inner and outer surfaces. *Neuron* **14,** 1055–1063.

Pennefather, P., Oliva, C., and Mulrine, N. (1992). Origin of the potassium and voltage dependence of the cardiac inwardly rectifying K-current. *Biophys. J.* **61,** 448–462.

Sakmann, B., and Trube, G. (1984). Voltage-dependent inactivation of inward-rectifying single-channel currents in the guinea-pig heart cell membrane. *J. Physiol.* **347,** 659–683.

Schwalbe, R. A., Wang, Z., Bianchi, L., and Brown, A. M. (1996). Novel sites of N-glycosylation in ROMK1 reveal the putative pore-forming segment H5 as extracellular. *J. Biol. Chem.* **271,** 24,201–24,206.

Silver, M. R., and DeCoursey, T. E. (1990). Intrinsic gating of inward rectifier in bovine pulmonary artery endothelial cells in the presence and absence of internal Mg^{2+}. *J. Gen. Physiol.* **96,** 109–133.

Stanfield, P. R., Davies, N. W., Shelton, P. A., Khan, I. A., Brammar, W. J., Standen, N. B., and Conley, E. C. (1994). The intrinsic gating of inward rectifier K^+ channels expressed from the murine IRK1 gene depends on voltage K^+ and Mg^{2+}. *J. Physiol.* **475,** 1–7.

Stanfield, P. R., Davies, N. W., Shelton, P. A., Sutcliffe, M. J., Khan, I. A., Brammar, W. J., and Conley, E. C. (1994). A single aspartate residue is involved in both intrinsic gating and blockage by Mg^{2+} of the inward rectifier IRK1. *J. Physiol.* **478,** 1–6.

Taglialatela, M., Wible, B. A., Caporaso, R., and Brown, A. M. (1994). Specification of pore properties by the carboxyl terminus of inward rectifying K^+ channels. *Science* **264,** 844–847.

Takumi, T., Ishii, T., Horio, Y., Morishige, K., Takahashi, N., Yamada, M., Yamashita, T., Kiyama, H., Sohmiya, K., Nakanishi, S. and Kurachi, Y. (1995). A novel ATP-dependent inward rectifier potassium channel expressed predominantly in glial cells. *J. Biol. Chem.* **270,** 16,339–16,346.

Vandenberg, C. A. (1987). Inward rectification of a potassium channel in cardiac ventricular cells depends on internal magnesium ions. *Proc. Natl. Acad. Sci. USA* **84,** 2560–2564.

Wible, B. A., Taglialatela, M., Ficker, E., and Brown, A. M. (1994). Gating of inwardly rectifying K^+ channels localized to a single negatively charged residue. *Nature* **371,** 246–249.

Yang, J., Jan, Y. N., and Jan, L. Y. (1995). Control of rectification and permeation by residues in two distinct domains in an inward-rectifier K^+ channel. *Neuron* **14,** 1047–1054.

Yang, J., Yu, M., Jan, Y. N., and Jan, L. Y. (1997). Stabilization of ion selectivity filter by pore loop ion-pairs in an inwardly rectifying potassium channel. *Proc. Natl. Acad. Sci. USA* **94,** 1568–1572.

Yellen, G., Jurman, M. E., Abramson, T., and MacKinnon, R. (1991). Mutations affecting internal TEA blockade identify the probable pore-forming region of a K^+ channel. *Science* **251,** 939–942.

CHAPTER 12

Potassium Channels with Two P Domains

Florian Lesage and Michel Lazdunski
Institut de Pharmacologie Moléculaire et Cellulaire, CNRS UPR 411, Sophia Antipolis, 06560 Valbonne, France

I. INTRODUCTION

K^+ channels are noncovalent polymers of pore-forming subunits generally associated with accessory subunits (for review see Jan and Jan, 1994; Chandy and Gutman, 1995; Isomoto *et al.,* 1997; Nichols and Lopatin, 1997). Four types of auxiliary subunits have been cloned to date (Kvβ, $K_{Ca}\beta$, SUR, and IsK), and only two pore-forming subunit types has been isolated until recently (*Shaker* and IRK). Except for IsK and K_{Ca}, several genes have

1063-5823/99 $30.00

been identified for each class of subunit, from 2 genes for SUR to more than 25 genes for *Shaker.* Adequate combination of coexpressed subunits has allowed the reconstitution of different K^+ currents previously identified by electrophysiology. Voltage-gated (Kv) and Ca^{2+}-sensitive (K_{Ca}) K^+ channels have been shown to correspond to associations of *Shaker* subunits with or without Kvβ, $K_{Ca}\beta$, or IsK subunits; inward rectifiers and G protein-coupled K^+ channels to associations of IRK subunits; and ATP-sensitive K^+ channels (K_{ATP}) to associations of IRK subunits with SUR subunits.

By computational "mining" of informatic DNA databases, our laboratory has recently identified a new class of pore-forming subunits (Fink *et al.,* 1996; Lesage *et al.,* 1996b; Duprat *et al.,* 1997). The major structural difference between these subunits and the *Shaker* and IRK subunits is the presence of two P domains instead of one, and four transmembrane segments (TMS) instead of six or two. The two P domain K^+ channel denomination is used to underline this feature. With the cloning and the functional expression of several members of this 4TMS/2P family, it is now clear that this unusual structure is associated with unique functional properties. These properties strongly suggest that the two P domain K^+ channels are responsible for the resting membrane K^+ conductances of many cell types. This chapter focuses on this peculiar class of K^+ channels. Their structural, electrophysiological, pharmacological, and regulatory properties as well as their distribution in mammals are described and their potential roles are discussed. Related channels identified in *Drosophila* (Goldstein *et al.,* 1996), nematode (Ketchum *et al.,* 1995; Lesage *et al.,* 1996b; Wei *et al.,* 1996), yeast (Ketchum *et al.,* 1995; Zhou *et al.,* 1995; Lesage *et al.,* 1996a; Reid *et al.,* 1996), and plant (Czempinski *et al.,* 1997) are also discussed.

II. TWIK: A NEW TYPE OF PORE-FORMING SUBUNIT

A. Structural Properties

1. Primary Structure

All the pore-forming K^+ channel subunits cloned from bacteria, animals, and plants contain a peculiar region of 15–25 residues, called H5 or P, that has been well conserved during evolution (Pongs, 1993; Jan and Jan, 1994; Mackinnon, 1995). Conservation of this domain is not surprising since it has been shown to form a major part of the K^+-selective pore of *Shaker* channels (Heginbotham *et al.,* 1994). We used this signature domain to search for homologous sequences in informatic DNA databases to identify part of a yet to be cloned K^+ channel complementary DNA (cDNA). A 298-bp fragment characterized in the frame of the systematic sequencing of expressed DNA in human brain was identified. It was amplified by polymerase chain reaction (PCR) and used to probe a human tissue Northern blot. A strong signal was found in the kidney and the probe was subsequently used to clone the corresponding full-length cDNA from a human kidney library. This cDNA contains a large open reading frame encoding a 336-amino-acid polypeptide. The protein does not display any significant sequence similarity with the other cloned K^+ channel subunits except in two regions that are highly homologous to the P domain signature sequence. These regions were termed P1 and P2. The channel was called TWIK [TWIK-1 in our original publication (Lesage *et al.*, 1996b)] for *T*andem of P domains in a *W*eak *I*nward rectifier *K^+* channel because of the presence of two P domains and its electrophysiological properties (Section IIB). Hydropathy analysis of TWIK predicts the presence of four TMS termed M1 to M4, M1 and M2 flanking P1 and M3 and M4 flanking P2 (Fig. 1). This overall structure is very different from the structure of other pore-forming K^+ channel subunit, which have a hydrophobic core containing six TMS and one P domain for the *Shakers* and two TMS and one P domain for the IRKs. Another important feature of TWIK is the presence of an extended domain between M1 and P1 (around 60–70 residues) that is never found in *Shaker* or IRK proteins. Finally, TWIK does not contain an S4-like domain, which is responsible for the voltage dependence of *Shaker* channels. The gene encoding human TWIK has been assigned to chromosome 1q42–q43 (Lesage *et al.*, 1996c).

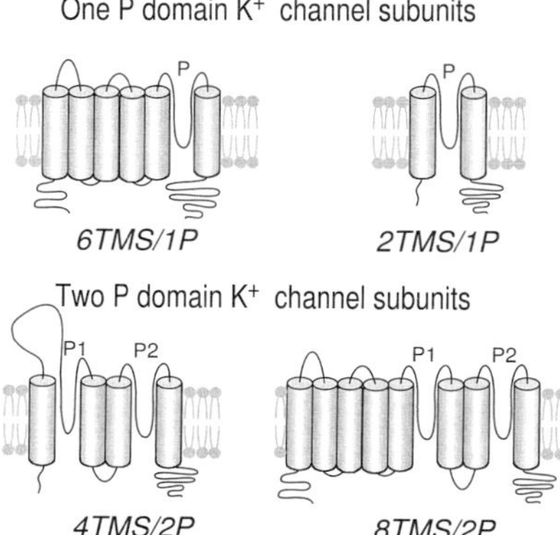

FIGURE 1 Representation of the four different classes of pore-forming K^+ channel subunits. K^+ channels with one P domain comprise the *Shaker* superfamily with six transmembrane segments (TMS) and the IRK family with only two TMS. K^+ channels with two P domains comprise the channels related to TWIK with four TMS and the unique TOK channel from yeast with eight TMS.

2. Disulfide-Bridged Dimerization

Membrane topologies of *Shaker* and IRK subunits are now well established. In both cases, carboxy and amino termini are cytoplasmic and the P domains are inserted into the membrane from the outside (Schwalbe *et al.*, 1997; Shih and Goldin, 1997). This suggested a similar model for TWIK wherein the extremities were cytoplasmic and the M1P1 interdomain extracellular, in order to respect the know orientation of P domains. We demonstrated the extracellular localization of the M1P1 loop by preparing antibodies directed against this region and by using them to detect the presence of the M1P1 loop at the surface of nonpermeabilized cells that expressed TWIK (Lesage *et al.*, 1996d). These antibodies were also used to carry out the biochemical characterization of TWIK from insect cells infected by a recombinant baculovirus. Western blot analysis of TWIK revealed the presence of an immunoreactive complex of 65–80 kDa in the absence of reducing agents. This complex was dissociated by the reducing agent β-mercaptoethanol to give a band of 40 kDa (Lesage *et al.*, 1996d). The same result was obtained by analyzing TWIK from mouse brain synaptic membranes (Lesage *et al.*, 1997). This suggested that TWIK was able to form homodimeric complexes containing an interchain disulfide bond. This hypothesis was verified by localizing the self-interacting domain responsible for the dimerization and by identifying the cysteine residue implicated in the formation of the disulfide bridge. A mutant of TWIK (C69S), in which the cysteine residue 69 is replaced by a serine, is unable to form the 68–80 kDa complex. This mutant TWIK/C69S also fails to produce active channels. On the other hand, a 34-amino-acid domain comprising C69 and localized in the extracellular M1P1 loop is sufficient to promote the self-dimerization of fusion proteins when expressed in bacteria. The secondary structure analysis of this domain predicts that it forms a hydrophilic α helix with a regular occurrence of hydrophilic charged residues and large apolar residues that are typical of the interdigitating helices. Finally, a second mutant (TWIK/N95A) was constructed, in which asparagine residue 95 localized in the M1P1 loop was mutated, and it was shown to be less glycosylated. Taken together, these different results lead us to propose a topological model where the functional TWIK channels are covalent homodimers with extracellular interacting domains that are glycosylated (Lesage *et al.*, 1997). This structure is very different from the known structures of *Shaker* and IRK channels, which are noncovalent tetramers (MacKinnon, 1991; Yang *et al.*, 1995) (Fig. 2). Assuming that four P domains are required to form an active K^+-selective pore, functional TWIK channels might be formed from the interaction of two monomers if both P domains are functional but a higher stoichiometry might be necessary if only one of the P domains is functional. A strong argument against this possibility

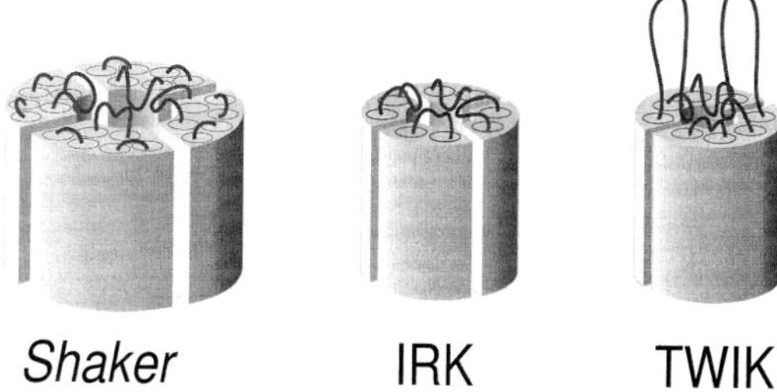

FIGURE 2 Representation of the quaternary structure of mammalian K^+ channels. Only the hydrophobic cores and the external domains of pore-forming subunits are shown. *Shaker* and IRK channels form noncovalent tetramers and TWIK forms homodimers with interacting extended M1P1 loops.

is the existence of many related proteins in a variety of species (mammals, nematode, and fly) (Sections V and VI). Although the overall sequence conservation is low among all these proteins (around 25% of amino acid identity), both P domains are always better conserved, with more than 50% amino acid identity (Section VB). This conservation over a long period of evolution between invertebrates and vertebrates is probably associated with a functional role for both P domains.

B. Functional Properties

In *Xenopus* oocytes, TWIK directs the expression of time-independent currents, that is, currents that are instantaneous and noninactivating (Lesage *et al.,* 1996b, 1997). The current–potential (I–V) relationships were established for different external K^+ concentrations ($[K^+]_{out}$). They show that the reverse potential of currents closely follows the potassium equilibrium potential (E_K), indicating that TWIK currents are K^+-selective and voltage-independent. TWIK currents were measured at all potentials. For normal external K^+ concentrations, the I–V relationship is nearly linear and saturates only for depolarizations positive to 0 mV. This weak inward rectification was also obtained for TWIK currents recorded in the inside out patch configuration in the presence of Mg^{2+} on the cytoplasmic side (3 or 10 mM). This rectification disappears in the absence of Mg^{2+}. In 3 mM internal Mg^{2+}, the unitary conductance was 19 pS at −80 mV and 34 pS at −80 mV in symmetrical 140 mM K^+. The behavior of TWIK channels was very flickering. The removal of internal Mg^{2+} did not influence the mean open times but the unitary conductance measured at +80 mV rose to 35 pS. This result suggests that a very rapid Mg^{2+}-induced flickering at +80 mV has led to an underestimation of the true conductance value. As expected for a time- and voltage-independent channel opened at all membrane potentials, TWIK activity is associated with a strong hyperpolarization of the resting membrane potential (E_m), with the E_m reaching a value close to E_K in TWIK-expressing oocytes.

III. A FAMILY OF TWIK-RELATED K+ CHANNELS IN HUMANS AND MICE

A. Cloning and Primary Structure

1. TREK

Both *Shaker* and IRK pore-forming subunits form large families of genes in mammals. This suggested that TWIK could be the first member of a novel family of new K^+ channel subunits. In order to clone such TWIK-related channels in mammals, we screened public DNA databases with the TWIK sequence. We found that the genome of *Caenorhabditis elegans* encoded several potential proteins with four TMS and two P domains. The sequence conservation between these sequences and TWIK is low (less than 25% amino acid identity) but some short amino acid stretches are better conserved, particularly in both P domains (P1 and P2) and in the second TMS (M2) (detailed in Section VB). Degenerate primers corresponding to M2 and P2 were designed and used to amplify DNA from mouse brain cDNA. A 280-bp fragment was isolated and its sequencing revealed significant sequence similarities with the M2P2 region of TWIK. This fragment was labeled and used to screen a mouse brain cDNA library under conditions of high stringency. A cDNA of 2.5 kb was isolated and sequenced. It contains an open reading frame of 1113 bp encoding a 370-amino-acid polypeptide. Despite a similar overall structure with two P domains and four potential TMS, the sequence identity with TWIK was only 25%. Clearly, this novel protein belongs to the TWIK superfamily. However, the low conservation suggests that this channel corresponds to a distinct subfamily, and it has been called TREK [TREK-1 in our original publication (Fink *et al.*, 1996)] for *T*WIK-*RE*lated *K*+ channel. TREK is 34 residues longer than TWIK because its amino and carboxy termini are more extended. Like TWIK, TREK contains an extended M1P1 loop and a cysteine residue at a position equivalent to cysteine 69 of TWIK. This suggests that the covalent homodimerization of TREK observed in transfected cells (unpublished results) could occur via the same mechanism as that for TWIK.

2. TASK

TASK is the third TWIK-related channel that we cloned in mice and humans (Duprat *et al.,* 1997). TWIK and TREK sequences were used to search for related sequences in the subset of expressed sequence tag (EST) database of Genbank. ESTs from mouse brain were identified and used to clone the corresponding cDNAs from human and mouse cDNA libraries. Human cDNA encodes a 395-amino-acid polypeptide sharing 25–28% sequence identity with TWIK and TREK, and the same overall structure with four TMS and two P domains. Considering its strong pH sensitivity (Section IIID1), this channel was called TASK for *T*WIK-related *A*cid-*S*ensitive K^+ channel. From a structural point of view, TASK differs from other TWIK-related channels by having a more extended carboxy terminus and no cysteine residue in the M1P1 loop. As expected from the latter point, TASK does not form covalent dimers in transfected cells (unpublished data). This result indicates that an interchain disulfide bridge dimerization is not the rule to form functional 4TMS/2P channels as earlier results on TWIK and TREK suggested. Another important feature of TASK is that its carboxy terminus contains a sequence (S/TXV) that has been shown to be sufficient to promote the association of some receptors and channels with synaptic proteins such as PSD95 or SAP90 (Sheng and Kim, 1996). In particular, such an association has been reported for K^+ channels belonging to the Kv1 and Kir2 subfamilies. TASK is the first 4TMS/2P channel subunit to share this characteristic and to be expected to have a synaptic localization.

B. Electrophysiological Properties

1. Outwardly Rectifying TREK Currents

TREK was expressed both in *Xenopus* oocytes and in transfected fibroblasts (Fink *et al.,* 1996). Like TWIK, TREK directs the expression of instantaneous and noninactivating currents that are selective for K^+. One of the major differences between both currents is the nature of their rectification. Unlike TWIK, which is an inward rectifier, TREK is an outward rectifier. Almost no inward current was recorded in an external medium containing 2 m*M* K^+. When $[K^+]_{out}$ was increased, an inward current was revealed that saturated upon hyperpolarization, the I–V curve was shifted rightward, and the reversal potential of the currents closely followed E_K. This behavior is very different from that of the "classical" outward rectifier voltage-gated K^+ channels (Kv) that belong to the *Shaker* superfamily. These channels are activated upon depolarization and opened from a fixed threshold potential whose value depends on the properties of an internal positively charged domain called S4 or the voltage sensor. The threshold potentials of Kv channels are always positive compared to E_K and, under physiological conditions, they only pass outward currents. The TREK structure does not have any domain similar to the voltage sensor of Kv channels. On the other hand, the TREK threshold activation potential is not fixed and closely follows E_K. This indicates that the molecular mechanisms of the outward rectification for TREK and for the Kv channels are different. A yeast two P domain K^+ channel shares this unusual behavior. In this case, we have shown that the preferential tendency of the channel to pass outward currents probably results from structural elements contained in the protein sequence itself (see Section VIA). Preliminary results with TREK indicate that this is also the case with this channel. In inside out patches, TREK currents are highly flickering with a single-channel conductance of 48 pS.

2. "GHK" Rectifying TASK Currents

As expected from the previous data on TWIK and TREK, TASK was found to produce K^+-selective and time-independent currents, that is, without activation and inactivation kinetics (Duprat *et al.,* 1997). However, TASK currents are neither inwardly rectifying as TWIK currents are, nor outwardly rectifying as TREK, but exhibit a novel type of behavior. TASK currents show an outward rectification when external $[K^+]$ is low ($[K^+]_{out}$ = 2 m*M*), but this rectification is not observed for high $[K^+]_{out}$ (98 m*M*). The rectification (indicated in the Table I as "GHK rectification") can be approximated by the Goldman–Hodgkin–Katz current equation that predicts a curvature of the *I*–V plot under asymmetric conditions. This indicates that TASK lacks intrinsic voltage sensitivity. To our knowledge, this behavior has never been observed in the case of K^+ channels with one P domain. Only ORK1, a 4TMS/2P K^+ channel cloned from *Drosophila,* exhibits such a strict voltage independence (Section VA) (Goldstein *et al.,* 1996). Like TWIK and TREK, TASK is highly flickering. As expected for a time- and voltage-independent K^+-selective current, TASK expression is associated with a strong polarization of the resting membrane potential, E_m reaching a value close to E_K.

C. Pharmacology

The pharmacology of the two P domain K^+ channels is summarized in Table I. Classical blockers of K^+ channels are not very efficient on these channels. TWIK, TREK, and TASK currents are insensitive to 4-aminopyridine (4-AP) and tetraethylammonium (TEA), except TWIK, which is

TABLE I

Properties of K^+ Channels with Two P Domains

		Electrophysiology		Pharmacology (external block)					
		Behavior	γ	TEA	A-AP (0.1 mM)	$BaCl_2$	Quinine (50 μM)	Regulations	References
TWIK	336 aa	Inward rectifier	34 pS	30%/10 mM	<10%	50%/0.1 mM	50%	$[H^+]_{in}$ PKC activators	Lesage *et al.* (1996a) Lesage *et al.* (1997)
TREK	370 aa	Outward rectifier	48 pS	<10%/1 mM	<10%	50%/0.1 mM	<10%	PKC activators PKA activators $[Na^+]_{out}$	Fink *et al.* (1996)
TASK	395 aa	GHK rectifier	nd	<10%/mM	<10%	<10%/0.1 mM	<10%	$[H^+]_{out}$	Duprat *et al.* (1997)
ORK1	618 aa	GHK rectifier	nd	<10%/25 mM	nd	<10%/1 mM	nd	nd	Goldstein *et al.* (1996)
KCO1	363 aa	Outward rectifier	64 pS	nd	nd	50%/3.8 mM	nd	$[Ca^{2+}]_{in}$	Czempinski *et al.* (1997)
TOK1	691 aa	Outward rectifier	26 pS	50%/5.5 mM	<10%	50%/3 mM	>50%	$[H^+]_{in}$ PKC activators	Lesage *et al.* (1996b)

slightly blocked by TEA (30% at 10 m*M*). TWIK is also inhibited by quinine (50% at 50 μM) while TREK and TASK are not. $BaCl_2$ blocks 50% of TWIK and TREK currents at 0.1 m*M*. Finally, these 4TMS/2P channels are insensitive to toxins that block Kv or K_{Ca} K^+ channels (charybdotoxin, apamin, and dendrotoxin), to the K_{ATP} channel blocker glibenclamide, and to the K_{ATP} channel opener pinacidil.

D. Regulation

1. TWIK and TASK Modulation by pH

The channel activity of TWIK is downregulated by internal acidification (Lesage *et al.,* 1996b). Intracellular acidification in oocytes has been achieved with two methods: (i) superfusion with a CO_2-enriched solution that produces acidification by a mechanism involving the bicarbonate transport system; and (ii) treatment with dinitrophenol (DNP), a metabolic inhibitor that uncouples the H^+ gradient in mitochondria and that induces internal acidosis. Both manipulations resulted in a marked decrease of the TWIK currents, up to 95% (CO_2) and 80% (DNP) of the control amplitude values. The DNP-induced inhibition of the single K^+ channel activity was still observed under cell-attached patch-clamp conditions. However, after excision of the patch, the channel activity became insensitive to acidification of the internal solution obtained either by modifying the Na_2HPO_4/NaH_2PO_4 buffer ratio or by bubbling CO_2. The effect of pH on the TWIK channel activity is probably indirect through an unknown mechanism.

TASK currents are not affected by internal pH modifications but are very sensitive to variations of extracellular pH in a narrow physiological range (Duprat *et al.,* 1997). As much as 90% of the maximum current is recorded at pH 7.66 and only 10% at pH 6.68. The external pH dependence of TASK, at +50 mV, indicates a pK value of 7.3 and a Hill coefficient of 1.6. The Hill coefficient of ~1.6 found for the H^+ concentration dependence of the TASK current is consistent with the idea that the channel is formed by the assembly of two subunits, as suggested by the biochemical characterization of TWIK (Section IIA2). Thus, these two subunits would have strong cooperative interactions with regard to H^+.

2. TREK Sensitivity to External Na^+ and K^+

TREK currents are sensitive to both extracellular K^+ and Na^+ (Fink *et al.,* 1996). In oocytes, the *I–V* relationship of TREK is outwardly rectifying when the external medium contains 2 m*M* K^+ and 96 m*M* Na^+. When external Na^+ is substituted by K^+, the *I–V* curve is shifted both rightward and downward. A decrease of the slope conductance upon depolarization in the K^+-rich solution and the activation of an inward current that saturates upon hyperpolarization are observed. To elucidate this unusual behavior the separate effects of external K^+ and external Na^+ concentrations ($[K^+]_{out}$ and $[Na^+]_{out}$) have been studied. *N-Methyl*-D-glucamine (NMDG) was used to substitute these ions. The currents are inhibited when Na^+ (96 m*M*) is substituted by NMDG, with no variations of the currents recorded upon hyperpolarization. This inhibition is clearly voltage-independent. The mean slope conductance between 0 and +50 mV is dose-dependent and linear between 0 and 96 m*M* Na^+ with a slope of 140 nS/m*M*. The same inhibition was observed when Na^+ was substituted by Li^+ instead of NMDG, suggesting a specific effect of Na^+ ions on the activity of TREK channels. The effects of external K^+ were then studied in a Na^+-free (96 m*M* NMDG) solution. Upon substitution of external NMDG by K^+ (98 m*M* external K^+), an inward current is revealed that saturates for negative potentials. The outward current slope conductance between 0 and +50 mV is clearly enhanced, indicating that TREK is activated by extracellular K^+ in controlled $[Na^+]_{out}$. This effect is unusual since an increase of $[K^+]_{out}$ lowers the chemical driving force for outward K^+ flux and would be expected to decrease rather than increase TREK outward currents. Nevertheless, such a $[K^+]_{out}$ dependence has been reported for many outward and inward currents both in heart and brain cells (Carmeliet, 1989; Pardo *et al.,* 1992; Sanguinetti and Jurkiewicz, 1992).

3. TWIK and TREK Modulation by Activators of Protein Kinases C and A

TWIK activity is upregulated by activation of protein kinase C (PKC) (Lesage *et al.,* 1996b, 1997). Activation of PKC by phorbol-12-myristate-acetate (PMA) increases the TWIK currents in oocytes. This effect was also observed in the cell-attached patch configuration upon application of PMA in the external medium. However, PKC probably does not act by direct phosphorylation of the channel protein since mutation of the only Thr residue (Thr161) in a PKC phosphorylation consensus motif did not abolish the upregulation. This unique PKC phosphorylation site is obligatory neither for intrinsic channel activity nor for regulation. PMA inhibition of another K^+ channel, Kv1.2, was recently shown to be indirect, involving the activation of a protein tyrosine kinase as an intermediate step (Lev *et al.,* 1995). A similar mechanism might apply for TWIK, except that, in this case, the involvement of a tyrosine protein kinase seems unlikely since the only consensus motif for a tyrosine kinase is located in the large extracellular loop M1P1.

Extensive inhibitions of TREK activity are observed after activation of protein kinases A (PKA) and C in oocytes (Fink *et al.,* 1996). Internal injection of GTPγS results in a decrease of TREK currents, while injection of GDPβS produces an increase of activity. These results suggest that TREK currents are inhibited via a pathway involving G-protein activation. A direct effect of G proteins on TREK channels can probably be excluded since effects of nucleotides were maximal only after 5 min of application. Activation of G proteins could lead to the activation of adenylate cyclase and/or phospholipase C and the resulting increase of cAMP and/or diacylglycerol levels could then lead to the activation of PKA and PKC, respectively. The possible modulation of TREK by activating PKA and PKC pathways was also examined. Perfusion of a mixture of 3-isobutyl-1-methylxanthine and forskolin to increase intracellular cAMP levels resulted in inhibition of currents, while a perfusion of PMA to activate PKC also resulted in an inhibition, with the two types of inhibition being additive. These results suggest that TREK activity is regulated via phosphorylation and dephosphorylation mechanisms implicating PKC as well as PKA. However, since applications of purified PKA and PKC on the internal side of channels in the excised patch configuration have had no effect, both mechanisms are probably indirect.

E. Distribution

1. Tissue Distribution by Northern Blot and Reverse Transcriptase PCR

In a first attempt to determine the tissue distributions of the 4TMS/2P channels, Northern blot analyses have been performed using human and mouse tissues for TWIK (Lesage *et al.,* 1996b, 1997) and TASK (Duprat *et al.,* 1997), and mouse tissues for TREK (Fink *et al.,* 1996). The tissues examined were brain, heart, skeletal muscle, kidney, lung and liver. To obtain a more detailed view, we also carried out simultaneous reverse transcriptase (RT)-PCR experiments on cDNAs prepared from an extended panel of mouse tissues. Experimental conditions were chosen in order to achieve semiquantitative PCR and, as expected, for each channel, the relative levels of expression in the tissues studied by Northern blot were similar to the relative levels in the same tissues studied by RT-PCR. This RT-PCR analysis is shown in Fig. 3. TWIK, TREK, and TASK are widely expressed and are found in excitable as well as nonexcitable tissues. It is also clear that despite a quasi-ubiquitous distribution, TWIK, TREK, and TASK each have a unique pattern of expression in terms of relative levels among the examined tissues. For example, TWIK is expressed at similar levels in the nervous system (brain, cerebellum, brainstem, and spinal cord) and the digestive tract (stomach, intestine, and colon) which is not the case for TASK and TREK. On the other hand, TWIK expression in heart is very low in comparison to TREK and TASK, and the only one

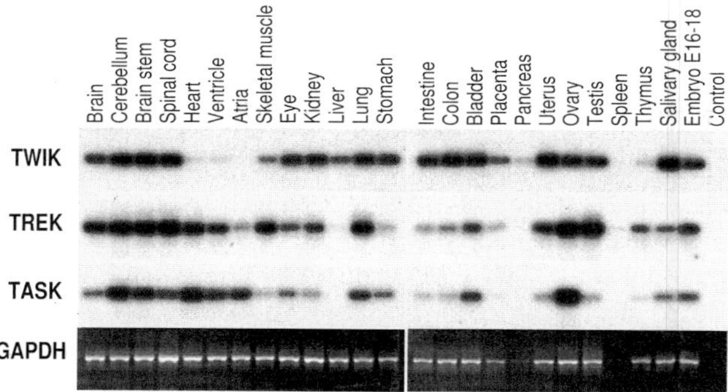

FIGURE 3 Tissue distribution of the TWIK, TREK, and TASK channels in the mouse. Transcript fragments were amplified by PCR using specific primers and analyzed by Southern blot using internal ^{32}P-labeled oligonucleotides.

to have an expression level in atria equivalent to the expression level in brain is TASK. Finally, it is worthwhile to note that TWIK, TASK, and TREK are all expressed in embryos.

2. Regional Localization by *in Situ* Hybridization

The brain localization of the four 4TMS/2P channels has been studied by *in situ* hybridization in the mouse (Fink *et al.*, 1996; Duprat *et al.*, 1997; Lesage *et al.*, 1997). Despite colocalizations in some regions of the brain, each channel has a unique pattern of expression. They are all expressed in the hippocampal formation with the highest level in the granule cell layer of the dentate gyrus and substantial levels in the pyramidal cells of the CA1, CA2, and CA3 fields of the Ammon' horn. The are also all expressed in the cerebellar cortex, in the granule cell layer as well as in the Purkinje cells, and in the cerebral cortex. However, the expression level of TASK is weak in the latter. In the other brain regions, expression of each channel is different. For instance, TREK is highly expressed in the olfactory bulb but not TWIK. TWIK transcripts are also absent in the basal ganglia while a low level of TREK and a high level of TASK were detected in this region. On the other hand, TASK is well expressed in the substantia nigra, TREK only faintly, and TWIK is absent. *In situ* hybridization studies have also been carried out in the mouse heart for TASK. In this tissue, TASK is highly expressed in the atria and faintly in the ventricle.

IV. WHAT ARE THE FUNCTIONAL ROLES FOR TWIK-RELATED K^+ CHANNELS?

TWIK, TREK, and TASK form a novel family of K^+ channels. These 4TMS/2P channels produce currents that are instantaneous and voltage-independent. TASK channels are strictly voltage-independent, with no rectification other than that predicted by the constant field theory, while both TWIK and TREK channels have a mild voltage sensitivity that is revealed by a weak rectification of currents. For TWIK, the inward rectification is due to a voltage-dependent blocking by internal Mg^{2+} as for the classical inward rectifiers belonging to the 2TMS/1P IRK class. For TREK, the outward rectification is probably due to a mechanism intrinsic to the channel itself. All these channels are opened at all membrane potentials and are able to drive the resting membrane potential close to E_K. This property, as well as their wide distribution in the mouse, indicates that they are probably the long sought after background or leak K^+ channels that play a major role in setting the resting membrane potential in many cell types.

The existence of background conductances was postulated originally in neurons (Hodgkin and Huxley, 1952; Baker *et al.,* 1987; Jones, 1989; Hille, 1992. Baker and colleagues (1987) have shown that in myelinated nerve, different K^+ conductances were successively removed by sequential application of TEA, 4-AP, and Cs^+. However, after this treatment, axons still exhibited a pronounced outward rectification. This result suggested that a residual K^+ background conductance, which was outward rectifying as expected from the constant field theory, was present in nerve to set the resting potential. K^+ currents displaying biophysical properties that fit well with this leak conductance have been recorded in many different preparations of neuronal cells: in invertebrates, in *Aplysia* sensory neurons (Siegelbaum *et al.,* 1982), in leech AP neurons (Pellegrini *et al.,* 1989), and in lobster stretch receptor neurons (Theander *et al.,* 1996); in vertebrates, in bullfrog sympathetic ganglia (Koyano *et al.,* 1992), in *Xenopus laevis* myelinated nerve (Koh *et al.,* 1992) and demyelinated axons (Wu *et al.,* 1993), in rat hippocampal (Premkumar *et al.,* 1990a,b; Kim *et al.,* 1995) and premotor respiratory neurons (Wagner and Dekin, 1997), and in guinea pig submucosal neurons (Shen *et al.,* 1992). Such background K^+ channels have also been recorded in other tissues: in *Bufo* smooth cells (Ordway *et al.,* 1991), in rat ventricular myocytes (Yue and Marban, 1988; Backx and Marban, 1993) and carotid body cells (Buckler, 1997), and in bovine adrenocortical cells (Enyeart *et al.,* 1996). All these channels are instantaneous and noninactivating. They are also voltage-independent and exhibit an outward rectification. When determined, their single-channel behavior is flickering. The majority of these currents are insensitive to TEA and 4-AP, and they are differentially affected by Ba^{2+}. TREK and TASK channels share all these properties, suggesting that these cloned 4TMS/2P K^+ channels, as well as other uncloned members of the same family, underly the recorded native background K^+ channels. Other results supporting this view are the regulation properties that are shared by some cloned and native channels. One of the best characterized background K^+ currents is the S-type current in *Aplysia* (Siegelbaum *et al.,* 1982; Buttner *et al.,* 1989). Like TREK, the S-type channel is time- and voltage-independent, outwardly rectifying, flickering, and TEA-resistant. Like the S-type channel, TREK is inhibited by increasing the internal cAMP concentration. Moreover, background K^+ currents with properties similar to TREK currents were also recorded from neurons that are activated by application of baclofen, which binds to the GABAb receptor that is negatively coupled to adenylate cyclase (Premkumar *et al.,* 1990a; Wagner and Dekin, 1997).

If the general role of background K^+ channels is clear, that is, the setting and the modulation of the resting membrane potential, the particular function that each native, as well as each cloned channel plays, is not well understood. In *Aplysia,* the S-type current is probably involved in synaptic depression by presynaptic inhibition of neurotransmitter release (Piomelli *et al.,* 1987), and in rat neurons, the GABAb-coupled background K^+ channel may be involved in the generation of the presynaptic inhibitory potentials (Wagner and Dekin, 1997). The very wide tissue distribution of the recently cloned channels suggests that the number of functional roles for these background channels should be much larger than previously suspected.

V. TWIK-RELATED K+ CHANNELS IN *DROSOPHILA* AND *CAENORHABDITIS ELEGANS*

A. ORK1

Functional complementation in *Saccharomyces cerevisiae* has been used to clone plant K^+ channels belonging to the 6TMS/1P family and was successfully used by Goldstein *et al.* (1996) to clone a 4TMS/2P K^+ channel from *Drosophila melanogaster.* When expressed in a yeast strain defective for K^+ uptake, the fly channel confers the ability to grow on low-K^+ medium (Goldstein *et al.,* 1996) upon these cells. This 698-amino-acid channel has the same structural features as mammalian TWIK-related channels, with which it shares around 25% amino acid identity. In *Xenopus* oocytes, it directs the expression of K^+ currents with electrophysiological properties very similar to TASK. This channel, called ORK1 for *O*pen *R*ectifier *K*$^+$ channel, produces K^+-selective currents that are instantaneous, noninactivating, and do not display any rectification other than the rectification predicted by the Goldmann–Hodgkin–Katz equation (GHK rectification). No information concerning the regulation and a possible sensitivity of this channel to pH and polyunsaturated fatty acids has been reported. ORK is blocked by Ba^{2+} but not by TEA. This block by the Ba^{2+} is voltage-dependent and affects only inward currents. ORK mRNA is found in the nervous system of *Drosophila* where it may be involved in setting the resting membrane potential of neurons.

B. *C. elegans* Proteins

Data accumulated from the systematic sequencing of the *Caenorhabditis elegans* genome are providing a unique opportunity to describe the K^+ channel gene repertoire expressed in a single model animal (Wei *et al.*, 1996). When this sequencing program has achieved its goal by the end of 1998, all the possible K^+ channel genes will be identified in this organism. Today, 70–80% of the genomic sequence is available and its analysis gives invaluable information. Genes potentially encoding proteins related to 6TMS/1P and 2TMS/1P subunits in mammals are identified in *C. elegans* as well as genes encoding 4TMS/2P subunits. A surprising result is the abundance of 4TMS/2P channels: More than 39 genes, in a total of 60 genes for the complete set of pore-forming K^+ channel subunits (Salkoff *et al.*, 1997). We do not know whether all these potential channel subunits are expressed and functional, but the conservation of their overall structure (four TMS and the M1P1 extracellular loop) as well as the conservation of their P domain sequences (detailed below) suggests that this is the case. We have aligned 32 of these potential TWIK-related channels together with the two P domain channels identified in other species (a part of this alignment is shown in Fig. 4). Two important points can be deduced from this alignment. First, the amino acid identity between the *C. elegans* subunits is around 25% with a few exceptions where this percentage is more important (30–40% and more than 80% between two channels that probably result from a recent event of gene duplication). Second, from this alignment it is not possible to identify orthologous genes for TWIK and TREK in *C. elegans*. Only one product seems to be more related to TASK than to the others and could correspond to a gene ortholog to the TASK gene. It may be that the orthologous genes of TWIK and TREK have not been sequenced yet (and are localized in the remaining 20% of the unsequenced part of the *C. elegans* genome). However, the easier interpretation of these different results is that the general structure of these channels has been well conserved during evolution but not the primary sequence. This case is very different from that of the 6TMS/1P channels. For these channels, the different subfamilies identified in mammals (for example *Shaker, Slo,* SK, KvLQT) have also been found in *C. elegans* (Wei *et al.*, 1996). This might be

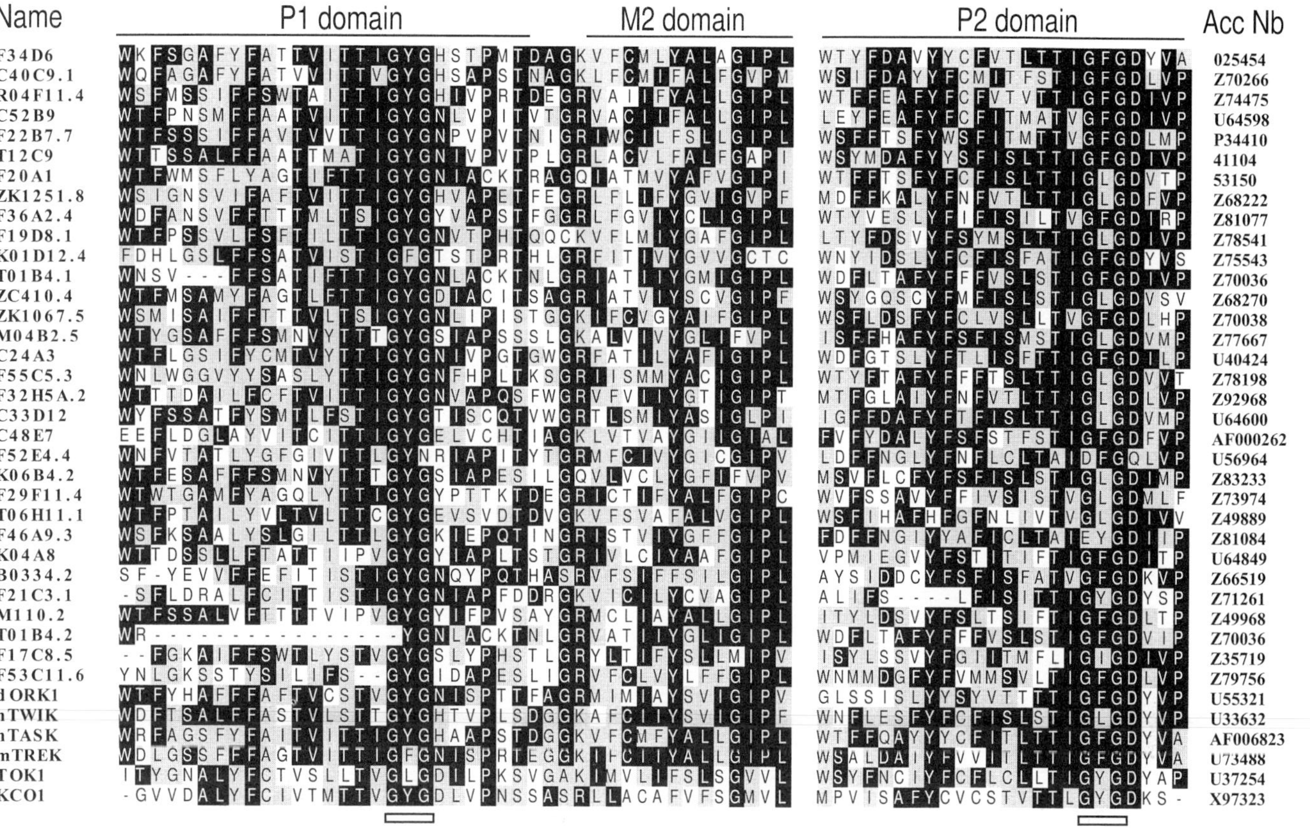

FIGURE 4 Multiple alignment of K^+ channels with two P domains. K^+ channels from *Caenorhabditis elegans* and other species were extracted from Genbank and aligned with ClustalW. P1domains, the amino terminus of the M2 segment, and the P2 domain were formatted with Boxshade. Identical and conserved residues are boxed in black and gray, respectively. The GXG motifs found in the P domains are underlined. Channel names and their Genbank accession numbers are indicated on the left and on the right of the figure, respectively.

due to the fact that the functioning of the 6TMS/1P channels is sophisticated (implicating voltage-dependence, activation, and inactivation mechanisms) and that the sequences that underly these mechanisms have undergone a more important selection pressure than the primary sequence of TWIK-related channels that are time- and voltage-independent.

A major structural feature of the TWIK-related channels is the presence of two P domains. As discussed previously, if both domains are functional, this implies that the active channels are dimers. The availability of the *C. elegans* sequences gives us a unique chance to compare a large number of P domains (P1 and P2) from many different TWIK-related subunits. Alignment of both P domains and a part of the M2 domain of the 32 *C. elegans* sequences and 6 channels identified in other species is shown in Fig. 4. The motif that is the best conserved in the P domains is GXG. In the K^+ channels with one P domain, the residue X flanked by glycines (G) is always a tyrosine (Y) or a phenylalanine (F). When this residue Y or F is replaced by another type of residue, the channel loses its selectivity for K^+ or is no longer functional (Heginbotham *et al.,* 1994). For the TWIK-related channels, the situation seems to be different. In P1, the residue X corresponds mainly to a tyrosine or a phenylalanine (in TREK), as expected, but also to a leucine (L) (in TOK). In P2, X can be a leucine or an isoleucine (I) as well as a phenylalanine or a tyrosine. This could mean that the particular structure of the 4TMS/2P channels can accommodate a larger variety of residues at this position, which is key for K^+ selectivity, than the K^+ channels with only one P domain. Two *C. elegans* proteins that are more closely related to each other (around 40% amino acid identity) than to the other channels (around 25%) are intriguing in this regard: They do not display the canonic GXG motifs, X being Y, F, L, or I. F52E4.4 has the motifs GYN and DFG and F46A9.3 has the motifs GYG and EYG, in the P1 and P2 domains, respectively. These substitutions observed in F52E4.4 and F46A9.3 might be associated with a loss of function. However, their overall structure has been conserved and the majority of the residues besides these motifs that are conserved in the family are also conserved in these channels. A more attractive hypothesis is that these changes are associated with a loss of selectivity and that these channels correspond to a new family of nonselective cationic channels issued from the TWIK-related K^+-selective channel family. The GXG motif is also absent in the nonselective cationic cyclic nucleotide-gated channels that probably have a common ancestor with the 6TMS/1P K^+ channels and modification of this motif in a 6TMS/1P K^+ channel has been shown to be sufficient to produce nonselective cationic channels (Heginbotham *et al.,* 1992).

VI. OTHER TWO P DOMAIN K^+ CHANNELS IN YEASTS AND PLANTS

A. *TOK1*

The yeast *S. cerevisiae* channel, called TOK1/YKC1/YORK/DUK (Ketchum *et al.,* 1995; Zhou *et al.,* 1995; Lesage *et al.,* 1996a; Reid *et al.,* 1996), is the only two P domain K^+ channel characterized to date to have eight potential TMS instead of four. It does not have the extended linker loop M1P1 characteristic of TWIK-related channels and does not share more than 20% amino acid identity with these channels. Currently, its phylogenic relation with the 4TMS/2P K^+ channels is difficult to establish. No 8TMS/2P channel gene has been detected in the large part of the *C. elegans* genome already sequenced. This suggests that this type of channel might be restricted to yeast and absent in higher organisms.

In terms of function, TOK1 channels produces outward rectifying currents whose activation threshold closely follows E_K as observed for TREK currents. However, unlike TREK, the rectification is "strict" and no inward currents can be recorded in high $[K^+]_{out}$. It has been suggested that divalent cations in the extracellular solution might be responsible for a voltage-dependent blocking of inward currents (Ketchum *et al.,* 1995). However, it is clear now that this is not the case and that this rectification is probably due to a mechanism intrinsic to the channel itself (Lesage *et al.,* 1996a; Loukin *et al.,* 1997). On the other hand, upon depolarization TOK1 induces a two-step current composed of instantaneous and delayed slow components, the ratio between both components being dependent on the holding potential. We have proposed a model that takes into account all of these properties (Lesage *et al.,* 1996a). In this model, an endogenous blocking particle such as an external domain of the channel enters the permeation pathway upon hyperpolarization. The presence of two binding sites is hypothesized to account for the two kinetically distinct current components. This view is supported by the fact that Ba^{2+} and quinine display two affinities for TOK1 blockade. TOK1 was also shown to be blocked by TEA. Like the mammalian 4TMS/2P channels, the activity of TOK1 is tightly regulated. TOK1 is inhibited by decreasing the internal pH and increased by activation of protein kinase C (Lesage *et al.,* 1996a).

B. *KCO1*

KCO1 (for *K^+* channel, *Ca^{2+}*-activated, *O*utward rectifying) is the first cloned plant K^+ channel with outward rectification properties (Czempinski *et al.,* 1997). KCO1 has recently been isolated from *Arabidopsis thaliana* where only 6TMS/2P inward rectifier K^+ channels were cloned. Heterologous expression of this channel in insect cells results in noninactivating and outward-rectifying K^+-selective currents. This rectification is of the same type as that of TOK1. A major functional difference between KCO1 and the other 2P channels characterized until now is the activation and inactivation kinetics displayed by the currents. Another feature of KCO1 is its steep Ca^{2+} dependence. Its activation is strongly dependent on the presence of nanomolar concentration of cytosolic free Ca^{2+} ($[Ca^{2+}]_{in}$). No current was detected when $[Ca^{2+}]_{in}$ was adjusted to <150 n*M*. KCO1 strongly activates at increasing $[Ca^{2+}]_{in}$ reaching a maximal value at $[Ca^{2+}]_{in}$, of 300 n*M*. In *Arabidopsis,* KCO1 might participate in repolarizing the membrane potential during signaling events associated with an increase of the internal calcium concentration.

In terms of structure, KCO1 is a 4TMS/2P K^+ channel of 363 residues. However, like TOK1, it seems to be only distantly related to the TWIK family. KCO1 has four TMS but it does not bear the extended M1P1 loop that is found in all other 4TMS/2P found in mammals, worm, and fly. Moreover, its sequence similarity with these channels is very low with less than 20% identity. With its structural and functional differences, KCO1 is an atypical member of the two P domain class of K^+ channels. This may correspond to a special adaptation of this class of channels in plants.

VII. CONCLUSION

Background K^+ currents that are believed to underly the resting membrane conductance of many cell types have been recorded in various animal species ranging from invertebrates to vertebrates. We have cloned several members of this functional family of K^+ channels. They form a novel structural class of pore-forming subunits with two P domains that is widely expressed through the living species. Closely related channel proteins have been isolated or identified in animals as diverse as nematode, fly, and mammals, and some more distantly related channels have been cloned from yeast and plants. The channels that we have isolated produce time- and voltage-independent K^+ currents. Their activity is highly regulated by various agents such as H^+, extracellular K^+ and Na^+, and activators of PKA and PKC.

Many questions concerning K^+ channels with two P domains remain to be answered. Are the homodimers we have identified the functional channel units? Is there a heteromeric dimerization between the different possible subunits? What levels of the K^+ channels are expressed in mammals, and are auxiliary subunits associated with channel subunits of this novel class?

In terms of function, the major goal will be to find a specific pharmacology for the cloned channels to better analyze their role in native cells. This return from cloning to physiology will certainly be the most difficult part of the work. However, success has recently been obtained in studying a variety of hereditary channelopathies concerning K^+ channels with one P domain (Sanguinetti and Spector, 1997). No doubt there will be rapid progress in this direction for K^+ channels with two P domains. This third class of K^+-selective pore-forming subunits is the latest discovered and could be the last one. It is unlikely that nature has created K^+ channels without any P domain. Thus, new K^+ channels that are cloned and expressed will probably fall into one of the three structural classes that are already known.

Acknowledgments

We thank all the investigators of the laboratory who have taken part in the study of K^+ channels with two P domains. We also thank Dr. A. Patel for careful reading of the manuscript, D. Doume for secretarial assistance, and F. Aguila for the artwork.

References

Backx, P. H., and Marban, E. (1993). Background potassium current active during the plateau of the action potential in guinea-pig ventricular myocytes. *Circ. Res.* **72,** 890–900.

Baker, M., Bostock, H., Grafe, P., and Martius, P. (1987). Function and distribution of three types of rectifying channel in rat spinal root myelinated axons. *J. Physiol.* **383,** 45–67.

Buckler, K. J. (1997). A novel oxygen-sensitive potassium current in rat carotid body type I cells. *J. Physiol. (London)* **498,** 649–662.

Buttner, N., Siegelbaum, S. A., and Volterra, A. (1989). Direct modulation of *Aplysia* S-K^+ channels by a 12-lipoxigenase metabolite of arachidonic acid. *Nature* **342,** 553–555.

Carmeliet, E. (1989). K^+ channels in cardiac cells: Mechanisms of activation, inactivation, rectification and K^+e sensitivity. *Pflügers Arch.* **414,** S88–S92.

Chandy, K. G., and Gutman, G. A. (1995). Voltage-gated potassium channel genes. *In* "Ligand and Voltage-Gated Ion Channels" (R. A. North, Ed.), pp. 1–71. CRC Press, Boca Raton, FL.

Czempinski, K., Zimmermann, S., Ehrhardt, T., and MullerRober, B. (1997). New structure and function in plant K^+ channels: KCO1, an outward rectifier with a steep Ca^{2+} dependency. *EMBO J.* **16,** 2565–2575.

Duprat, F., Lesage, F., M. Fink, R. R., Heurteaux, C., and Lazdunski, M. (1997). TASK, a human background K^+ channel to sense external pH variations near physiological pH. *EMBO J.* **16,** 5464–5471.

Enyeart, J. J., Mlinar, B., and Enyeart, J. A. (1996). Adrenocorticotropic hormone and cAMP inhibit noninactivating K^+ current in adrenocortical cells by an a-kinase-independent mechanism requiring ATP hydrolysis. *J. Gen. Physiol.* **108,** 251–264.

Fink, M., Duprat, F., Lesage, F., Reyes, R., Romey, G., Heurteaux, C., and Lazdunski, M. (1996). Cloning, functional expression and brain localization of a novel unconventional outward rectifier K^+ channel. *EMBO J.* **15,** 6854–6862.

Goldstein, S. A. N., Price, L. A., Rosenthal, D. N., and Pausch, M. H. (1996). ORK1, a potassium-selective leak channel with two pore domains cloned from *Drosophila melano-*

gaster by expression in *Saccharomyces cerevisiae. Proc. Natl. Acad. Sci. USA* **93,** 13,256–13,261.

Heginbotham, L., Abramson, T., and Mackinnon, R. (1992). A functional connection between the pores of distantly related ion channels as revealed by mutant K^+ channels. *Science* **258,** 1152–1155.

Heginbotham, L., Lu, Z., Abramson, T., and Mackinnon, R. (1994). Mutations in the K^+ channel signature sequence. *Biophys. J.* **66,** 1061–1067.

Hille, B. (1992). "Ionic Channels in Excitable Membranes." Sinauer Associates, Sunderland, MA.

Hodgkin, A. L., and Huxley, A. F. (1952). A quantitative description of membrane current and its application to conduction and excitation in nerve. *J. Physiol.* **117,** 500–544.

Isomoto, S., Kondo, C., and Kurachi, Y. (1997). Inwardly rectifying potassium channels: Their molecular heterogeneity and function. *Jpn. J. Physiol.* **47,** 11–39.

Jan, L. Y., and Jan, Y. N. (1994). Potassium channels and their evolving gates. *Nature* **371,** 119–122.

Jones, S. W. (1989). On the resting potential of isolated frog sympathetic neurones. *Neuron* **3,** 153–161.

Ketchum, K. A., Joiner, W. J., Sellers, A. J., Kaczmarek, L. K., and Goldstein, S. A. N. (1995). A new family of outwardly rectifying potassium channel proteins with two pore domains in tandem. *Nature* **376,** 690–695.

Kim, D. H., Sladek, C. D., Aguadovelasco, C., and Mathiasen, J. R. (1995). Arachidonic acid activation of a new family of K^+ channels in cultured rat neuronal cells. *J. Physiol. (London)* **484,** 643–660.

Koh, D. S., Jonas, P., Bräu, M. E., and Vogel, W. (1992). A TEA-insensitive flickering potassium channel active around the resting potential in myelinated nerve. *J. Membr. Biol.* **130,** 149–162.

Koyano, K., Tanaka, K., and Kuba, K. (1992). A patch-clamp study on the muscarine-sensitive potassium channel in bullfrog sympathetic ganglion cells. *J. Physiol. (London)* **454,** 231–246.

Lesage, F., Guillemare, E., Fink, M., Duprat, F., Lazdunski, M., Romey, G., and Barhanin, J. (1996a). A pH-sensitive yeast outward rectifier K^+ channel with two pore domains and novel gating properties. *J. Biol. Chem.* **271,** 4183–4187.

Lesage, F., Guillemare, E., Fink, M., Duprat, F., Lazdunski, M., Romey, G., and Barhanin, J. (1996b). TWIK-1, a ubiquitous human weakly inward rectifying K^+ channel with a novel structure. *EMBO J.* **15,** 1004–1011.

Lesage, F., Lauritzen, I., Duprat, F., Reyes, R., Fink, M., Heurteaux, C., and Lazdunski, M. (1997). The structure, function and distribution of the mouse TWIK-1 K^+ channel. *FEBS Lett.* **402,** 28–32.

Lesage, F., Mattei, M. G., Fink, M., Barhanin, J., and Lazdunski, M. (1996c). Assignment of the human weak inward rectifier K^+ channel TWIK-1 gene to chromosome 1q42–q43. *Genomics* **34,** 153–155.

Lesage, F., Reyes, R., Fink, M., Duprat, F., Guillemare, E., and Lazdunski, M. (1996d). Dimerization of TWIK-1 K^+ channel subunits *via* a disulfide bridge. *EMBO J.* **15,** 6400–6407.

Lev, S., Moreno, H., Martinez, R., Canoll, P., Peles, E., Musacchio, J. M., Plowman, G. D., Rudy, B., and Schlessinger, J. (1995). Protein tyrosine kinase PYK2 involved in Ca^{2+}-induced regulation of ion channel and MAP kinase functions. *Nature* **376,** 737–745.

Loukin, S. H., Vaillant, B., Zhou, X. L., Spalding, E. P., Kung, C., and Saimi, Y. (1997). Random mutagenesis reveals a region important for gating of the yeast K^+ channel YKC1. *EMBO J.* **16,** 4817–4825.

Mackinnon, R. (1991). Determination of the subunit stoichiometry of a voltage-activated potassium channel. *Nature* **350,** 232–235.

Mackinnon, R. (1995). Pore loops: An emerging theme in ion channel structure. *Neuron* **14,** 889–892.

Nichols, C. G., and Lopatin, A. N. (1997). Inward rectifier potassium channels. *Annu. Rev. Physiol.* **59,** 171–191.

Ordway, R. W., Singer, J. J., and Walsh, J. V., Jr. (1991). Direct regulation of ion channels by fatty acids. *Trends Neurosci.* **14,** 96–100.

Pardo, L. A., Heinemann, S. H., Terlau, H., Ludewig, U., Lorra, C., Pongs, O., and Stümer, W. (1992). Extracellular K^+ specifically modulates a rat brain potassium channel. *Proc. Natl. Acad. Sci. USA* **89,** 2466–2470.

Pellegrini, M., Simoni, A., and Pellefrino, M. (1989). Two types of K^+ channels in excised patches of somatic membrane of leech AP neuron. *Brain Res.* **483,** 294–300.

Piomelli, D., Volterra, A., Dale, N., Siegelbaum, S. A., Kandel, E. R., Schwarz, J. H., and Belardetti, F. (1987). Lipoxygenase metabolites of arachidonic acid as second messengers for presynaptic inhibition of *Aplysia* sensory neurons. *Nature* **328,** 38–43.

Pongs, O. (1993). Structure-function studies on the pore of potassium channels. *J. Membr. Biol.* **136,** 1–8.

Premkumar, L. S., Chung, S. H., and Gage, P. W. (1990a). GABA-induced potassium channels in cultured neurons. *Proc. R. Soc. Lond. B* **241,** 153–158.

Premkumar, L. S., Gage, P. W., and Chung, S. H. (1990b). Coupled potassium channels induced by arachidonic acid in cultured neurons. *Proc. R. Soc. Lond. B* **242,** 17–22.

Reid, J. D., Lukas, W., Shafaatian, R., Bertl, A., Scheurmannkettner, C., Guy, H. R., and North, R. A. (1996). The *S. cerevisiae* outwardly-rectifying potassium channel (DUK1) identifies a new family of channels with duplicated pore domains. *Recept. Chan.* **4,** 51–62.

Salkoff, L., Butler, A., Nonet, M., and Wei, A. (1997). The impact of the *C. elegans* genome-sequencing project on K^+ channel biology. *Pflügers Arch. Eur. J. Physiol.* R79.

Sanguinetti, M. C., and Jurkiewicz, N. K. (1992). Role of external Ca^{2+} and K^+ in gating of of cardiac delayed rectifier K^+ currents. *Pflügers Arch.* **420,** 180–186.

Sanguinetti, M. C., and Spector, P. S. (1997). Potassium channelopathies. *Neuropharmacology* **36,** 755–762.

Schwalbe, R. A., Bianchi, L., and Brown, A. M. (1997). Mapping the kidney potassium channel ROMK1. *J. Biol. Chem.* **272,** 25,217–25,223.

Shen, K. Z., North, R. A., and Surprenant, A. (1992). Potassium channels opened by noradrenaline and other transmitters in excised membrane patches of guinea-pig submucosal neurones. *J. Physiol.* (*London*) **445,** 581–599.

Sheng, M., and Kim, E. (1996). Ion channel associated proteins. *Curr. Opin. Neurobiol.* **6,** 602–608.

Shih, T. M., and Goldin, A. L. (1997). Topology of the Shaker potassium channel probed with hydrophilic epitope insertions. *J. Cell Biol.* **136,** 1037–1045.

Siegelbaum, S. A., Camardo, J. S., and Kandel, E. (1982). Serotonin and cyclic AMP close single K^+ channels in *Aplysia* sensory neurones. *Nature* **229,** 413–417.

Theander, S., Fahraeus, C., and Grampp, W. (1996). Analysis of leak current properties in the lobster stretch receptor neurone. *Acta Physiol. Scand.* **157,** 493–509.

Wagner, P. G., and Dekin, M. S. (1997). cCAMP modulates an S-type K^+ channel coupled to GABA(B) receptors in mammalian respiratory neurons. *Neuroreport* **8,** 1667–1670.

Wei, A., Jegla, T., and Salkoff, L. (1996). Eight potassium channel families revealed by the *C. elegans* genome project. *Neuropharmacology* **35,** 805–829.

Wu, J. V., Rubinstein, T., and Shrager, P. (1993). Single channel characterization of multiple types of potassium channels in demyelinated *Xenopus* axons. *J. Neurosci.* **13,** 5153–5163.

Yang, J., Jan, Y. N., and Jan, L. Y. (1995). Determination of the subunit stoichiometry of an inwardly rectifying potassium channel. *Neuron* **15,** 1441–1447.

Yue, T. Y., and Marban, E. (1988). A novel cardiac potassium channel that is active and conductive at depolarized potentials. *Pflügers Arch. Eur. J. Physiol.* **413,** 127–133.

Zhou, X. L., Vaillant, B., Loukin, S. H., Kung, C., and Saimi, Y. (1995). YKC1 encodes the depolarization-activated K^+ channel in the plasma membrane of yeast. *FEBS Lett.* **373,** 170–176.

CHAPTER 13

Intracellular Regulation of Inwardly Rectifying Potassium Channels

J. P. Ruppersberg,*,† J. Schultz,* U. Brändle,† B. Fakler,* and U. Schulte*

*Institute of Physiology and †Department of Otolaryngology, University of Tübingen, 72076 Tübingen, Germany

I. INTRODUCTION

Inwardly rectifying K^+ channels encoded by the genes of the Kir superfamily (Doupnik *et al.,* 1995) determine resting potential and membrane excitability in many types of cells (Hille, 1992). Regulation of these channels by intracellular second messengers is therefore assumed to be of great physiological relevance for control of excitability . This is particularly well established for the ATP dependence of Kir6.0 (KATP) channels (Inagaki *et al.,* 1995a,b; Sakura *et al.,* 1995) that control insulin secretion (Ashcroft *et al.,* 1984) and for the regulation of Kir3.0 (GIRK) K^+ channels by G

1063-5823/99 $30.00

proteins (Krapivinsky *et al.*, 1995; Kubo *et al.*, 1993; Lesage *et al.*, 1994) that determine vagal control of the heart rate (Wickman and Clapham, 1995). These two modes of intracellular regulation are addressed in other chapters of this book. This chapter focuses on the less subunit-specific mechanisms of regulation of Kir channels by protein kinases (Cohen *et al.*, 1996; Fakler *et al.*, 1994; McNicholas *et al.*, 1994; Wischmeyer and Karschin, 1996), ATP hydrolysis (Fakler *et al.*, 1994), changes in intracellular pH (*Doi et al.*, 1996; Fakler *et al.*, 1996b; Tsai *et al.*, 1995), and intracellular redox reactions (Ruppersberg and Fakler, 1996). These types of regulation are supposed to control various Kir channels and potentially play a role also in the Kir3 and Kir6 subtypes. Some underlying mechanisms are fairly well understood, others are controversially discussed, and some are still unclear.

The mechanism underlying pH dependence in renal outer medulla (ROMK) Kir1.1 channels is well established: Regulation by intracellular hydrogen ions has been explained by titration of an N-terminal lysine residue whose pK_A is shifted to the neutral pH range (Fakler *et al.*, 1996b). A lysine is also present at the homologous position in Kir4.1 channels (Bond *et al.*, 1994). We therefore address the question of whether this residue plays a similar role in pH regulation of Kir4.1 channels as is known for Kir1.1. Surprisingly, the same site also controls the pH dependence of other subtypes of Kir channels, such as Kir2.1 and Kir6.2, although these subunits exhibit amino acid residues at the homologous positions that are not titratable.

There is still some controversy about intracellular regulation of Kir channels by phosphorylation. Upregulation by protein kinase A (PKA) was first described for Kir1.1 and Kir2.1 channels (Fakler *et al.*, 1994; McNicholas *et al.*, 1994). For Kir2.1 channels this regulation seems to depend on the expression system (Wischmeyer and Karschin, 1996). Moreover, PKA regulation might control binding of the Kir channel subunits to the PDZ domains of chapsyns like PSD-95 (Cohen *et al.*, 1996).

Some types of intracellular regulation of Kir channels are rather complex: One example is dependence on ATP. While KATP channels are blocked by ATP without hydrolyzing it (Ashcroft and Rorsman, 1989), in Kir2.1 ATP hydrolysis reverses the spontaneous rundown (Fakler *et al.*, 1994). A similar process has also been described for Kir6.2 channels where it is independent of the presence of the regulatory SUR subunit (Tucker *et al.*, 1997). Chimeric channels constructed from Kir2.1 and Kir1.1 are analyzed here to define domains involved in this regulation.

Even more complex is the regulation by oxidation and the effects of antioxidants that have been studied in Kir2.1 channels (Ruppersberg and

Fakler, 1996). It is shown here that Kir1.1 also shows a dependence on oxidation that is comparable to that of Kir2.1.

This chapter gives an overview of the multiple modes of intracellular regulation of Kir channels and provides some new insights into mechanisms and subunit specificity of the underlying molecular processes.

II. RESULTS

A. *Regulation by Intracellular pH*

1. Lysine Titration in the N-Terminus of Kir1.1 Channels at Neutral Intracellular pH

Kir1.1 channels that have been localized to the apical membrane of renal epithelial cells of the cortical collecting duct (Lee and Hebert, 1995; Wang *et al.,* 1990) show strong dependence on intracellular pH (Tsai *et al.,* 1995). These channels are fully active around pH 7.4 but almost completely closed at pH values below 7.0. Fitting a dose–response curve to this strong pH dependence yields large Hill coefficients, suggesting that the subunits of the channel interact in a cooperative manner during pH-induced inactivation (Fakler *et al.,* 1996b). In a recent study addressing the molecular determinant of this type of regulation, we showed that a lysine residue in the Nterminus at position 80 is critical for this pH effect (Fakler *et al.,* 1996b). Mutating this residue in Kir1.1 to methionine or threonine removes the strong pH dependence of the resulting channels. In contrast, introducing a lysine residue at the homologous position in Kir2.1 resulted in an identically pronounced pH dependence of the resulting Kir2.1(M84K) channels. Mutating this lysine to arginine irreversibly inactivates the channel whereas mutating it to histidine results in a channel that is as strongly pH dependent as the wild-type channel but at more acidic pH values around pH 5.0. Mutating a cysteine at the same position and modifying it with the positively charged cysteine reagent ethylammonium methanethiosulfonate blocks the channel in a way reminiscent of the arginine mutation. Putting these results together strongly suggested that abnormal titration of lysine 80 at neutral pH underlies block of Kir1.1 channels in response to intracellular acidification (Fakler *et al.,* 1996b).

2. Kir4.1: A Strong Inward Rectifier Regulated by Intracellular pH

Kir4.1 channels (Bond *et al.,* 1994) have not been described in kidney but are highly expressed in many brain regions. On the cellular level they were shown to be expressed in Müller glia cells in the retina (Ishii *et al.,* 1997), in hair cells of the inner ear (Glowatski *et al.,* 1995), and in marginal

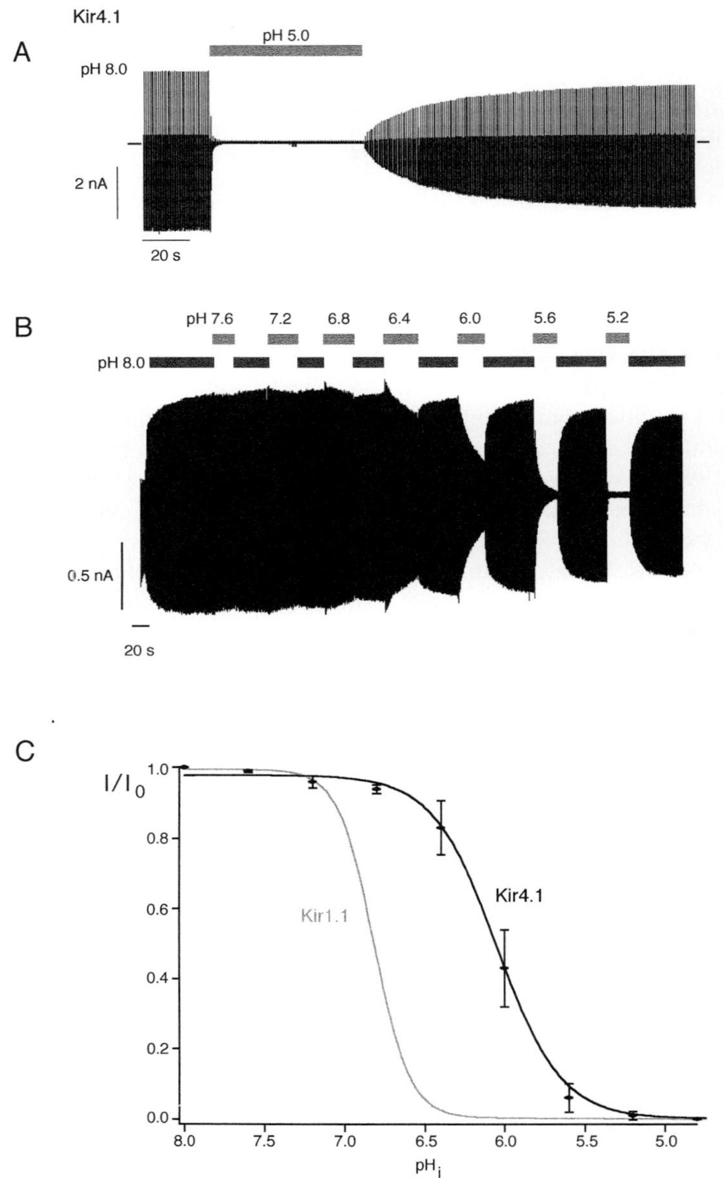

FIGURE 1 Regulation of Kir4.1 channels by intracellular pH. (A) Currents mediated by Kir4.1 channels in response to alternating voltage steps of 100 msec from −80 to 0 or +50 mV. No run down at intracellular pH 8.0. Complete block of the current at intracellular pH 5.0. (B) Currents as in A but at various pH values as indicated. (C) Dose–response curve

cells of stria vascularis (Hibino *et al.*, 1997). Because channels of this type also carry a lysine residue at the position homologous to the pH-regulation site in Kir1.1 channels, a similar pH dependence might be expected. Regulation of these K^+ channels at physiological pH values might be of physiological relevance for pH and K^+ homeostasis in brain. We therefore examined the pH dependence of Kir4.1 channels in giant inside out patches from *Xenopus laevis* oocytes. Figure 1A shows current responses to alternating voltage jumps of 100 msec duration going from -80 to either 0 or 50 mV every 500 msec recorded under symmetrical K^+ concentrations in the absence of intracellular Mg^{2+} or polyamines. At the start, the patch is kept at an intracellular pH 8.0, at which the current remains stable around 3 nA in positive and negative directions. Subsequently, the intracellular pH is switched to 5.0, which results in a complete current decline within 3 sec. The patch is kept at pH 5.0 for 1 min before the intracellular pH is switched back to 8.0, which leads to almost complete recovery within 2 min. This result suggests the existence of a pH-dependent downregulation in Kir4.1 reminiscent of the pH effect on Kir1.1 (for comparison see Fig. 4A). Regulation of Kir4.1 showed even higher reversibility upon realkalinization, as found for Kir1.1; this topic which will be addressed later. Measurements as in Fig. 1A were carried out at various pH values (example in Fig. 1B), which allowed the construction of the dose–response curve (from $n = 11$ patches) shown in Fig. 1C (in which the dose–response curve of Kir1.1 is illustrated in gray for comparison). It is obvious from Fig. 1C that the dose–response curve obtained for Kir4.1 is shifted to acidic values compared to Kir1.1 by approximately one pH unit. This result is surprising because it suggests that the corresponding lysine residue in Kir4.1 titrates at pH values that differ even more from the pK_A value of free lysine (free lysine titrates at pH 10.5). The yet unknown mechanism causing the shift of this pK_A to acidic values must thus be even more effective in Kir4.1 than it is in Kir1.1. In contrast, it has been found for chimeric Kir channels that the titration of the respective lysine is shifted to values closer to the pK_A of free lysine; this was interpreted as a disturbance of channel structure in the close vicinity of the pH-regulation site (Fakler *et al.*, 1996b). The pH dependence of Kir4.1 shown in Fig. 1C shows almost no channel inhibition at pH values between pH 7.4 and 6.8, the range where the pH may be under

---

constructed from measurements as in A and B. Block values were determined by monoexponential fits to the time course of block upon change in pH and normalized to the respective amplitude at pH 8.0. The dose–response curve was fitted with a modified Hill equation: $I(X)/I_0 = 1/\{1 + (X/K_{0.5})^N\}$; with $K_D = K_{0.5}{}^N$, where X is the concentration of intracellular H^+ ions ($[H^+]_i$), $K_{0.5}$ is $[H^+]_i$ at half maximal inhibition, N is the Hill coefficient, and I_0 is the asymptotic value of $I(X)$. The fit resulted in a $pK_{0.5}$ of 6.06 and an N of 2.3.

physiological conditions. This indicates that although the pH dependence of Kir4.1 channels was suggested to be similar to Kir1.1 channels, as concluded from the existence of a lysine residue at the pH-regulation site, they are effectively pH independent in the physiological range. Under pathophysiological conditions, where very acidic pH values may occur, the pH dependence of Kir4.1 might be again relevant, however.

3. KATP Channels of the Kir6.2 Type Show a pH Optimum

Dependence on intracellular pH is also found in some Kir channels that do not carry a lysine residue at the pH-regulation site. One example is Kir6.2, which forms functional channels if coassembled with the sulfonylurea receptor SUR1/2 and which represents the Kir subunit underlying KATP channels (Inagaki *et al.,* 1995a). The current responses displayed in Fig. 2A start with the excision of the giant patch into the inside out configuration. Washout of ATP rapidly activates the channels and a large and fairly stable current is observed while the patch is kept at pH 8.0. Stepping pH to 7.6, 7.2, and 6.8 shows successive increases in current compared to pH 8.0. In contrast, further acidification to 6.4 and 6.0 leads to an obvious current decline, occurring on a slower time scale, however. Thus Kir6.2 channels show a pH-dependent current maximum around neutral pH. The independence of the two underlying processes is best displayed in the step from pH 8.0 to pH 6.0, which shows two phases: a small but rapid increase followed by a slower current decrease. The data in Fig. 2A therefore suggest that there are two inversely pH-dependent processes superimposed on each other: one rather rapid current increase upon acidification and a slower process downregulating the channels in response to the same stimulus of acidification. In order to investigate whether there is any relation between the pH regulation site determined in Kir1.1 and Kir4.1 and one of these effects, we mutated the Kir6.2 subunit, at the homologous position 71, where the wild-type Kir6.2 carries a threonine, to methionine as it is in the rather pH-insensitive Kir2.1 channels (Kir6.2(T71M)). To our surprise this led to an enhanced amplitude of the pH-dependent current optimum, as displayed in Fig. 2B. The absolute position of the optimum was also slightly shifted to more acidic values, suggesting a strengthening of the hydrogen ion-dependent upregulation. At pH 6.0 a pronounced current decline, slightly faster but comparable to the pH 6.0 effect in the wild-type channels, was observed. Although it was unexpected for this mutation to enhance the process that increases current upon acidification, this result clearly demonstrates that the complex pH dependence of Kir6.2 channels is influenced by the same site that mediates pH dependent inactivation of Kir1.1 channels. Mutations at this site might therefore also have an

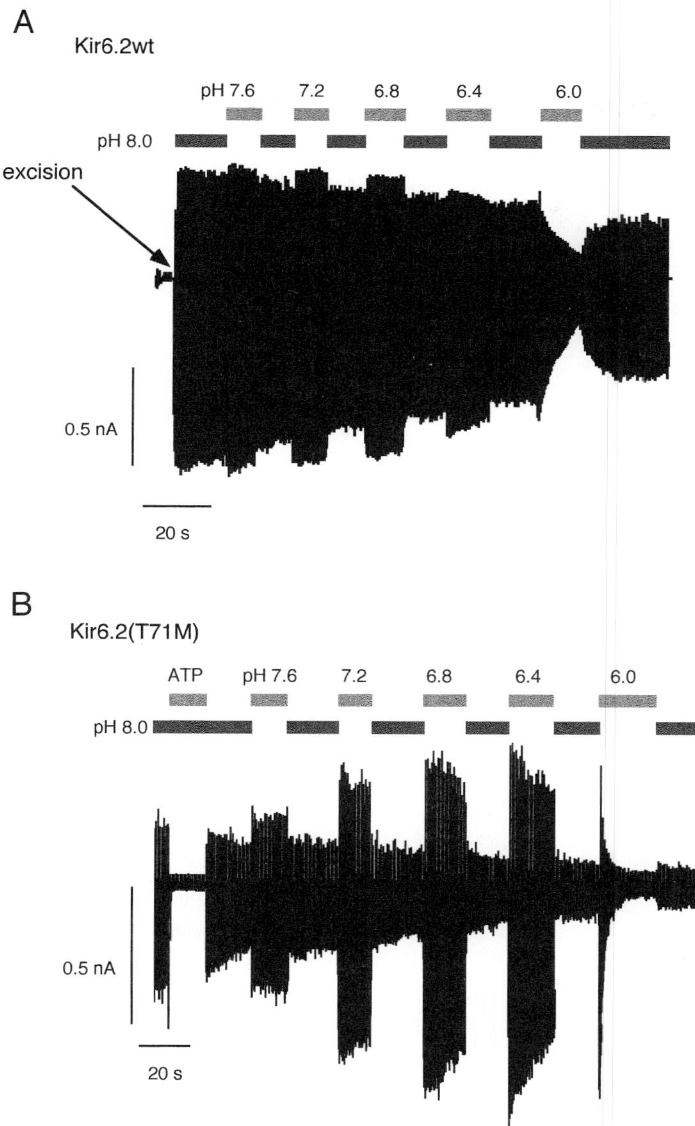

FIGURE 2 Regulation of Kir6.2 channels by intracellular pH. (A) Currents mediated by Kir6.2/SUR1 channels measured as in Fig. 1. Current appears upon excision of the patch. Slow run down at intracellular pH 8.0. Current optimum at intracellular pH 6.8. (B) Currents mediated by Kir6.2(T71M)/SUR1 channels as in A. The pH optimum of the current at an intracellular pH 6.8 is more pronounced. Initial application of ATP visualizes sensitivity to ATP block in this mutant.

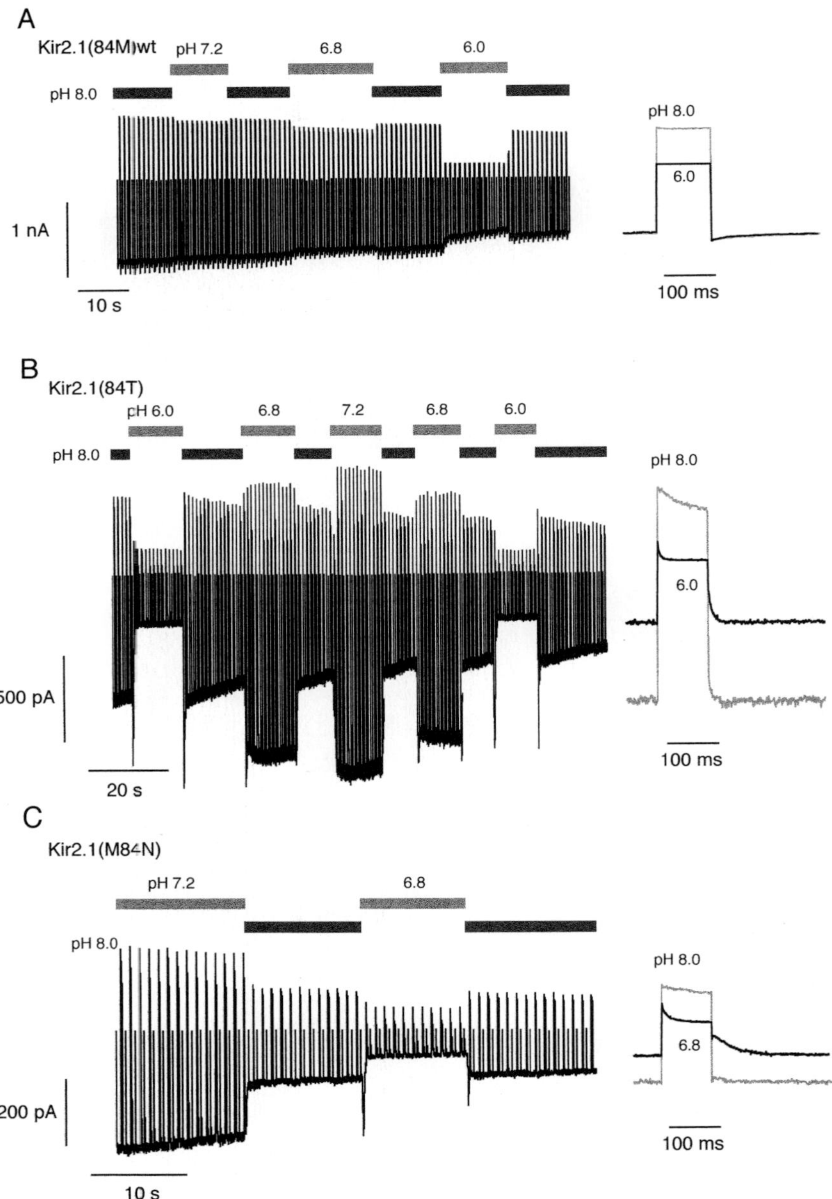

FIGURE 3 Regulation of Kir2.1(84T) channels by intracellular pH. (A) Kir2.1 wild-type channels show insensitivity to changes in intracellular pH. Only outward current is inhibited by voltage-dependent hydrogen ion block. The inset on the right shows that hydrogen ion

effect on channels such as Kir2.1, which are pH insensitive in their wild-type form.

4. Role of the M84T Mutation in Kir2.1 for pH Regulation and Block by Hydrogen Ions

As we have shown previously (Fakler *et al.,* 1996a), Kir2.1 wild-type channels that carry a methionine residue at the position homologous to the pH-regulatory site of Kir1.1 are rather insensitive to changes in intracellular pH. Only at pH values around 6.0 does a voltage-dependent block by hydrogen ions occur, as displayed by the difference of inward and outward current block in Fig. 3A. There is a variant of Kir2.1, however, which we cloned independently from rat skeletal muscle (Fakler *et al.,* 1994) that carries a threonine at this site (Kir2.1(84T)). This variant could not be reproducibly found in rat skeletal muscle and therefore it may be assumed that it might represent a rare allele of Kir2.1. This variant shows significant differences in intracellular block by magnesium ions (Mg^{2+}) and spermine (Fakler *et al.,* 1996b). Since, as shown above, the corresponding site influenced the pH dependence of Kir6.2, we examined the pH dependence of Kir2.1(84T). Surprisingly, the pH dependence of Kir2.1(84T), shown in Fig. 3B, largely differed from that of Kir2.1 wild-type channels (Fig. 3A) and was more similar to the behavior of the Kir6.2 channels shown in Fig. 2A. There is a clear pH optimum around pH 7.2 while the current at pH 8.0 is only one-half and the current at pH 6.0 is only about one-third of the current amplitude at pH 7.2. Moreover, inhibition in inward and outward directions is fairly similar in Kir2.1(84T) on the first view, suggesting that the inhibition at pH 6.0 does not show a similarly pronounced voltage dependence as do the Kir2.1 wild-type channels. This is explained by the higher time resolution visualized in the insets on the right side of Fig. 3, which show that there is a time-dependent closing of the Kir2.1(84T) channels at positive potentials (inset Fig. 3B), suggesting voltage-dependent hydrogen block similar to Kir2.1 wild type but on a slower time scale. For the wild-type channels this voltage-dependent block seems to be faster than the resolution of our measurements and appears to be instantaneous (inset Fig. 3A), effectively reducing the peak amplitude. Voltage-dependent hydrogen block became more effective but developed

---

block occurs instantaneously on an enlarged time scale. (B) Kir2.1(84T) channels show a pH optimum of inward and outward currents at pH 7.2 similar to Kir6.2 channels in Figs. 2A and 2B. The inset displays that a voltage-dependent hydrogen ion block exists also in these channels but develops much slower. (C) The pH dependence of Kir2.1(84N) channels is stronger than that of Kir2.1(84T) channels. The voltage-dependent hydrogen block develops slower (inset).

even slower than in Kir2.1(84T) channels when we mutated an asparagine at position 84 in Kir2.1 (Kir2.1(M84N)). Asparagine is found at the homologous position in Kir3.0 channels. As shown in Fig. 3C, for this mutant both inward and outward current were already markedly smaller at pH 6.8 than at pH 8.0. Because for both Kir2.1(84T) and Kir2.1(M84N) the regulatory effects affect not only the outward but also the inward currents at −80 mV, we conclude that these channel subtypes are regulated in their activity by pH similar to Kir6.2 channels and that this effect is additional to the voltage-dependent block also seen in wild-type channels. However, the mechanism of this pH-dependent regulation leading to an pH optimum is not yet understood.

B. Regulation by Protein Phosphorylation

1. Regulation of Kir2.1 Channels by PKA Depends on the Expression System

Several reports investigated the regulation of Kir channels by protein kinases. Stimulators of protein kinase C such as phorbol esters or SC10 induce downregulation of Kir channels expressed in *Xenopus* oocytes (McNicholas *et al.*, 1994; Fakler *et al.*, 1994; Cohen *et al.*, 1996 Di Magno *et al.*, 1996; Henry *et al.*, 1996; Wischmeyer and Karschin, 1996). It is not known, however, whether the channel protein itself is a target for protein kinase C. Regarding PKA, the published data are somewhat contradictory. Depending on the expression system, either up- or downregulation of Kir channels by PKA, has been reported. Upregulation was observed upon expression of Kir1.1 and Kir2.1 in *Xenopus* oocytes (Fakler *et al.*, 1994 McNicholas *et al.*, 1994) while downregulation was found for Kir2.1 channels expressed in COS7 cells after transient transfection (Wischmeyer and Karschin, 1996). For the inhibitory effect in the transfected cells, a target phosphorylation site at the end of the C terminus was identified, suggesting that this effect may involve direct phosphorylation of the channel protein whereas the upregulatory effect might be mediated by a more complex mechanism. In inside out patches from *Xenopus* oocytes, phosphorylation of this site seems to have no effect, however, since a mutation of the critical serine to alanine results in indistinguishable regulatory behavior (data not shown). Recently it has been pointed out that this phosphorylation site is part of the consensus sequence for binding the C terminus of a channel protein to chapsyns like PSD-95, which have been found to interact with members of the Kir2 family as well as with Kir4.1 (Cohen *et al.*, 1996; Horio *et al.*, 1997). Moreover, it was shown that phosphorylation of this C-terminal serine residue prevents binding to the PDZ domains of the chapsyn . It is unclear at present how much of a relation there is between the observed downregulation by protein kinase A and chapsyn binding.

C. Regulation by Oxidation

1. Regulation of Kir2.1 by Antioxidants

The most complicated intracellular regulation of Kir channels seems to be the effect of oxidants and antioxidants. In inside out patch experiments with Kir2.1 channels, both up- and downregulation of Kir2.1 channels have been observed (Ruppersberg and Fakler, 1996). In some cases both processes run in sequence through an intermediate equilibrium. These results suggest the involvement of more than one regulatory process in this type of regulation. The underlying mechanism is not yet characterized. Upregulation by dithiothreitol (DTT) was found to be not distinct from the effect of the catalytic subunit of PKA (Fakler *et al.*, 1994; Rupperberg and Fakler, 1996). One possible assumption may be that DTT rescues the function of a patch-adherent kinase. The opposite effect then might be the rescue of the corresponding patchadherent protein phosphatase. The latter speculation is supported by the observation that the effect of G-protein stimulation on Kir2.1, which decreases the current, is antagonized by the phosphatase inhibitor microcystin (Ruppersberg and Fakler, 1996). This suggested the possible involvement of a patch-adherent G-protein-controlled phosphatase that might be sensitive to oxidation.

2. pH Dependence of Oxidation in Kir1.1

Regulation of Kir2.1 by DTT varies from patch to patch and from cell to cell and therefore might be based on patch-adherent enzymes rather than cysteine residues of the channel protein itself. As shown below, redox regulation of Kir1.1 channels is very reliable, which might suggest the involvement of cysteine residues of the channel protein. Repetitive acidic inactivation of Kir1.1 channels leads to an increasing loss of channels and there is a nonrecoverable loss in the presence of alkalic pH values (Fig. 4A). Because it was suggested that Kir1.1 channels may be upregulated by ATP (Ho *et al.*, 1993) we tried, without success, to recover activity of these channels by application of 1 m*M* Mg-ATP (Fig. 4A). Because Kir2.1 channels as well as ATP-dependent processes are also upregulated by the antioxidant DTT, we tested the effect of DTT on Kir1.1 channels. As shown in Fig. 4B (left trace), DTT did not influence the current amplitude when it was initially applied at pH 8.0. Subsequent acidification and realkalinization did not lead to any permanent loss of channel activity in the presence

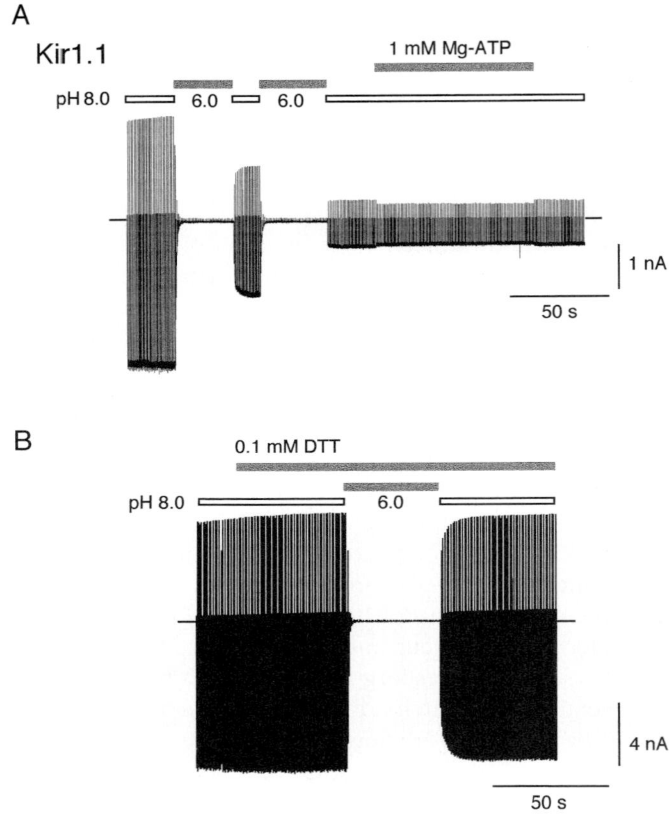

FIGURE 4 Effect of oxidants and antioxidants on pH regulation of Kir1.1 channels. (A) Incomplete recovery from pH-dependent block of Kir1.1 channels. No run down at pH 8.0. Channels are repetitively blocked by application of pH 6.0. Incomplete recovery is not reversed by 1 m*M* Mg-ATP. (B) Complete recovery from pH-dependent block of Kir1.1 channels in the presence of 0.1 m*M* DTT. DTT has no initial effect on the current at pH 8.0.

of 0.1 m*M* DTT. These results suggest that oxidation of Kir1.1 or a related protein occurs predominantly at acidic pH.

D. Regulation by ATP Hydrolysis

1. Regulation of Kir2.1

It was recently shown that Kir2.1 channels show a spontaneous rundown in the absence of Mg-ATP that is reversed by application of Mg-ATP to the internal side of the patch. This effect is most likely mediated by an ATPase and not by a protein kinase since it does not activate the channel permanently upon the use of ATPβS (Fakler *et al.,* 1994). In the absence of Mg-ATP neither DTT nor PKA plus ATP-βS were sufficient to recover channels from rundown. As mentioned above, pH dependence could be transferred to Kir2.1 channels by mutating the methionine residue at position 84 to lysine. As already seen in Kir1.1 channels, transient acidification results in a loss of channels upon realkalinization (Fig. 5A; left trace). This loss of channels induced by acidification was even more pronounced than in Kir1.1. We therefore tried to rescue channels by adding 0.1 m*M* DTT (Fig. 5A; right trace) or 1 m*M* Mg-ATP (Fig. 5B; left trace) in separate experiments. Neither experiments led to substantial recovery of channels upon realkalinization. In contrast, when both DTT and Mg-ATP were coapplied with the acidic solution the result was a complete recovery of the current. This suggests that the pH-dependent inactivation conferred to Kir2.1 channels by the mutation of a lysine residue at position 84 might represent an inactivating transition similar to the spontaneous rundown of Kir2.1 wild-type channels, which is also reversed by DTT and Mg-ATP.

2. ATP Regulation Is Transferred to Kir1.1 Channels by Exchange of the C-Terminus with Kir2.1

Since recovery of the current after acidification required Mg-ATP in Kir2.1(M84K) channels but not in Kir1.1 channels, we were interested in the question of which region of the channel protein mediates this differential Mg-ATP sensitivity. We therefore used chimeric channels previously constructed to investigate the localization of the pH-regulation site (Fakler *et al.,* 1996). As shown in Fig. 5B the C-terminal part of the Kir2.1 channel seems to be involved in the ATP effect. Linking the C terminus of Kir2.1 to the end of the second putative transmembrane segment of Kir1.1 results in channels that do not recover from acidification with 0.1 m*M* DTT alone (Fig. 5B; left trace). Complete recovery is achieved, however, when both DTT and Mg-ATP are coapplied with the acidic solution, as shown above for the Kir2.1(M84K) channels. This strongly suggests that the C terminus of Kir2.1 is involved in the mechanism restoring channel activity by ATPase activity.

E. Regulation by Yet Unknown Intracellular Factors

1. Patch-Cram Experiments with Kir3.2 Channels Suggest Unknown Intracellular Regulatory Factors

Functional expression of K+ channels of Kir3.2 and Kir3.4 in *X. laevis* oocytes results in spontaneous active channels while expression of Kir3.1

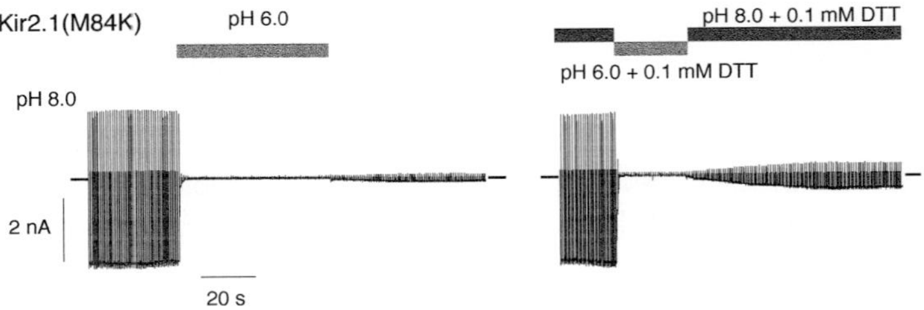

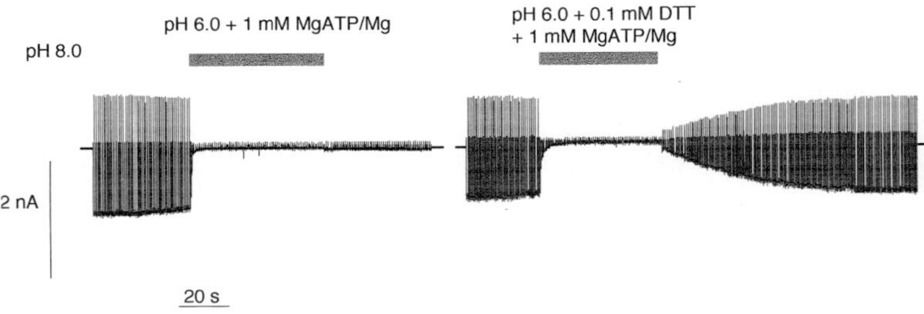

B

Kir1.1-2.1c

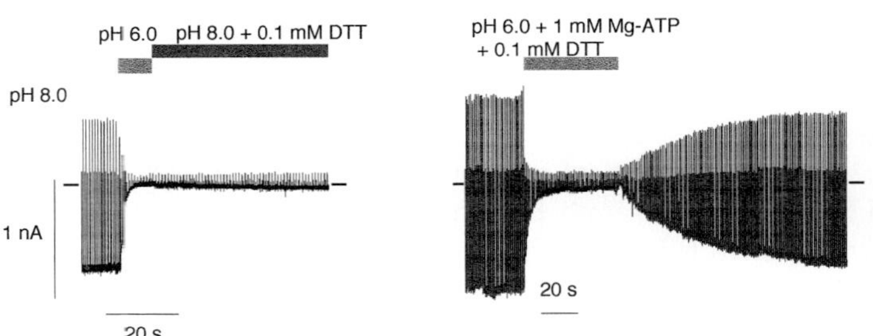

FIGURE 5 Effect of Mg-ATP and antioxidants on pH regulation of Kir2.1(M84K) channels. (A) Left trace: Almost no recovery from pH-dependent block of Kir2.1(M84K) channels. No run down at pH 8.0. Right trace: Incomplete recovery is not reversed by 0.1 m*M* DTT alone. (B) Left trace: Almost no recovery from pH-dependent block of Kir2.1(M84K) channels if 1 m*M* Mg-ATP is coapplied. Right trace: Complete recovery is induced by coapplication of 0.1 m*M* DTT plus 1 m*M* Mg-ATP. (C) Left trace: Incomplete recovery from pH-dependent block of Kir1.1 channels with Kir2.1 C terminus in the presence of 0.1 m*M* DTT. Right trace: Complete recovery is again induced by coapplication of 0.1 m*M* DTT plus 1 m*M* Mg-ATP, indicating a role of the C-terminus in the Mg-ATP effect.

does not result in homomeric channels. Coexpression of Kir3.1 channels with either of the two other subunits leads to the formation of channels with low basal activity and receptor-dependent stimulation via G-protein activation. It is yet unclear, however, whether Kir3.2 and Kir3.4 themselves are regulated by G protein or other intracellular second messengers. We therefore tried to investigate regulation of homomeric Kir3.2 channels in giant inside out patches from *Xenopus* oocytes. Figure 6 shows an experiment in which a giant inside out patch from a *Xenopus* oocyte is excised and the internal side is exposed to standard high K^+ intracellular solution. The current declines spontaneously and does not recover upon any combination of DTT, ATP, GTPγS, and catalytic subunit of PKA (data not shown). Interestingly, as is also visible in Fig. 6, some current is recovered with a physiological concentration of 5 m*M* intracellular Na^+. This corresponds to the results of a recent study that showed that Kir3.1/3.4 channels are upregulated by high concentrations of internal Na^+ (Sui *et al.,* 1996). Because oocytes are known to have low internal Na^+ concentration, it seems unlikely that the rapid current decline upon patch excision is solely due to a washout of Na^+. Almost complete recovery is achieved after

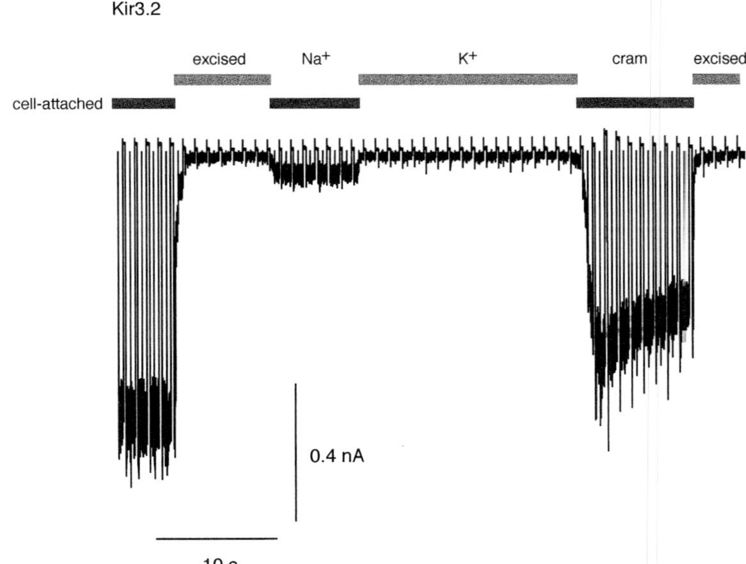

FIGURE 6 Effect of an unknown cytoplasmic factor on Kir3.2 channels. Current in a cell-attached patch with Kir3.2 channels is rapidly lost upon patch excision. Some current is restored on application of 5 m*M* Na^+. There is almost complete recovery of the current after patch cram.

reexposure of the inside of the patch to the cytoplasm of the oocyte. This is performed by cramming the inside out patch back into the oocyte. This experiment suggests that in addition to Na^+ and ATP other yet unknown intracellular factors may be required for maintenance of the currents mediated by Kir3.2.

III. DISCUSSION

At first glance, this chapter addresses a whole zoo of intracellular regulatory effects on Kir channels. By looking more carefully into the interrelations of the effects reported here, a common principle emerges. The different types of pH dependence discussed here all depend on the particular amino acid residue located at the pH-regulation site of Kir1.1 or at its homologous location in the various channel subunits. The role of this site can be direct titration of a lysine residue present at this site but there are also effects on intracellular pore block by hydrogen ions and on an as yet unclear mechanism generating a pH optimum around pH 7. For the lysine titration, this site is shown to induce a state transition of the channels that closely interacts with other regulatory effects. It allows reversible or irreversible inactivation by oxidation or makes some channels dependent on Mg-ATP.

In view of these interactions, it seems likely that the channels in all experiment discussed here undergo very similar conformational changes from an open to an inactive state. It might furthermore be speculated that the type of state transition discussed here is independently superimposed on the more specific types of regulation by ATP in Kir6 and by G proteins in Kir3 channels. State-dependent oxidation of Kir1.1 channels might serve to further elucidate the structural basis of the underlying transition in future mutagenesis studies.

Besides this possible common principle of Kir inactivation, these channels have, of course, their individuality with respect of this type of regulation. The first example is that Kir4.1 channels, although carrying a lysine residue at the pH-regulation site, are not pH regulated at neutral but at more acidic pH values. This result is surprising since it implies that this lysine residue titrates at even more acidic pH values although a physiological necessity for this is not immediately obvious. Since heteromultimer formation between Kir1.1 and Kir4.1 has been shown (Glowatzki *et al.,* 1995) it might be possible that nature generates K^+ channels with tuned pH sensitivity by varying the subunit composition. Such physiological relevance needs to be shown, however.

Another issue of individuality, not fully understood at present, is the differential effect of the various amino acid residues at the pH-regulation site of Kir6.2 and Kir2.1 channels. While a methionine at that site makes Kir2.1 channels completely insensitive to changes in internal pH, the same residue increases pH dependence in Kir6.2 channels. In contrast, the threonine residue that mediates pH sensitivity of the putative Kir2.1(84T) allele and that is present in wild-type Kir6 channels at this position makes the latter less sensitive to changes in pH.

Sensitivity to Mg-ATP was originally described in Kir1.1 channels (Ho *et al.*, 1993) and is also present in Kir6.2 channels (Tucker *et al.*, 1997). Nevertheless, Mg-ATP reverses the downregulatory effect of acidification only in Kir2.1(M84K) channels and not in Kir1.1 wild-type channels, while an antioxidant like DTT is required for current recovery in both channel types. It remains, therefore, to be elucidated whether the mechanisms of the Mg-ATP and DTT effect are the same in all channel subtypes. Similarly unclear is the effect of protein kinases, although there seems to be a well-defined downregulatory action of protein kinase C on Kir2.3 channels (Henry *et al.*, 1996).

Finally, there are some reports of intracellular regulation by metal ions like Mg^{2+} (Chuang *et al.*, 1997) and Na^+ (Sui *et al.*, 1996) regulating Kir channels. These regulatory mechanisms are not yet understood but may be important links between ion transport and K^+ channel activity.

IV. MATERIALS AND METHODS

A. Mutagenesis and cRNA Synthesis

All mutants were prepared according to Herlitze and Koenen (1990), subcloned into pBF1/2 vector, and the mutation verified by sequencing. The chimeric construct Kir1.1–2.1c corresponds to RO-IR(C) of Doi *et al.* (1996).

B. cRNA Injection and Preparation of *Xenopus* Oocytes

Capped cRNA was synthesized *in vitro* using SP6 polymerase (Promega, Heidelberg, FRG) and stored in stock solutions at −70°C. *Xenopus* oocytes were surgically removed from adult females and manually dissected. About 50 nl of cRNA solution was injected into Dumont Stage VI oocytes. Oocytes were treated with collagenase Type II (Sigma, 0.5 mg/ml) and incubated at 19°C for 1–3 days prior to use.

C. Electrophysiology

All voltage- and patch-clamp experiments were performed at room temperature (approximately 22°C) 3–7 days after injection. Expression of Kir channels was first tested by two-microelectrode recordings, and oocytes providing inwardly rectifying currents of more than 10 μA at a membrane potential of -100 mV were used for patch-clamp experiments. Giant patch pipets (Hilgemann *et al.,* 1991) were made from thick-wall borosilicate glass, had resistances of 0.3 to 0.6 MΩ (tip diameter of 20 to 30 μm), and were filled with (in m*M*) 120 KCl, 10 Hepes, and 1.8 $CaCl_2$. Currents were recorded and corrected for capacitive transients using an EPC9 amplifier (HEKA Electronics, Lamprecht, FRG), whose analog filter was set to 3 kHz (-3 dB). Leakage correction was not performed. Data were sampled at about 0.3 kHz and stored on optical disk.

Solutions were applied to the cytoplasmic side of the excised patches via three motor-driven double-barrel pipets arranged in parallel. Solutions had the following composition (in m*M*, pH adjusted to the values indicated with KOH) with the concentration of free Mg^{2+} ions calculated according to Fabiato (1988): standard intracellular solution without ATP (120 KCl, 10 HEPES, 10 EGTA), which was also used as bathing solution; and standard intracellular solution with Mg-ATP (120 KCl, 10 Hepes, 10 EGTA, 2.44 $MgCl_2$, 1.13 K_2ATP); DTT and DTNB were added in the concentrations indicated. DTNB, DTT, and ATP were purchased from Sigma (St. Louis, MO).

Acknowledgments

We thank F. Ashcroft for providing Kir6.2 and SUR1, J. P. Adelman for Kir4.1, and C. Stansfeld for reading the manuscript. This work was supported by the Deutsche Forschungsgemeinschaft Ru535/4-1.

References

Ashcroft, F. M., Harrison, D. E., and Ashcroft, S. J. H. (1984). Glucose induces closure of single potassium channels in isolated rat pancreatic β-cells. *Nature* **312,** 446–448.

Ashcroft, F. M., and Rorsman, P. (1989). Electrophysiology of the pancreatic beta-cell. *Prog. Biophys. Mol. Biol.* **54,** 87–143.

Bond, C. T., Pessia, M., Xia, X. M., Lagrutta, A., Kavanaugh, M. P., and Adelman, J. P. (1994). Cloning and expression of a family of inward rectifier potassium channels. *Recept. Chan.* **2,** 183–191.

Chuang, H.-h., Jan, Y. N., and Jan, L. Y. (1997). Regulation of IRK3 inward rectifier K^+ channel by m1 acetylcholine receptor and intracellular magnesium. *Cell,* **89,** 1121–1132.

Cohen, N. A., Brenman, J. E., Snyder, S. H., and Bredt, D. S. (1996). Binding of the inward rectifier K^- channel Kir2.3 to PSD-95 is regulated by protein kinase a phosphorylation. *Neuron* **17,** 759–767.

Di Magno, L., Dascal, N., Davidson, N., Lester, H. A., and Schreibmayer, W. (1996). Serotonin and protein kinase C modulation of a rat brain inwardly rectifying K^+ channel expressed in xenopus oocytes. *Pflugers Arch.* **431,** 335–340.

Doi, T., Fakler, B., Schultz, J. H., Schulte, U., Brändle, U., Weidemann, S., Zenner, H.-P., Lang, F., and Ruppersberg, J. P. (1996). Extracellular K^+ and intracellular pH allosterically regulate renal Kir1.1 channels. *J. Biol. Chem.* **271,** 17,261–17,266.

Doupnik, C. A., Davidson, N., and Lester, H. A. (1995). The inward rectifier potassium channel family. *Curr. Opin. Neurobiol.* **5,** 268–277.

Fabiato, A. (1988). Computer programs for calculating total from specified free or free from specified total concentrations in aqueous solutions containing multiple metals and ligands. *Methods Enzymol.* **157,** 378–417.

Fakler, B., Bond, C. T., Adelman, J. P., and Ruppersberg, J. P. (1996a). Heterooligomeric assembly of inward-rectifier K^+ channels from subunits of different subfamilies: Kir2.1 (IRK1) and Kir4.1 (BIR10). *Pflügers. Arch.* **433,** 77–83.

Fakler, B., Brändle, U., Glowatzki, E., Zenner, H.-P., and Ruppersberg, J. P. (1994). Kir2.1 inward rectifier K^+ channels are regulated independently by protein kinases and ATP hydrolysis. *Neuron,* **13,** 1413–1420.

Fakler, B., and Ruppersberg, J. P. (1996). Functional and molecular diversity classifies the family of inward-rectifier K^+ channels. *Cell. Physiol. Biochem.* **6,** 195–209.

Fakler, B., Schultz, J., Yang, J., Schulte, U., Brändle, U., Zenner, H. P., Jan, L. Y., and Ruppersberg, J. P. (1996b). Identification of a titratable lysine residue that determines sensitivity of kidney potassium channels (ROMK) to intracellular pH. *EMBO J.* **15,** 4093–4099.

Glowatzki, E., Fakler, G., Brändle, U., Rexhausen, U., Zenner, H. P., Ruppersberg, J. P., and Fakler, B. (1995). Subunit-dependent assembly of inward-rectifier K^+ channels. *Proc. R. Soc. B* **261,** 251–261.

Henry, P., Pearson, W. L., and Nichols, C. G. (1996). Protein kinase C inhibition of cloned inward rectifier (HRK1/KIR2.3) K^+ channels expressed in *Xenopus* oocytes. *J. Physiol.* **495,** 681–688.

Herlitze, S., and Koenen, M. (1990). A general and rapid mutagenesis method using polymerase chain reaction. *Gene* **91,** 143–147.

Hilgemann, D. W., Nicoll, D. A., and Philipson, K. D. (1991). Charge movement during Na^+ translocation by native and cloned cardiac Na^+/Ca^{2+} exchanger. *Nature* **352,** 715–718.

Hille, B. (1992). "Ionic Channels in Excitable Membranes," 2nd ed. Sinauer Associates, Sunderland, MA.

Ho, K., Nichols, C. G., Lederer, W. J., Lytton, J., Vassilev, P. M., Kanazirska, M. V., and Hebert, S. C. (1993). Cloning and expression of an inwardly rectifying atp-regulated potassium channel. *Nature* **362,** 31–38.

Horio, Y., Hibino, H., Inanobe, A., Yamada, M., Ishii, M., Tada, Y., Satoh, E., Hata, Y., Takai, Y., and Kurachi, Y. (1997). Clustering and enhanced activity of an inwardly rectifying potassium channel, Kir4.1, by an anchoring protein, PSD-95/SAP90. *J. Biol. Chem.* **272,** 12,885–12,888.

Inagaki, N., Gonoi, T., Clement, J. P. T., Namba, N., Inazawa, J., Gonzalez, G., Aguilar-Bryan, L., Seino, S., and Bryan, J. (1995a). Reconstitution of IKATP: An inward rectifier subunit plus the sulfonylurea receptor. *Science* **270,** 1166–1170.

Inagaki, N., Tsuura, Y., Namba, N., Masuda, K., Gonoi, T., Horie, M., Seino, Y., Mizuta, M., and Seino, S. (1995b). Cloning and functional characterization of a novel ATP-sensitive potassium channel ubiquitously expressed in rat tissues, including pancreatic islets, pituitary, skeletal muscle, and heart. *J. Biol. Chem.* **270,** 5691–5694.

Ishii, M., Horio, Y., Tada, Y., Hibino, H., Inanobe, A., Ito, M., Yamada, M., Gotow, T., Uchiyama, and Kurachi, Y. (1997). Expression and clustered distribution of an inwardly rectifying potassium channel, KAB-2/Kir4.1, on mammalian retinal Müller cell membrane: Their regulation by insulin and laminin signals. *J. Neurosci.* **17,** 7725–7735.

Krapivinsky, G., Gordon, E. A., Wickman, K., Velimirovic, B., Krapivinsky, L., and Clapham, D. E. (1995). The G-protein-gated atrial K^+ channel IKACh is a heteromultimer of two inwardly rectifying K^+-channel proteins. *Nature* **374,** 135–141.

Kubo, Y., Reuveny, E., Slesinger, P. A., Jan, Y. N., and Jan, L. Y. (1993). Primary structure and functional expression of a rat G-protein-coupled muscarinic potassium channel. *Nature* **364,** 802–806.

Lee, W.-S., and Hebert, S. (1995). ROMK inwardly rectifying ATP-sensitive K^+ channel. I. Expression in rat distal nephron segments. *Am. J. Physiol.* **268,** F1124–F1131.

Lesage, F., Duprat, F., Fink, M., Guillemare, E., Coppola, T., Lazdunski, M., and Hugnot, J.-P. (1994). Clonign provides evidence for a family of inward rectifier and G-protein coupled K^+ channels in the brain. *FEBS Lett.* **353,** 37–42.

McNicholas, C. M., Wang, W., Ho, K., Hebert, S. C., and Giebisch, G. (1994). Regulation of ROMK1 K^+ channel activity involves phosphorylation processes. *Proc. Natl. Acad. Sci. USA* **91,** 8077–8081.

Ruppersberg, J. P., and Fakler, B., (1996). Complexity of the regulation of Kir2.1 K^+ channels. *Neuropharmacology* **35,** 887–893.

Sakura, H., Ämmälä, C., Smith, P. A., Gribble, F. M., and Ashcroft, F. M. (1995). Cloning and functional expression of the cDNA encoding a novel ATP-sensitive potassium channel expressed in pancreatic β-cells, brain, heart and skeletal muscle. *FEBS Let.* **377,** 338–344.

Sui, J. L., Chan, K. W., and Logothetis, D. E. (1996). Na^+ activation of the muscarinic K^+ channel by a G-protein-independent mechanism. *J. Gen. Physiol.* **108,** 381–391.

Tsai, T. D., Shuck, M. E., Thompson, D. P., Bienkowski, M. J., and Lee, K. S. (1995). Intracellular H^+ inhibits a cloned rat kidney outer medulla K^+ channel expressed in *Xenopus* oocytes. *Am. J. Physiol.* **268,** C1173–C1178.

Tucker, S. J., Gribble, F. M., Zhao, C., Trapp, S., and Ashcroft, F. M. (1997). Truncation of Kir6.2 produces ATP-sensitive K^+ channels in the absence of the sulphonylurea receptor. *Nature* **387,** 179–183.

Wang, W., Schwab, A., and Giebisch, G. (1990). Regulation of small-conductance K^+ channel in apical membrane of rat cortical collecting tubule. *Am. J. Physiol.* **259,** F494–F502.

Wickman, K., and Clapham, D. E. (1995). Ion channel regulation by G proteins. *Physiol. Rev.* **75,** 865–885.

Wischmeyer, E., and Karschin, A., (1996). Receptor stimulation causes slow inhibition of IRK1 inwardly rectifying K^+ channels by direct protein kinase A-mediated phosphorylation. *Proc. Nat. Acad. Sci. USA* **93,** 5819–5823.

CHAPTER 14

Regulation of Ion Channels by Membrane Proteins

Hiroshi Hibino,* Andre Terzic,† Atsushi Inanobe,* Yoshiyuki Horio,* and Yoshihisa Kurachi*

*Department of Pharmacology II, Faculty of Medicine, Osaka University, Osaka 565, Japan; and †Department of Pharmacology and Medicine, Mayo Clinic, Rochester, Minnesota 55905

Current Topics in Membranes, Volume 46

1063-5823/99 $30.00

I. INTRODUCTION

A variety of cells including neuronal cells, cardiac myocytes, and skeletal muscle, possess inwardly rectifying K^+ (Kir) channels through which currents flow more readily in the inward than in the outward direction (Isomoto *et al.,* 1997). These K^+ channels play pivotal roles in many physiological functions such as maintenance of the resting membrane potential, regulation of the action potential duration, receptor- and metabolism-dependent regulation of cellular excitability, and secretion and absorption of K^+ ions across cell membranes. Recent molecular biological dissection shows that the DNAs encoding Kir channels constitute a new family of K^+ channels, whose subunit contains two putative transmembrane domains and one pore-forming region. So far, more than 10 cDNAs of Kir channel subunits have been isolated and may be classified into five subfamilies: (1) IRK subfamily (IRK1-3/Kir2.1–2.3), (2) GIRK subfamily (GIRK1-4/Kir3.1–3.4), (3) ATP-dependent Kir subfamily (ROMK1/Kir1.1, K_{AB}-2/Kir4.1), (4) ATP-sensitive Kir subfamily (uK_{ATP}-1/Kir6.1, BIR/Kir6.2), and (5) Kir5.1 subfamily (Fig. 1). The function of Kir5.1 has not yet been examined well.

Xenopus oocytes injected with cRNAs of IRKs elicit the classical Kir channel currents. GIRKs as heteromultimers compose the G-protein-gated Kir (K_G) channels, which are regulated by a variety of G_i/G_o-coupled inhibitory neurotransmitter receptors, such as m_2-muscarinic, serotonergic (5-HT_{1A}), $GABA_B$, somatostatin, and opioid (μ, δ, κ) receptors. ROMK1 and K_{AB}-2 are characterized by a Walker Type A ATP-binding motif in their carboxyl termini, and may be involved in K^+ transport in renal epithelial and

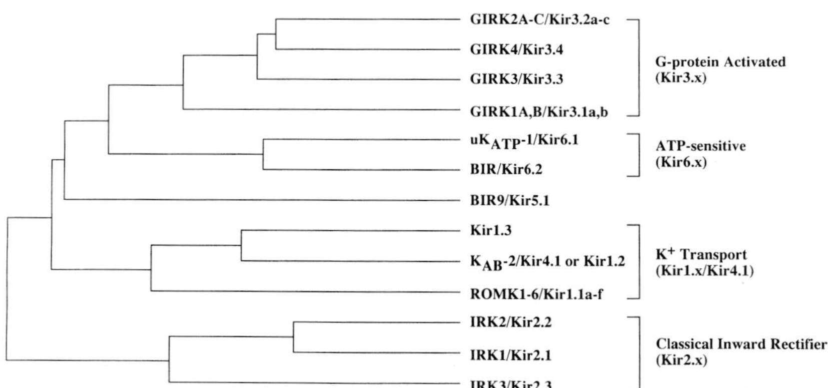

FIGURE 1 The evolutionary tree of the inwardly rectifying K^+ channel family with two membrane-spanning domains.

brain glial cells. uK_{ATP}-1 and BIR form, with sulfonylurea receptors, the so-called ATP-sensitive K^+ channels. The Na^+-activated Kir channels identified electrophysiologically in neurons and cardiac myocytes have not yet been cloned, but may belong to this family of K^+ channels. Thus, each subfamily can be characterized as being responsible for one specific and important physiological function.

The Kir subunits, as described above, exhibit the simplest structure as ion channels. This simple structure is shared by the P_{2X}-receptor channel, the epithelial amiloride-sensitive Na^+ channel, and proton-sensing nonselective cation channels (Surprenant *et al.,* 1995; Cannessa *et al.,* 1993; Waldmann *et al.,* 1997). The simple structure of these ion channels may indicate that they are ancestors of ion channels and may have long histories in evolution. Thus, these channels may have evolved during their long histories to be involved in various essential cellular functions. Control of the function of Kir subunits as ion channels and their localization in the cell membrane by various other proteins may be the keys to understanding how the cells have evolved to utilize the subunits for their physiological roles.

II. MODULATION OF Kir CHANNEL FUNCTION

Ion channels are usually classified into two major categories (Hille, 1992): one is voltage dependent and the other is ligand (Table I; gated Hille, 1992). If strictly adopted, this classification may not include Kir channels. The Kir channels are not regulated purely by voltage across the cell membrane, because the voltage-sensing mechanism, which is usually the re-

TABLE I

Regulation of Ion Channels

Regulation of channel function
1. Voltage; primary regulator of ion channels
2. Extracellular ligands; neurotransmitters such as ACh, glutamic acid, GABA
Intracellular substances
3. Intracellular ligands; ATP, cyclic nucleotides, Ca^2, Mg^{2+}
4. Phosphorylation
Proteins
5. β subunits
6. G proteins
7. ABC proteins; sulfonyl urea receptors, CFTR, p-glycoproteins (?)
8. Anchoring proteins; PSD and SAP family

Note. See the text for further details and abbreviations.

peated positively charged amino acids located in the S4 segment of voltage-dependent K^+, Na^+, and Ca^{2+} channels, is lacking in Kir subunits (Papazian *et al.,* 1991). The Kir subunits themselves do not possess ligand-binding sites for neurotransmitters and hormones, and thus are not directly regulated by ligands. The control of Kir channel activity can be divided into three categories: the first is the direct modulation of Kir subunits by ions, phosphorylation, and substances, the second is control of Kir subunit function by other proteins, and the third is control of Kir channel function by cytoskeleton and related proteins (Fig. 2). The first category includes (1) Mg^{2+} and polyamine interaction causing the inwardly rectifying property of the Kir channels (Falker *et al.,* 1994; Lopatin *et al.,* 1994; Ficker *et al.,* 1994; Yamada and Kurachi, 1995); (2) intra- and extracellular proton inhibition of Kir channel activity (Coulter *et al.,* 1995; Doi *et al.,* 1996), and (3) phosphorylation-induced inhibition of Kir channel activity (Wischmeyer and Karschin, 1996). In some Kir channels, Mg^{2+} regulates not only the rectifying property but also the channel activity, and thus can be a second messenger of several neurotransmitters (Chuang *et al.,* 1997). These subunits are discussed in other chapters of this book. In the second category, are examples featuring Kir subunit function. For example, the trimeric GTP-binding proteins couple G-protein-gated K^+ channel and inhibitory membrane receptors (Kurachi, 1996). Recently it has been clarified that an ATP-binding cassette protein (i.e., sulfonylurea receptors) controls the activity of the Kir6.0 subfamily Kir channel and forms so-called ATP-sensitive K^+ channels (Inagaki *et al.,* 1995). The third category, which has recently been shown to be involved in control of Kir channel function, is the cytoskeleton and related proteins, including actin filaments and the PSD family of anchoring proteins (Terzic and Kurachi, 1996; Cohen *et al.,* 1996; Horio *et al.,* 1997). Evidence has been accumulating rapidly that cytoskeleton effects on ion channels including Kir may be an essential component in regulation of function and localization of ion channels in various tissues. Because this subject has been studied well in ion channels other than Kir, such as the nicotinic acetylcholine receptor and glycine receptor channels, we summarize the current knowledge of cytoskeleton effects not only on Kir channels but also on other ion channels.

III. CYTOSKELETON EFFECTS ON ION CHANNELS

The cytoskeleton forms fibrillar network structures throughout the cytosol, including the microenvironment surrounding ion channel proteins within the plasma membrane. The cytoskeleton is made of microfilaments, microtubules, intermediate filaments, and associated proteins (Gallo-Payet and Payet, 1995). Microfilaments are composed of G-actin monomers, which

Mechanisms for Kir Channel Control of Cellular Function

(1) Control of Channel Properties
- (a) Inward-rectification — Mg^{2+}_i, polyamines
- (b) Channel Activity — ions (H^+_i, Mg^{2+})
- (c) Regulatory Proteins — Sulfonylurea receptors (SURs); G proteins (βγ subunits)

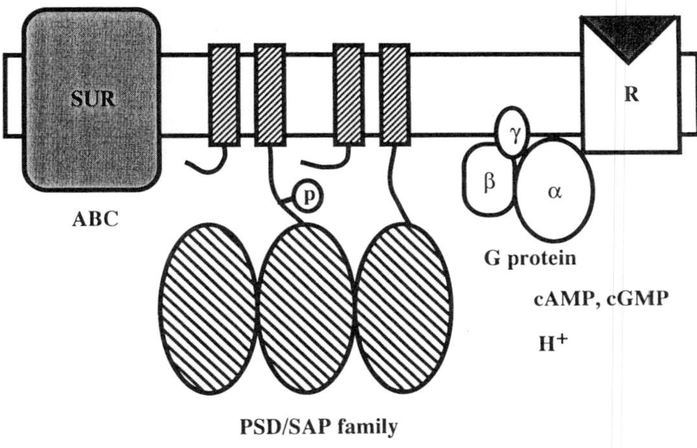

(2) Control of Channel Localization
- (a) PSD/SAP anchoring proteins
- (b) other unidentified mechanisms

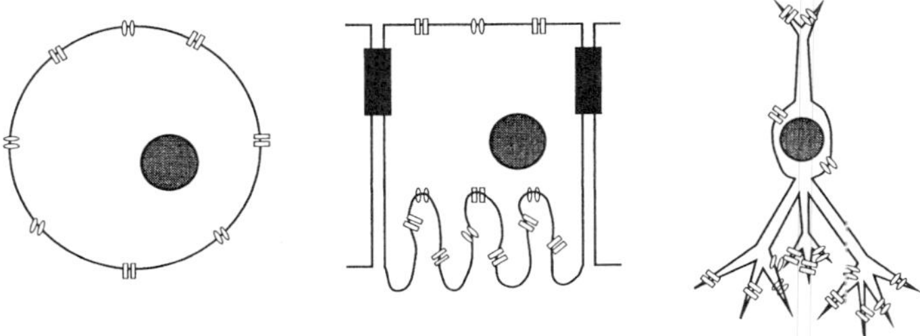

FIGURE 2 Mechanisms for Kir channel control of cellular function.

assemble into actin polymers (F-actin). Actin polymers link into three-dimensional frameworks, which crosslink myosin or form cortical networks at the cell periphery. There is a dynamic equilibrium between polymeric forms of actin and G-actin monomers (Fig. 3A). The status of actin and the state of myofilament organization are set by diverse actin-binding proteins,

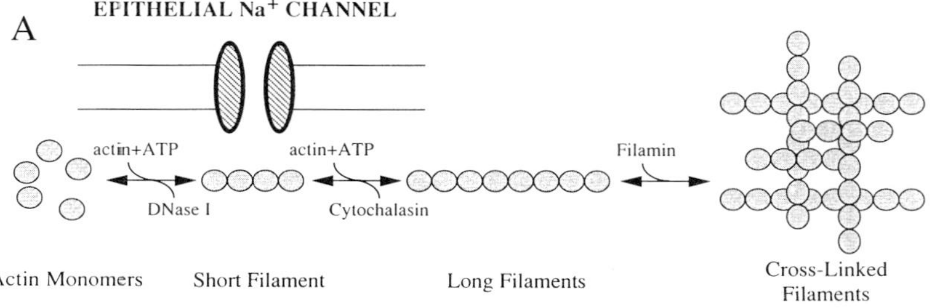

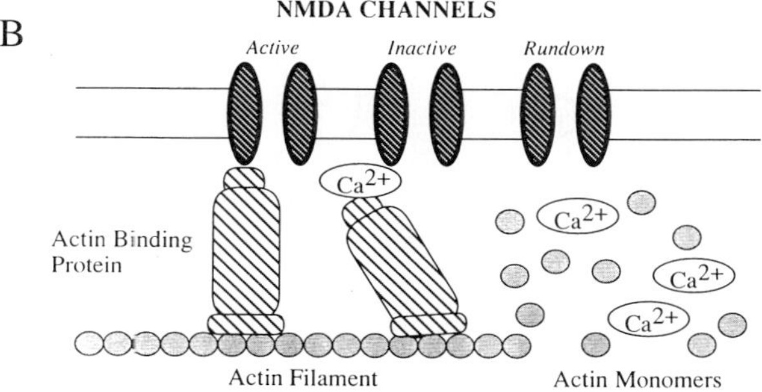

FIGURE 3 (A) A simplified scheme of dynamic interactions among various conformations of actin. In the presence of MgATP, monomeric G-actin nucleates into short actin filaments (e.g., tetramers) that anneal to produce long actin filaments. DNase I inhibits polymerization by binding with G-actin. Cytochalasins favor the transition from long to short filaments. Crosslinking proteins, such as filamin, induce the gelation of actin filaments. The length of actin filaments appears to be a determinant of channel activation. For example, an increase in the availability of short actin filaments enhances the probability of renal epithelial Na^+ channels to being open. Modified from Cantiello (1995) (see also Gallo-Payet and Payet, 1995). (B) Proposed model of actin-dependent regulation of NMDA channels. This model (Rosenmund and Westbrook, 1993) includes an actin-binding protein that dissociates from the NMDA channel in a Ca^{2+}-dependent manner, leading the channel to inactivation. Calcium-dependent actin filament depolymerization results in removal of the actin-binding protein and channel rundown.

including profilin (which binds G-actin), gelsolin (which caps F-actin), and filamin and α-actinin (which crosslink microfilaments). Microtubules are composed of α- and β-tubulin heterodimers that form tubular filaments. Polymerization, stabilization, and modulation of microtubule function are regulated by microtubule-associated proteins. The cytoskeleton is essential not only in the maintenance of cell shape and motility, but also in the distribution, stability, and function of integral membrane proteins (Bennett and Gilligan, 1993; Hitt and Luna, 1994).

Interactions between various ion channel proteins and cytoskeletal structures are numerous and have been implicated in mediating spatial sorting of surface channel proteins and in the regulation of channel activity (Cantiello, 1995; Gomperts, 1996; Smith and Benos, 1996).

IV. SITE-DIRECTED ION CHANNEL DISTRIBUTION BY CYTOSKELETON-ASSOCIATED PROTEINS

A critical role for the cytoskeleton, and associated proteins, lies in governing ion channel distribution within specialized regions of plasma membranes. Such regulated distribution of ion channels at the cell surface is necessary for proper intra- and intercellular signaling, in particular within and between excitable cells, such as neurons and epithelial cells (Sheng, 1996). In the nervous system, electrical signaling is driven by the synchronized functioning of ion channels, which are typically localized at specific locations, such as the neuromuscular junction, node of Ranvier, or postsynaptic sites. In addition to the nervous system, association of cytoskeletal proteins, such as ankyrin and spectrin, with ion channels and ion transporters, including the Cl/HCO_3 exchanger and the α subunit of the Na–K pump, has been reported in other tissues such as the erythrocyte or epithelial cells (Cantiello, 1995; Smith and Benos, 1996). More recently, various K^+ channels including Kir1.1 and Kir4.1 have been shown to be localized at the apical and basolateral sides of renal tubular epithelial cells, respectively (Lee and Hebert, 1995; Ito *et al.*, 1996). This section provides an overview of the interaction between cytoskeletal proteins and the ion channel responsible for site-directed ion channel distribution in the nervous system (Table II).

A. Rapsyn and Clustering of Nicotinic Acetylcholine Receptors at the Neuromuscular Junction

It is well established that site-specific ion channel clustering of nicotinic acetylcholine receptors (nAChRs) at the neuromuscular junction (NMJ) is essential for synaptic efficacy during neurotransmission. At the neuromuscular junction, nAChRs are localized at the motor end plate opposite the presynaptic nerve terminal. In this specialized membrane domain, nAChRs are tightly clustered at a density of $10{,}000/\mu m^2$, in contrast to the thousand-fold lower density of nAChRs found just outside of the motor end plate (Froehner, 1993). The spatially localized aggregation of nACh receptor/

TABLE II

Ion Channel Distribution Directed by Cytoskeleton-Associated Proteins

Channel	Location	Cytoskeleton-associated protein
Na^+ channel	Node of Ranvier	Ankyrin
Glycine receptor	Postsynaptic membrane	Gephyrin
NMDA receptor	Postsynaptic density	PSD-95 α_2-actinin NF-L
AMPA receptor	Postsynaptic membrane	GRIP
mGlu receptor	Postsynaptic membrane	Homer
Shaker type K^+ channel (Kv1.4)	Various neural microdomains	PSD-95
Inwardly rectifying K^+ channels		
Kir2.3	Forebrain	PSD-95
Kir4.1	Retinal Muller cells, glial cells	SAP97

Note. See the text for further details and abbreviations.

channel complexes ensures a rapid and robust response to released acetylcholine molecules, while a change in the cluster density of nAChRs strongly impacts postsynaptic response (Gomperts, 1996). The 43-kDa protein rapsyn, which is associated with the inner face of the postsynaptic membrane, clusters and localizes nAChRs and links them to the subsynaptic cytoskeleton (Phillips *et al.,* 1991; Froehner, 1993). Mutational studies on rapsyn have identified its binding domains for nAChRs, as well as for the cytoskeleton, within the primary structure of rapsyn. The targeting of nAChR and rapsyn to the NMJ is further orchestrated by the nerve-derived factor agrin (see later; Wallace, 1992).

Rapsyn can also cluster dystroglycan, a member of the dystrophin–glycoprotein complex (DGC) found along the sarcolemma of skeletal muscle and associated with the cytoskeleton, with the extracellular matrix, and with synapse-specific proteins, utrophin and β_2-syntrophin (Apel and Merlie, 1995). By binding to DGC, rapsyn may, in turn, localize nAChR clusters to the synapse. Targeted disruption of the rapsyn gene abolishes nAChRs clustering in the neuromuscular junction, directly implicating rapsyn as essential for the immobilization of nAChRs in the postsynaptic membrane (Gautam *et al.,* 1995). The importance of protein association with the DGC is further underscored by the demonstration that several muscular dystrophies, which cause improper function of the neuromuscular junction with progressive muscle weakness and premature death, derive from genetic errors within members of the DGC and associated loss of the cytoskeleton–extracellular matrix linkage (Campbell, 1995).

B. Ankyrin and Na^+ Channels in the Node of Ranvier

Segregation of ion channels to specialized regions of neurons is also crucial for the propagation of action potentials (APs). Studies of lateral mobility and topography indicate that Na^+ channels are freely mobile on the neuronal cell body, but are immobile at the axon hillock, presynaptic terminal, and focal points along the axon. In particular, a high density of voltage-gated Na^+ channels is found at the node of Ranvier, a specialized membrane domain in myelinated axons essential for fast propagation of nerve impulses. Such domain-dependent expression of Na^+ channels has been ascribed to an interaction of channel proteins with a specific isoform of ankyrin, a cytoskeletal linker protein (Srinivasan *et al.,* 1988). Specifically, during the process of myelinization, ankyrin clusters at the ends of Schwann cell processes and flanks the site of nodal formation. As myelinization proceeds, clustered zones fuse to form mature nodes of Ranvier. During nodal development, Na^+ channels and adhesion molecules (i.e., neurofascin) colocalize with ankyrin cluster zones and define the molecular composition of the node of Ranvier. Thus, the multivalent properties of ankyrin (i.e., a capacity to bind not only to membranes but also to ion channel proteins and adhesion molecules) play an important role in the coordinated recruitment and targeted distribution of Na^+ channels to the node of Ranvier, and thereby in AP conduction (Davis *et al.,* 1996).

C. Gephyrin and the Postsynaptic Ion Channel/Receptor Mosaic

The molecular mechanisms underlying the postsynaptic ion channel/receptor mosaic within central nervous system neurons have been related, at least in part, to cytoskeleton-binding protein. In particular, clustering of inhibitory glycine receptors (GlyR) at postsynaptic membranes, underneath glycine-releasing nerve terminals, has been ascribed to the associations of these pentameric receptor/ion channel complexes with gephyrin, a 93-kDa tubulin-binding protein (Kuhse *et al.*, 1995). Specifically, a gephyrin-binding domain has been identified in the cytoplasmic loop of the β-subunit of GlyR, between the third and fourth transmembrane segment (Meyer *et al.,* 1995). Antisense inhibition of the expression of gephyrin prevents GlyR accumulation at postsynaptic membrane specialization sites (Kirsch *et al.,* 1993). Furthermore, depolymerization of microtubules in special neurons reduces the percentage of cells with postsynaptic gephyrin clusters and disperses postsynaptic GlyR clusters. This supports the notion that postsynaptic localization of GlyR is regulated by the gephyrin-mediated anchoring of receptor polypeptides to the subsynaptic cytoskeleton through interactions with microtubules and possibly microfilaments (Kirsch and Betz, 1995).

The essential role for gephyrin in setting postsynaptic receptor/channel topology may not be restricted to GlyR. Gephyrin may also anchor another inhibitory receptor–ion channel complex, the $GABA_A$ receptor/channel, to postsynaptic membranes in hippocampal neurons (Craig *et al.*, 1996).

D. Postsynaptic Density Proteins and Glutamate Receptor/Channels

At central excitatory synapses, the *N*-methyl-D-aspartate (NMDA) receptor is highly concentrated at postsynaptic sites. These ionotropic neurotransmitter-gated receptor/channels are actually imbedded in the postsynaptic density (PSD), a specialized compartment of the submembrane cytoskeleton. Recently, it has been established that one of the subunits of the NMDA receptor complex (NR2) specifically interacts with the postsynaptic density protein PSD-95 (Kornau *et al.*, 1995). PSD-95 (also known as the synapse-associated protein 90 kDa, or SAP90) is a 95-kDa, cytoskeleton-associated protein abundant in the postsynaptic synaptosomal fraction (Gomperts, 1996). It has been proposed that members of the PSD-95 family serve to anchor NMDA receptors to the submembrane cytoskeleton and to aid in the assembly of signal transduction complexes at postsynaptic sites (Ehlers *et al.*, 1996; Gomperts, 1996).

Other types of glutamate receptors, that is, α-amino-3-hydroxy-5-methyl-4-isoxazole propionic acid (AMPA) and metabotrophic glutamate (mGlu) receptors, are also regulated by membrane-anchoring proteins other than PSD-95. They are coexpressed with each other or with NMDA receptors in glutamatergic synapses (Nusser *et al.*, 1994; Sheng, 1997). However, these three types of glutamate receptors differ in their subsynaptic distribution, mGlu receptors being located at the periphery, whereas NMDA and AMPA receptors are more central (Nusser *et al.*, 1994). AMPA receptors, do not bind to PSD-95 (Sheng, 1997). Recently, GRIP (glutamate receptor interacting protein) and Homer, novel anchoring proteins, have been cloned and identified to bind to AMPA receptors and mGlu receptors, respectively (Dong *et al.*, 1997; Brakeman *et al.*, 1997). Thus, GRIP, Homer, and PSD-95 may be candidates for differential distribution of AMPA, mGlu, and NMDA receptors within the same glutamatergic synapses (Sheng, 1997). Furthermore, it is suggested that these anchoring proteins function as adaptor proteins that physically couple the different classes of glutamate receptors to distinct downstream signaling pathways.

E. PSD-95-Related Proteins and Voltage-Gated K^+ Channels

This class of channel-associated proteins, collectively named the PSD-95/SAP90 family (which in addition to PSD-95 also includes related proteins such as hdlg/SAP97), also interacts with subunits of several voltage-dependent K^+ (KV) channels (Kim *et al.,* 1995; Sheng, 1996). The voltage-dependent K^+ channels are concentrated at various neuronal microdomains, including presynaptic terminals, nodes of Ranvier, and dendrites, where they regulate local membrane excitability. Evidence has been obtained that cell-surface clustering of Shaker-type K^+ channels, such as Kv1.4, is mediated by the PSD-95 family of membrane- and cytoskeleton-associated proteins. This occurs through direct and specific binding of the C-terminal cytoplasmic tails of K^+ channel subunits to two characteristic, so-called PDZ, domains in the PSD-95 protein (see Table III).

F. PSD-95 and Inwardly Rectifying K^+ Channels

In addition to promoting clustering of NMDA receptors and voltage-dependent K^+ channels, the cytoskeletal protein PSD-95 also binds inwardly rectifying K^+ channels, including Kir2.1, Kir2.3, and Kir4.1. Kir2.3 colocalizes with PSD-95 in neuronal populations in the forebrain, whereas a PSD-95/Kir2.3 complex occurs in the hippocampus (Cohen *et al.,* 1996).

Kir4.1, a glial cell inwardly rectifying K^+ channel (Takumi *et al.,* 1995), is colocalized with SAP97 in retinal Muller cells (Horio *et al.,* 1997) and possibly in renal tubular epithelium and the marginal cells in cochlear stria vascularis (Ito *et al.,* 1996; Hibino *et al.,* 1997). These anchoring proteins seem not only to cluster Kir4.1 on the cell membrane, but also to stimulate the channel current by increasing the functional channel number in the cell membrane (Horio *et al.,* 1997). Recently, it was also found that a cardiac two-pore K^+ channel (cTBAK) possesses the C-tail domain interacting with PSD family proteins (Kim *et al.,* 1998). Therefore, many other K^+ channels might also be under control of this family of anchoring proteins.

Based on the multiple protein–protein interactions, and the observation that PSD-95 and related proteins form oligomers, a scaffolding role for PSD-95 in organizing signaling cascades at the PSD has been proposed (Cohen *et al.,* 1996; Gomperts, 1996).

G. PDZ Domains and Protein–Protein Interaction

PDZ domains are viewed as modular protein-binding sites that recognize a short consensus peptide sequence (Fig. 4), making a small domain of

TABLE III

Ion Channels and Receptors Carry a T/SXV Motif

	E-T/SXV motif		Hydrophobic T/SXV motif	
NMDA receptors	NR2A	-ESDV		
	NR2B	-ESDV		
	NR2C	-ESDV		
	NR2D-2	-ESDV		
Voltage-gated K^+ channels	Kv1.4	-ETDV	Kv1.1	-LTDV
	Kv1.5	-ETDV	Kv1.2	-LTDV
			Kv1.3	-FTDV
			Kv1.6	-LTEV
			Kv3.2b	-PSIL
			Kv3.3b	-PSIL
			Kv4.1	-ISSL
			Kv4.2	-VSAL
			Kv4.3	-VSAL
Inwardly rectifying K^+ channels	IRK1/Kir2.1	-ESEI		
	IRK2/Kir2.2	-ESEI		
	IRK3/Kir2.3	-ESRI		
	GIRK2C/Kir3.2c	-ESKV		
K^+ channels with two-pore domains			cTBAK (TASK)	-RSSV
G-protein-coupled receptors	β1 receptor	-ETVV	5HT2A	-VSCV
	MAS oncogene	-ETVV	5HT2C	-ISSV
			VIP	-VSLV

Note. S, possible phosphorylation site by PKA. Various kinds of channels and receptors can bind to the PDZ domains of PSD/SAP family proteins via a T/SXV motif in their C termini.

a much large protein responsible for specific protein–protein association (Sheng, 1996). PDZ domains are 90-amino-acid repeats in the N-terminal half of PSD-95-related proteins (PSD-95, hdlg/SAP97, chapsyn-110/PSD-93). In addition to the presence of three PDZ domains, these proteins, which also belong to the membrane-associated guanylate-kinase (MAGUK) superfamily of proteins, are characterized by the existence of Src 3 homology (SH3) and guanylate kinase-like domains within the C-terminal region (Fig. 4; Gomperts, 1996; Sheng, 1996). The NR2 subunit of the NMDA receptor and the C-terminal region of Shaker K^+ channels and Kir channels possess four highly conserved amino acids (–E–S/T–D–V motif) that are specifically recognized by the PDZ do-

A The structure of PSD-95/SAP90 family proteins

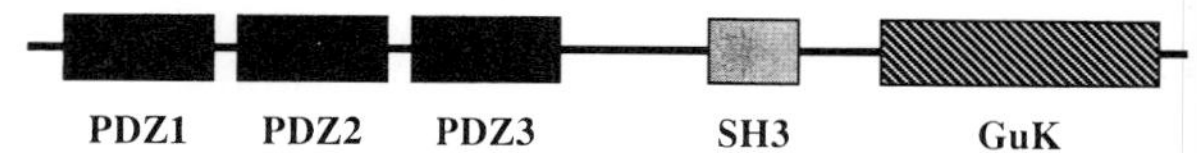

B Clustering of channels by PSD-95/SAP90 family

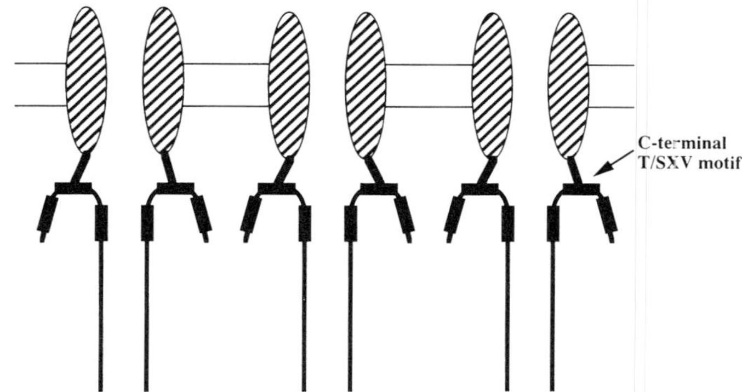

FIGURE 4 The structure of PSD-95/SAP90 family proteins (A) and clustering of channels by PSD-95/SAP90 family proteins (B). GuK, guanylate kinase-like domain.

mains in PSD-95 (see Table III; Doyle *et al.,* 1996; Gomperts, 1996; Sheng, 1996).

The ability of PDZ domains to function as independent modules for protein–protein interaction suggests that PDZ-domain-containing polypeptides may be widely involved in the organization of proteins at sites of membrane specialization (Kim *et al.,* 1995; Sheng, 1996). While Shaker, Kir, and NR2 proteins do not cluster in the absence of PSD-95, coexpression of Shaker-type K^+, Kir, or NR2 subunits of NMDA receptors with PSD-95, SAP-97, or chapsyn-110 results in the coclustering of both proteins (Kim *et al.,* 1995, 1996; Sheng, 1996; Horio *et al.,* 1997). This emphasizes the importance of PSD-95 proteins in directing the distribution of NMDA, voltage-dependent K^+, and Kir channels.

H. α_2-Actinin and Ion Channel Expression in Postsynaptic Domains

Other cytoskeleton-related proteins, such as α_2-actinin, have also been related to the regulation of NMDA receptor expression within the postsynaptic domains. The integrity of postsynaptic actin filaments is, indeed, important for NMDA receptor function. A specific biochemical association between α_2-actinin and the cytoplasmic tail of the NMDA receptor subunits NR1A, NR1C, and NR2B has recently been demonstrated. It has, therefore, been suggested that the actin-binding protein, α_2-actinin, could serve as an anchor protein to mediate NMDA receptor association with the postsynaptic actin cytoskeleton (Wyszynski *et al.*, 1997). The interaction of α_2-actinin with NMDA receptor proteins may, in turn, affect the plasticity of excitatory synapses.

I. Neurofilament and NMDA Receptors in Postsynaptic Membrane

A recent study has shown that NMDA receptors can bind not only cytoskeleton-associated proteins, but also cytoskeleton. Ehlers and colleagues (1998) identified the direct interaction of 68-kDa neurofilament subunit NF-L with NR1A using the yeast two-hybrid system. This interaction occurs between the C terminus of NR1A and the rod domain of NF-L. Furthermore, they found that NR1 and NF-L coexist in the dendrites of cultured hippocampal neurons. As described above, the C termini of NR1A and NR2B can interact with α_2-actinin, which binds actin. Because NF-L binds to β-spectrin, one of the actin-binding proteins (Frappier *et al.*, 1991, 1992), the complex of NR1A, NR2B, β-spectrin, NF, and actin can be constructed in the postsynaptic membrane. Thus, the cellular distribution and functional properties of NMDA receptors may be regulated by the cytoskeleton-associated and cytoskeletal proteins.

In summary, evidence has been obtained to indicate that cytoskeleton-associated proteins (including rapsyn, ankyrin, gephyrin, PSD-95, α_2-actinin) and cytoskeleton (NF-L) direct and maintain the site-specific distribution of several ion channels (such as nAChRs and Na^+ and K^+ channels, as well as glycine and NMDA receptor/channel complexes) within microdomains of the plasma membrane (see Table II).

V. ROLE FOR PHOSPHORYLATION IN CYTOSKELETAL PROTEIN-DIRECTED CLUSTERING OF ION CHANNELS

Targeting of ion channels to discrete plasma membrane sites is a dynamic process and is believed to be commonly regulated by additional enzymatic processes. Most commonly, evidence for protein phosphorylation has been obtained.

A. *Tyrosine Kinase Activity and Clustering of Nicotinic Acetylcholine Receptors*

During development, nAChRs are clustered to postsynaptic muscular junctions in response to release of the nerve-derived factor agrin (Wallace, 1992). Agrin-mediated signal transduction is mediated by receptor protein tyrosine kinase (Glass *et al.*, 1996). Targeted disruption of the gene encoding MuSK, a receptor tyrosine kinase selectively localized to the postsynaptic muscle surface, disrupts neuromuscular synapse formation (DeChiara *et al.*, 1996). Thus, nAChR clustering requires not only the cytoskeleton-associated protein rapsyn, as discussed earlier, but also protein–tyrosine kinase activity. Moreover, it has also been reported that rapsyn may induce autophosphorylation of MuSK. This could, in turn, lead to a MuSK-specific phosphorylation of the β subunit of nAChRs (Gillespie *et al.*, 1996). Thus, rapsyn may mediate the synaptic localization of MuSK in muscle, which may play an important role in the rapsyn-induced clustering of nAChRs within the neuromuscular junction (Ruegg and Bixby, 1998).

B. *Protein Kinase A Activity and Synaptic Channel Density*

It has recently been established that phosphorylation by protein kinase A (PKA) is important for the interaction between inwardly rectifying K^+ channels and the cytoskeletal protein, PSD-95. Stimulation of PKA in intact cells causes rapid dissociation of the inwardly rectifying K^+ channel, Kir2.3, from PSD-95 (Cohen *et al.*, 1996). A serine residue (Ser-440), located within the C-terminal tail of the inwardly rectifying K^+ channel, Kir2.3, is not only critical for interaction with PSD-95, but also serves as a substrate for phosphorylation by PKA. In turn, phosphorylation and dephosphorylation of such amino acid residues may regulate the dynamic interaction between K^+ channels and the cytoskeleton. Rapid cyclic AMP-mediated changes in the structure of the PSD may, in turn, determine the postsynaptic channel density and mediate synaptic plasticity (Cohen *et al.*, 1996). A similar PKA phosphorylation site exits in the C-terminal tail of cTBAK, which can interact with PSD-95 (Kim *et al.*, 1998). Thus, the interaction between PSD-95-anchoring proteins and PKA-dependent phosphorylation might be involved in regulation of channel function more widely than recognized currently.

Anchoring of PKA also appears important in the regulation of synaptic function. Specifically, it has been shown that anchoring of PKA by A-kinase-anchoring proteins (AKAPs) is required for the modulation of AMPA/kainate channels. Intracellular perfusion of hippocampal neurons with peptides derived from the conserved kinase-binding region of AKAPs prevents PKA-mediated regulation of AMPA/kainate currents as well as fast excitatory synaptic currents. Thus, positioning of kinases by anchoring proteins near their substrates, including ion channel complexes, may be essential in regulation of the electrical properties of a cellular membrane (Rosenmund *et al.,* 1994).

VI. REGULATION OF ION CHANNEL FUNCTION BY CYTOSKELETAL PROTEINS

As described earlier, ion channels and other ion transport molecules are integral to the plasma membrane and are surrounded by cytoskeletal strands, in particular actin filaments (Ruknudin *et al.,* 1991; Horber *et al.,* 1995). Actin accounts for more than 20% of total cell proteins, and actin and actin-binding proteins couple to several ion channels and ion transport molecules (Cantiello, 1995). This is of importance in epithelia and neurons, where maintenance of ion channels and transporters to specific membrane domains is vital for the normal function of epithelial and neuronal tissues (Smith and Benos, 1996). In addition to structural interactions, there is growing evidence for functional interactions between ion channels and the adjacent actin microfilament network in both epithelia and nervous tissues, as well as in the heart (Cantiello and Prat, 1996).

A. *Epithelial Ion Channel Function*

1. Regulation of Epithelial Na^+ Channels by the Cytoskeleton

It has been established that epithelial Na^+ channels are linked to cytoskeleton structures, including ankyrin, spectrin, and actin (Smith and Benos, 1996). Using antibodies generated against the purified renal epithelial Na^+ channel, the Na^+ channels were found to colocalize to the apical membrane with actin and apically associated isoforms of ankyrin and spectrin (Smith *et al.,* 1991; Cantiello and Prat, 1996). It is a proline-rich region in the epithelial Na^+ channel that mediates binding to an SH3 regions of a-spectrin, which in turn maintains the polarized distribution of the channel to the apical membrane of renal epithelia (Rotin *et al.,* 1994). While such interaction may serve to determine the spatial distribution of Na^+ channels, colocalization of actin filaments with apical Na^+ channels has also been related to the functional regulation of Na^+ channel activity (Cantiello, 1995). Agents that depolymerize actin filaments, such as cytochalasin D (see Fig. 3A), enhance the open probability of a 9-pS Na^+ channel activity in renal epithelial cells (Cantiello *et al.,* 1991). In contrast, DNase I, which stabilizes the pool of monomeric actin (see Fig. 3A), lacked such effect. The length of the actin filaments appears to be a determinant of channel activation. Addition of short actin filaments to excised membrane patches enhanced the probability for Na^+ channels to be open, yet whenever actin was added after being polymerized to achieve predominantly long filaments, no Na^+ channel activation was observed (Cantiello *et al.,* 1991; Cantiello, 1995). This suggests that short actin filaments, in contrast to G-actin or long actin filaments, are responsible for channel activation (Cantiello and Prat, 1996).

The actin-dependent regulation of channel activity may participate in the stretch-dependent activation of the renal epithelial Na^+ channel (Awayda *et al.,* 1995). Under basal conditions, stabilized actin filaments may contribute to maintaining Na^+ channels in the closed state, whereas actin depolymerization, by stretch or a hormone (e.g., vasopressin), may result in channel activation, either by affecting the membrane environment or by interacting with other membrane–cytoskeleton proteins associated with the channel (Cantiello, 1995; Cantiello and Prat, 1996; Smith and Benos, 1996). The effect of actin on Na^+ channels is modulated by phosphorylation through PKA (Prat *et al.,* 1993). Actin filaments have also been implicated in hormone (vasopressin)-mediated insertion of Na^+ channels into the apical membrane (Verrey *et al.,* 1995).

The amiloride-sensitive renal Na^+ channels are composed of three subunits of α-, β-, γ-ENaC (Cannessa *et al.,* 1994). Mutations in β and γ subunits of ENaC are responsible for Liddle syndrome (Shimkets *et al.,* 1994; Hansson *et al.,* 1995). These mutations cause deletion or dysfunction of PY motifs (PPPXY) of the subunits, resulting in loss of interaction of the Na^+ channel with Nedd4 protein through their WW domain. Because Nedd4 facilitates downregulation of the Na^+ channel in normal conditions, the Liddle mutation causes an increase of the number of Na^+ channels in the membrane and its high activity.

2. Regulation of Epithelial Chloride Channels by the Cytoskeleton

Chloride (Cl^-) channels play an important role in fluid movement across epithelia. The actin cytoskeleton appears to regulate the behavior of the cystic fibrosis transmembrane regulator (CFTR), a low-conductance Cl^- channel predominantly expressed in the apical membrane of epithelia (Cantiello and Prat, 1996; Smith and Benos, 1996). Severing of the endogenous actin cytoskeleton by cytochalasin D, or direct addition of exogenous actin, induces activation of Cl^- current in adenocarcinoma cells transfected with the human CFTR gene (Prat *et al.,* 1995). Thus, increasing the availability of "short" filaments (see Fig. 3A) by either disruption of preexisting filaments or *de novo* formation of new ones activates CFTR. In contrast, decreasing the number of "short" actin filaments by preventing actin polymerization using DNase I, for example, or by bundling filaments with filamin inhibits CFTR-associated channel activity (Cantiello and Prat, 1996). In this regard the actin-dependent regulation of CFTR activity appears similar to that of epithelial Na^+ channels (see earlier discussion; Cantiello, 1995), since long, filamentous actin maintains the channels in the closed state while severing of actin into "short" filaments activates channels (Cantiello, 1995; Cantiello and Prat, 1996; Smith and Benos, 1996). Comparison of amino acid sequences of CFTR with those of known actin-binding proteins, such as severin and filamin, revealed putative actin-binding domain(s) within the nucleotide-binding folds of CFTR (Prat *et al.,* 1995).

It should be pointed out, however, that the effect of the F-actin network-dependent regulation of Cl^- channels may not be uniform. For example, opening of a different Cl^- conductance, the 33−pS Cl^- channel present in real proximal tubule epithelia or of Cl^- conductances in bronchial epithelia, is actually inhibited by cytochalasin D (Suzuki *et al.,* 1993; Hug *et al.,* 1995). Signaling pathways that include disruption of F-actin have been suggested to mediate the activation of a 305-pS Cl^- channel, during cell swelling, in renal collecting duct cells (Schwiebert *et al.,* 1994). Regardless of the outcome on channel activity resulting from the modification of actin microfilament structure, these experimental data point toward a functional interaction between the actin cytoskeleton and epithelial Cl^- channels.

Further evidence for an interaction between actin and a Cl^- channel or associated protein has been obtained from coimmunoprecipitation studies. Actin does not coimmunoprecipitate with a 27–kDa protein, named pI_{Cln}, which gives rise to an outwardly rectifying Cl^- channel activity (Paulmichl *et al.,* 1992; Krapivinsky *et al.,* 1994). Thus, in epithelia the actin network interacts with channel-related proteins and regulates Cl^- channel behavior.

3. Regulation of Epithelial K^+ Channels by the Cytoskeleton

The membrane cytoskeleton is involved in the modulation of the low-conductance K^+ channel present in the apical membrane of the cortical collecting duct (Wang *et al.,* 1994). This K^+ channel is inactivated by application of known disrupters of actin filaments, such as cytochalasins. Phalloidin, however, which stabilizes actin filaments, prevents cytochalasin-induced K^+ channel inactivation. Based on such findings, it has been proposed that the actin cytoskeleton is critically involved in the interaction between epithelial

K^+ channel proteins and the lipid phase of the cell membrane (Wang *et al.*, 1994).

Taken together, these findings indicate that the activity of various epithelial Na^+, K^+, and Cl^- channels is modulated by agents acting on cytoskeletal structures (Table IV). This, in turn, suggests a role for the submembrane cytoskeleton in the regulation of ion channel function in epithelial tissues.

B. Neuronal Ion Channel Function

In addition to compartmenting and anchoring integral membrane proteins, it has also been suggested that the neuronal cytoskeleton modulates neuronal excitability and synaptic plasticity through regulation of ion channel function (Fukuda *et al.*, 1981; Rosenmund and Westbrook, 1993). The initial observation was that cytoskeleton breakdown decreases AP upstroke in dorsal root ganglion neurons and axons, apparently through regulation of Na^+ and Ca^{2+} channels (Matsumoto and Sakai, 1979; Fukuda *et al.*, 1981). Thereafter, cytoskeletal breakdown in neuronal tissue was also shown to affect membrane excitability through regulation of Ca^{2+} and NMDA channels (Fukuda *et al.*, 1981; Johnson and Byerly, 1993; Rosenmund and Westbrook, 1993). Here we summarize the studies that relate to the cytoskeleton-dependent regulation of ion channel activity.

1. Regulation of Neuronal Ca^{2+} Channels by the Cytoskeleton

It was first shown that the cytoskeletal disrupter colchicine causes a reduction of the upstroke velocity of action potentials of cultured neurons. From these data, it was inferred that Ca^{2+} channels in neurons interact with microtubules (Fukuda *et al.*, 1981). The metabolic dependence and-

TABLE IV
Effects on Actin Cytoskeleton on Selected Epithelial Channels

Channel	Tissue	Effect
9-pS Na^+ channel	Renal epithelial (A6) cells	Activation by short actin filaments
CFTR	Transfected adenocarcinoma cell line	Activation by long actin filaments
33-pS Cl^- channel	Proximal tubular cell line	Activation by long actin filaments
30-pS K^+ channel	Principal cells in cortical collecting duct	Inactivation by actin filament disrupters

inactivation by intracellular Ca^{2+} of Ca^{2+} channels were found to be mediated by an allosteric interaction between channel proteins and the cytoskeleton (Johnson and Byerly, 1993). Cytoskeletal disruption (by colchicine and cytochalasin B) prevents ATP from preserving Ca^{2+} channel activity, whereas cytoskeletal stabilizers (taxol and phalloidin) reduce both channel dependence on ATP and inactivation by intracellular Ca^{2+}. An allosteric interaction between the cytoskeleton and Ca^{2+} might trigger a conformational change in the cytoskeleton, rapidly closing the adjacent Ca^{2+} channels (Johnson and Byerly, 1993). Cytoskeletal stabilizers would reduce Ca^{2+}-induced channel inactivation by restricting the Ca^{2+}-dependent conformational change in the cytoskeleton. Thus, it is proposed that Ca^{2+}-dependent inactivation of Ca^{2+} current in neurons may be related to cytoskeleton integrity (Johnson and Byerly, 1993).

2. Regulation of Neuronal NMDA Channels by the Cytoskeleton

F-actin is a major component of the cytoskeleton in postsynaptic densities and dendritic spines, and is under dynamic regulation of both Ca^{2+}, which rapidly induces depolymerization, and adenosine triphosphate (ATP), which promotes repolymerization. Actin depolymerization influences NMDA channel activity in whole-cell recordings of cultured hippocampal neurons (Rosenmund and Westbrook, 1993). Specifically, the ATP- and Ca^{2+}-dependent "rundown" of NMDA channels (a parameter used to probe channel regulation) was prevented when actin depolymerization was blocked by phalloidin. This agent binds to F-actin and shifts the equibrium between F-actin and actin monomers (G-actin) toward the polymerized state. Cytochalasins, which enhance actin–ATP hydrolysis, induced NMDA channel rundown, whereas taxol or colchicine, which stabilize or disrupt microtubule assembly, had no effect (Rosenmund and Westbrook, 1993). These results were interpreted to suggest that Ca^{2+} and ATP can influence NMDA channel activity by altering the state of actin polymerization (Rosenmund and Westbrook, 1993; see Fig. 3B). Thus, actin dynamics may contribute to calcium-dependent postsynaptic events, such as long-term depression.

C. Cardiac Ion Channel Function

In addition to epithelial and neuronal ion channels, more recently, indications were obtained for a functional interaction between the cytoskeleton and Na^+, Ca^{2+}, as well as ATP-sensitive K^+ channels expressed in cardiac myocytes. These data are based primarily on the ability of agents known to affect the cytoskeleton and modulate ion channel activity.

1. Cytoskeleton Modulates Gating of Voltage-Dependent Cardiac Na^+ Channels

Agents that interfere with actin polymerization, such as cytochalasin D, reduce whole-cell peak Na^+ current and slow current decay in ventricular cardiac myocytes (Undrovinas *et al.,* 1995). Application of cytochalasin on the cytoplasmic side of inside out patches results in reduction of peak open probability, accompanied with long bursts of Na^+ channel openings. These results were interpreted to indicate that cytochalasin D, through effects on the cytoskeleton, induces cardiac Na^+ channels to enter a mode characterized by a lower peak open probability but a greater persistent activity as if the inactivation rate were slowed.

2. Cytoskeleton Disrupters Regulate L-Type Cardiac Ca^{2+} Channels

Initially, it was observed that cardiac excitability can be modulated by agents that target microtubules, such as tubulin, a depolymerizing agent (Lampidis *et al.,* 1992). Colchicine, which dissociates microtubules into tubulin, and taxol, which dissociates microtubules, strongly influence the kinetics of L-type Ca^{2+} channels in intact cardiac cells (Galli and DeFelice, 1994). Colchicine increases the probability that Ca^{2+} channels are in the closed state, whereas taxol increases the probability that Ca^{2+} channels are in the open state. Moreover, taxol lengthens the mean open time of Ca^{2+} channels. Neither taxol nor colchicine affects the number of Ca^{2+} channels (Galli and DeFelice, 1994). Several interpretations were proposed for these findings, including a direct interaction of tubulin with Ca^{2+} channels, or alternatively an action of taxol and/or colchicine through the buffering ability of the cytoskeleton to regulate the effective concentration of inactivating ions near the mouths of channels (Galli and DeFelice, 1994). This relates to the concept that the dynamics of current-induced inactivation are dictated by restricted and heterogeneous spaces near the membrane, as well as by transient local buffering within cells. In this regard, the structure of the cytoskeleton surrounding the mouths of channels could contribute to both compartmentalization and buffering. Thus, alterations in the structure of the cytoskeleton within the Ca^{2+} channel's microenvironment could participate in the phenomenon of channel inactivation, and thereby in the regulation of cell excitability (Galli and DeFelice, 1994).

3. Actin Filaments Regulate Cardiac ATP-Sensitive K^+ Channel Activity

The defining property of ATP-sensitive K^+ (K_{ATP}) channels is their inhibition by intracellular ATP, whereby these channels are viewed as a link between the metabolic state and electrical excitability of a cardiac cell (Terzic *et al.,* 1995). Opening of K_{ATP} channels in the myocardium is sensitive to the mechanical distortion of the membrane (Van Wagoner, 1993), suggesting that the integrity of the microenvironment surrounding K_{ATP} channels may play a role in modulating channel activity. Indeed, cytoskeletal disrupters, DNase I (Fig. 5) and cytochalasin B (but not antimicrotubule agents), were found to antagonize the ATP-induced inhibition of cardiac K_{ATP} channels, that is, they produced an apparent decrease in the sensitivity of K_{ATP} channel toward ATP-induced inhibition that was partially restored by addition of purified actin subunits (Terzic and Kurachi, 1996). Taken together, these findings may fulfill the established criteria for a disrupter of actin microfilaments to regulate a specific ion channel (Cantiello, 1995), and support the notion that DNase I acts on actin filaments to modulate K_{ATP} channel activity (Terzic and Kurachi, 1996).

The subsarcolemmal actin microfilament network may be of importance in governing not only the ATP-dependent gating of the channel, but also the sulfonylurea-dependent K_{ATP} channel regulation. In addition to ATP, a major pharmacological property of K_{ATP} channels is their sensitivity to sulfonylurea drugs that are considered among the more specific K_{ATP} channel ligands to inhibit channel activity. DNase I, when applied to the internal surface of excised membrane patches, impaired the action of sulfonylurea drugs on myocardial K_{ATP} channel activity (Brady *et al.,* 1996). Specifically, this high-affinity actin-sequestering protein, which depolymerizes actin fil-

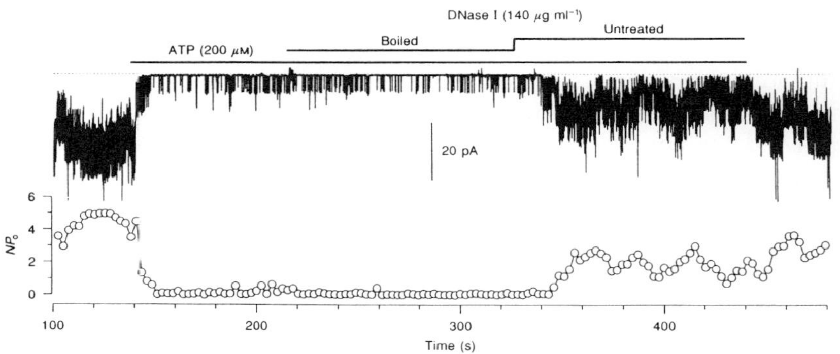

FIGURE 5 The actin microfilament disrupter, DNase I, enhances K_{ATP} channel opening. Untreated DNase I antagonized ATP-induced K_{ATP} channel inhibition. By contrast, DNase that had been denatured by boiling had no effect. Upper trace: Original trace record from an inside out patch excised from a guinea pig ventricular cardiac cell. Lower trace: Channel open probability calculated over 2.5-sec intervals. Reproduction, with permission, from Terzic and Kurachi (1996).

aments, decreased the apparent sensitivity of K_{ATP} channels toward inhibition by a prototype sulfonylurea, glyburide. The effect of DNase appeared to be mediated through binding to actin molecules since cytoskeletal strands are present in excised membrane patches, whereas other known targets of DNase are absent (Ruknudin *et al.,* 1991; Horber *et al.,* 1995). Denatured DNase I could not antagonize glyburide-induced K_{ATP} channel inhibition, which is consistent with the requirement that the native structure of the DNase protein be intact for it to form complexes with actin molecules and interfere with actin filament formation. Coincubation of DNase I with excess purified actin, which forms 1:1 molar complexes with DNase, prevented DNase action on K_{ATP} channels, suggesting that unoccupied binding sites for actin binding on the DNase molecule are important for the modulation of K_{ATP} channel regulation (Terzic and Kurachi, 1996; Brady *et al.,* 1996). The finding that the intraburst kinetic properties and conductance of the channel were not affected by DNase could suggest that the loss of responsiveness of the channel to sulfonylurea, as well as ATP, was secondary to alterations remote from the pore region of the channel.

The coupled impairment in ligand sensitivity induced by DNase supports the notion that sulfonylurea drugs and ATP bind to a common subunit, within the so-called sulfonylurea-binding protein, distinct from the presumed pore-forming components of the cardiovascular K_{ATP} channel (Inagaki *et al.,* 1996; Isomoto *et al.,* 1996). Further evidence for a functional linkage of K_{ATP} channels to the actin cytoskeleton was obtained from the observation that phalloidin, an actin filament stabilizing agent, could maintain channel activity and partially restore rundown channel activity (Furukawa *et al.,* 1996). It was thus proposed that for "fully activated" channels long, polymerized F-actin filaments are required. "Partially rundown" channels were associated with short actin filaments capped by actin-binding proteins. "Completely rundown" channels were related to depolymerized G-actin (Furukawa *et al.,* 1996). Taken together, these results could be interpreted to indicate that cardiac K_{ATP} channels can be regulated by the assembly and disassembly of the actin cytoskeleton network (Terzic and Kurachi, 1996; Brady *et al.,* 1996; Furukawa *et al.,* 1996).

VII. MECHANOSENSITIVE GATING OF ION CHANNELS AND CYTOSKELETON

The interaction of protein channels with the cytoskeleton has also been suggested to be involved in the gating of mechanosensitive and transduction channels. This is the case, for example, in certain specialized hair cells (Sach, 1988). Also, in skeletal muscle, an absence of normal dystrophin (a spectrin-like component of the cortical cytoskeleton) is associated with altered mechanosensitive gating, thus implicating cortical cytoskeleton in the mechanism of stretch sensitivity (Morris, 1995).

Gating of stretch-activated channels is thought to rely on forces between the cytoskeleton and the attached membrane channels. Although the biochemical basis of this interaction is uncertain, disruption of actin by cytochalasin alters behavior of mechanosensitive channels (Sachs, 1986). In the case of vertebrate hair cells, adaptation of the transduction current involves a Ca^{2+}- and actin-dependent mechanism. Ca^{2+} influx is believed to activate a molecular motor that maintains gating spring tension by moving along the actin core of the stereocilia (Hudspeth, 1989).

VIII. SUMMARY

The cytoskeleton regulates ion channel function through integrated interactions with cytoskeleton-associated proteins and dynamic regulation of its own state. Two main roles for the cytoskeleton have been recognized in the regulation of ion channel function: (1) targeted distribution of ion channel proteins within specialized domains of the plasma membranes, and (2) modulation of ion channel activity.

It is now established that cytoskeleton-associated proteins, such as rapsyn, ankyrin, gephyrin, PSD-95, and α_2-actinin, target the distribution of Na^+ and K^+ channels, nAChRs, glycine, and NMDA receptor/channel complexes to specialized membrane domains including the postsynaptic membranes or the nodes of Ranvier (see Table II). Cytoskeleton-dependent targeting and maintenance of ion channels at discrete plasma membrane sites, in turn, are regulated by catalytic processes, including protein phosphorylation by tyrosine kinases and PKA. Thus, structural interactions between the cytoskeleton, cytoskeleton-associated proteins, and channel/receptor subunits determine the highly specialized distribution of ion channel proteins within domains of plasma membranes. Such site-directed distribution and anchoring of ion channels are required for proper intra- and intercellular signaling.

In addition to structural interactions, functional interactions between ion channel and colocalized subplasmalemmal cytoskeletal networks have been described in epithelia, neurons, and cardiac cells. Apparently, the open probability of epithelial Na^+, K^+, and Cl^- channels is regulated by the length of actin filament networks (see Table IV). Also the cytoskeleton has been suggested to modulate neuronal excitability and synaptic plasticity through modulation of ion channel activity, in particular through regulation of the open probability and kinetics of Ca^{2+} and NMDA channels. In the heart, the activities of Na^+, Ca^{2+}, and K_{ATP} channels have been shown to depend on the integrity of the cytoskeleton. Therefore, based on the current understanding of the relationship between the cytoskeleton and ion channels, it has become apparent that modulation of the cytoskeleton and associated proteins may represent important means of regulating the physiology of ion channels, and thereby cellular functions, including signaling and excitability.

Moreover, disturbances of the cytoskeleton or associated proteins can occur under disease conditions, including muscular dystrophies (Campbell, 1995), as well as under pathophysiological conditions, such as ischemia and hypoxia (Ganote and Armstrong, 1993). Thus, it is conceivable that under such conditions the distribution and behavior of ion conductances, which depend on the integrity of cytoskeleton networks, could be dramatically altered. The effects of cytoskeleton and ion channel function in disease states await elucidation.

References

Apel, E. D., and Merlie, J. P. (1995). Assembly of the postsynaptic apparatus. *Curr. Opin. Neurobiol.* **5,** 62–67.

Awayda, M. S., Ismailov, I. I., Berdiev, B. K., and Benos, D. J. (1995). A cloned renal epithelial Na^+ channel proteins display stretch activation in planar lipid bilayers. *Am. J. Physiol.* **268,** C1450–C1459.

Bennett, V., and Gilligan, D. M. (1993). The spectrin-based membrane skeleton and micron-scale organization of the plasma membrane. *Annu. Rev. Cell Biol.* **9,** 27–66.

Brady, P. A., Alekseev, A. E. A., Aleksandrova, L. A., Gomez, L. A., and Terzic, A. (1996). A disrupter of actin microfilaments impairs sulfonylurea-inhibitory gating of cardiac K_{ATP} channels. *Am. J. Physiol.* **271,** H2710–H2716.

Brakeman, P. R., Lanahan, A. A., O'Brien, R., Roche, K., Barnes, C. A., Huganir, R. L., and Worley, P. F. (1997). Homer: A protein that selectively binds metabotropic glutamate receptors. *Nature* **386,** 284–287.

Campbell, K. P. (1995). Three muscular dystrophies: Loss of cytoskeleton-extracellular matrix linkage. *Cell* **80,** 675–679.

Cannessa, C. M., Horisberger, J.-D., and Rossier, B. C. (1993). Epithelial sodium channel related to proteins involved in neurodegeneration. *Nature* **361,** 467–470.

Cannessa, C. M., Schild, L., Gary, B., Thorens, B., Gautschi, I., Horisberger, J.-D., and Rossier, B. C. (1994). Amiloride-sensitive epithelial Na^+ channel is made of three homologous subunits. *Nature* **367,** 463–467.

Cantiello, H. F. (1995). Role of the actin cytoskeleton on epithelial Na^+ channel regulation. *Kidney Int* **48,** 970–984.

Cantiello, H. F., and Prat, A. G. (1996). Role of actin filament organization in ion channel activity and cell volume regulation. *Curr. Top. Membr.* **43,** 373–396.

Cantiello, H. F., Stow, J. L., Prat, A. G., and Ausiello, D. A. (1991). Actin filaments regulate epithelial Na^+ channel activity. *Am. J. Physiol.* **261,** C882–C888.

Chuang, H., Jan, Y. N., and Jan, L. Y. (1997). Regulation of IRK3 inward rectifier K^+ channel by m1 acetylcholine receptor and intracellular magnesium. *Cell* **89,** 1121–1132.

Cohen, N. A., Brenman, J. E., Snyder, S. H., and Bredt, D. S. (1996). Binding of the inward rectifier K^+ channel Kir 2.3 to PSD-95 is regulated by protein kinase A phosphorylation. *Neuron* **17,** 759–767.

Coulter, K. L., Perier, F., Radele, C. M., and Vandenberg, C. A. (1995). Identification and molecular localization of a pH-sensing domain for the inward rectifier potassium channel HIR. *Neuron* **15,** 1157–1168.

Craig, A. M., Banker, G., Chang, W., McGrath, M. E., and Serpinskaya, A. S. (1996). Clustering of gephyrin at GABAergic but not glutamatergic synapses in cultured rat hippocampal neurons. *J. Neurosci.* **16,** 3166–3177.

Davis, J. Q., Lambert, S., and Bennett, V. (1996). Molecular composition of the node of Ranvier—Identification of ankyrin-binding cell adhesion molecules neurofascin and NrCAM at nodal axon segments. *J. Cell Biol.* **135,** 1355–1367.

DeChiara, T. M., Bowen, D. C., Valenzuela, D. M., Simmons, M. V., Poueymirou, W. T., Thomas, S., Kinetz, E., Compton, D. L., Rojas, E., Park, J. S., Smith, C., DiStefano, P. S., Glass, D. J., Burden, S. J., and Yancopoulos, G. D. (1996). The receptor tyrosine kinase MuSK is required for neuromuscular junction formation *in vivo. Cell* **85,** 501–512.

Doi, T., Fakler, B., Schultz, J. H., Schulte, U., Brandle, U., Weidemann, S., Zenner, H. P., Lang, J., and Ruppersberg, J. P. (1996). Extracellular K^+ and intracellular pH allosterically regulate renal Kir1.1 channels. *J. Biol. Chem.* **271,** 17,261–17,266.

Dong, H., O'Brien, R. J., Fung, E. T., Lanhan, A. A., Worley, P. F., and Huganir, R. L. (1997). GRIP: A synaptic PDZ domain-containing protein that interacts with AMPA receptors. *Nature* **386,** 279–284.

Doyle, D. A., Lee, A., Lewis, J,. Kim, E., Sheng, M., and MacKinnon, R. (1996). Crystal structures of a complexed and peptide-free membrane domain—Molecular basis of peptide recognition by PDZ. *Cell* **85,** 1067–1076.

Ehlers, M. D., Fung, E. T., O'Brien, R. J., and Huganir, R. L. (1998). Splice variant-specific interaction of the NMDA receptor subunit NR1 with neuronal intermediate filaments. *J. Neurosci.* **18,** 720–730.

Ehlers, M. D., Mammen, A. L., Lau, L. F., and Huganir, R. L. (1996). Synaptic targeting of glutamate receptors. *Curr. Opin. Cell Biol.* **8,** 484–489.

Falker, B., Brandle, U., Glowatzki, E., Weldemann, S., Zenner, H. P., and Ruppersberg, J. P. (1994). Strong voltage dependent inward rectification of inward rectifier K^+ channels is caused by intracellular spermine. *Cell* **80,** 149–154.

Ficker, E., Taglialatela, M., Wible, B. A., Henley, C. M., and Brown, A. M. (1994). Spermine and spermidine as gating molecules for inward rectifier K^+ channels. *Science* **266,** 1068–1072.

Frappier, T., Derancourt, J., and Pradel, L. A. (1992). Actin and neurofilament binding domain of brain spectrin beta subunit. *Eur. J. Biochem.* **205,** 85–91.

Frappier, T., Stetzkowski-Marden, F., and Pradel, L. A. (1991). Interaction domains of neurofilament light chain and brain spectrin. *Biochem. J.* **275,** 521–527.

Froehner, S. C. (1993). Regulation of ion channel distribution at synapses. *Annu. Rev. Neurosci.* **16,** 347–368.

Fukuda, J., Kameyama, M., and Yamaguchi, K. (1981). Breakdown of cytoskeletal filaments selectively reduces Na and Ca spikes in cultured mammal neurons. *Nature* **294,** 82–85.

Furukawa, T., Yamane, Y., Terai, Y., Katayama, Y., and Hiraoka, M. (1996). Functional linkage of the cardiac ATP-sensitive K^+ channel to the actin cytoskeleton. *Pflugers Arch.* **431,** 504–512.

Galli, A., and DeFelice, L. J. (1994). Inactivation of L-type Ca channels in embryonic chick ventricle cells: Dependence on the cytoskeletal agents colchicine and taxol. *Biophys. J.* **67,** 2296–2304.

Gallo-Payet, N., and Payet, M. D. (1995). Excitation-secretion coupling. *In* "Cell Physiology" (N. Sperelakis, Ed.), pp. 465–482. Academic Press, San Diego.

Ganote, C., and Armstrong, S. (1993). Ischaemia and the myocyte cytoskeleton: Review and speculation. *Cardiovasc. Res.* **27,** 1387–1403.

Gautam, M., Noakes, P. G., Mudd, J., Nichol, M., Chu, G. C., Sanes, J. R., and Merlie, J. P. (1995). Failure of postsynaptic specialization to develop at neuromuscular junctions of rapsyn-deficient mice. *Nature* **377,** 195–196.

Gillespie, S. K., Balasusramanian, S., Fung, E. T., and Huganir, R. L. (1996). Rapsyn clusters and activates the synapse-specific receptor tyrosine kinase MuSK. *Neuron* **16,** 953–962.

Glass, D. J., Bowen, D. C., Stitt, T. N., Radziejeski, C., Bruno, J., Ryan, T. E., Gies, D. R., Shah, S., Mattsson, K., Burden, S. J., DiStefano, P. S., Valenzuela, D. M., Dechiara, T. M., and Yancopoulos, G. D. (1996). Agrin acts via a MuSK receptor complex. *Cell* **85,** 513–523.

Gomperts, S. N. (1996). Clustering membrane proteins: It's all coming together with the PSD-95/SAP90 protein family. *Cell* **84,** 659–662.

Graig, A. M., Banker, G., Chang, W., McGrath, M. E., and Serpinskaya, A. S. (1996). Clustering of gephyrin at GABAergic but not glutamatergic synapses in cultured rat hippocampal neurons. *J. Neurosci.* **16,** 3166–3177.

Hansson, J. H., Nelson-Williams, C., Suzuki, H., Schild, L., Shimkets, R., Lu, Y., Canessa, C., Iwasaki, T., Rossier, B., and Lifton, R. P. (1995). Hypertension caused by a truncated epithelial sodium channel subunit: Genetic heterogeneity of Liddle syndrome. *Nature Genet.* **11,** 76–82.

Hibino, H., Horio, Y., Inanobe, A., Doi, K., Ito, M., Yamada, M., Gotow, T., Uchiyama, Y., Kawamura, M., Kubo, T., and Kurachi, Y. (1997). An ATP-dependent inwardly rectifying potassium channel, K_{AB}-2 (Kir4.1), in cochlear stria vascularis of inner ear: Its specific subcellular localization and correlation with the formation of endocochlear potential. *J. Neurosci.* **17,** 4711–4721.

Hille, B. (1992). Potassium channels and chloride channels. *In* "Ionic Channels of Excitable Membrane" (B. Hille, Ed.), pp. 115–139. Sinauer Associates, Sunderland, MA.

Hitt, A. L., and Luna, E. A. (1994). Membrane interactions with the actin cytoskeleton. *Curr. Opin. Cell Biol.* **6,** 120–130.

Horber, J. K. H., Mosbacher, J., Haberele, W., Ruppersberg, J. P., and Sackmann, B. (1995). A look at membrane patches with a scanning force microscope. *Biophys. J.* **68,** 1687–1693.

Horio, Y., Hibino, H., Inanobe, A., Yamada, M., Ishii, M., Tada, Y., Sato, E., Hata, Y., Takai, Y., and Kurachi, Y. (1997). Clustering and enhanced activity of an inwardly rectifying potassium channel, Kir4.1, by an anchoring protein, PSD-95/SAP90. *J Biol. Chem.* **272,** 12,885–12,888.

Hudspeth, A. J. (1989). How the ear's works work. *Nature* **341,** 397–401.

Hug, T., Koslowsky, T., Ecke, T., Greger, R., and Kunzelmann, K. (1995). Actin-dependent activation of ion conductances in bronchial epithelial cells. *Pflugers Arch* **429,** 682–690.

Inagaki, N., Gonoi, T., Clement, J. P., Namba, N., Inazawa, J., Gonzalez, G., Aguilar-Bryan, L., Seino, S., and Bryan, J. (1995). Reconstitution of I_{KATP}: An inward rectifier subunit plus the sulfonylurea receptor. *Science* **270,** 1166–1170.

Inagaki, N., Gonoi, T., Clement, J. P., Wang, C. Z., Aguilar-Bryan, L., Bryan, J., and Seino, S. (1996). A family of sulfonylurea receptors determines the pharmacological properties of ATP-sensitive K^+ channels. *Neuron* **16,** 1011–1017.

Isomoto, S., Kondo, C., and Kurachi, Y. (1997). Inwardly rectifying potassium channels: their molecular heterogeneity and function. *Jpn. J. Physiol.* **47,** 11–39.

Isomoto, S., Kondo, C., Yamada, M., Matsumoto, S., Higashiguchi, O., Horio, Y., Matsuzawa, Y., and Kurachi, Y. (1996) A novel sulfonylurea receptor forms with BIR (Kir 6.2) a smooth muscle type ATP-sensitive K^+ channel. *J. Biol. Chem.* **271,** 24321–24324.

Ito, M., Inanob, A., Horio, Y., Hibino, H., Isomoto, S., Ito, H., Mori, K., Tonosaki, A., Tomoike, H,, and Kurachi, Y. (1996). Immunolocalization of an inwardly rectifying K^+ channel, K_{AB}-2 (Kir4.1), in the basolateral membrane of renal distal tubular epithelia. *FEBS Lett.* **388,** 11–15.

Johnson, B. D., and Byerly, L. (1993). A cytoskeletal mechanism for Ca^{2+} channel metabolic dependence and inactivation by intracellular Ca^{2+}. *Neuron* **10,** 797–804.

Kim, D., Fujita, A., Horio, Y., and Kurachi, Y. (1998). Cloning and functional expression of a novel cardiac two pore background K^+ channel (cTBAK-1). *Circ. Res.,* **82,** 513–518.

Kim, E., Niethammer, M., Rothschild, A., Jan, Y. N., and Sheng, M. (1995). Clustering of Shaker- type K^+ channels by interaction with a family of membrane-associated guanylate kinases. *Nature* **378,** 85–88.

Kim, E., Cho, K.-O., Rothschild, A., and Sheng, M. (1996). Heteromultimerization and NMDA receptor-clustering activity of chapsyn-110, a member of the PSD-95 family of proteins. *Neuron* **17,** 103–113.

Kirsch, J., and Betz, H. (1995). The postsynaptic localization of the glycine receptor-associated protein gephyrin is regulated by the cytoskeleton. *J. Neurosci.* **15,** 4148–4156.

Kirsch, J., Wolters, I., Triller, A., and Betz, H. (1993). Gephyrin antisense oligonucleotides prevent glycine receptor clustering in spinal neurons. *Nature* **366,** 745–748.

Kornau, H. C., Schenker, L. T., Kennedy, M. B., and Seeburg, P. H. (1995). Domain interaction between NMDA receptor subunits and the postsynaptic density protein PSD-95. *Science* **269,** 1737–1740.

Krapivinsky, G. B., Ackerman, M., Gordon, E., Krapivinsky, L., and Clapham, D. E. (1994). Molecular characterization of a swelling-induced chloride conductance regulatory protein, pI_{Cln}. *Cell* **76,** 439–448.

Kuhse, J., Betz, H., and Kirsch, J. (1995). The inhibitory glycine receptor: Architecture, synaptic localization and molecular pathology of a postsynaptic ion-channel complex. *Curr. Opin. Neurobiol.* **5,** 318–323.

Kurachi, Y. (1996). Muscarinic and purinergic regulation of cardiac K^+ channels. *In* "Molecular Physiology and Pathophysiology of Cardiac Ion Channels and Transporters" (M. Morad, S. Ebashi, W. Trautwein, and Y. Kurachi, Eds.), pp. 177–186. Kluwar, Dordrecht.

Lampidis, T. J., Kolonias, D., Savara, J. N., and Rubin, R. (1992). Cardiostimulatory and antiarrhythmic activity of tubulin-binding agents. *Proc. Natl. Acad. Sci. USA* **86,** 1256–1260.

Lee, W.-S., and Hebert, S. C. (1995). ROMK inwardly rectifying ATP-sensitive K^+ channel. I. Expression in rat distal nephron segments. *Am. J. Physiol.* **268,** F1124–F1131.

Lopatin, A. N., Makhina, E. N., and Nichols, C. G. (1994). Potassium channel block by cytoplasmic polyamines as the mechanism of intrinsic rectification. *Nature* **372,** 366–369.

Matsumoto, G., and Sakai, H. (1979). Microtubules inside the plasma membrane of squid giant axons and their possible physiological function. *J. Membr. Biol.* **50,** 1–14.

Meyer, G., Kirsch, J., Betz, H., and Langosch, D. (1995). Identification of a gephyrin binding motif on the glycine receptor beta subunit. *Neuron* **15,** 563–572.

Morris, C. E. (1995). Stretch-sensitive ion channels. *In* "Cell Physiology" (N. Sperelakis, Ed.), pp. 483–489. Academic Press, San Diego.

Nusser, Z., Mulvihill, E., Streit, P., and Somogyi, P. (1994). Subsynaptic segregation of metabotropic and ionotropic glutamate receptors as revealed by immunogold localization. *Neuroscience* **61,** 421–427.

Papazian, D. M., Timpe, L. C., Jan, Y. N., and Jan, L. Y. (1991). Alteration of voltage-dependence of *Shaker* potassium channel by mutations in the S4 sequence. *Nature* **349,** 305–310.

Paulmichl, M., Li, Y., Wickman, K., Acherman, M., Peralta, E., Clapham, D. (1992). New mammalian chloride channel identified by expression cloning. Nature 356: 238–241.

Phillips, W. D., Kopta, C., Blount, P., Gardner, P. D., Steinbach, J. H., and Merlie, J. P. (1991). ACh receptor-rich membrane domains organized in fibroblasts by recombinant 43-kilodalton protein. *Science* **251,** 568–570.

Prat, A. G., Bertorello, A. M., Ausiello, D. A., and Cantiello, H. F. (1993). Activation of epithelial Na^+ channels by protein kinase A requires actin filaments. *Am. J. Physiol.* **265,** C224–C233.

Prat, A. G., Xiao, Y.-F., Ausiello, D. A., and Cantiello, H. F. (1995). c-AMP-independent regulation of CFTR by the actin cytoskeleton. *Am. J. Physiol.* **268,** C1522–C1561.

Rosenmund, C., Carr, D. W., Bergeson, S. E., Nilaver, G., Scott, J. D., and Westbrook, G. L. (1994). Anchoring of protein kinase A is required for modulation of AMPA/kainate receptors on hippocampal neurons. *Nature* **368,** 853–856.

Rosenmund, C., and Westbrook, G. L. (1993). Calcium-induced actin depolymerization reduces NMDA channel activity. *Neuron* **10,** 805–814.

Rotin, D., Bar-Sagi, D., O'Brodovich, H., Merilainen, J., Lehto, V. P., Canessa, C. M., Rossier, B. C., and Downey, G. P. (1994). An SH3 binding region in the epithelial Na^+ channel (αrENaC) mediates its localization at the apical membrane. *EMBO J.* **13,** 4440–4450.

Ruegg, M. A., and Bixby, J. L. (1998). Agrin orchestrates synaptic differentiation at the vertebrate neuromuscular junction. *Trends Neurosci.* **21,** 22–27.

Ruknudin, A., Song, M. J., and Sachs, F. (1991). The ultrastructure of patch-clamped membranes: A study using high voltage electron microscopy. *J. Cell Biol.* **112,** 125–134.

Sachs, F. (1986). Biophysics of mechanoreception. *Membr. Bio. Chem.* **6,** 173–195.

Sachs, F. (1988). Mechanical transduction in biological system. *Crit. Rev. Biomed. Engineer.* **16,** 141–169.

Schwiebert, E. M., Mills, J. W., and Stanton, B. A. (1994). Actin-based cytoskeleton regulates a chloride channel and cell volume in a renal cortical collecting duct cell line. *J. Biol. Chem.* **269,** 7081–7089.

Sheng, M. (1996). PDZs and receptor/channel clustering: Rounding up the latest suspects. *Neuron* **17,** 575–578.

Sheng, M. (1997). Glutamate receptors put in their place. *Nature* **386,** 221–223.

Shimkets, R. A., Warnock, D. G., Bositis, C. M., Nelson-Williams, C., Hansson, J. H., Schambelan, M., Gill, J. R. Jr., Ulick, S., Milora, R. V., Findling, J. W., Canessa, C. M., Rossier, B. C., and Lifton, R. P. (1994). Liddle's syndrome: heritable human hypertension caused by mutation in the β subunit of the epithelial sodium channel. *Cell* **79,** 407–414.

Smith, P. R., and Benos, D. J. (1996). Regulation of epithelial ion channel activity by the membrane-cytoskeleton. *Curr. Top. Membr.* **43,** 345–372.

Smith, P. R., Saccomani, G., Joe. E.-H., Angelides, K. J., and Benos, D. J. (1991). Amiloride-sensitive sodium channel is linked to the cytoskeleton in renal epithelial cells. *Proc. Natl. Acad. Sci. USA* **88,** 6971–6975.

Srinivasan, Y., Elmer, L., Davis, J., Bennett, V., and Angelides, K. (1988). Ankyrin and spectrin associate with voltage-dependent sodium channels in brain. *Nature* **333,** 177–180.

Surprenant, A., Buell, G., and North, R. A. (1995). P2X receptors bring new structure to ligand-gated ion channels. *Trends Neurosci.* **18,** 224–229.

Suzuki, M., Miyazaki, K., Ikeda, M., Kawaguchi, Y., and Sakai, O. (1993). F-actin network may regulate a Cl^- channel in proximal tubule cells. *J. Membr. Biol.* **134,** 31–39.

Takumi, T., Ishii, T., Horio, Y., Morishige, K.-I., Takahashi, N., Yamada, M., Yamashita, T., Kiyama, H., Sohmiya, K., Nakanishi, S., and Kurachi, Y. (1995). A novel ATP-dependent inward rectifier potassium channel expressed predominantly in glial cells. *J. Biol. Chem.* **270,** 16,339–16,346.

Terzic, A., Jahangir, A., and Kurachi, Y. (1995). Cardiac ATP-sensitive K^+ channels: Regulation by intracellular nucleotides and K^+ channel opening-drugs. *Am. J. Physiol.* **269,** C525–C545.

Terzic, A., and Kurachi, Y. (1996). Actin microfilament disrupters enhance K_{ATP} channel opening in patches from guinea-pig cardiomyocytes. *J. Physiol.* **492,** 395–404.

Terzic, A., and Kurachi, Y. (1998). Cytoskeleton effects on ion channels. *In* "Cell Physiology Source Book" (N. Sperelakis, Ed.), pp. 532–543. Academic Press, San Diego.

Undrovinas, A. I., Shander, G. S., and Makielski, J. C. (1995). Cytoskeleton modulates gating of voltage-dependent sodium channel in heart. *Am. J. Physiol.* **269,** H203–H214.

Van Wagoner, D. R.(1993). Mechanosensitive gating of atrial ATP-sensitive potassium channels. *Circ. Res.* **72,** 973–983.

Verrey, F., Groscurth, P., and Bolliger, U. (1995). Cytoskeletal disruption of A6 kidney cells: Impact on endo/exocytosis and NaCl transport regulation by antidiuretic hormone. *J. Membr. Biol.* **145,** 193–294.

Waldmann, R., Champigny, G., Bassilana, F., Heurteaux, C., and Lazdunski, M. (1997). A proton-gated cation channel involved in acid-sensing. *Nature* **386,** 173–177

Wallace, B. G. (1992). Mechanism of agrin-induced acetylcholine receptor aggregation. *J. Neurobiol.* **23,** 592–604.

Wang, W.-H., Cassola, A., and Giebisch, G. (1994). Involvement of actin cytoskeleton in modulation of apical K channel activity in rat collecting duct. *Am. J. Physiol.* **267,** F592–F598.

Wischmeyer, E., and Karschin, A. (1996). Receptor stimulation causes slow inhibition of IRK1 inwardly rectifying K^+ channels by direct protein kinase A-mediated phosphorylation. *Proc. Natl. Acad. Sci. USA* **93,** 5819–5823.

Wyszynski, M., Lin, J., Rao, A., Nigh, E., Beggs, A. H., Craig, A. M., and Morgan, S. (1997). Competitive binding of α-actinin and calmodulin to the NMDA receptor. *Nature* **385,** 439–442.

Yamada, M., and Kurachi, Y. (1995). Spermine gates inward-rectifying muscarinic but not ATP-sensitive K^+ channels in rabbit atrial myocytes. *J. Biol. Chem.* **270,** 9289–9294.

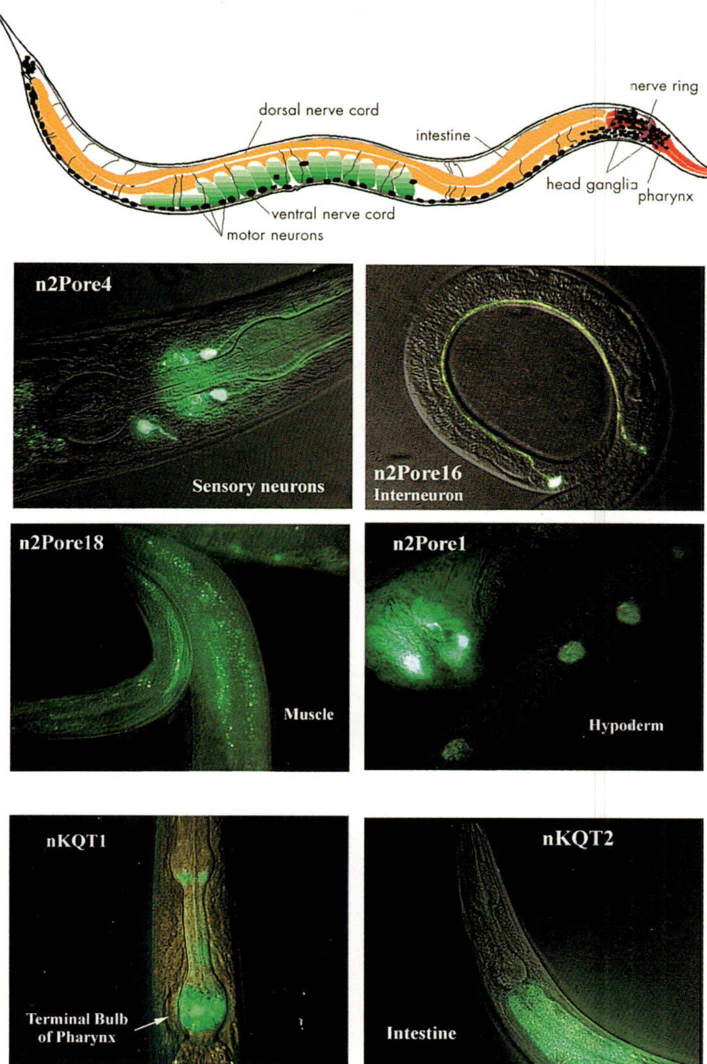

FIGURE 2-2 GFP–promotor transformation experiments for several *C. elegans* K^+ channel genes. n2Pore indicates a gene encoding a 4TM channel. nKQT indicates a gene encoding a channel of the 6TM KQT gene family (see text and Wei *et al.*, 1996). The labeled tissue type where expression is manifest is indicated on the figures. The tissue-specific patterns of expression for the six channel genes shown do not overlap. The general morphology of *C. elegans* (a hermaphrodite) is illustrated at the top; this figure also shows the outline of the nervous system. Prominent features include the cephalized "brain" ganglia and motor neuron cell bodies located along a ventral nerve cord. The body of *C. elegans* is organized similarly to higher animals in having a "brain," complex sensory receptors, and a food intake system in its anterior "head." The digestive tract, composed of only 20 cells is, nevertheless, quite long. There is also a complex reproductive tract (see Wood, 1988).

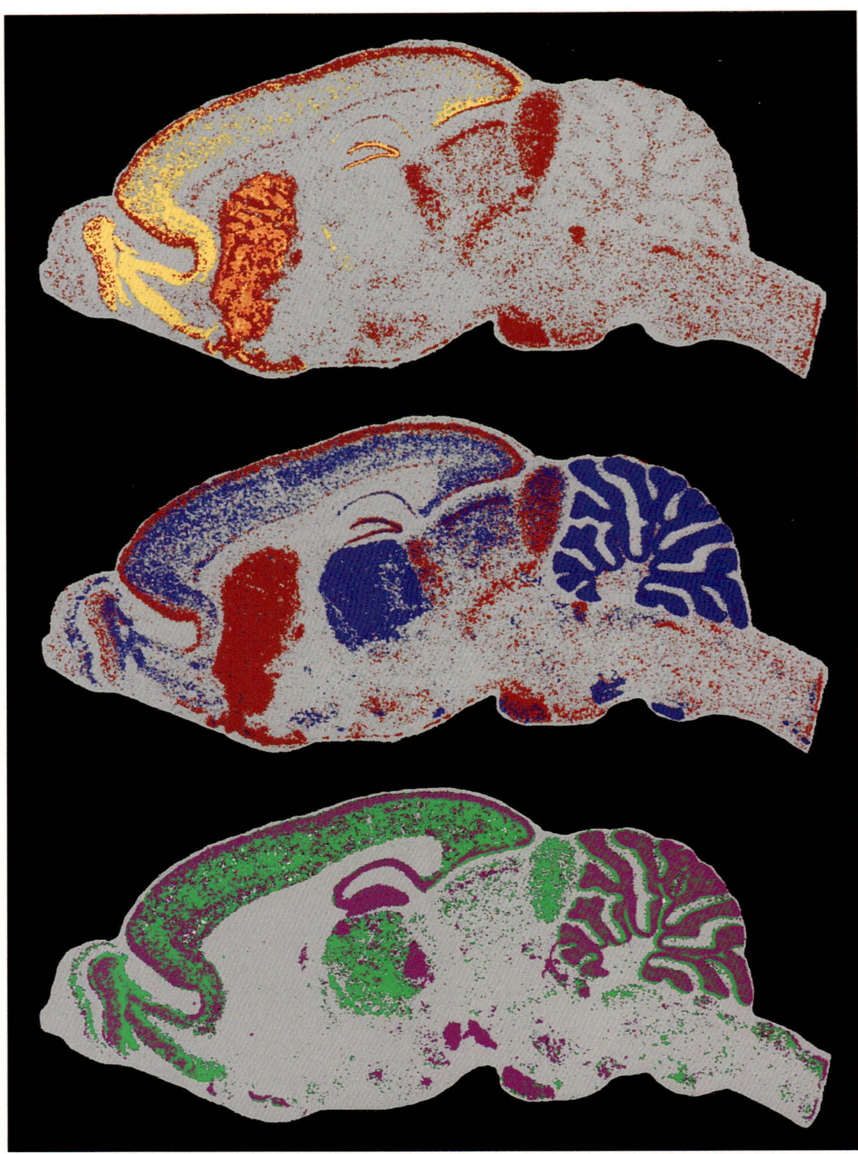

FIGURE 15-2 Cellular expression of mRNA transcripts of inwardly rectifying K^+ (Kir) channels in the developing rat brain at Postnatal Day 21. The figure depicts false-color transformations of digitized mRNA expression patterns as revealed by radiolabeled oligonucleotide *in situ* hybridization. Representative expression patterns of Kir channel subunits in adjacent 16-μm sagittal sections have been superimposed to demonstrate (i) forebrain and midbrain expression (Kir2.1 in red, Kir2.3 in yellow; at top), (ii) differential expression (Kir2.1 in red, Kir2.2 in blue; in the middle), and (iii) overlapping expression (Kir3.1 in green, Kir3.2 in magenta; at the bottom).

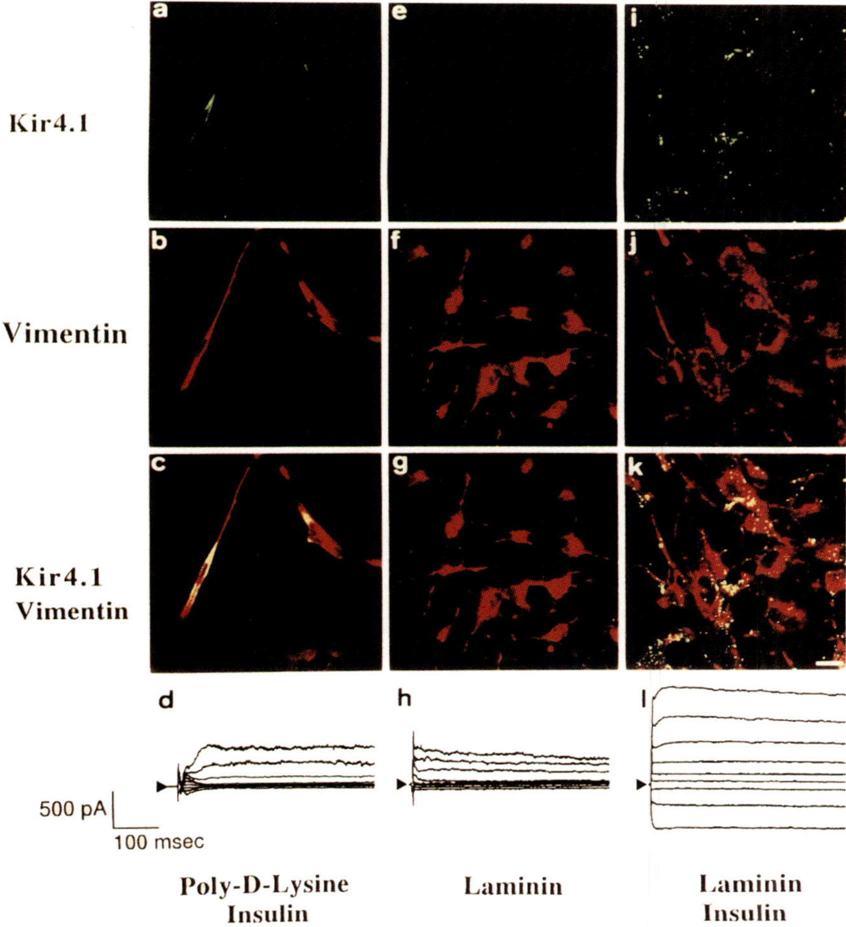

FIGURE 25-4 Expression and distribution of Kir4.1 in cultured cells from rat retina. Immunostaining of Kir4.1 (green) and vimentin, a marker of Müller cells (red), and whole-cell currents of cultured cells. Dissociated cells were cultured for 4 days on poly-D-lysine with 1 μ*M* insulin (a–d), on laminin without insulin (e–h), and on laminin with 1μ*M* insulin (i–l). In i and k, Kir4.1 clustered on the membrane of cells. Scale bar (shown in k): 20 μm. Whole-cell current recordings were performed in these cells. The holding potential was –70 mV, and traces were elicited with voltage steps from –120 to +40 mV in 20-mV increments (inset). In l, large inward currents were recorded, while no inward currents were recorded in d and h. Both insulin and laminin were needed for the expression of Kir4.1 on the membrane of Müller cells.

CHAPTER 15

Distribution of Inwardly Rectifying Potassium Channels in the Brain

Christine Karschin and Andreas Karschin
Molecular Neurobiology of Signal Transduction, Max-Planck-Institute of Biophysical Chemistry, 37070 Göttingen, Germany

I. INTRODUCTION

In mammals, molecular cloning to date has revealed the existence of 14 unique members (plus alternatively spliced Kir1.1b–f, Kir3.1, and Kir3.2 transcripts) of the gene family of inwardly rectifying K^+ (Kir) channels. Based on sequence similarities and functional properties a new nomenclature has recently been established to group Kir subunits into five subfamilies (for references see reviews in Doupnik *et al.*, 1995; Fakler and Ruppersberg, 1996; Isomoto *et al.*, 1997; as well as chapters in this volume). Kir1 channels are ATP sensitive and "mildly" rectifying and include kidney Kir1.1 (alternatively termed ROMK or K_{AB}-1), Kir1.2 (BIRK1, BIR10, Kir4.1, K_{AB}-2)

1063-5823/99 $30.00

present in the central nervous system (CNS), and the recently identified Kir1.3 subunits (Shuck *et al.,* 1997) primarily expressed in kidney, lung, and pancreas. "Strongly" rectifying Kir2 channels (IRK, BIR) comprise Kir2.1 (IRK1, HHBIRK1), Kir2.2 (IRK2, BIR8), Kir2.3 (IRK3, BIR11, HIR, BIRK2, hIRK2, HRK1, BIK), and the novel Kir2.4 isoforms (IRK4; Töpert *et al.,* 1998). Four isoforms of the Kir3 subfamily are strongly expressed in brain and heart and are likely to be activated directly by $G\beta\gamma$ subunits: Kir3.1 (GIRK1, KGA), Kir3.2 (GIRK2, BIR1, K_{ATP}-2), Kir3.3 (GIRK3), and Kir3.4 (GIRK4, rcKATP-1, CIR). Kir5.1 (BIR9) subunits differ significantly in their primary amino acid sequence from all other known Kir channels. Finally, the ubiquitously distributed Kir6.1 (uK_{ATP}-1) and Kir6.2 subunits (BIR) that are inhibited by cellular ATP interact with sulfonylurea receptors (SUR) in the formation of classic K_{ATP} channels.

In spite of some conflicting data on their tissue distribution, Kir1.2, Kir2.1, Kir2.2, Kir2.3, Kir2.4, Kir3.1, Kir3.2, Kir3.3, Kir3.4, Kir6.1, and Kir6.2 subunits are clearly expressed in the mammalian brain, as indicated by Northern blot, *in situ* mRNA, and more recently, immunocytochemical analyses. Other subunits appear to be predominantly present in peripheral tissues and are not or are only weakly expressed in brain (Kir1.1, Kir1.3). These, and where only insufficient information is presently available (Kir5.1), will not be considered in this overview of Kir channel distribution in the rodent brain.

II. DISTRIBUTION OF Kir CHANNEL SUBUNITS

Expression and localization of Kir subunit mRNAs in the mammalian brain has mainly been demonstrated using *in situ* hybridization (Figs. 1, 2). Because mRNAs are predominantly found in somata, this analysis defines brain nuclei and cell populations expressing a given channel subunit. Relative amounts of observed hybridization signals also reflect strong or weak expression in these brain nuclei (Table I). More recent immunocytochemical studies that evaluate the localization of Kir2.1, Kir3.1, Kir3.2, and Kir3.4 proteins using specific antibodies reach beyond the information provided by mRNA localization. Both light and electron microscopic studies have identified Kir subunits at high cellular resolution, allowing localization at pre- or postsynaptic sites, or defining neuronal compartments with several subunits coexpressed.

A. Kir1 Subfamily

1. Kir1.2

Abundant Kir 1.2 subunit mRNA expression was first detected in the rat brain by Bredt *et al.* (1995) and was primarily localized in neurons

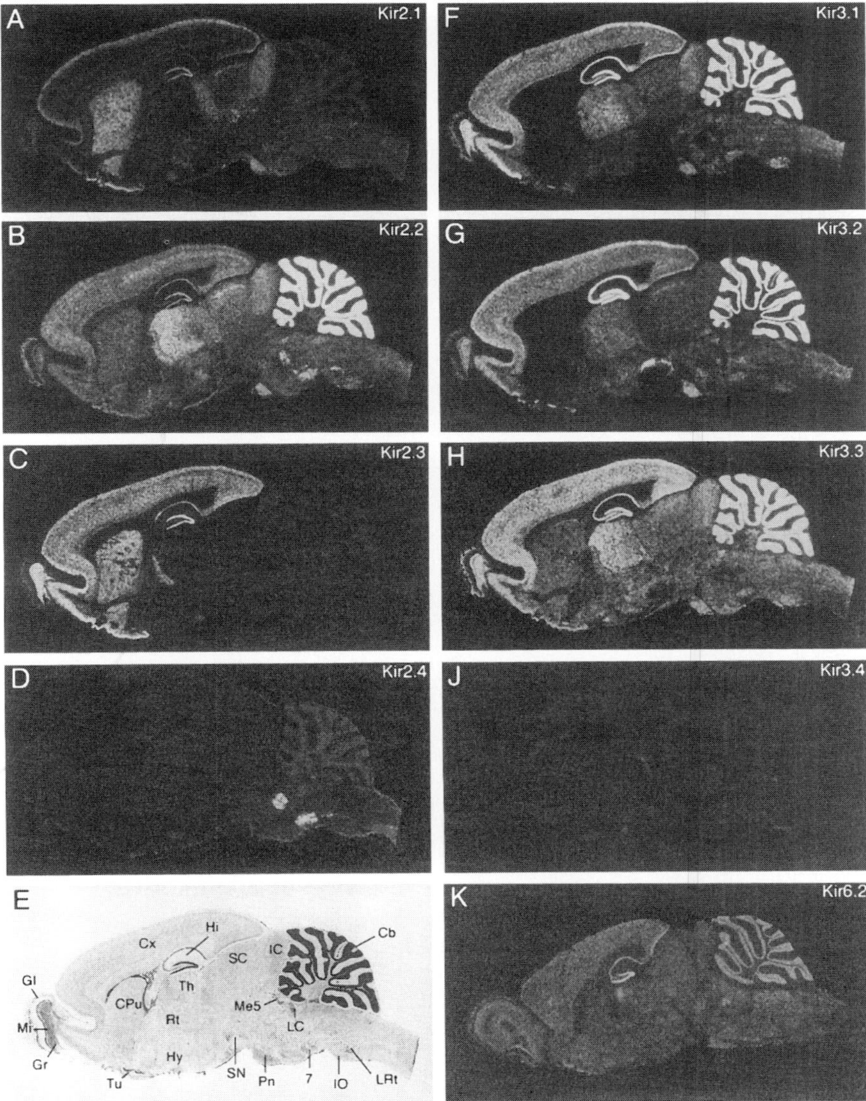

FIGURE 1 Kir 2 and Kir3 mRNA subunit expression in the P21 rat brain and Kir6.2 in adult mouse brain (K). Shown are a Nissl-stained rat 16-μm sagittal section in brightfield (E) and X-ray images of adjacent brain sections (A–D, F–J) hybridized with specific 3′ ^{35}S-endlabeled 45–50 bp antisense oligonucleotide probes. Cb, cerebellum; Cpu, caudate putamen; Cx, cortex; Gl, glomerular cell layer; Gr, granule cell layer olfacory bulb; Hi, hippocampus; Hy, hypothalamus; IC, inferior colliculus; IO, inferior olive; LC, locus coeruleus; LRt, lateral reticular nucleus; Me5, mesencephalic trigeminal nucleus; Mi, mitral cell layer olfactory bulb; Pn, pontine nucleus; Rt, thalamic reticular nucleus; SC, superior colliculus; SN, substantia nigra; Th, thalamus; Tu, olfactory tubercle; 7, facial nucleus. Exposure time, 25 days.

TABLE I

Distribution of Kir Channel Subunits in the Rodent Brain

Brain region	Kir1.2	Kir2.1	Kir2.2	Kir2.3	Kir2.4	Kir3.1	Kir3.2	Kir3.3	Kir3.4	Kir6.2
Forebrain										
Olfactory bulb										
Granule cell layer		++	+	++	0	+++	++	+++	(+)	+
Mitral cell layer		+/+++*	+	++	0	+++	++	+++	(+)	+++
Periglomerular cell layer		0/+++*	++	0	0	++	+	+++	(+)	++
Anterior olfactory nucleus		+	+	+++	0	++++	+++	++++	(+)	++
Olfactory tubercle		+++	+	+++	0	+	+	+++	(+)	++
Piriform cortex		++	++	+++	0	++++	+++	+++	(+)	+++
Neocortex										
Layer II, III	+++	+++	++	+++	0	++++	+++	++++	(+)	+++
Layer IV-VI	+++	++	++	++	0	++++	+++	++++	++	+++
Subiculum		+	++	++	0	+++	++	+++	(+)	++
Entorhinal cortex		++	++	++	0	++++	++	++++	(+)	+++
Hippocampus										
Dentate gyrus granule cells	+++	+++	++	+++	0	++++	++++	++++	(+)	+++
CA1–CA2 pyramidal cells	+++	+	++	++	0	+++	++++	++++	(+)	++
CA3 pyramidal cells	+++	0/++*	0/++*	0/++*	0	+++	+++	+++	(++)	
Tenia tecta		+	±	+++	0	++++	+++	+++	(±)	+++
Indusium griseum		0	+	++	0	+++	+++	+	(++)	+++
Septum										
Bed nuclei of stria terminalis		0	0	0	0	++/0*	++	++	0	+
Lateral septal nucleus		±	±	0	0	+++	++	+	(++)	+
Nuclei of the diagonal band		0	+++	+	0	+	+	+++	0	+++
Basal ganglia										
Caudate putamen		+++	++	+++	0	+	0	++	0/++*	++
Nucleus accumbens		+++	++	+++	0	+	0	+	0/++*	++
Globus pallidus		0	0	0	0	+	0	+	0/++*	++

Amygdala		+	+	±	0	+++	+++	+++	0	+
Lateral olfactory tract nucleus		+++	0	0	0	+++	++	+++		+++
Thalamus										
Reticular nucleus		0	++++	+++	0	+++	++	+++	(+)	+++
Geniculate nuclei		0	++++	0	0	+++	+++	++++	(+)	+
Anterior dorsal nucleus		0	++++	0	+	+++	++	++++	(+)	+++
Lateral nuclei		0	++++	0	0	+++	+++	++++	(+)	+
Ventroposterior nuclei		0	++++	0	0	+++	++	++++	(+)	++
Hypothalamus										
Hypothalamic area		+	+	0	0	±	+	++	(+)	+
Preoptic area		0	±	0	0	±	+	++	(0)	+
Magnocellular preoptic nucleus		0	++	0	0	+	+	++		+++
Supraoptic nucleus		0	++	0	0	0	0	++	0	
Mammillary nuclei		0	++	0/++*	0	++	+	+	0	
Epithalamus										
Medial habenula		+	++++	0/++*	0	++/±*	++	0	+++	
Lateral habenula		0	0	0	0	0	0	++	0	
Midbrain										
Subthalamic nucleus	+++	+	++	0	++	+	++	++		+++
Anterior pretectal nucleus	+++	+++	+	0	0	++	0	++		
Ventral lateral lemniscus	++++	0	+	0	0	+++	++	+++		
Superior colliculus	+++	+++	+++	0	0	++	+	+++	++	++
Inferior colliculus	+++	++	++	0	0	+++	±	+++	0	++
Central gray		+	+	0	0	+	+	++	0	++
Deep mesencephalic nucleus		+++	++	0	0	+	+	++	++	
Oculomotor nucleus (III)		0	++	0	+++	+++	++	++	0	
Red nucleus		0	+	0	0	+++	++	+++	0	++
Substantia nigra										
Pars compacta		0	+	0	0	0	++++	++	0	++
Pars reticulata		±	+	0	0	0	0	++	0	++

(continues)

TABLE I (*Continued*)

Brain region	Kir1.2	Kir2.1	Kir2.2	Kir2.3	Kir2.4	Kir3.1	Kir3.2	Kir3.3	Kir3.4	Kir6.2
Ventral tegmental area	+++	0	0	0	0	0/++*	++++	+	0	++
Interpeduncular nucleus			+		0	+	0	++	0	
Hindbrain										
Cerebellum										
Deep nuclei	+++	0/+*	+	0	0	+++	0	+++	0	+++
Molecular layer		0/++*	0	0	0	±	0	++	0	+
Granule cell layer	++	0/++*	++++	0	0	+++	+++	+++	(++)	+
Purkinje cells		0/++*	0	0	0	0	0	+++	(+)	+++
Pontine nucleus	++++	++	+	0/++*	0	+++	++	+++	0	++
Superior olive		±	±	0	0	+	+	++	0	
Trapezoid body		0	++	0	++	+++	0	+	0	
Locus coeruleus		0	+	0	0	++	++	+++		+
Mesencephalic trigeminal nucleus		0	++	0	0	++	++	++	0	
Motor trigeminal nucleus (V)	++++	0	++++	0	++++	++	++	+++	0	
Facial nucleus (VII)		+	++++	0	++++	++	++	+++	0	+
Ambiguus nucleus		+	+++	0	++++	+++	+	++	0	
Inferior olivary nuclei		0	++++	0	0	0++*	0	++	++	+

Accessory nucleus										
Raphe nuclei		0	+	0	0	+	0	++		+
Pontine reticular formation	++++	+	++	0	0	++	++	++	0	++
Dorsal tegmental nucleus		0	+	0	0	+	++	+++	0	
Parabrachial nucleus		0	0	0	0	+/+++*	++	++++	0	
Vestibular nucleus	++++	0	+	0	0	++	+++	+++	++	+++
Cochlear nuclei	++++	0	++++	0	0	+++	+++	+++	0	+++
Cuneate nucleus		0	0	0	0	++		++	0	
Solitary nucleus		0	0	0	0	0/+++*		++	0	
Hypoglossal nucleus (XII)		0	+++	0	++++	+++		++	0	+++
Spinal trigeminal nucleus	++++	0	0	0	0	+++	++	+++	0	++
	Kir1.2	Kir2.1	Kir2.2	Kir2.3	Kir2.4	Kir3.1	Kir3.2	Kir3.3	Kir3.4	Kir6.2

Note. Data are compiled from investigations using both radioactive *in situ* hybridization mRNA analysis and immunohistochemical detection of Kir protein in rat or mouse brain sections with subunit-specific antibodies. mRNA levels (rated as the relative silver grain density) and protein expression levels represent a consensus from published data: +++, abundant; ++, moderate; +, low; 0, no expression. Kir1.2: Bredt *et al.*, 1995 (rat); Kir2.1: Morishige *et al.*, 1993 (mouse), Horio *et al.*, 1996 (mouse), Karschin *et al.*, 1996 (rat); Kir2.2: Horio *et al.*, 1996 (mouse), Karschin *et al.*, 1996 (rat); Kir2.3: Bredt *et al.*, 1995 (rat), Horio *et al.*, 1996 (mouse), Karschin *et al.*, 1996 (rat); Kir2.4: Töpert *et al.*, 1998 (rat); Kir3.1: DePaoli *et al.*, 1994 (rat), Karschin, C. *et al.*, 1994, 1996 (rat), Bausch *et al.*, 1995 (rat); Kobayashi *et al.*, 1995 (mouse); Liao *et al.*, 1996 (rat); Ponce *et al.*, 1996 (rat); Miyashita and Kubo, 1997 (rat); Kir3.2: Dixon *et al.*, 1995 (rat), Karschin, C. *et al.*, 1996 (rat), Kobayashi *et al.*, 1995 (mouse), Liao *et al.*, 1996 (rat); Kir3.3: Dissmann *et al.*, 1996 (rat), Karschin *et al.*, 1996 (rat), Kobayashi *et al.*, 1995 (mouse); Kir3.4: Karschin *et al.*, 1996 (rat), Spauschus *et al.*, 1996 (rat), Iizuka *et al.*, 1997; Kir6.2: Karschin *et al.*, 1997 (rat). Values in parentheses are for cRNA probes. An asterisk indicates significantly conflicting data from different investigators; a blank indicates no data available to the authors.

of the neocortex, hippocampus, superior and inferior colliculus, lateral lemniscus, cerebellar deep nuclei and granule cells, and most intense in brainstem nuclei (spinal trigeminal, vestibular, cochlear, pontine, motor trigeminal nuclei, and area postrema). Controversially, both the mRNA and protein of this subunit were described by another group to be predominantly expressed in white matter oligodendroglia, cerebellar Bergmann glia, and retinal Müller cells (Takumi *et al.,* 1995; Ishii *et al.,* 1997; see below). Strong Kir1.2 signals on RNA blots of corpus callosum support these latter data (Shuck *et al.,* 1997). A discrepancy between expression in neurons and nonneuronal elements (e.g., also marginal epithelial cells in the inner ear; Hibino *et al.,* 1997) might eventually be explained by the existence of a second brain-specific transcript, as indicated on Northern blots (Shuck *et al.,* 1997).

B. Kir2 Subfamily

1. Kir2.1

Kir2.1 mRNAs are widely distributed in both the mouse brain (Morishige *et al.,* 1993; Horio *et al.,* 1996) and rat brain (Karschin *et al.,* 1996; Fig. 1). Northern blot (Kubo *et al.,* 1993; Ishii *et al.,* 1994; Takahashi *et al.,* 1994) and *in situ* hybridization signals were strongest in forebrain structures, for example, olfactory bulb, piriform cortex, neocortex Layers II–III, hippocampus, and striatum. In the midbrain, high levels were restricted to the anterior pretectal nucleus, mesencephalic deep nucleus, and optic nerve layer of the superior colliculus. In the brainstem, only lower levels of transcript were found in pontine and facial nuclei. While mRNA signals were undetectable in the cerebellum of both species (Karschin *et al.,* 1996), Kir2.1 proteins were found in the molecular layer, granule cell dendrites, and Purkinje cell somata and dendrites (Liao *et al.,* 1996, Miyashita and Kubo, 1997; Signorini *et al.,* 1997). Analogously, mRNA levels were low in CA1–3 hippocampal pyramidal cells, but protein was detected in somata and proximal dendrites in stratum oriens, stratum radiatum and stratum lacunosum-moleculare (Liao *et al.,* 1996; Miyashita and Kubo, 1997).

2. Kir2.2

There is general agreement on the wide distribution of Kir2.2 mRNA in both mouse (Horio *et al.,* 1996) and rat brain (Karschin *et al.,* 1996) with very high levels in cerebellar granule cells, the large motoneurons of the facial and motor trigeminal nuclei, and other brainstem nuclei (e.g., inferior olive, hypoglossal, and cochlear nuclei). Strong expression also in the large aspiny cholinergic neurons in caudate putamen, periglomerular cells in the olfactory bulb, and neurons in the epithalamic medial habenula stands out from a generally more moderate expression in the forebrain that is supported by Northern blot analysis (Takahashi *et al.,* 1994). The observation of high Kir2.2 mRNA levels in thalamic nuclei of the rat brain, but not the mouse brain, may be attributed to a species difference.

3. Kir2.3

Accumulating data from Northern blot and *in situ* hybridization analyses demonstrate that Kir2.3 mRNA is indeed almost exclusively limited to the forebrain (Morishige *et al.,* 1994; Bredt *et al.,* 1995; Horio *et al.,* 1996; Karschin *et al.,* 1996) and not expressed ubiquitously throughout the rat brain (Falk *et al.,* 1995; Fink *et al.,* 1996). Noticeably highest mRNA levels were present in the whole olfactory system (i.e., olfactory bulb, anterior olfactory nucleus, olfactory tubercle, and piriform cortex) as well as the caudate putamen/nucleus accumbens complex and all neocortical layers. Dentate gyrus granule cells and the CA2 cells of the hippocampal region were more strongly positive than other pyramidal neurons of the CA field. Finally, in the midbrain, solely neurons in the thalamic reticular nucleus were strongly labeled.

4. Kir2.4

In contrast to Kir2.3 the recently described Kir2.4 subunits are virtually absent from the forebrain, but are strongly expressed in several distinct nuclei of the medulla and brainstem (Töpert *et al.,* 1998; Fig. 1). These include nuclei related to the special visceral motor column such as the motor trigeminal, facial, ambiguus, and accessory nuclei, as well as other cranial motor nerve nuclei (e.g., abducens, hypoglossus, oculomotoric). In these nuclei Kir2.4 was solely expressed in the large cholinergic motoneurons and in none of the smaller (inter-) neurons and glial cells. Surprisingly, the choroid plexus of the fourth ventricle (not the third and lateral ventricle) was strongly labeled, suggesting expression in nonneuronal cells. Kir2.4 mRNA was furthermore detected in the trapezoid body, in the anterodorsal thalamic nucleus, and in the subthalamic nucleus, but at much lower levels.

C. *Kir3 Subfamily*

1. Kir3.1

Kir3.1 mRNA is abundantly expressed throughout the brain of both mouse (Kobayashi *et al.,* 1995) and rat (De Paoli *et al.,* 1994; C. Karschin *et al.,* 1994, 1996; Ponce *et al.,* 1996) and protein is found mainly on somata and dendrites with a distribution matching the transcript pattern (Bausch *et al.,* 1995; Liao *et al.,* 1996; Ponce *et al.,* 1996; Miyashita and Kubo, 1997). Prominent expression occurs in the olfactory system, hippocampus, dentate gyrus, amygdala, neocortex, lateral septal nuclei, thalamus, red nucleus, cerebellum granule cells, deep cerebellar nuclei, and several brainstem nuclei (e.g., pontine nucleus, cochlear, vestibular nuclei, and spinal trigeminal nuclei). Kir3.1 mRNAs are virtually absent from the hypothalamus and the basal ganglia.

2. Kir3.2

The Kir3.2 mRNA distribution is generally very similar to that of Kir3.1 in that mRNA is widely expressed and protein is prominent in neuronal somata and dendrites in the hippocampus, dentate gyrus, amygdala, neocortex, lateral septal nuclei, cerebellar granule cells, and several brainstem nuclei (Liao *et al.,* 1996; Adelbrecht *et al.,* 1997; Murer *et al.,* 1997). Significant differences exist only (i) in the dopaminergic neurons of the substantia nigra pars compacta and ventral tegmental area where Kir3.1 is absent (Fig. 2); (ii) in the caudate-putamen complex where Kir3.2 is absent and Kir3.1 is moderately expressed; and (iii) in cerebellar deep nuclei and most thalamic nuclei where Kir3.1 is more strongly expressed.

3. Kir3.3

Of all subunits, Kir3.3 is the most widely distributed with moderate to strong expression levels in the majority of brain regions of the rat (Dissmann *et al.,* 1996; Karschin *et al.,* 1996). Strongest signals are present in the olfactory system, all neocortical layers, hippocampus, and thalamus. As one of the few exceptions, cerebellar Purkinje cells express Kir3.3, but neither Kir3.1 or Kir3.2, mRNA. In mouse brain, Kir3.3 subunits are also widely distributed, but found only in lower levels in olfactory bulb, caudate putamen, and septum (Kobayashi *et al.,* 1995). Immunohistochemical data to evaluate subcellular Kir3.3 distribution are not yet available.

4. Kir3.4

With both oligonucleotide and RNA probes, Kir3.4 mRNA showed the most restricted distribution of all Kir3 subunits (Karschin *et al.,* 1996; Spauschus *et al.,* 1996). Moderate mRNA expression levels were only observed in the medial habenula, superior colliculus, and cerebellar Purkinje cells, structures that were also immunopositive with α-Kir3.4-specific antibodies (Murer *et al.,* 1997). In the globus pallidus, ventral pallidum, ventral striatum, and nucleus of the diagonal band, Kir3.4 protein is present, but no significant mRNA level has been detected. Also neocortical Layer VI neurons (adjacent to white matter) were immunopositive, but contained high mRNA levels only during postnatal development (Karschin and Karschin, 1997).

Contrary to these reports are the findings of Iizuka *et al.* (1997). They used digoxygenin-labeled RNA probes and antibodies and detected both strong Kir3.4 mRNA and protein expression throughout the brain with highest levels in cortex, hippocampus, olfactory system, basal ganglia, cerebellar Purkinje and basket cells, numerous brainstem nuclei, and the nonneuronal cells of the choroid plexus.

D. Kir6 Subfamily

1. Kir6.1

The mRNA of this subunit is found throughout the brain, but only in small, scattered cells that may represent a subpopulation of glial cells (see below). Preliminary observations indicate that the mRNA of the SUR2B isoform is distributed similarly, suggesting that both components possibly interact in the formation of a nonneuronal type of K_{ATP} channel.

2. Kir6.2

The ubiquitous distribution of Kir6.2 mRNA in the rodent brain largely overlaps with the SUR1 mRNA pattern. Expression levels are generally moderate except for elevated signals in olfactory bulb mitral cells, cortex, hippocampus, nuclei of the diagonal band, subthalamic nucleus, nuclei of the lateral olfactory tract, thalamic anterior dorsal nucleus, cerebellar Purkinje cells and deep nuclei, and several brainstem nuclei (e.g., vestibular, cochlear, hypoglossal nuclei). No protein localization studies are yet available for either Kir6.2 or SUR1 subunits to demonstrate their subcellular colocalization.

III. THE POSSIBLE FORMATION OF HETEROMULTIMERIC Kir CHANNELS IN THE BRAIN

The first evidence for the formation of heteromeric Kir channels was reported by Krapivinsky *et al.* (1995) in the heart, where Kir3.1 and Kir3.4 (CIR, GIRK4) subunits together give rise to G-protein-activated K_{ACh} channels. Assembly into functional Kir3 channels has now been demonstrated electrophysiologically and biochemically for combinations of Kir3.1/Kir3.2, Kir3.1/Kir3.3, Kir3.1/Kir3.4, Kir3.2/Kir3.3, Kir3.2/Kir3.4, Kir3.3/Kir3.4 (see Chapter 9 by Tinker and Jan in this volume). From their vastly overlapping distribution in the CNS, it is likely that Kir3.1, Kir3.2, and

Kir3.3 subunits (Kir3.4 express only poorly in the adult mammalian CNS) in many regions are present in the same cell populations (Figs. 1, 2). High-resolution immunohistochemical subcellular localization, however, has been shown so far only for Kir3.1 (Bausch *et al.*, 1995; Ponce *et al.*, 1996; Drake *et al.*, 1997; Miyashita and Kubo, 1997) and Kir3.2 (Kofuji *et al.*, 1996; Slesinger *et al.*, 1996; Lauritzen *et al.*, 1997) initiated by the great interest in the Kir3.2-defective *weaver* mouse. Using subunit-specific antibodies, Liao and colleagues (1996) also showed aggregation by coimmunoprecipitation of Kir3.1/3.2 gene products from the mouse cerebrum, cerebellum, and hippocampus, suggesting that this combination constitutes a major form of neuronal G-protein-gated Kir channel.

In contrast to Kir3 subunits, which unfavorably assemble into homomeric proteins, members of all other Kir subfamilies (except Kir1.3, Kir5.1, and Kir6.2 subunits, which require association with sulfonylurea receptors) give rise to homomeric channels when expressed heterologously. Only one Kir2 subunit is expressed in several neuronal populations; in other cells possible heteromerization among Kir2 subunits is still under debate. There are conflicting reports as to whether Kir2.1 subunits coassemble with other Kir2 subunits (Fink *et al.*, 1996) or not (Tinker *et al.*, 1996). The cellular distribution of Kir2.1 and Kir2.3 transcripts overlap considerably in the rat forebrain (Fig. 2; see color plate). Of special interest are also many brainstem cranial nerve nuclei that highly express both Kir2.2 and the novel Kir2.4 subunits, suggesting interaction at the protein level (Töpert *et al.*, 1998). Coassembly of Kir2.1 across subunit borders with Kir1.2 (Kir4.1) subunits has also been well established (Fakler *et al.*, 1996). Recombinant Kir4.1 subunits also form heteromeric channels with Kir1.1 (Glowatzki *et al.*, 1995), Kir3.4 (Tucker *et al.*, 1996), and Kir5.1 subunits (Pessia *et al.*, 1996), which might be of relevance in the brain. No consent, however, has yet been reached on the precise localization of the Kir1.2 isoform(s).

IV. SUBCELLULAR LOCALIZATION OF Kir CHANNEL SUBUNITS

Corresponding to their potential role in postsynaptic inhibition, light and electron microscopic analyses of antibody staining patterns localized Kir2.1, Kir3.1, Kir3.2, and Kir3.4 subunits postsynaptically on cell somata and primary dendrites. However, in some cases they were also found in remote synaptic regions and presynaptically on axon terminals.

A. Postsynaptic Localization

In most regions of the brain where strong mRNA signals were present, channel proteins were predominantly expressed on cell bodies and dendrites, and only occasionally on varicose fibers (Bausch *et al.,* 1995; Liao *et al.,* 1996; Ponce *et al.,* 1996; Adelbrecht *et al.,* 1997; Drake *et al.,* 1997; Iizuka *et al.,* 1997; Miyashita and Kubo, 1997; Murer *et al.,* 1997). Kir2.1 subunits have been demonstrated only postsynaptically in somata and primary dendrites of Purkinje cells, cortical Layer II, III, and V pyramidal cells, and hippocampal neurons (Liao *et al.,* 1996; Miyashita and Kubo, 1997; Signorini *et al.,* 1997). In cerebellar granule cell layer glomeruli (composed of clustered granule cell dendrites surrounding a mossy fiber) Kir3.1 and Kir3.2 protein were solely detected postsynaptically on dendrites, but not on mossy fiber axons (Liao *et al.,* 1996; Ponce *et al.,* 1996; Miyashita and Kubo, 1997; Murer *et al.,* 1997). Both proteins are also strongly expressed in the hippocampus stratum lacunosum moleculare (slm, the terminal fields of the perforant pathway where thalamic fibers synapse onto pyramidal cell dendrites). At higher resolution Kir3.1 was detected predominantly on the pyramidal cell spiny dendrites, often juxtaposed to asymmetric postsynaptic densities, but not on axons (Drake *et al.,* 1997; but see Ponce *et al.,* 1996, who attributed punctate staining in slm to terminal fields of thalamic projections).

B. Presynaptic Localization

On the contrary, several findings of presynaptic localization may indicate that Kir3 subunits play a role in modulation of transmitter release. (i) Kainate lesions of thalamic input to the cortex were found to reduce the high Kir3.1 protein levels in the whisker barrels of the primary somatosensory cortex, suggesting a presynaptic localization on thalamocortical projection fibers or axons from Layer IV neurons (Liao *et al.,* 1996; Ponce *et al.,* 1996; Miyashita and Kubo, 1997). (ii) Both Kir3.1 and Kir3.2 proteins are also present on axon terminals in the lateral septum (Liao *et al.,* 1996; Murer *et al.,* 1997) and likely originate from the major input neurons in the hippocampus. (iii) Presynaptic localization of Kir3.1 has been found in varicose terminals, but not in dendrites in the paraventricular hypothalamic nucleus (PVN). Several PVN projection areas such as amygdala, subiculum, and lateral septum contain high levels of Kir3.1 mRNA. (iv) Kir3.4 immunoreactivity is present in axon terminals of basket cells in cerebellum and hippocampus (Iizuka *et al.,* 1997).

V. Kir CHANNEL SUBUNITS IN GLIA AND NONNEURONAL ELEMENTS

Glial cells in the central and peripheral nervous system contain a large number of functionally different Kir channels that serve to control K^+

homoeostasis, glia–neuron signaling, or myelinogenesis (e.g., Barres *et al.,* 1990; A. Karschin *et al.,* 1994; Sontheimer, 1994). However, remarkably little is known about their molecular identity. Most investigators using *in situ* hybridization or immunhistochemical techniques describe Kir subunit expression restricted to neuronal cell populations, but absent from glia. Possibly expression in glial cells is (i) overlooked due to the small cell size, (ii) below the threshold of detection, (iii) reduced due to low turnover rates of Kir transcripts, and (iv) yet unknown glia-specific Kir subunits may exist (see also Chapter 25 by Horio and Kurachi in this volume).

Within the Kir2 family reverse transcriptase polymerase chain reaction detection of Kir2.1 subunits from single cerebral oligodendrocytes (Karschin and Wischmeyer, 1995) and immunohistochemical localization of Kir2.1 or Kir2.3 proteins in the nodal microvilli of Schwann cells ensheathing myelinating nerves remain an exception (Mi *et al.,* 1996). On the basis of these channels Schwann cells and oligodendrocytes may provide for the accumulation or spatial buffering of K^+ ions from the extracellular space. In CNS, white matter oligodendrocytes, but also cerebellar Bergmann glia and the abundant Müller cells in the retina glial cell-specific Kir1.2 (Kir4.1, K_{AB}-2) subunits likely play a similar role (Takumi *et al.,* 1995; Ishii *et al.,* 1997). Most interestingly, Kir1.2 channels were found clustered in the endfeet facing the vitreous and the distal portions of the Müller cells, where they are crucial for establishing a K^+ gradient in the mammalian retina (e.g., Newman, 1993; Newman and Reichenbach, 1996). Other nonneuronal elements of Kir1.2 expression include the endolymph-facing marginal cells of the stria vascularis and satellite cells surrounding spiral ganglion neurons in the mammalian cochlea. In the inner ear, Kir1.2 channels may be involved in the generation of the endocochlear potential and thus are essential for the sense of hearing (Hibino *et al.,* 1997).

Northern blot analysis demonstrated moderate mRNA levels of Kir6.1 subunits in the brain (Inagaki *et al.,* 1995) which alternatively to Kir6.2 isoforms may combine with sulfonylurea receptors in the formation of K_{ATP} channels. Analysis of transcript expression at the cellular level using oligonucleotide probes revealed that Kir6.1 subunits were present in a distinct cell population throughout the brain (Karschin *et al.,* 1997). Positive cells that were also labeled with SUR2-specific probes had very small somata (<7 μm in diameter), were sparsely distributed, and likely represent a glioblast precursor population (C.K., unpublished observations).

VI. Kir CHANNEL EXPRESSION IN THE DEVELOPING CNS

From their function in controlling neuronal excitability and susceptibility to external stimuli it is evident that Kir channels play an important role during synaptogenesis and in earlier developmental events of nonsynaptic communication and trophic interactions. An ontogenic role of Kir channel activity was suspected when mutant Kir3.2 genes were found responsible for the defective differentiation of cerebellar and substantia nigra neurons. Both Kir3.2 and associated Kir3.1 subunits are present in the mouse cerebellar anlage as early as Embryonic Day E14.5 and continue to be expressed in the cerebellum at high levels during development (Kofuji *et al.,* 1996; Slesinger *et al.,* 1996).

Using *in situ* hybridization, we recently supplied comprehensive localization profiles of all Kir2 and Kir3 subunits during embryonic and postnatal development of the rat nervous system and determined the onset of channel expression as well as the ontogenic transitions leading to adult expression (Karschin and Karschin, 1998; Fig. 3). No data are available yet for the expression of Kir1.2, Kir5.1, and Kir6.1/Kir6.2 subunits that have been detected in the adult rodent brain. In the rat, mRNAs of all subunits are already detectable at E12, when the neural tube is closed, but most neurons are not born yet. Only Kir2.3 mRNA is absent from the brain until E21 together with Kir3.4 from the embryonic spinal cord. All other subunits appear with distinct and mainly nonoverlapping expression patterns (e.g., Kir3.2 transcripts that in the adult together form a major type of G-protein-gated Kir channel are absent from the developing cortex and thalamus). During further development, expression in the CNS becomes more widespread leading to widely overlapping mRNA patterns as in the adult rat. In general, subunits are mainly found in regions that are also positive in the adult.

Altogether in the CNS there is neither a specific subunit nor a brain region that exhibits "early" or "late" expression; that is, appearance of transcripts is not generally related to the progressive and recessive phases during neurogenesis, but rather regulated differentially for each subunit and any specific set of neurons. It is important to note that together with mRNA expression channel protein can be detected in immature, yet undifferentiated neurons (e.g., Kir3.1 in cerebellar and hippocampal granule cells; Miyashita and Kubo, 1997). For each subunit (except Kir2.3) there are cells in which (i) early expression starts soon after neurogenesis; (ii) expression is late, long after cells are born and have already migrated and differentiated (e.g., Kir3.4 subunits except in the ventromedial hypothalamic nucleus are absent from the embryonic brain); (iii) expression is found in the proliferating embryonic neuroepithelium that lines the ventricles and gives rise to all central neural elements (except Kir2.2 and Kir3.4), i.e., high during proliferative phases; and (iv) expression during ontogeny is transient and virtually absent in the adult (in particular Kir3.2 and Kir3.4).

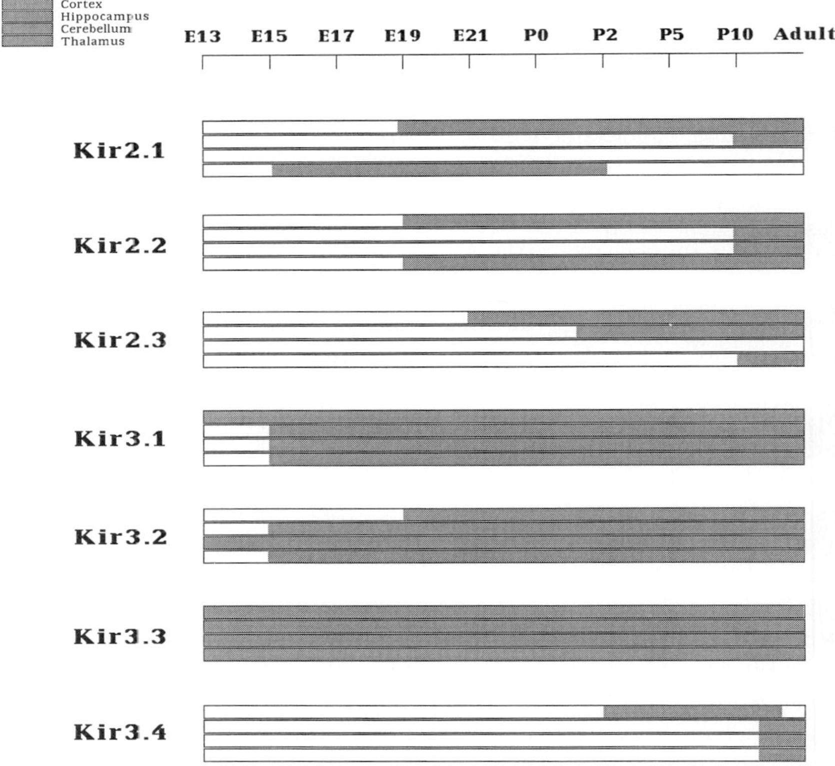

FIGURE 3 Schematic representation of Kir subunit mRNA expression throughout the developing rat cortex, hippocampus, cerebellum, and thalamus. Differences in the onset of expression between substructures, e.g., cortical layers, cerebellar cell types, or thalamic nuclei, have not been considered.

The distinct and strong expression of Kir3.2 mRNA in neuroepithelia and subventricular zones where cerebellar granule cells and future dopaminergic neurons of the substantia nigra are generated may provide a partial answer why in *weaver* mutant mice preferentially neurons in these two regions degenerate (even though mutated Kir3.2 subunits are widely expressed throughout the brain). Accordingly, in brain areas that appear unaffected by the *weaver* mutation (e.g., cortex and thalamus), Kir3.2 expression starts after initial events in cell differentiation have taken place and is not observed in neuroepithelium. Thus it appears that the sites of strong Kir3.2 expression in neuronal precursors may define cell populations most vulnerable to the devastating effect of excessive Na^+ influx caused by the defective Kir3.2 gene.

Acknowledgments

We thank all members of the MNS group for helpful discussions and unpublished information, the graphics department for excellent technical assistance, and Prof. W. Stühmer for generous support. A.K. is supported in part by the Deutsche Forschungsgemeinschaft.

References

Adelbrecht, C., Murer, M. G., Lauritzen, I., Lesage, F., Lazdunski, M., Agid, Y., and Raisman-Vozari, R. (1997). An immunocytochemical study of a G-protein gated inward rectifier K^+ channel (GIRK2) in the *weaver* mouse mesencephalon. *NeuroReport* **8,** 969–974.

Barres, B. A., Chun, L. L. Y., and Corey, D. P. (1990). Ion channels in vertebrate glia. *Annu. Rev. Neurosci.* **13,** 441–474.

Bausch, S. B., Patterson, T. A., Ehrengruber, M. U., Lester, H. A., Davidson, N., and Chavkin, C. (1995). Colocalization of Mu opioid receptors with GIRK1 potassium channels in the rat brain: An immunocytochemical study. *Recept. Chan.* **3,** 221–241.

Bredt, D. S., Wang, T.-L., Cohen, N. A., Guggino, W. B., and Snyder, S. H. (1995). Cloning and expression of two brain-specific inwardly rectifying potassium channels. *Proc. Natl. Acad. Sci. USA* **92,** 6753–6757.

Dascal, N., Schreibmayer, W., Lim, N. F., Wang, W., Chavkin, C., DiMagno, L., Labarca, C., Kieffer, B. L., Gaveriaux-Ruff, C., Trollinger, D., Lester, H. A., and Davidson, N. (1993). Atrial G protein-activated K^+ channel: Expression cloning and molecular properties. *Proc. Natl. Acad. Sci. USA* **90,** 10,235–10,239.

DePaoli, A. M., Bell, G. I., and Stoffel, M. (1994). G protein-activated inwardly rectifying potassium channel (GIRK1/KGA) mRNA in adult rat heart and brain by *in situ* hybridization histochemistry. *Mol. Cell. Neurosci.* **5,** 515–522.

Dissmann, E., Wischmeyer, E., Spauschus, A., v. Pfeil, D., Karschin, C., and Karschin, A. (1996). Functional expression and cellular mRNA localization of a G protein-activated K^+ inward rectifier isolated from rat brain. *Biochem. Biophys. Res. Commun.* **223,** 474–479.

Dixon, A. K., Gubitz, A. K., Ashford, M. L. J., Richardson, P. J., and Freeman, T. C. (1995). Distribution of mRNA encoding the inwardly rectifying K^+ channel, BIR1 in rat tissues. *FEBS Lett.* **374,** 135–140.

Doupnik, C. A., Davidson, N., and Lester, H. A. (1995). The inward rectifier potassium channel family. *Curr. Opin. Neurobiol.* **5,** 268–277.

Drake, C. T., Bausch, S. B., Milner, T. A., and Chavkin, C. (1997). GIRK1 immunoreactivity is present predominantly in dendrites, dendritic spines, and somata in the CA1 region of the hippocampus. *Proc. Natl. Acad. Sci. USA* **94,** 1007–1012.

Fakler, B., Bond, C. T., Adelman, J. P., and Ruppersberg, J. P. (1996). Heterooligomeric assembly of inward-rectifier K^+ channels from subunits of different subfamilies: Kir2.1 (IRK1) and Kir4.1 (BIR10). *Pflügers Arch.* **433,** 77–83.

Fakler, B., and Ruppersberg, J. P. (1996). Functional and molecular diversity classifies the family of inward-rectifier K^+ channels. *Cell. Physiol. Biochem.* **6,** 195–209.

Falk, T., Meyerhof, W., Corrette, B. J., Schäfer, J., Bauer, C. K., Schwarz, J. R., and Richter, D. (1995). Cloning, functional expression and mRNA distribution of an inwardly rectifying potassium channel protein. *FEBS Lett.* **367,** 127–131.

Fink, M., Duprat, F., Heurteaux, C., Lesage, F., Romey, G., Barhanin, J., and Lazdunski, M. (1996). Dominant negative chimeras provide evidence for homo and heteromultimeric assembly of inward rectifier K^+ channel proteins via their N-terminal end. *FEBS Lett.* **378,** 64–68.

Glowatzki, E., Fakler, G., Brändle, U., Rexhausen, U., Zenner, H.-P., Ruppersberg, J. P., and Fakler, B. (1995). Subunit-dependent assembly of inward-rectifier K^+ channels. *Proc. R. Soc. Lond. B* **261,** 252–261.

Hibino, H., Horio, Y., Inanobe, A., Doi, K., Ito, M., Yamada, M., Gotow, T., Uchiyama, Y., Kawamura, M., Kubo, T., and Kurachi, Y. (1997). An ATP-dependent inwardly rectifying potassium channel, K_{AB}-2 (Kir4.1), in cochlear stria vascularis of inner ear: Ist specific subcellular localization and correlation with the formation of endocochlear potential. *J. Neurosci.* **17,** 4711–4721.

Horio, Y., Morishige, K.-I., Takahashi, N., and Kurachi, Y. (1996). Differential distribution of classical inwardly rectifying potassium channel mRNAs in the brain: Comparison of IRK2 with IRK1 and IRK3. *FEBS Lett.* **379,** 239–243.

Iizuka, M., Kubo, Y., Tsunenari, I., Pan, C. X., Akiba, I., and Kono, T. (1995). Functional characterization and localization of a cardiac-type inwardly rectifying K^+ channel. *Recept. Chan.* **3,** 299–315.

Iizuka, M., Tsunenari, I., Momota, Y., Akiba, I., and Kono, T. (1997). Localization of a G-protein-coupled inwardly rectifying K^+ channel, CIR, in the rat brain. *Neuroscience* **77,** 1–13.

Inagaki, N., Gonoi, T., Clement, IV J. P., Namba, N., Inazawa, J., Gonzalez, G., Aguilar-Bryan, L., Seino, S., and Bryan, J. (1995). Reconstitution of I_{KATP}: an inward rectifier subunit plus the sulfonylurea receptor. *Science* **270,** 1166–1170.

Ishii, M., Yamahishi, T., and Taira, N. (1994). Cloning and functional expression of a cardiac inward rectifier K^+ channel. *FEBS Letts.* **338,** 107–111.

Ishii, M., Horio, Y., Tada, Y., Hibino, H., Inanobe, A., Ito, M., Yamada, M., Gotow, T., Uchiyama, Y., and Kurachi, Y. (1997). Expression and clustered distribution of an inwardly rectifying potassium channel, K_{AB}-2/Kir4.1, on mammalian retinal Müller cell membrane: Their regulation by insulin and laminin signals. *J. Neurosci.* **17,** 7725–7735.

Isomoto, S., Kondo, C., and Kurachi, Y. (1997). Inwardly rectifying potassium channels: Their molecular heterogeneity and function. *Jpn. J. Physiol.* **47,** 11–39.

Karschin, A., and Wischmeyer, E. (1995). Identification of G protein-regulated inwardly rectifying K^+ channels in rat brain oligodendrocytes. *Neurosci. Lett.* **183,** 135–138.

Karschin, A., Wischmeyer, E., Davidson, N., and Lester, H. A. (1994). Fast inhibition of inwardly rectifying K^+ channels by multiple neurotransmitter receptors in oligodendroglia. *Eur. J. Neurosci.* **6,** 1756–1764.

Karschin, C., Dissmann, E., Stühmer, W., and Karschin, A. (1996). IRK(1-3) and GIRK(1-4) inwardly rectifying K^+ channel mRNAs are differentially expressed in the adult rat brain. *J. Neurosci.* **16,** 3559–3570.

Karschin, C., Ecke, C., Ashcroft, F. M., and Karschin, A. (1997). Overlapping distribution of K_{ATP} channel-forming Kir6.2 subunit and the sulfonylurea receptor SUR1 in rodent brain. *FEBS Lett.* **401,** 59–64.

Karschin, C., and Karschin, A. (1997). Ontogeny of gene expression of Kir channel subunits in the rat. *Mol. Cell. Neurosci.* **10,** 131–148.

Karschin, C., Schreibmayer, W., Dascal, N., Lester, H., Davidson, N., and Karschin, A. (1994). Distribution and localization of a G protein-coupled inwardly rectifying K^+ channel in the rat. *FEBS Lett.* **348,** 139–144.

Kobayashi, T., Ikeda, K., Ichikawa, T., Abe, S., Togashi, S., and Kumanishi, T. (1995). Molecular cloning of a mouse G-protein-activated K^+ channel (mGIRK1) and distinct distributions of three GIRK (GIRK1, 2 and 3) mRNAs in mouse brain. *Biochem. Biophys. Res. Commun.* **208,** 1166–1173.

Kofuji, P., Hofer, M., Millen, K. J., Millonig, J. H., Davidson, N., Lester, H. A., and Hatten, M. E. (1996). Functional analysis of the *weaver* mutant GIRK2 K^+ channel and rescue of *weaver* granule cells. *Neuron* **16,** 941–952.

Krapivinsky, G., Gordon, E. A., Wickman, K., Velimirovic, B., Krapivinsky, L., and Clapham, D. E. (1995). The G-protein-gated atrial K^+ channel I_{KACh} is a heteromultimer of two inwardly rectifying K^+-channel proteins. *Nature* **374,** 135–141.

Kubo, Y., Baldwin, T. J., Jan, Y. N., and Jan, L. Y. (1993). Primary structure and functional expression of a mouse inward rectifier potassium channel. *Nature* **362,** 127–133.

Lauritzen, I., De Weille, J. D., Adelbrecht, C., Lesage, F., Murer, M. G., Raisman-Vozari, R., and Lazdunski, M. (1997). Comparative expression of the inward rectifier K^+ channel GIRK2 in the cerebellum of normal and *weaver* mutant mice. *Brain Res.* **753,** 8–17.

Liao, Y. J., Jan, Y. N., and Jan, L. Y. (1996). Heteromultimerization of G-protein-gated inwardly rectifying K^+ channel proteins GIRK1 and GIRK2 and their altered expression in *weaver* brain. *J. Neurosci.* **16,** 7137–7150.

Mi, H., Deerinck, T. J., Jones, M., Ellisman, M. H., and Schwarz, T. L. (1996). Inwardly rectifying K^+ channels that may participate in K^+ buffering are localized in microvilli of Schwann cells. *J. Neurosci.* **16,** 2421–2429.

Miyashita, T., and Kubo, Y. (1997). Localization and developmental changes of the expression of two inward rectifying K^+-channel proteins in the rat brain. *Brain Res.* **750,** 251–263.

Morishige, K.-I., Inanobe, A., Takahashi, N., Yoshimoto, Y., Kurachi, H., Miyake, A., Tokunaga, Y., Maeda, T., and Kurachi, Y. (1996). G protein-gated K^+ channel (GIRK1) protein is expressed presynaptically in the paraventricular nucleus of the hypothalamus. *Biochem. Biophys. Res. Commun.* **220,** 300–305.

Morishige, K.-I., Takahashi, N., Findlay, I., Koyama, H., Zanelli, J. S., Peterson, C., Jenkins, N.A., Copeland, N. G., Mori, N., and Kurachi, Y. (1993). Molecular cloning, functional expression and localization of an inward rectifier potassium channel in the mouse brain. *FEBS Lett.* **336,** 375–380.

Morishige, K.-I., Takahashi, N., Jahangir, A., Yamada, M., Koyama, H., Zanelli, J. S., and Kurachi, Y. (1994). Molecular cloning and functional expression of a novel brain-specific inward rectifier potassium channel. *FEBS Lett.* **346,** 251–256.

Murer, G., Adelbrecht, C., Lauritzen, I., Lesage, F., Lazdunski, M., Agid, Y., and Vozari-Raisman, R. (1997). An immunocytochemical study on the distribution of two G-protein-gated inward rectifier potassium channels (GIRK2 and GIRK4) in the adult rat brain. *Neuroscience* **80,** 345–357.

Newman, E. A. (1993). Inward-rectifying potassium channels in retinal glial (Müller) cells. *J. Neurosci.* **13,** 3333–3345.

Newman, E., and Reichenbach, A. (1996). The Müller cell: A functional element of the retina. *Trends Neurosci.* **19,** 307–312.

Pessia, M., Tucker, S. J., Lee, K., Bond, C. T., and Adelman, J. P. (1996). Subunit positional effects revealed by novel heteromeric inwardly rectifying K^+ channels. *EMBO J.* **15,** 2980–2987.

Ponce, A., Bueno, E., Kentros, C., Vega-Saenz de Miera, E., Chow, A., Hillman, D., Chen, S., Zhu, L., Wu, M. B., Rudy, B., and Thorenhill, W. B. (1996). G-protein-gated inward rectifier K^+ channel proteins (GIRK1) are present in the soma and dendrites as well as in nerve terminals of specific neurons in the brain. *J. Neurosci.* **16,** 1990–2001.

Shuck, M. E., Piser, T. M., Bock, J. H., Slightom, J. L., Lee, K. S., and Bienkowski, M. J. (1997). Cloning and characterization of two K^+ inward rectifier (Kir) 1.1 potassium channel homologs from human kidney (Kir1.2 and Kir1.3). *J. Biol. Chem.* **272,** 586–593.

Signorini, S., Liao, Y. J., Duncan, S. A., Jan, L. Y., and Stoffel, M. (1997). Normal cerebellar development but susceptibility to seizures in mice lacking G protein-coupled, inwardly rectifying K^+ channel GIRK2. *Proc. Natl. Acad. Sci. USA* **94,** 923–927.

Slesinger, P. A., Patil, N., Liao, Y. J., Jan, Y. N., Jan, L. Y., and Cox, D. R. (1996). Functional effects of the *weaver* mouse mutation on G protein-gated inwardly rectifying K^+ channels. *Neuron* **16,** 321–331.

Sontheimer, H. (1994). Voltage-dependent ion channels in glial cells. *Glia* **11,** 156–172.

Spauschus, A., Lentes, K.-U., Wischmeyer, E., Dissmann, E., Karschin, C., and Karschin, A. (1996). A G-protein-activated inwardly rectifying K^+ channel (GIRK4) from human hippocampus associates with other GIRK channels. *J. Neurosci.* **16,** 930–938.

Takahashi, N., Morishige, K.-I., Jahangir, A., Yamada, M., Findlay, I., Koyama, H., and Kurachi, Y. (1994). Molecular cloning and functional expression of cDNA encoding a second class of inward rectifier potassium channels in the mouse brain. *J. Biol. Chem.* **269,** 23,274–23,279.

Takumi ,T., Ishii, T., Horio, Y., Morishige, K.-I., Takahashi, N., Yamada, M., Yamashita, T., Kiyama, H., Sohmiya, K., Nakanishi, S., and Kurachi, Y. (1995). A novel ATP-dependent inward rectifier potassium channel expressed predominantly in glial cells. *J. Biol. Chem.* **270,** 16,339–16,346.

Tinker, A., Jan, Y. N., and Jan, L. Y. (1996). Regions responsible for the assembly of inwardly rectifying potassium channels. *Cell* **87,** 857–868.

Töpert, C., Döring, F., Wischmeyer, E., Karschin, C., Brockhaus, J., Ballanyi, K., Derst, C., and Karschin, A. (1998). Kir2.4: A novel K^+ inward rectifier channel associated with motoneurons of cranial nerve nuclei. *J. Neurosci.* 18**(11),** 4096–4109.

Tucker, S. J., Bond, C. T., Herson, P., Pessia, M., and Adelman, J. P. (1996). Inhibitory interactions between two inward rectifier K^+ channel subunits mediated by the transmembrane domains. *J. Biol. Chem.* **271,** 5866–5870.

PART III

G-Protein-Gated Potassium Channels

CHAPTER 16

G-Protein-Gated Potassium Channels: Implication for the *weaver* Mouse

Betsy Navarro,† Shawn Corey,* Matthew Kennedy,† and David E. Clapham†

*Neuroscience Program, Mayo Foundation, Rochester, Minnesota 55905; and
†Cardiovascular Division, Children's Hospital and Harvard Medical School, Boston, Massachusetts 02115

I. INTRODUCTION

In 1964, Lane *et al.* discovered a spontaneous mutation in mice known as the *weaver* mutation. These mice displayed an instability of gait and a fine rapid tremor. The brains of *weaver* mice had cytoarchitectural defects

1063-5823/99 $30.00

in both the striatal dopaminergic and cerebellar neurons, concurrent with their abnormalities in motor performance (Rakic and Sidman, 1973). Specifically, in the cerebellum, granule cells failed to migrate from the external granular cell layer to the internal granule cell layer and died (Goldowitz and Mullen, 1982). This led to initial speculation that a cell adhesion protein or migration molecule was responsible for the *weaver* phenotype (Hatten *et al.,* 1984, 1986). Surprisingly, Patil *et al.* (1995) found that the *weaver* mice contained a missense mutation in the G-protein-gated inwardly rectifying potassium channel subunit (GIRK2, Fig. 1). This is the first ion channel to be linked to a neurodegenerative disorder and could provide key insights into the role of GIRKs in brain and human disease. Ion channels containing GIRK2 are distributed ubiquitously throughout the brain (Kobayashi *et al.,* 1995; Karschin *et al.,* 1996; Liao *et al.,* 1996; Murer *et al.,* 1997) and testis (Harrison and Roffler-Tarlov, 1994; Verina *et al.,* 1995) and are thought to play an important role in receptor-mediated regulation of cell excitability (Wickman and Clapham, 1995). Here we describe the properties of GIRKs, the properties of wild-type GIRK2 and its potential roles in brain, and the pathophysiology of the GIRK2 mutation in *weaver* mice.

II. THE G-PROTEIN-ACTIVATED INWARDLY RECTIFYING POTASSIUM CHANNELS

A. $G_{\beta\gamma}$ Regulation and Physiological Role

Heterotrimeric G proteins ($G_{\alpha\beta\gamma}$) play a critical role in signal propagation from the cell surface to the cell interior (Neer and Clapham, 1988; Wickman *et al.,* 1997). They mediate the intracellular effects of many extracellular signals, working as a network to simultaneously coordinate multiple intracellular regulatory pathways. The binding of a ligand to its receptor catalyzes the exchange of GDP for GTP on the G_α subunit; the G_α subunit in turn dissociates from the $G_{\beta\gamma}$ subunit (Clapham and Neer, 1997). This allows the G_α and/or $G_{\beta\gamma}$ subunits to regulate downstream effectors such as adenylyl cyclase, phospholipase C (PLC), the muscarinic potassium channel I_{KACh}, cGMP-dependent phosphodiesterase, β-adrenergic receptor kinase (BARK), and phosphoinositol 3-kinase (PI3-kinase).

A large number of G-protein-coupled receptors hyperpolarize cells by activating inwardly rectifying K^+ channels (GIRKs). GIRKs are activated primarily by direct $G_{\beta\gamma}$ binding (Clapham, 1994). A prototypical example of a GIRK current (K_G) is I_{KACh} (Logothetis *et al.,* 1987). I_{KACh} is responsible for the acetylcholine (ACh)-induced deceleration of heart rate (Wickman *et al.,* 1998). Briefly, ACh is released from the vagus nerve and activates

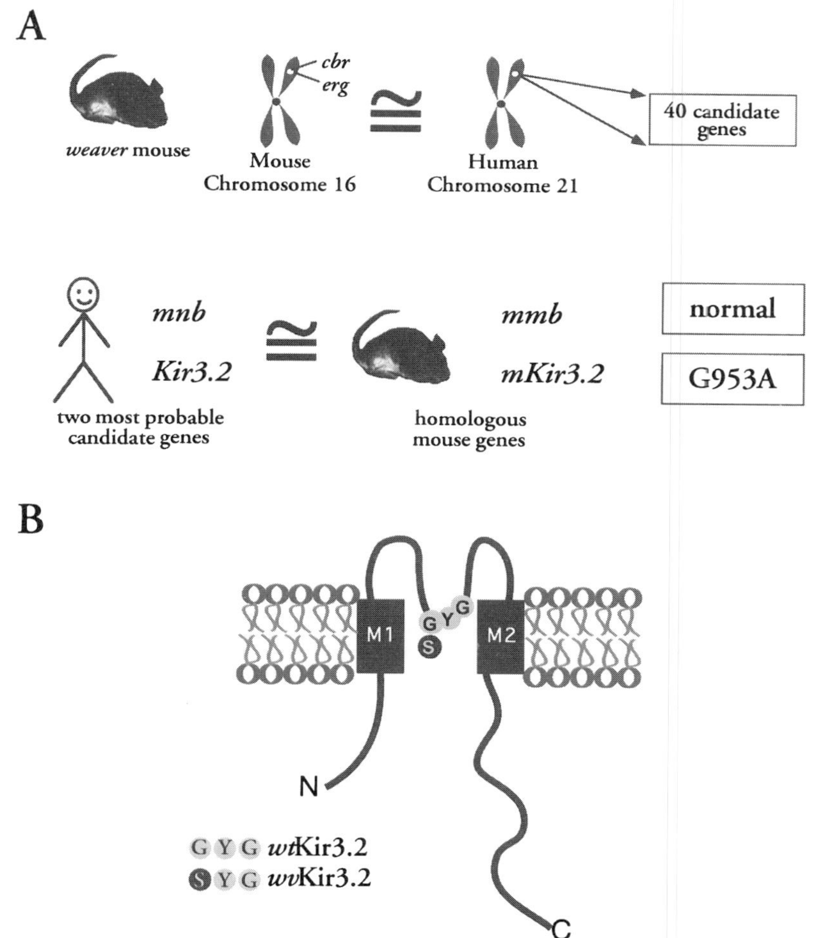

FIGURE 1 Cloning of the *weaver* mutation *wv*Kir 3.2. (A) The weaver mutation was originally mapped to mouse chromosome 16. This locus of mouse chromosome 16 is equivalent to a better defined locus of human chromosome 21. From this locus of human chromosome 21, 40 candidate genes were selected. Of the 40 candidate genes, 2 were considered to be the most likely candidates, Kir 3.2 and *mmb*. Cloning of the mouse homolog to Kir 3.2 revealed a single point mutation (G953A) in *wv*Kir 3.2. (B) The G953A base pair substitution creates a serine for glycine substitution (G156S) in a critical pore region of Kir 3.2.

the m2-muscarinic receptor. The m2 receptor in turn catalyzes the release of $G_{\beta\gamma}$ from G_{α}. $G_{\beta\gamma}$ is then free to directly bind I_{KACh} and consequently increase the open probability of the channel (Fig. 2A,B) (Logothetis *et al.*,

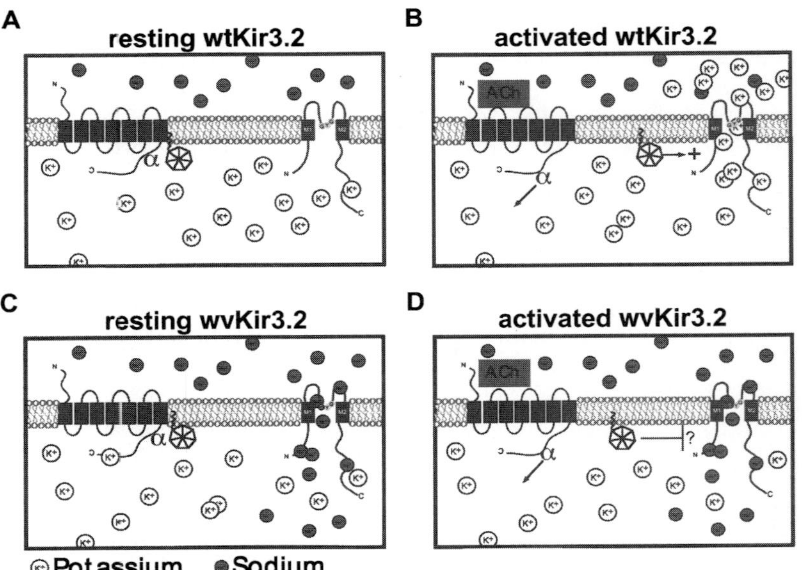

FIGURE 2 Activation and pore properties of *wv*Kir 3.2 versus *wt*Kir 3.2. (A) Resting *wt*Kir.3.2 is not permeant to cations or anions. (B) Binding of an agonist to its G-protein-coupled receptor results in the $G_{\beta\gamma}$-induced activation of *wt*Kir 3.2. This allows K^+ to flow in the outward direction, driving the cell toward E_K. (C) Resting *wv*Kir 3.2 forms a nonselective cation channel allowing Na^+ to "leak" into the cell. (D) Binding of an agonist to its G-protein-coupled receptor does not result in any additional activation of *wv*Kir 3.2.

1987; Kurachi *et al.,* 1989; Clapham, 1994; Wickman *et al.,* 1994). The increased I_{KACh} current drives the cell toward E_K, which in turn decreases heart rate. In brain, GIRK channels are coupled to the 5-HT1A serotoninergic, D2 dopaminergic, α_2-adrenergic, m2-muscarinic, α-aminobutyric acid ($GABA_B$, somatostatin and opioid receptors (Table I). GIRKs are particularly important in the receptor-mediated regulation of cell excitability. Activation of K_G current following the stimulation of postsynaptic receptors decreases neuronal excitability, whereas stimulation of presynaptic receptors inhibits neurotransmitter release (Nicoll *et al.,* 1988; North, 1989). GIRK channels may modulate neuronal excitability by regulating firing threshold, firing frequency, membrane potential, and neurotransmitter release. Thus, in brain, GIRKs play a crucial role in neurotransmitter-mediated changes in inhibitory and slow excitatory synapses (North, 1989; Grigg *et al.,* 1996).

TABLE I
Brain Localization of Kir 3.0 Currents and the Receptors to Which They Are Coupled

Brain area	K_G	m2	D2	$GABA_B$	A-1	5-HT1A	SOM	κ,μ-Opioid
Cerebral cortex	Yes			Yes				
Caudate putamen	Yes		Yes					
Hippocampus	Yes					Yes		Yes
CA3	Yes			Yes	Yes			
Thalamus	?[a]			Yes				
Hypothalamus	Yes			Yes				Yes
Substantia nigra compacta	Yes		Yes	Yes				
Locus coeruleus	Yes						Yes	Yes
Raphe nucleus	Yes					Yes		
VTA	?		Yes					
Pontine reticular formation	Yes	?						
Cerebellar granulle cells	Yes						Yes	

[a] Potassium currents that were not tested for G-protein activation.

B. Nomenclature

The GIRKs form the Kir 3.0 subfamily of inwardly rectifying K^+ channels (Doupnik *et al.,* 1995). There are four mammalian members of the Kir 3.0 subfamily. The first member of the subfamily, Kir 3.1 (GIRK1, or KGA), was cloned by homology to Kir1.1 (Kubo *et al.,* 1993) and by expression cloning (Dascal *et al.,* 1993). Kir 3.2 (GIRK2) and Kir 3.3 (GIRK3) were cloned from mouse brain (Lesage *et al.,* 1994), whereas Kir 3.4 (GIRK4) was cloned from rat heart (Krapivinsky *et al.,* 1995a). The Kir 3.0 nomenclature for GIRKs will be used throughout the rest of this chapter.

C. Heterotetrameric Structure of the Kir 3.0 Subfamily

It was originally believed that Kir 3.1 and Kir 3.4 were the sole components of two separate ion channels, I_{KACh} and I_{KATP}, respectively (Kubo *et al.,* 1993; Ashford *et al.,* 1994). We now know that I_{KACh} is a heterotetrameric complex (Corey *et al.,* 1998) composed of two Kir 3.1 and two Kir 3.4 subunits (Krapivinsky *et al.,* 1995a,b). In contrast, I_{KATP} is composed of a G-protein-independent subunit, Kir 6.2, and the sulfonylurea receptor

(Inagaki *et al.,* 1995). Likewise in brain, Kir 3.1 combines with Kir 3.2 to form a heteromultimeric channel (Liao *et al.,* 1996). The heteromultimeric structure of Kir 3.0 subfamily members has been demonstrated repeatedly by coimmunoprecipitation (Krapivinsky *et al.,* 1995a; Chan *et al.,* 1996) and functional coexpression (Duprat *et al.,* 1995; Hedin *et al.,* 1996; Kofuji *et al.,* 1995; Krapivinsky *et al.,* 1995a; Lesage *et al.,* 1995; Chan *et al.,* 1996; Velimirovic *et al.,* 1996). In addition the heteromultimeric structure of the Kir 3.0 subfamily members has been implied by immunocytochemistry (Karschin *et al.,* 1994) and *in situ* hybridization (Kobayashi *et al.,* 1995; Ponce *et al.,* 1996; Spauschus *et al.,* 1996). It remains to be determined whether homotetrameric Kir 3.0 channels exist *in vivo.* In heterologous expression systems Kir 3.2 and Kir 3.4 form homotetramers (Corey *et al.,* 1998) that are transported to the cell membrane (Kennedy *et al.,* 1996) yet produce only atypical flickery currents (Krapivinsky *et al.,* 1995a; Velimirovic *et al.,* 1996). Kir 3.1, on the other hand, forms homotetramers (Corey *et al.,* 1998) that are not transported to the cell membrane (Kennedy *et al.,* 1996).

D. Kir 3.0 Electrophysiology

Inward rectifier K^+ channels are characterized by their ability to pass K^+ ions in the inward direction (from the extracellular to the intracellular environment) better than in the outward direction (Hille, 1992). Kir 3.0 heterotetramers exhibit strong inward rectification (Krapivinsky *et al.,* 1995a; Velimirovic *et al.,* 1996). This inward rectification is a direct result of an intracellular Mg^{2+}-mediated blockade and an intrinsic gating process (Logothetis *et al.,* 1987; Matsuda *et al.,* 1987; Aleksandrov *et al.,* 1996; Kurachi *et al.,* 1989). Despite the fact that Kir channels rectify inwardly, it is the smaller outward K^+ current which is physiologically relevant.

Heterologous expression of the m2 receptor and Kir 3.1 with either Kir 3.2 or Kir 3.4 produces an inwardly rectifying K^+ current on activation of the $G_{\alpha i}$-linked receptor (m2). The m2 receptor can be bypassed by the overexpression of G-protein subunits $G_{\beta 1}$ and $G_{\gamma 2}$. This confirms that the $G_{\beta\gamma}$ subunit, not the G_{α} subunit, is responsible for increasing the channel open probability (Logothetis *et al.,* 1987; Kubo *et al.,* 1993; Wickman *et al.,* 1994; Krapivinsky *et al.,* 1995a). Activation of Kir 3.1/Kir (3.2 or 3.4) with either $G_{\beta\gamma}$ or GTPγS produces channels that are nearly identical to native neuronal and cardiac Kir 3.0 channels: 32–35 pS single-channel conductance in symmetrical 140 m*M* K^+ and a mean open time of 1 msec (Logothetis *et al.,* 1987; Krapivinsky *et al.,* 1995a; Grigg *et al.,* 1996; Kofuji *et al.,* 1996; Velimirovic *et al.,* 1996). This again demonstrates that Kir 3.0 channels are responsible for the native $G_{\beta\gamma}$-activated, inwardly rectifying K^+ currents in brain and heart. In addition to being activated by $G_{\beta\gamma}$, Kir 3.0 channels may be activated by intracellular Na^+ (millimolar concentration range; Sui *et al.,* 1996). The physiological significance of Na^+ activation remains to be determined.

Despite the high homology between Kir 3.0 family members, the channel openings of Kir 3.2 and Kir 3.4 homomultimers are flickery, whereas Kir 3.1 homomultimers are nonfunctional (Krapivinsky *et al.,* 1995a; Navarro *et al.,* 1996; Velimirovic *et al.,* 1996). The ability—of Kir 3.0 subunits to form heteromultimers with other Kir 3.0 family members, to be gated by G proteins, and to interact with other proteins adds considerable complexity to their behavior.

III. WILD-TYPE Kir 3.2

The *weaver* mutation in Kir 3.2 is pleiotropic leading to diverse and complicated mutant phenotypes. Thus, it is worth examining the normal genomic structure, expression pattern, electrophysiologic properties, and potential functions of wild-type Kir 3.2 (*wt*Kir 3.2), before examining *weaver* Kir 3.2 (*wv*Kir 3.2) in detail.

A. Cloning and Genomic Structure of Kir 3.2

Initially, *Kir 3.2* was cloned from rat (Lesage *et al.,* 1994) and mouse (Isomoto *et al.,* 1996) brain based on their homology to Kir 2.1. Patil *et al.* (1995) mapped the *Kir 3.2* gene to mouse chromosome 16. More recently, the human *Kir 3.2* cDNA (KCNJ7, genetic denomination) was cloned from pancreas (Tsaur *et al.,* 1995). It was mapped to chromosome 21q22.19 (the human equivalent of mouse chromosome 16). The entire open reading frame of *Kir 3.2* (alternatively known as $K_{ATP\text{-}2}$ or BR1) is contained in three exons. The transcripts encode a 423 amino acid (48.4 kDa) protein. At least four different functional isoforms of mouse Kir 3.2 (Lesage *et al.,* 1994; Stoffel *et al.,* 1995; Isomoto *et al.,* 1996) are formed by alternative splicing (Fig. 3). The predicted primary structure of the alternatively spliced transcripts reveals a conserved amino terminus (amino acids 1–94) and central core [M1 (95–117), H5 (142–158), and M2 (167–191)] domains. The carboxyl termini of the alternatively spliced transcripts are divergent (Stoffel *et al.,* 1995; Isomoto *et al.,* 1996). The functions of these splice variants is largely unknown.

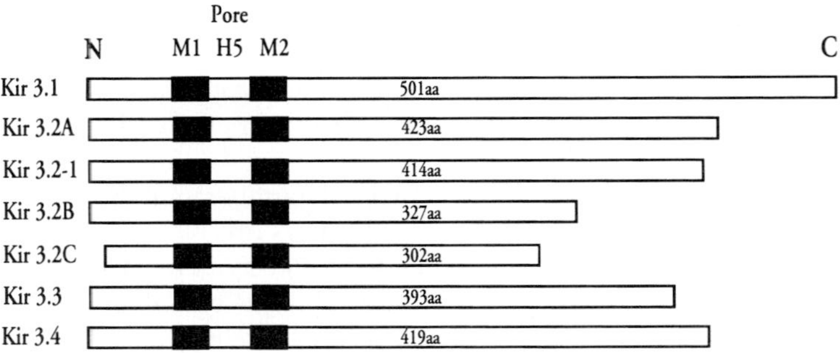

FIGURE 3 Kir 3.2 has numerous splice variants.

B. Expression Pattern of Kir 3.2

Kir 3.2 mRNA is expressed mainly in brain, with lower levels expressed in pancreas and in testis (Lesage *et al.,* 1994). The specific localization patterns of the splice variants are unknown. For region-specific expression patterns of Kir 3.2 protein in brain, see Table II. Interestingly, in the superior colliculus, the hypothalamus, the medial habenula, and the amygdala, Kir 3.2 protein is expressed in the absence of Kir 3.1 protein. This raises the possibility that in these regions Kir 3.2 exists as either a homomultimeric channel or is associated with an undetermined subunit. At the subcellular level, it is likely that Kir 3.2 is present both presynaptically and postsynaptically (Liao *et al.,* 1996).

C. Physiological Relevance of Kir 3.2 in Brain

Several electrophysiological studies have reported Kir 3.0-like currents in brain (Table I). In addition, immunostaining studies show that Kir 3.2 protein is present both presynaptically and postsynaptically in neurons. Because Kir channels are important in setting the membrane potential and regulating action potential duration (Hille, 1992), it is likely that Kir 3.2-containing channels are important in regulating neuronal excitability. In brain, Kir 3.0 channels may be a major downstream effector of inhibitory neurotransmitters. Thus, Kir 3.2 channels probably influence the generation of slow postsynaptic potentials (North, 1989; Velimirovic *et al.,* 1996) as well as the synaptic integration of signals. Activation of Kir 3.2-containing channels presynaptically may be involved in the inhibition of neurotransmitter release. In addition, Kir 3.2-containing channels may contribute the susceptibility to neurodegenerative disorders, as suggested by *weaver* mice. Overall, the role of Kir 3.2 in brain remains undefined.

IV. *weaver* MICE Kir 3.2

An important channelopathy was discovered when it was found that a defect in Kir 3.2 was responsible for the *weaver* phenotype. This was the first time that an ion channel had been linked to a neurodegenerative disease. Thus, a detailed study of *weaver* mice could provide clues about the role of Kir 3.0 channels in brain, the pathophysiology of human neurodegenerative diseases, and the functional importance of certain neuronal cell types.

A. Genotype

The autosomal recessive *weaver* mutation was mapped to mouse chromosome 16 by Reeves *et al.* (1989). Patil *et al.* (1995) then identified the *weaver* gene by using a technique for direct cDNA selection. Briefly, Patil *et al.* took advantage of the conserved synteny between the *weaver* locus on chromosome 16 and the equivalent, and more well-defined, locus on human chromosome 21. On closer examination of the human locus, they identified two candidate genes that mapped to this region of chromosome 21 (*mmb* and *Girk2*). After cloning the corresponding *mmb* and *Girk2* genes from wild-type mice, they looked for mutations in either of these genes in *weaver* mice. Although *weaver* mice contained a wild-type copy of the *mmb* gene, they carried a single point mutaton (G953A) in the *Girk2* gene (Fig. 1A). This base substitution in Kir 3.2 creates a substitution of glycine for serine (G156S) in a critical pore region of Kir 3.2 (Fig. 1B).

B. Phenotype

The single point mutation in *weaver* mice is pleiotropic, with abnormalities being present at birth. The most distinguishing feature of *weaver* mice is their weaving gait (origin of the *weaver* designation). In addition to the abnormalities in their motor function, *weaver* mice display sporadic seizures (Eisenberg and Messer, 1989) and male infertility, which has been largely ignored (Vogelweid *et al.*, 1993; Verina *et al.*, 1995). Mice heterozygous for the weaver gene display sporadic seizures and eventually become infertile.

TABLE II

Brain Immunolocalization of Some G-Protein-Coupled Receptors and Kir 3.0 Channels

Brain area	Kir 3.1	Kir 3.2	Kir 3.4	m2	D2	$GABA_B$	A-1	5-HT1A	SOM	δ-Opioid	κ-Opioid	μ-Opioid
Cerebral cortex				Yes			Yes		Yes	Yes	Yes	Yes
Layer I	Yes	Yes			Yes							
Layer VI	Yes	Yes	Yes		Yes							
Pyramidal cells		Yes				Yes						
Caudate putamen	Yes	Yes	Yes	Yes	Yes		Yes	Yes	Yes	Yes	Yes	Yes (sd)[a]
Globus pallidus	Yes		Yes (sd)	Yes	Yes	Yes	Yes			Yes	Yes	Yes
Substantia nigra compacta		Yes (sd)			Yes	Yes	Yes				Yes	Yes
Hippocampus				Yes		Yes			Yes (sd)			Yes
CA1–CA3	Yes	Yes			Yes		Yes	Yes (ps)[b]		Yes		
Dentate gyrus	Yes	Yes			Yes		Yes	Yes				
Amygdala		Yes (sd)		Yes		Yes	Yes	Yes	Yes (sd)	Yes	Yes	Yes
Olfactory bulb	Yes	Yes	Yes		Yes	Yes	Yes					Yes
Olfactory tubercle			Yes	Yes	Yes	Yes	Yes				Yes	
Islands of Calleja			Yes	Yes	Yes				Yes			

Thalamus	Yes	Yes		Yes		Yes	Yes		Yes		Yes	
Medial habenula		Yes	Yes		Yes				Yes (sd)		Yes	
Hypothalamus		Yes	Yes	Yes	Yes				Yes		Yes	Yes
Midbrain												
Superior colliculus		Yes	Yes (sd)	Yes		Yes	Yes		Yes		Yes	Yes
Inferior colliculus	Yes				Yes					Yes	Yes	Yes
Brain stem												
Reticular formation	Yes											
Pontine nuclei	Yes	Yes	Yes	Yes		Yes	Yes			Yes		
Inferior olive	Yes				Yes	Yes			Yes			
Cerebellum				Yes	Yes	Yes	Yes		Yes[c]		Yes	Yes
Granular layer	Yes(d)[c]	Yes(d)[c]										
Purkinje cell layer			Yes (sd)					Yes				
Basket cell			Yes (sd)									

[a] sd, somatodendritic.
[b] ps, postsynaptic.
[c] d, dendritic.
[d] Transitory receptor expression starts at embryonic day 16 (E16), peaks at birth, and disappears after postnatal day 18 (P18).

1. Behavior

The behavior of *weaver* mice has been described in detail. The hallmark of *weaver* mice is their impaired motor activity. The motor abnormalities include less activity in the hole-board test and a loss of balance patterns. In addition, *weaver* mice exhibit an instability of gait, fine rapid tremor in their trunks, and hyperactivity. Alterations in movement occur very early in development (postnatal day 3), concurrently with the earliest biochemical and neuroanatomical deficits (Bolivar *et al.*, 1996). Interestingly, +/*wv* heterozygous mice have no obvious motor problems.

2. Neuroanatomy

a. Cerebellar Defects. The primary neuroanatomical defect in *weaver* mice is a loss of cerebellar granule cells (Rakic and Sidman, 1973). By embryonic day 19, the size of the cerebellum (Wei *et al.*, 1994) and fissurization are reduced (Bayer *et al.*, 1996). By postnatal day 14 (P14) the external granular layer, including the proliferative and premigratory compartments, is absent. Consequently, the major source of excitatory input to the Purkinje cells is eliminated, since the external granular layer is the source of the basket, stellate and granule cells which form critical synapses with the Purkinje cells. Hence, the cerebellums of *weaver* mice undergo an extensive reorganization of cell interactions secondary to the absence of mature granule cells. This reorganization of cerebellar circuitry causes physiological and morphological alterations in the terminal domains of the GABAergic cerebellar corticovestibular projections (Purkinje cell projections; Sotelo, 1975, 1990).

Granule cell precursors undergo apoptosis during the first 2 weeks of postnatal life, prior to normal granule cell migration (which begins near P4) and synaptogenesis (Migheli *et al.*, 1995). The *weaver* mice may have defects even earlier in granule cell development, such as during the exit of neuroblasts from the cell cycle or during axogenesis (Smeyne and Goldowitz, 1989). The mechanisms by which the *weaver* gene inhibits neurite extension and/or induces death of the granule neurons in the cerebellums of *weaver* mice are not presently understood. However, recent work suggests that the *weaver* mutation accelerates the intrinsic cell death machinery (Migheli *et al.*, 1995).

b. Ventral Forebrain Defects. In addition to their cerebellar defects, *weaver* mice have defects in their nigrostriatal dopaminergic systems. The pars compacta portion of the substantia nigra is hypocellular (Privat, 1978) due to a loss of up to 70% of the mesostriatal dopaminergic neurons (Chou *et al.*, 1991; Roffler-Tarlov *et al.*, 1996). Wei *et al.* (1997), using tyrosine hydroxylase immunocytochemistry, have shown that dopaminergic neurons die between P7 and P20. By three months of age, *weaver* mice exhibit nerve cell losses in all 3 areas of the mesencephalic (A8, A9, and A10) dopaminergic cell systems (Wei *et al.,* 1997). Such losses account for the failed dopaminergic striatal, limbic, and cortical projection fields (Triarhou *et al.,* 1986; Chou *et al.,* 1991; Simon and Ghetti, 1994; Simon *et al.,* 1994; Verney *et al.,* 1995). Striatal tyrosine hydroxylase and dopamine content are reduced by 60–70%, and dopamine uptake is reduced by up to 95% (Simon *et al.,* 1994). Thus, *weaver* mice have a dramatic failure of the nigrostriatal pathway. This defect is so severe that certain adaptive/compensatory mechanisms must be masking a more severe phenotype.

In summary, in cerebellum, the *weaver* mutation is lethal to a specific subpopulation of proliferating and migrating cells. However, in the ventral midbrain the *weaver* gene causes the death of dopaminergic neurons after their proliferation and migration (Bayer *et al.,* 1996). Mice heterozygous for the *weaver* mutation do show some pathology at the cellular level, albeit much less severe than in homozygous mice.

3. Neurobiochemistry

In *weaver* mice, there is a massive postnatal loss of dopamine in the caudate putamen, the target of the nigrostriatal system, yet there is a relative preservation of dopamine in the ventral striatum, a target of the mesolimbic system. Changes in dopaminergic uptake, dopamine content, and tyrosine hydroxylase activity start as early as postnatal day 3 (Simon *et al.,* 1994). There is also a concomitant death of catecholaminergic neurons in the substantia nigra, with much less cell death in the limbic midbrain area. Despite this, striatal cholinergic and GABAergic interneurons of *weaver* mice are normal (Simon *et al.,* 1994). However, Stotz *et al.* (1993, 1994) reported that the dorsal portion of the striatum is hyperinnervated by serotonin fibers. This may be an adaptive or compensatory response to the decreased dopamine content of the striatum. The aberrant cellular distribution in the cerebellar cortex of *weaver* mice alters glutamate uptake (Fischer-Bovenkerk *et al.,* 1988) and the GABAergic system (Roffler-Tarlov and Turey, 1982). This suggests an interdependence between the granule cell (glutaminergic neuron) and its main postsynaptic target, the Purkinje cell (GABAergic neuron).

V. THE PATHOPHYSIOLOGY OF *weaver* MICE

A. Electrophysiological Properties of weaver *Kir 3.2*

1. Decreased K^+ Conductance of *wv*Kir 3.2-Containing Heteromultimeric Channels

Since 1996, our understanding of *wv*Kir 3.2 electrophysiology has greatly increased (Kofuji *et al.,* 1995; Navarro *et al.,* 1996; Velimirovic *et al.,* 1996).

Coexpression of Kir 3.1 and *wv*Kir 3.2 subunits in Chinese hamster ovary cells gives rise to a markedly reduced GTPγS-activated Kir current, when compared to Kir 3.1/Kir 3.2 cotransfected cells (Fig. 4A). This suggests that Kir 3.1/*wv*Kir 3.2 heteromultimers are functionally impaired. These results were confirmed by similar findings in oocytes (Kofuji *et al.,* 1996). Surprisingly, *wv*Kir 3.2 homomultimers produce a larger whole-cell current than Kir 3.2 homomultimers. A possible explanation for these findings is

A. **Symmetrical 140 mM [K^+]**

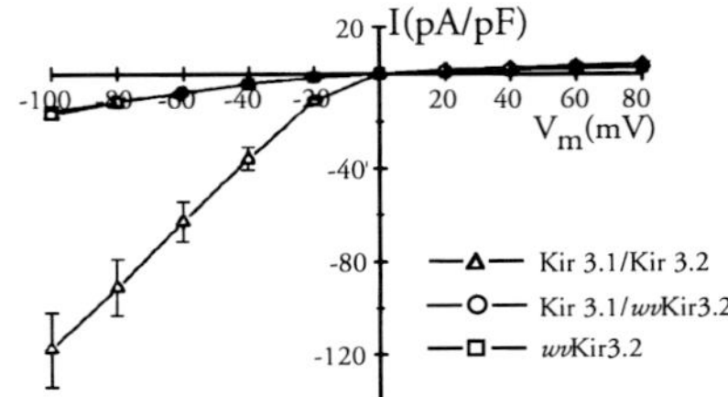

B. **138 mM [Cs^+]$_o$ + 2 mM [K^+]$_o$**

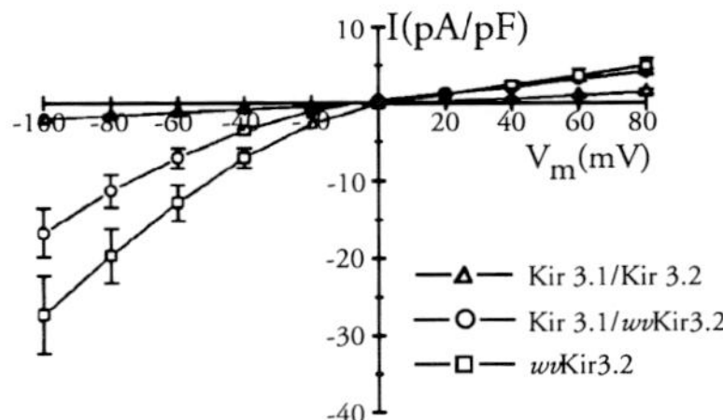

C. **138 mM [Na^+]o + 2 mM [K^+]o**

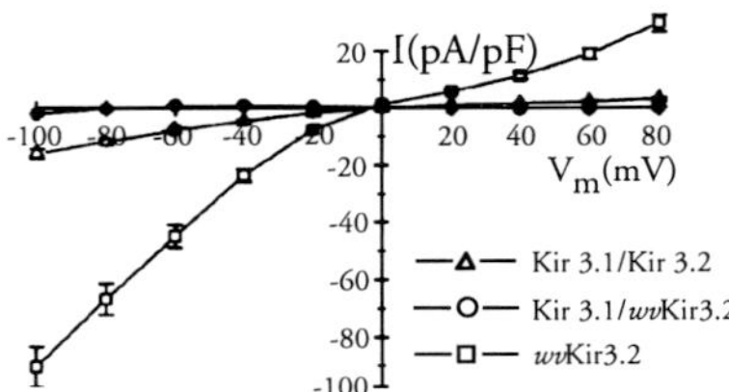

FIGURE 4 Loss of selectivity of *wv*Kir 3.2-containing channels. Glycine 156 in the pore region is critical for maintaining ion selectivity. Modifed from Navarro *et al.* (1996).

that *wv*Kir 3.2 has acquired novel functional properties, while still retaining its susceptibility to inhibition by external Ba^{2+}.

2. Loss of Selectivity in *wv*Kir 3.2-Containing Channels

Permeability measurements show that both *wv*Kir 3.2 -containing heteromultimers and homomultimers are nonselective cation channels (Fig. 4B,C). *wv*Kir 3.2 homomultimers conduct sodium better than Kir 3.1/*wv*Kir 3.2 heteromultimers. The loss of *wv*Kir 3.2 K^+ selectivity can be explained by the GYG to SYG mutation that occurs in the channel signature sequence (TxxTxGYG) of the channel. This signature sequence is a region common to all K^+ channels and is required for K^+ channel selectivity (MacKinnon and Yellen, 1990; Heginbotham *et al.,* 1992, 1994). In fact, when a *wv*Kir 3.2-equivalent mutation was made in *Shaker* K^+ channels, nonselective cation channels were produced (Heginbotham *et al.,* 1994).

The loss of selectivity and the resulting Na^+ conductance through *wv*Kir 3.2 channels could have disastrous consequences for the neuron. First, Na^+ could enter through the pore of *wv*Kir 3.2, further activating the channel (Na^+-dependent activation), thus leading to an ever-increasing cycle of channel activation (Silverman *et al.,* 1996). Finally, increased Na^+ would lead to increased Na^+/Ca^{2+} exchange, which in turn would lead to increased intracellular calcium levels and cell death (Fig. 5). Even a smaller basal *wv*Kir 3.2 Na^+ conductance could challenge cell viability. First, the extra Na^+ conductance could lead to chronic depolarization of the neuron. Second, the chronic depolarization could lead to the constant activation of L-type calcium channels and finally Ca^{2+}-induced cell death (Liesi and Wright, 1996). Interestingly, if *wv*Kir 3.2 homomultimers existed in discrete regions of the brain, one would predict that these regions would be most severely affected, all other factors being equal. As previously mentioned, *wv*Kir 3.2 homomultimers conduct Na^+ better than Kir 3.1/*wv*Kir 3.2 heteromultimers.

3. *wv*Kir 3.2 Channels Are $G_{\beta\gamma}$ Insensitive

When Kir 3.1/*wv*Kir 3.2 or *wv*Kir 3.2 is expressed in CHO cells, they are not activated by $G_{\beta\gamma}$ (Navarro *et al.,* 1996). This is a consequence either of the inability of $G_{\beta\gamma}$ to activate *wv*Kir 3.2 channels or of constitutive activation of *wv*Kir 3.2 channels. In any case, channels that contain *wv*Kir 3.2 are no longer tightly linked to the G-protein-coupled receptor pathways.

4. Native Kir 3.2 Currents in Cerebellar Granule Cells

It is likely that there are Kir 3.2 currents in granule cells and that in *weaver* mice these currents are altered (Table I). The G-protein-activated currents in normal cerebellar granule cells are potassium-selective, show strong inward rectification, and are blocked by Ba^{2+}. These properties are

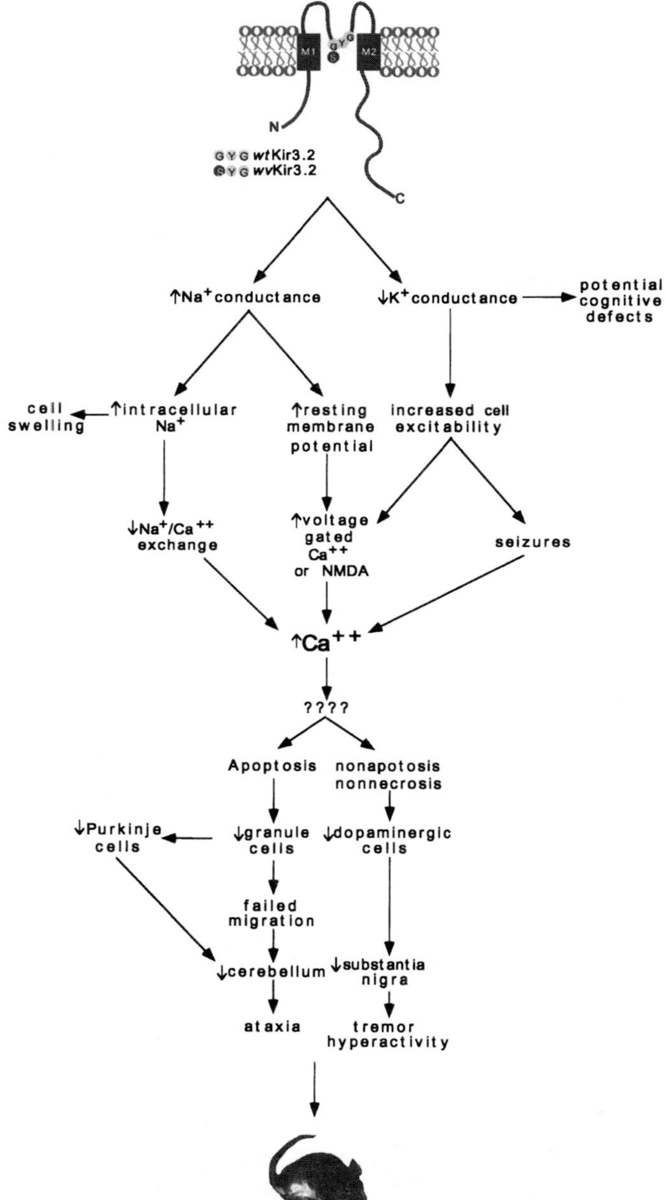

FIGURE 5 Speculative schematic of *weaver* pathophysiology.

similar to channels containing Kir 3.2 subunits in heterologous expression systems (Section II,D).

The type and extent of current alterations in *weaver* cerebellar granule cells are debated. Surmeier *et al.* (1996) and Lauritzen *et al.* (1997) report that Kir 3.0 currents are severely reduced, whereas Kofuji *et al.* (1996) report that there is an increase in a G-protein-independent Na^+ "leakage" conductance. These findings are not mutually exclusive. Both the decrease in granule cell Kir 3.0 conductance and the increase in granule cell Na^+ "leakage" are compatible with electrophysiological properties of *wv*Kir 3.2 expressed in heterologous systems (Sections V,A,1 and V,A,2).

B. Potential Pathophysiological Consequences of weaver Kir 3.2

It is well established that *weaver* mice have a selective defect in cell differentiation (Rakic and Sidman, 1973; Sotelo and Changeux, 1974; Hatten *et al.*, 1984, 1986; Goldowitz, 1989; Sotelo, 1990). Prior to the identification of *wv*Kir 3.2, several mechanisms for the defective cell differentiation were suggested. These included aberrant proteolytic enzymes (aprotinin), extracellular matrix molecules (B2 chain of laminin), and intracellular calcium levels. The unanticipated discovery of *wv*Kir 3.2 dramatically altered our understanding of *weaver* pathophysiology. Kir 3.2 is not directly involved in any of the previously mentioned processes. *wv*Kir 3.2 causes both a new function (increased Na^+ conductance) and a loss of function (reduction in the functional expression of Kir 3.1/Kir 3.2 heterotetrameric channels). Thus, the *wv*Kir 3.2 allele is neomorphic for some phenotypes and hypomorphic for other phenotypes.

1. New Function: Increased Na^+ Conductance

The pore mutation in Kir 3.2 suggests one of two etiologies for *weaver* cerebellar granule cell pathophysiology. Either K_G currents are required for normal granule cell differentiation/migration (loss of function), or the increased *wv*Kir 3.2 Na^+ conductance leads to cell death prior to cell differentiation/migration. There is considerable evidence for the second etiology. The viability of CHO cells expressing *wv*Kir 3.2 is dramatically decreased. In transfected CHO cells 90% of wild-type Kir 3.2-expressing cells versus only 22% of *wv*Kir 3.2-expressing cells were alive 24 hr following transfection (Navarro *et al.*, 1996). Our efforts directed at studying the physical interaction of *wv*Kir 3.2 and Kir 3.1 subunits were hampered by extremely low expression levels of *wv*Kir 3.2 compared to either wild-type Kir 3.1 or Kir 3.2. Functional expresssion studies also suggested that *wv*Kir 3.2-expressing cells were not as viable as wild-type Kir 3.2-expressing cells.

We substituted the majority of the Na^+ ion in the culture media with *N*-methyl-D-glucamine (NMDG), a nonpermeant molecule, and found a significant increase in the amount of expressed *wv*Kir 3.2 subunit without changing the expression levels of wild-type Kir 3.1 and Kir 3.2 (Navarro et al., 1996). The mechanism by which low Na^+ media enhanced *wv*Kir 3.2 subunit expression is not known, but we would suggest based on our assessment of cell survival and immunopreciptiation experiments that there is a decrease in the cell death when *wv*Kir 3.2-expressing cells were cultured in low Na^+ media. Furthermore, the ability of *weaver* granule cells to differentiate *in vivo* was restored by blocking *wv*Kir 3.2 Na^+ influx with QX314 (Kofuji *et al.,* 1996).

2. Loss of Function: Decreased K^+ Conductance of *wv*Kir 3.2-Containing Heteromultimers

The absence of *wt*Kir 3.2 could be contributing to the *weaver* phenotype in addition to the new ability of *wv*Kir 3.2 channels to conduct Na^+. Kir 3.2 −/− mice are grossly indistinguishable from wild-type mice (Signorini *et al.,* 1997). Most importantly, they do not display ataxia or significant cerebellar degeneration as seen in *weaver.* The simplest interpretation of these data is that the acquisition of Na^+ conductance by Kir 3.2 is more important for cerebellar defects than the decrease in K^+ conductance. One exception, however, is that Kir 3.2 −/− mice develop spontaneous seizures similar to weaver mice. In brain regions where receptor-driven hyperpolarization is coupled to Kir 3.2 containing channels, the loss of K^+ conductance and added Na^+ conductance in *wv*Kir3.2 could cause hyperexcitability possibly leading to seizures. These observations suggest that the weaver phenotype is largely, but not completely, due to a new function, rather than loss of normal *wt*Kir 3.2 function. It is important to note that Kir3.2 −/− mice show several abnormalities at the cellular level, many of which are still being elucidated (see Signorini *et al.,* 1997).

The phenotype of heterozygous *weaver* mice is much less severe in terms of ataxia and cerebellar degeneration and is later in onset compared to homozygous *weaver* mice. The reason for this is unknown, but it could be due to the presence of one wild-type allele of Kir 3.2. If both alleles are expressed in a single cell, then heterotetrameric channels containing at least one each of Kir 3.1, Kir 3.2, or *wv*Kir 3.2 subunits may exist and display intermediate conductance and ion permeability. If only one allele is active in a given cell, then one would expect the cell's fate to be determined by the presence or absence of *wv*Kir 3.2 subunits. In transfected CHO cells the whole-cell K^+ current is increased when wild-type Kir 3.2 is coexpressed with Kir 3.1 and *wv*Kir 3.2 (Fig. 6A), but it is still much lower than that seen for wild-type Kir 3.1/Kir 3.2. Therefore, the presence

A. CHO cells

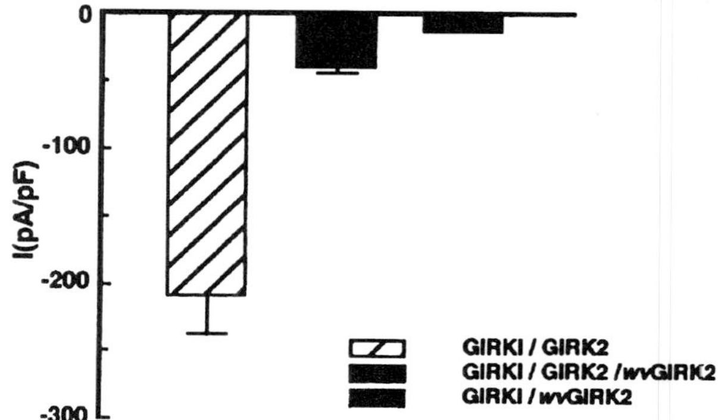

B. Oocytes

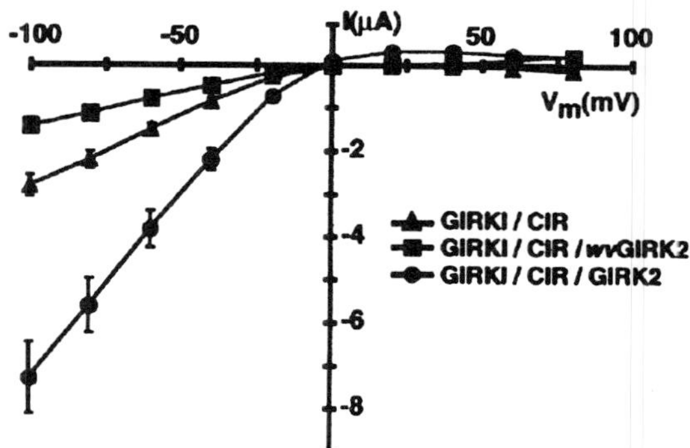

FIGURE 6 *wv*Kir 3.2 competes with *wt*Kir 3.2 and Kir 3.4 for Kir 3.1. (A) The fourfold reduction in whole-cell current observed after transfection with Kir 3.1/Kir 3.2/*wv*Kir 3.2 compared to Kir 3.1/Kir 3.2 suggests that *wv*Kir 3.2 is competing with *wt*Kir 3.2 for Kir 3.1. (B) Similar effects were observed in oocytes.

of all three subunits in a given cell leads to an intermediate whole-cell current. This same effect is seen when *wv*Kir 3.2 is coexpressed with Kir 3.1/Kir 3.4 subunits in oocytes (Fig. 6B). *wv*Kir 3.2 interacts with Kir 3.1 indistinguishably from wild-type Kir 3.2 (Navarro *et al.,* 1996), suggesting that, when all three subunits are coexpressed, channels containing all three subunits could form.

3. Mechanism of Cell Death: Apoptosis versus Necrosis

Cell death in *weaver* mice is most likely cell type-dependent. Cerebellar granule cells appear to die by apoptosis, whereas dopaminergic cells appear to die by a novel nonapoptotic, nonnecrotic mechanism (Gillardon *et al.*, 1995; Migheli *et al.*, 1995, 1997; Owens *et al.*, 1995). Identification of cell-specific factors that contribute to the variability in cell death could lead to a better understanding of human neurodegenerative disease. The death of other cell types such as Purkinje cells may be secondary to granule cell loss (Section IV,B,2,a).

VI. CONCLUSION

During the 1990s, our understanding of neurodegenerative disorders has greatly increased. Advances in molecular biology, cell biology, and electrophysiology have allowed us to begin to probe the inner workings of the nervous system. Yet, to date, very few animal models of neurodegenerative disease exist. The *weaver* mouse is one of the very first animal models of neurodegenerative disease and is proving invaluable to neuroscientists. These meandering, hyperactive, and Parkinsonianlike mice are teaching us about the role of Kir 3.0 channels in brain, the pathophysiology of human neurodegenerative diseases, and the functional importance of certain neuronal cell types.

We now know that *weaver* mice are a direct result of a mutation in a ion channel (Kir 3.2), rather than an adhesion or migratory molecule as originally speculated. This mutation affects both striatal dopaminergic and cerebellar granule cells (Rakic and Sidman, 1973), ultimately leading to cell death. This, in turn, indirectly affects many other cell types and systems. The cellular defects in weaver mice correlate well with their motor and gross anatomical abnormalities.

*wv*Kir 3.2 contains a mutation in a critical pore region of a G-protein-regulated inwardly rectifying K^+ channel subunit (Fig. 1). Kir 3.0 channels regulate membrane excitability and are presumably important for controlling presynaptic and postsynaptic inhibition. The GYG to SYG pore mutation leads to a nonselective cation (K^+, Na^+) channel. It is the increase in Na^+ conductance which is largely responsible for the neuronal cell death (gain of function). The absence of normally functional Kir 3.2 currents results in several defects measurable at the cellular level, as demonstrated by Kir 3.2 knockout mice (Kir 3.2 −/−). The Kir 3.2 −/− mice, unlike Kir 3.2 *wv/wv* mice, appear grossly normal. Although the exact pathway to cell death in *weaver* mice is unknown, the deaths of cerebellar granule cells and dopaminergic neurons are likely to have different etiologies. Granule neurons die by apoptosis, whereas dopaminergic neurons appear to die by a novel nonnecrotic, nonapoptotic mechanism.

Already *weaver* mice are providing some insights into human neuropathophysiology. The anomalous and repetitive behavior of *weaver* mice is similar to human compulsive spectrum disorders, such as Parkinson's disease. Thus, Bandmann *et al.* (1996) have assessed the possibility of a common genetic defect between *weaver* mice and Parkinson's patients. However, so far no family with a history Parkinson's disease has been found to have mutation in Kir 3.2 (Tanizawa *et al.,* 1996). Deciphering the exact mechanism of cell death in *weaver* mice, and determining what makes the cell's response to *wv*Kir 3.2 cell-type specific, will undoubtedly be important for our understanding of a wide variety of neurodegenerative diseases.

References

Aleksandrov, A., Velimirovic, B., and Clapham, D. E. (1996). Inward rectification of the IRK1 K^+ channel reconstituted in lipid bilayers. *Biophys. J.* **70,** 2680–2687.

Ashford, M. L. J., Bond, C. T., Blair, T. A., and Adelman, J. P. (1994). Cloning and functional expression of a rat heart K_{ATP} channel. *Nature (London)* **370,** 456–459.

Bandmann, O., Davis, M. B., Marsden, C. D., and Wood, N. W. (1996). The human homologue of the weaver mouse gene in familial and sporadic Parkinson's disease. *Neuroscience (Oxford)* **72,** 877–879.

Bayer, S. A., Wills, K. V., Wei, J., Feng, Y., Dlouhy, S. R., Hodes, M. E., Verina, T., and Ghetti, B. (1996). Phenotypic effects of the weaver gene are evident in the embryonic cerebellum but not in the ventral midbrain. *Brain Res. Dev. Brain Res.* **96,** 130–137.

Bolivar, V. J., Manley, K., and Fentress, J. C. (1996). The development of swimming behavior in the neurological mutant weaver mouse. *Dev. Psychobiol.* **29,** 123–137.

Chan, K. W., Sui, J. L., Vivaudou, M., and Logothetis, D. E. (1996). Control of channel activity through a unique amino acid residue of a G protein-gated inwardly rectifying K^+ channel subunit. *Proc. Natl. Acad. Sci. U.S.A.* **93,** 14193–14198.

Chou, H., Ogawa, N., Asanuma, M., Hirata, H., and Mori, A. (1991). Dopamine deficiency in the weaver mutant mouse: An animal model of olivopontocerebellar atrophy. *Res. Commun. Chem. Pathol. Pharmacol.* **74,** 117–120.

Clapham, D. E. (1994). Direct G protein activation of ion channels? *Annu. Rev. Neurosci.* **17,** 441–464.

Clapham, D. E., and Neer, E. J. (1997). G protein $\beta\gamma$ subunits. *Annu. Rev. Pharmacol.* **37,** 167–204.

Corey, S., Krapivinsky, G., Krapivinsky, L., and Clapham, D. E. (1998). Number and stoichiometry of subunits in the native atrial G-protein-gated K^+ channel, IKACh. *J. Biol. Chem.* **273,** 5271–5278.

Dascal, N., Schreibmayer, W., Lim, N. F., Wang, W., Chavkin, C., DiMagno, L. Labarca, C., Kiefer, B. L., Gaveriaux-Ruff, C., Trollinger, D., Lester, H. A., and Davidson, V. (1993). Atrial G-Protein-activated K^+ channel: Expression cloning and molecular properties. *Proc. Natl. Acad. Sci. U.S.A.* **90,** 10235–10239.

Doupnik, C. A., Lim, N. F., Kofuji, P., Davidson, N., and Lester, H. A. (1995). Intrinsic gating properties of a cloned G protein-activated inward rectifier K^+ channel. *J. Gen. Physiol.* **106,** 1–23.

Duprat, F., Lesage, F., Guillemare, E., Fink, M., Hugnot, J. P., Bigay, J., Lazdunski, M., Romey, G., and Barhanin, J. (1995). Heterologous multimeric assembly is essential for K^+ channel activity of neuronal and cardiac G-protein-activated inward rectifiers. *Biochem. Biophys. Res. Commun.* **212,** 657–663.

Eisenberg, B., and Messer, A. (1989). Tonic/clonic seizures in a mouse mutant carrying the weaver gene. *Neurosci. Lett.* **96,** 168–172.

Fischer-Bovenkerk, C., Kish, P. E., and Ueda, T. (1988). ATP-dependent glutamate uptake into synaptic vesicles from cerebellar mutant mice. *J. Neurochem.* **51,** 1054–1059.

Gillardon, F., Bäurle, J., Grusser-Cornehls, U., and Zimmermann, M. (1995). DNA fragmentation and activation of c-Jun in the cerebellum of mutant mice (weaver, Purkinje cell degeneration). *Neuroreport* **6,** 1766–1768.

Goldowitz, D. (1989). The weaver granuloprival phenotype is due to intrinsic action of the mutant locus in granule cells: Evidence from homozygous weaver chimeras. *Neuron* **2,** 1565–1575.

Goldowitz, D., and Mullen, R. J. (1982). Granule cell as a site of gene action in the weaver mouse cerebellum: Evidence from heterozygous mutant chimeras. *J. Neurosci.* **2,** 1474–1485.

Grigg, J. J., Kozasa, T., Nakajima, Y., and Nakajima, S. (1996). Single-channel properties of a G-protein-coupled inward rectifier potassium channel in brain neurons. *J. Neurophysiol.* **75,** 318–328.

Harrison, S. M., and Roffler-Tarlov, S. (1994). Male-sterile phenotype of the neurological mouse mutant weaver. *Dev. Dyn.* **200,** 26–38.

Hatten, M. E., Liem, R. K., and Mason, C. A. (1984). Defects in specific associations between astroglia and neurons occur in microcultures of weaver mouse cerebellar cells. *J. Neurosci.* **4,** 1163–1172.

Hatten, M. E., Liem, R. K., and Mason, C. A. (1986). Weaver mouse cerebellar granule neurons fail to migrate on wild-type astroglial processes in vitro. *J. Neurosci.* **6,** 2676–2683.

Hedin, K., Nancy, L., and Clapham, D. (1996). Cloning of a *Xenopus laevis* inwardly rectifying K^+ channel subunit that permits GIRK1 expression of I_{KACH} currents in oocytes. *Neuron* **16,** 423–429.

Heginbotham, L., Abramson, T., and MacKinnon, R. (1992). A functional connection between the pores of distantly related ion channels as revealed by mutant K^+ channels. *Science* **258,** 1152–1155.

Heginbotham, L., Lu, Z., Abramson, T., and MacKinnon, R. (1994). Mutations in the K^+ channel signature sequence. *Biophys. J.* **66,** 1061–1067.

Hille, B. (1992). "Ionic Channels of Excitable Membranes," 2nd Ed. Sinauer, Sunderland, Massachusetts.

Inagaki, N., Gonoi, T., Clement, J. P. T., Namba, N., Inazawa, J., Gonzalez, G., Aguilar-Bryan, L., Seino, S., and Bryan, J. (1995). Reconstitution of IKATP: An inward rectifier subunit plus the sulfonylurea receptor [see comments]. *Science* **270,** 1166–1170.

Isomoto, S., Kondo, C., Takahashi, N., Matsumoto, S., Yamada, M., Takumi, T., Horio, Y., and Kurachi, Y. (1996). A novel ubiquitously distributed isoform of GIRK2 (GIRK2B) enhances GIRK1 expression of the G-protein-gated K^+ current in *Xenopus* oocytes. *Biochem. Biophys. Res. Commun.* 218, 286–291.

Karschin, C., Schreibmayer, W., Dascal, N., Lester, H., Davidson, N., and Karschin, A. (1994). Distribution and localization of a G protein-coupled inwardly rectifying K^+ channel in the rat. *FEBS Lett.* **348,** 139–144.

Karschin, C., Dibmann, E., Stuhmer, W., and Karschin, A. (1996). IRK1(1–3) and GIRK(1–4) inwardly rectifying K^+ channel mRNAs are differentially expressed in the adult rat brain. *J. Neurosci.* **16,** 3559–3570.

Kennedy, M., Nemec, J., and Clapham, D. E. (1996). Localization and interaction of epitope-tagged GIRK1 and CIR inward rectifier K^+ channel subunits. *Neuropharmacology* **35,** 831–839.

Kobayashi, T., Kazutaka, I., Ichikawa, T., Abe, S., Togashi, S., and Kumanishi, T. (1995). Molecular cloning of a mouse G-protein-activated K^+ channel (mGIRK1) and distinct distributions of three GIRK (GIRK1, 2 and 3) mRNAs in mouse brain. *Biochem. Biophys. Res. Commun.* **208,** 1166–1173.

Kofuji, P., Davidson, N., and Lester, H. (1995). Evidence that neuronal G-protein-gated inwardly rectifying K^+ channels are activated by $G\beta\gamma$ subunits and function as heteromultimers. *Proc. Natl. Acad. Sci. U.S.A.* **92,** 6542–6546.

Kofuji, P., Hofer, M., Millen, K. J., Millonig, J. H., Davidson, N., Lester, H. A., and Hatten, M. E. (1996). Functional analysis of the weaver mutant GIRK2 K^+ channel and rescue of weaver granule cells. *Neuron* **16,** 941–952.

Krapivinsky, G., Gordon, E. A., Wickman, K., Velimiroviy, B., Krapivinsky, L., and Clapham, D. E. (1995a). The G-protein-gated atrial K^+ channel IKACh is a heteromultimer of two inwardly rectifying $K(^+)$-channel proteins. *Nature (London)* **374,** 135–141.

Krapivinsky, G., Krapivinsky, L., Velimirovic, B., Wickman, K., Navarro, B., and Clapham, D. E. (1995b). The cardiac inward rectifier K^+ channel subunit, CIR, does not comprise the ATP-sensitive K^+ channel, IKATP. *J. Biol. Chem.* **270,** 28777–28779.

Kubo, Y., Reuveny, E., Slesinger, P. A., Jan, Y. N., and Jan, L. Y. (1993). Primary structure and functional expression of a rat G-protein-coupled muscarinic potassium channel. *Nature (London)* **364,** 802–806.

Kurachi, Y., Ito, H., Sugimoto, T., Katada, T., and Ui, M. (1989). Activation of atrial muscarinic K^+ channels by low concentrations of beta gamma subunits of rat brain G protein. *Pfluegers Arch.* **413,** 325–327.

Lane, P. W. (1964). *Mouse News Letter,* 30–32.

Lauritzen, I., De Weille, J., Adelbrecht, C., Lesage, F., Murer, G., Raisman-Vozari, R., and Lazdunski, M. (1997). Comparative expression of the inward rectifier K^+ channel GIRK2 in the cerebellum of normal and weaver mutant mice. *Brain Res.* **753,** 8–17.

Lesage, F., Duprat, F., Fink, M., Guillemare, E., Coppola, T., Lazdunski, M., and Hugnot, J. P. (1994). Cloning provides evidence for a family of inward rectifier and G-protein coupled K^+ channels in the brian. *FEBS Lett.* **353,** 37–42.

Lesage, F., Guilemere, E., Fink, M., Duprat, F., Heurteaux, C., Fosset, M., Romey, G., Barhanin, J., and Lazdunski, M. (1995). Molecular properties of neuronal G-protein-activated inwardly rectifying K^+ channels. *J. Biol. Chem.* **270,** 28660–28667.

Liao, Y. J., Jan, Y. N., and Jan, L. Y. (1996). Heteromultimerization of G-protein-gated inwardly rectifying K^+ channel proteins GIRK1 and GIRK2 and their altered expression in weaver brain. *J. Neurosci.* **16,** 7137–7150.

Liesi, P., and Wright, J. M. (1996). Weaver granule neurons are rescued by calcium channel antagonists and antibodies against a neurite outgrowth domain of the B2 chain of laminin. *J. Cell Biol.* **134,** 477–486.

Logothetis, D. E., Kurachi, Y., Galper, J., Neer, E. J., and Clapham, D. E. (1987). The beta gamma subunits of GTP-binding proteins activate the muscarinic K^+ channel in heart. *Nature (London)* **325,** 321–326.

MacKinnon, R., and Yellen, G. (1990). Mutations affecting TEA blockade and ion permeation in voltage-activated K^+ channels. *Science* **250,** 276–279.

Matsuda, H., Saigusa, A., and Irisawa, H. (1987). Ohmic conductance through the inwardly rectifying K channel and blocking by internal Mg^{2+}. *Nature (London)* **325,** 156–159.

Migheli, A., Attanasio, A., Lee, W. H., Bayer, S. A., and Ghetti, B. (1995). Detection of apoptosis in weaver cerebellum by electron microscopic in situ end-labeling of fragmented DNA. *Neurosci Lett.* **199,** 53–56.

Migheli, A., Piva, R., Wei, J., Attanasio, A., Casolino, S., Hodes, M. E., Dlouhy, S. R., Bayer, S. A., and Ghetti, B. (1997). Diverse cell death pathways result from a single missense mutation in weaver mouse. *Am. J. Pathol.* **151,** 1629–1638.

Murer, G., Adelbrecht, C., Lauritzen, I., Lesage, F., Lazdunski, M., Agid, Y., and Raisman-Vozari, R. (1997). An immunocytochemical study on the distribution of two G-protein-gated inward rectifier potassium channels (GIRK2 and GIRK4) in the adult rat brain. *Neuroscience (Oxford)* **80,** 345–357.

Navarro, B., Kennedy, M., Velimirovic, B., Bhat, D., Peterson, A., and Clapham, D. (1996). Nonselective and $G_{\beta\gamma}$-insensitive *weaver* K^+ channels. *Science* **272,** 1950–1953.

Neer, E. J., and Clapham, D. E. (1988). Roles of G protein subunits in transmembrane signalling. *Nature (London)* **333,** 129–134.

Nicoll, R. A., Kauer, J. A., and Malenka, R. C. (1988). The current excitement in long-term potentiation. *Neuron* **1,** 97–103.

North, R. A. (1989). Twelfth Gaddum memorial lecture. Drug receptors and the inhibition of nerve cells. *Br. J. Pharmacol.* **98,** 13–28.

Owens, G. P., Mahalik, T. J., and Hahn, W. E. (1995). Expression of the death-associated gene RP-8 in granule cell neurons undergoing postnatal cell death in the cerebellum of weaver mice. *Brain Res. Dev. Brain Res.* **86,** 35–47.

Patil, N. Cox, D. R., Bhat, D., Myers, R. M., and Peterson, A. S. (1995). A potassium channel mutation in *weaver* mice implicates membrane excitability in granule cell differentiation. *Nat. Genet.* **11,** 126–129.

Ponce, A., Bueno, E., Kentros, C., Vega-Saenz de Miera, E., Chow, A., Hillman, D., Chen, S., Zhu, L., Wu, M. B., Wu, X., Rudy, B., and Thornhill, W. B. (1996). G-protein-gated inward rectifier K^+ channel proteins (GIRK1) are present in the soma and dendrites as well as in nerve terminals of specific neurons in the brain. *J. Neurosci.* **16,** 1990–2001.

Privat, A. (1978). Dendro-dendritic pentalaminar junctions in the weaver mouse cerebellum. *Acta Neuropathol.* **42,** 137–140.

Rakic, P., and Sidman, R. L. (1973). Weaver mutant mouse cerebellum: Defective neuronal migration secondary to abnormality of Bergmann glia. *Proc. Natl. Acad. Sci. U.S.A.* **70,** 240–244.

Reeves, R. H., Crowley, M. R., Lorenzon, N., Pavan, W. J., Smeyne, R. J., and Goldwitz, D. (1989). The mouse neurological mutant *weaver* maps within the region of chromosome 16 that is homologous to human chromosome 21. *Genomics* **5,** 522–526.

Roffler-Tarlov, S., and Turey, M. (1982). The content of amino acids in the developing cerebellar cortex and deep cerebellar nuclei of granule cell deficient mutant mice. *Brain Res.* **247,** 65–73.

Roffler-Tarlov, S., Martin, B., Graybiel, A. M., and Kauer, J. S. (1996). Cell death in the midbrain of the murine mutation weaver. *J. Neurosci.* **16,** 1819–1826.

Signorini, S., Liao, Y. J., Duncan, S. A., Jan, L. Y., and Stoffel, M. (1997). Normal cerebellar development but susceptibility to seizures in mice lacking G protein-coupled, inwardly rectifying K^+ channel GIRK2. *Proc. Natl. Acad. Sci. U.S.A.* **94,** 923–927.

Silverman, S. K., Kofuji, P., Dougherty, D. A., Davidson, N., and Lester, H. A. (1996). A regenerative link in the ionic fluxes through the weaver potassium channel underlies the pathophysiology of the mutation. *Proc. Natl. Acad. Sci. U.S.A.* **93,** 15429–15434.

Simon, J. R., and Ghetti, B. (1994). The weaver mutant mouse as a model of nigrostriatal dysfunction. *Mol. Neurobiol.* **9,** 183–189.

Simon, J. R., Richter, J. A., and Ghetti, B. (1994). Age-dependent alterations in dopamine content, tyrosine hydroxylase activity, and dopamine uptake in the striatum of the weaver mutant mouse. *J. Neurochem.* **62,** 543–548.

Smeyne, R. J., and Goldowitz, D. (1989). Development and death of external granular layer cells in the weaver mouse cerebellum: A quantitative study. *J. Neurosci.* **9,** 1608–1620.

Sotelo, C. (1975). Anatomical, physiological and biochemical studies of the cerebellum from mutant mice. II. Morphological study of cerebellar cortical neurons and circuits in the weaver mouse. *Brain Res.* **94,** 19–44.

Sotelo, C. (1990). Cerebellar synaptogenesis: What we can learn from mutant mice. *J. Exp. Biol.* **153,** 225–249.

Sotelo, C., and Changeux, J. P. (1974). Bergmann fibers and granular cell migration in the cerebellum of homozygous weaver mutant mouse. *Brain Res.* **77,** 484–491.

Spauschus, A., Lentes, K. U., Wischmeyer, E., Dissmann, E., Karschin, C., and Karschin, A. (1996). A G-protein-activated inwardly rectifying K^+ channel (GIRK4) from human hippocampus associates with other GIRK channels. *J. Neurosci.* **16,** 930–938.

Stoffel, M., Tokuyama, Y., Trabb, J. B., German, M. S., Tsaar, M. L., Jan, L. Y., Polonsky, K. S., and Bell, G. I. (1995). Cloning of rat KATP-2 channel and decreased expression in pancreatic islets of male Zucker diabetic fatty rats. *Biochem. Biophys. Res. Commun.* **212,** 894–899.

Stotz, E. H., Triarhou, L. C., Ghetti, B., and Simon, J. R. (1993). Serotonin content is elevated in the dopamine deficient striatum of the weaver mutant mouse. *Brain Res.* **606,** 267–272.

Stotz, E. H., Palacios, J. M., Landwehrmeyer, B., Norton, J., Ghetti, B., Simon, J. R., and Triarhou, L. C. (1994). Alterations in dopamine and serotonin uptake systems in the striatum of the weaver mutant mouse. *J. Neural Transm. Gen. Sect.* **97,** 51–64.

Sui, J. L., Chan, K. W., and Logothetis, D. E. (1996). Na^+ activation of the muscarinic K^+ channel by a G-protein-independent mechanism. *J. Gen. Physiol.* **108,** 381–391.

Surmeier, D. J., Mermelstein, P. G., and Goldowitz, D. (1996). The weaver mutation of GIRK2 results in a loss of inwardly rectifying K^+ current in cerebellar granule cells [see comments]. *Proc. Natl. Acad. Sci. U.S.A.* **93,** 11191–11195.

Tanizawa, Y., Matsubara, A., Ueda, K., Katagiri, H., Kuwano, A., Ferrer, J., Permutt, M. A., and Oka, Y. (1996). A human pancreatic islet inwardly rectifying potassium channel: cDNA cloning, determination of the genomic structure and genetic variations in Japanese NIDDM patients. *Diabetologia* **39,** 447–452.

Triarhou, L. C., Low, W. C., and Ghetti, B. (1986). Transplantation of ventral mesencephalic anlagen to hosts with genetic nigrostriatal dopamine deficiency. *Proc. Natl. Acad. Sci. U.S.A.* **83,** 8789–8793.

Tsaur, M. L., Menzel, S., Lai, F. P., Espinosa, R. R., Concannon, P., Spielman, R. S., Hanis, C. L., Cox, N. J., Le Beau, M. M., German, M. S., Jan, L. Y., Bell, G. I., and Stoffel, M. (1995). Isolation of a cDNA clone encoding a KATP channel-like protein expressed in insulin-secreting cells, localization of the human gene to chromosome band 21q22.1, and linkage studies with NIDDM. *Diabetes* **44,** 592–596.

Velimirovic, B. M., Gordon, E. A., Lim, N. F., Navarro, B., and Clapham, D. E. (1996). The K^+ channel inward rectifier subunits form a channel similar to neuronal G protein-gated K^+ channel. *FEBS Lett.* **379,** 31–37.

Verina, T., Tang, X., Fitzpatrick, L., Norton, J., Vogelweid, C., and Ghetti, B. (1995). Degeneration of Sertoli and spermatogenic cells in homozygous and heterozygous weaver mice. *J. Neurogenet.* **9,** 251–265.

Verney, C., Febvret-Muzerelle, A., and Gaspar, P. (1995). Early postnatal changes of the dopaminergic mesencephalic neurons in the weaver mutant mouse. *Brain Res. Dev. Brain Res.* **89,** 115–119.

Vogelweid, C. M., Verina, T., Norton, J., Harruff, R., and Ghetti, B. (1993). Hypospermatogenesis is the cause of infertility in the male weaver mutant mouse. *J. Neurogenet.* **9,** 89–104.

Wei, J., Dlouhy, S. R., Zhu, J., Ghetti, B., and Hodes, M. E. (1994). Analysis of region-specific library constructed by sequence-independent amplification of microdissected fragments surrounding weaver (wv) gene on mouse chromosome 16. *Somatic Cell Mol. Genet.* **20,** 401–408.

Wei, J., Dlouhy, S. R., Bayer, S., Piva, R., Verina, T., Wang, Y., Feng, Y., Dupree, B., Hodes, M. E., and Ghetti, B. (1997). In situ hybridization analysis of GIRK2 expression in the developing central nervous system in normal and weaver mice. *J. Neuropathol. Exp. Neurol.* **56,** 762–771.

Wickman, K., and Clapham, D. E. (1995). Ion channel regulation by G proteins. *Physiol. Rev.* **75,** 865–885.

Wickman, K. D., Iniguez-Luhi, J. A., Davenport, P. A., Taussig, R., Krapivinsky, G. B., Linder, M. E., Gilman, A. G., and Clapham, D. E. (1994). Recombinant G-protein beta gamma-subunits activate the muscarinic-gated atrial potassium channel. *Nature* (*London*) **368,** 255–257.

Wickman, K., Hedin, K., Perez, C., Krapivinsky, G., Stehno-Bittel, L., Velimirovic, B., and Clapham, D. E. (1997). The switch and assembly paradigms of transmembrane signalling. *In* "Handbook of Physiology; Section 14: Cell Physiology" (J. Hoffman and J. D. Jamieson, eds.). pp. 689–742. Oxford University Press, New York.

Wickman, K., Nemec, J., Gendler, S., and Clapham, D. E. (1998). Abnormal heart rate regulation in GIRK4 knockout mice. *Neuron* **20,** 103–114.

CHAPTER 17

G-Protein Control of G-Protein-Gated Potassium Channels

Eitan Reuveny* and Lily Yeh Jan†

*Department of Biological Chemistry, Weizmann Institute of Science, Rehovot 76100, Israel; and †Howard Hughes Medical Institute, Department of Physiology and Biochemistry, University of California, San Francisco, San Francisco, California 94143

I. INTRODUCTION

G-protein-activated potassium channels (GIRK) belong to the inwardly rectifying potassium channel superfamily. These channels are opened following the stimulation of $G_{i/o}$-coupled serpentine receptors to cause hyperpolarization of the cell. GIRK channels are involved in many physiological responses. In the heart, GIRK channels are responsible for the bradycardia induced by vagal nerve stimulation (parasympathetic), through the activation of m2 muscarinic receptors by acetylcholine (Loewi, 1921; Trautwein and Dudel, 1958). In the brain, these channels are widely distributed (Drake *et al.,* 1997; Karschin *et al.,* 1996; Liao *et al.,* 1996; Ponce *et al.,* 1996; Spauschus *et al.,* 1996) and are involved in the postsynaptic inhibitory action of many neurotransmitters such as γ-aminobutyric acid (through $GABA_B$ receptors), adenosine (through A1), and serotonin (through 5HT1A) (Lüscher *et al.,* 1997). Their function in the pancreas is less well understood,

1063-5823/99 $30.00

but suggestions have been made to include modulation of insulin secretion from islet β cells (Ferrer *et al.,* 1995).

II. PRIMARY STRUCTURE

GIRK channels belong to the large superfamily of inwardly rectifying potassium channels. This superfamily includes channel subunits with two predicted transmembrane domains, M1 and M2, and a putative pore region, H5, in between (Fig. 1) (Kubo *et al.,* 1993a,b). The absence of signal peptide suggests that the two transmembrane domains are flanked by cytoplasmic N and C termini. (The C-terminal domain is three to four times larger than the N-terminal domain.) Sequence comparison of all known inwardly rectifying potassium channels reveals that the most conserved region is the H5 pore region and the M1 and M2 transmembrane domains (Doupnik *et al.,* 1995). Some high homology is also found at the beginning (5′) of the C-terminal domain of the molecule (the first 150 amino acids following the end of M2). The N-terminal domain, by contrast, is less conserved among the various family members. The family of G-protein-coupled potassium channels (GIRK, Kir3.x), so far, is the only family of inwardly rectifying channels that are known to be activated by the $\beta\gamma$ subunits of G proteins.

Five members of the GIRK family are known to date (GIRK1–GIRK5 or Kir3.1–Kir3.5) (Dascal *et al.,* 1993; Hedin *et al.,* 1996; Krapivinsky *et al.,* 1995a; Kubo *et al.,* 1993b; Lesage *et al.,* 1994). They share 65–80% identity in their amino acid sequence. Molecular identification of various GIRK1–4 channels have been done for many mammalian species and one avian species; however, GIRK5, so far, has been isolated only from *Xenopus* oocytes (Hedin *et al.,* 1996).

A. GIRK Channels Are Heteromultimers

Using antibodies against GIRK1, Krapivinsky *et al.* (1995a) have been able to coimmunoprecipitate GIRK4 (also known as CIR) and to demonstrate that in heart, and in other heterologous expression systems, GIRK1 and GIRK4 are forming heteromultimers with electrophysiological characteristics identical to the well-characterized muscarinic potassium channel, $I_{K(ACh)}$ (Krapivinsky *et al.,* 1995a). Later it was shown that other members of the GIRK family can also form heteromultimers when coexpressed in *Xenopus* oocytes and various mammalian cell lines (Duprat *et al.,* 1995; Kofuji *et al.,* 1995; Lesage *et al.,* 1995; Velimirovic *et al.,* 1996). Indeed,

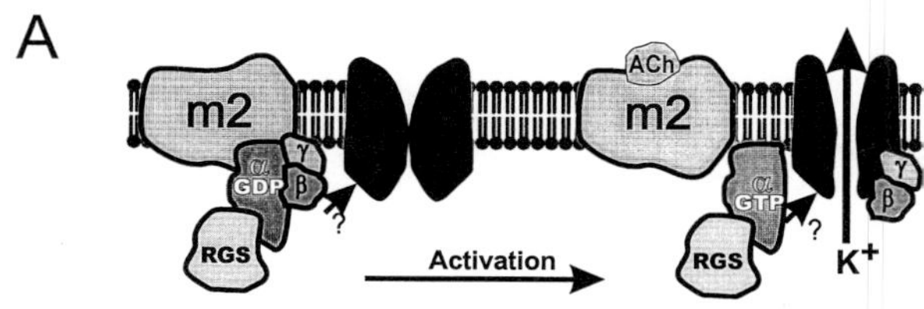

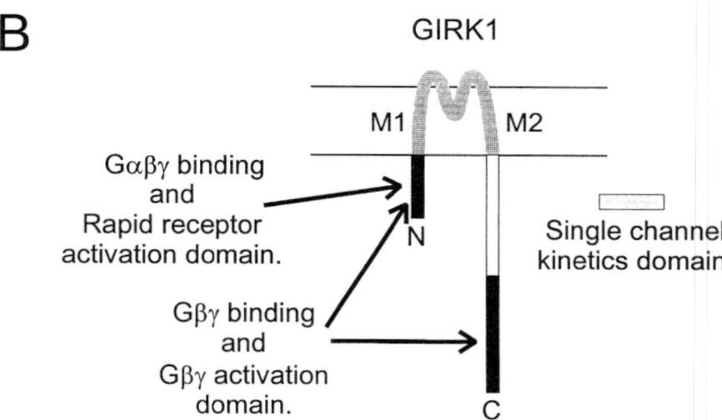

FIGURE 1 Proposed mechanism for G-protein activation of the GIRK channel. (A) Schematic representation of the working model on the activation of GIRK1 channel following muscarinic stimulation. At rest, the muscarinic receptor, inactive G-protein heterotrimer ($G_{\alpha\beta\gamma}$), and the GIRK1 channel may form a local complex which is specified, in part, by the binding interaction of the G protein with the GIRK1 channel. The interaction of the G-protein regulatory protein (RGS) with the G_α accelerates the GDP to GTP exchange and GTPase activity of the G_α–GTP, promoting rapid activation and deactivation at the end of receptor stimulation. (B) Summary of the functional domains implicated in G-protein activation of GIRK. The hydrophobic domain (M1–H5–M2) of GIRK1 contains a gate governing the single-channel openings. The N- and C-terminal hydrophilic domains of GIRK, on the other hand, contain regions important for the $G_{\beta\gamma}$ binding and $G_{\beta\gamma}$ activation of the GIRK channel. The N-terminal domain of GIRK1 also facilitates the rapid activation following receptor stimulation and binds the $G_{\alpha\beta\gamma}$ heterotrimer.

heteromultimers of GIRK1 and GIRK2 have been found in the cortex, hippocampus, and cerebellum of the mammalian brain (Liao *et al.,* 1996). These results suggest that GIRK family members can form functional homo- and heteromultimers except that GIRK1 has only been found to form functional heteromultimers. This feature of GIRK1 has been found to be due to one residue in the putative H5 pore domain (phenylalanine at position 137), which does not allow the formation of functional homomultimeric channels. Mutating this residue to the corresponding amino acid present in all other members of the IR superfamily (phenylalanine 137 to serine) allows the formation of functional homomultimers (Chan *et al.,* 1996).

Splice variants of GIRK1 and GIRK2 channels have been identified in brain tissue. In the case of GIRK2, these splice variants include structures such as PDZ binding domain, which allows the localization and clustering of these channels in specific regions similar to what have been shown for other channels (Sheng and Kim, 1996). Splice variants of GIRK1 are mainly characterized by exon omissions at the C-terminal end of the molecule (Nelson *et al.,* 1997). These splice variants, however, lack the C-terminal putative $G_{\beta\gamma}$ binding domains (see below), and their functions have not been determined yet.

B. GIRK Channel as a Tetramer

Studies of the voltage-gated potassium channels have revealed that the functional molecule is of a tetrameric structure (Liman *et al.,* 1992; MacKinnon, 1991). Because of similarities between voltage-gated potassium channels and inwardly rectifying potassium channels in their pore structure, it is not surprising that inwardly rectifying potassium channels form a similar functional tetrameric assembly (Glowatzki *et al.,* 1995; Silverman *et al.,* 1996; Tucker *et al.,* 1996; Yang *et al.,* 1995; but see Omori *et al.,* 1997). Construction of tandemly linked GIRK1 and GIRK4 channel subunits, in various positions and stoichiometries, has revealed that a heteromultimeric channel is likely to contain two subunits of each GIRK type in an alternating array within the tetramer (Silverman *et al.,* 1996; Tucker *et al.,* 1996). In addition, by using biochemical means, it was found that, in mouse forebrain preparations, antibodies against GIRK1 immunoprecipitated a macroprotein complex composed of GIRK1 and several associated proteins with molecular masses ranging from 50 to 62 kDa on SDS–polyacrylamide gel electrophoresis. On size fractionation by both sucrose density gradient centrifugation and gel filtration, the solubilized GIRK complex channel proteins migrated into a single peak, which suggests that the component subunits form a single macromolecule. On the basis of the values of size fractionation, the molecular mass of the GIRK channel macromolecule could be estimated at approximately 231,000 kDa. These results also suggest that the GIRK channel in the brain is most likely of a tetrameric structure (Inanobe *et al.,* 1995).

III. GIRK CHANNELS ARE ACTIVATED BY $G_{\beta\gamma}$

Following the demonstration that G proteins are responsible for the muscarinic activation of the $I_{K(ACh)}$ channel in atrial myocytes (Breitwieser and Szabo, 1985; Kurachi *et al.,* 1986; Pfaffinger *et al.,* 1985), heated debate concerning the role of the various G-protein subunits responsible for the activation was launched. Several groups were claiming that the activation of $I_{K(ACh)}$ was mediated via G_α–GTP, and other groups were suggesting that $G_{\beta\gamma}$ dimers are responsible for this action (Birnbaumer and Brown, 1987; Codina *et al.,* 1987; Yatani *et al.,* 1987; Ito *et al.,* 1992; Logothetis *et al.,* 1987; 1988). Progress in the biochemistry and molecular biology of G proteins and potassium channels has made possible two major technical advances which allowed further demonstration in support of the conclusion that the $G_{\beta\gamma}$ dimer is responsible for the activation of $I_{K(ACh)}$: (1) Active recombinant $G_{\beta\gamma}$ subunits activate $I_{K(ACh)}$; this activation could not have been due to contamination by active G_α–GTP. (2) The cloning and molecular identification of $I_{K(ACh)}$ has made it possible to heterologously express the cloned $I_{K(ACh)}$ (GIRK) and verify that $G_{\beta\gamma}$ subunits can activate these channels. Wickman *et al.* (1994) generated various recombinant $G_{\beta\gamma}$ subunit isoforms from Sf9 cells and showed that all six combinations tested have an apparent half-maximal activation constant (EC_{50}) of 3–30 n*M* for the native $I_{K(ACh)}$ in atrial myocytes. This important observation demonstrated the lack of specificity among the various $G_{\beta\gamma}$ subunits to activate the channel (with the exception of $G_{\beta1\gamma1}$ which lacks the isoprenylation linkage to the plasma membrane) and raised an important issue of receptor signal sorting and specificity. In addition, the ability of the inactive form of G_α, G_α–GDP, to reverse the $G_{\beta\gamma}$ activation of the channel, by presumably sequestering $G_{\beta\gamma}$, further demonstrated the involvement of $G_{\beta\gamma}$ in the activation process (Logothetis *et al.,* 1988; Reuveny *et al.,* 1994; Wickman *et al.,* 1994).

Using the cloned GIRK1 expressed in *Xenopus* oocytes, it was demonstrated that coexpression of GIRK1 (heteromultimerized with endogenous GIRK5) with $G_{\beta1\gamma2}$ but not $G_{\alpha i2}$ in *Xenopus* oocytes resulted in channel activity that persisted in the absence of cytoplasmic GTP (Reuveny *et al.,* 1994). This activity was reduced by fusion proteins of the β-adrenergic receptor kinase (βARK) C-terminal domain (Reuveny *et al.,* 1994), which is known to bind the $G_{\beta\gamma}$ dimer (Inglese *et al.,* 1994). In native tissue the βARK fragment was also able to inhibit $I_{K(ACh)}$ activity (Nair *et al.,* 1995). This effect of the C-terminal fragment of βARK1 on channel activity was likely due to reduced availability of $G_{\beta\gamma}$ to activate the channel, similar to what has been shown with G_α–GDP. In addition, the GIRK channel was activated by isoproterenol in *Xenopus* oocytes when coexpressed with β_2–adrenergic receptor (β_2–AR) and $G_{\alpha s}$. Since $G_{\alpha s}$ by itself does not activate the channel, it was concluded that $G_{\beta\gamma}$ released from the $G_{\alpha s\beta\gamma}$ complex due to receptor activation causes channel activation (Lim *et al.,* 1995). These results demonstrate the lack of receptor signaling specificity in *Xenopus* oocytes and suggest that the receptor specificity *in vivo* may be due to some unknown mechanisms, such as channel-receptor compartmentalization in specific membrane microdomains (Parton, 1996).

IV. REGIONS INVOLVED IN BINDING AND ACTIVATION OF GIRK CHANNELS

Studies have demonstrated that $G_{\beta\gamma}$ subunits directly bind the GIRK1, GIRK2, GIRK3, and GIRK4 proteins and have identified separate regions of the channel subunits involved in this binding. By using an antibody against the N terminus of GIRK4, Krapivinsky *et al.* (1995b) were able to show that the $G_{\beta\gamma}$ dimer binds to immunoprecipitated cardiac $I_{K(ACh)}$ as well as to recombinant GIRK1 and GIRK4 subunits, with dissociation constants (K_d) of 55, 50, and 125 n*M*, respectively. Huang *et al.* (1995), Kunkel and Peralta (1995), and Inanobe *et al.* (1995) have shown that purified fragments of the C and N termini of GIRK1 directly interact with $G_{\beta\gamma}$ dimers *in vitro.* Huang *et al.* (1997) further narrowed the required sequences for $G_{\beta\gamma}$ binding to one region in the N terminus (residues 34–86) and two regions in the C terminus (318–374 and 390–460) of GIRK1. The authors also found that the C-terminal fragments of other GIRK family members (GIRK2–GIRK4) exhibited weaker binding to $G_{\beta\gamma}$ than the GIRK1 C terminus. The stronger $G_{\beta\gamma}$ binding to the C-terminal domain of GIRK1 was probably due to amino acids 390–462, which are unique to GIRK1. Binding of $G_{\beta\gamma}$ to the N-terminal domains of GIRK1–4 was of similar potency. In addition, there appears to be cooperativity between the N terminus of GIRK1 or GIRK4 and the C-terminal domains of GIRK1 or GIRK4 in terms of $G_{\beta\gamma}$ binding (Huang *et al.,* 1997). These results are consistent with the idea that the N- and C-terminal domains of the cardiac $I_{K(ACh)}$ channel subunits may interact with each other to form a higher affinity binding site(s) for $G_{\beta\gamma}$.

In addition to demonstrating binding of the $G_{\beta\gamma}$ dimer to the N and C termini of GIRK1, Huang *et al.* (1995) showed binding of the trimeric $G_{\alpha\beta\gamma}$ and G_{α}–GDP to the N terminus but not to the C terminus of GIRK1. G_{α}–GDP was able to antagonize $G_{\beta\gamma}$ binding to the C-terminal domain and to the whole solubilized channel protein (Huang *et al.,* 1995; Inanobe *et al.,* 1995; Krapivinsky *et al.,* 1995b). In contrast, G_{α}–GDP together with $G_{\beta\gamma}$ were coimmuoprecipated with a fusion protein of the N terminus of GIRK1. The physiological significance of binding $G_{\beta\gamma}$ and $G_{\alpha\beta\gamma}$ to the N terminus remains to be determined. One possibility may be to anchor the G protein close to the channel macromolecule so as to ensure a rapid and efficient coupling (see below) (Huang *et al.,* 1995; Slesinger *et al.,* 1995).

Doupnik *et al.* (1996) estimated the rate of association and dissociation of $G_{\beta\gamma}$ from the C terminus of GIRK4 (amino acids 186–419) immobilized to the surface of a biosensor chip by measuring the refractive index detected by surface plasmon resonance. The estimated equilibrium dissociation binding constant (K_d) was approximately 800 nM, somewhat higher than that for the observed half-maximal channel activation in atrial myocytes (Krapivinsky *et al.*, 1995b). This discrepancy may be due to the requirement of cooperative interaction of N and C termini for $G_{\beta\gamma}$ binding (Huang *et al.*, 1997).

The $G_{\beta\gamma}$ binding to GIRK1 N- and C-terminal domains as demonstrated in biochemical assays has been shown to be important for channel gating by $G_{\beta\gamma}$ (Huang *et al.*, 1995; Slesinger *et al.*, 1995; Kunkel and Peralta, 1995). Peptides derived from the sequences of the N or the C terminus of GIRK1 partially inhibit both $G_{\beta\gamma}$ binding and channel activation (Huang *et al.*, 1995). By constructing chimeric channels between GIRK1 and IRK1 or IRK2, the latter being G-protein-insensitive inward rectifiers, it was shown that chimeric channels could be activated by $G_{\beta\gamma}$ or m2 muscarinic receptor stimulation if they contained either the N-terminal or part of the C-terminal hydrophilic domain of GIRK1 (amino acids 325–501 in Slesinger *et al.*, 1995, or 290–501 in Kunkel and Peralta, 1995). The ability of GIRK4 to form a G-protein-sensitive heteromultimeric channel with a IRK/GIRK1 chimeric channel that lacks the cytoplasmic domains of GIRK1 suggests that GIRK4 contains structural elements which gate by closing the channel in the absence of G proteins (Slesinger *et al.*, 1995). Qualitatively similar results were obtained using homomeric channel assembly of either GIRK1 (the homomultimer-forming mutant channel GIRK1F137S) or GIRK4, suggesting that the $G_{\beta\gamma}$ gating machinery is within a region that is conserved between GIRK1 and GIRK4 (Vivaudou *et al.*, 1997). A chimera of GIRK1 and IRK1 that contains the C-terminal $G_{\beta\gamma}$ binding domain and the hydrophobic core (M1 through M2) of GIRK1 exhibits 6 and 3.5 times slower activation and deactivation kinetics, respectively, compared to a similar chimera that contains the N-terminal domain of GIRK1 (Slesinger *et al.*, 1995). The rate of activation of heteromultimeric channels that contain the same chimera as well as GIRK4 was also slow (P. A. Slesinger, E. Reuveny, Y. N. Jan, and L Y. Jan, 1996). Thus, correlated with the ability of the N-terminal domain of GIRK1 to bind the $G_{\beta\gamma}$ dimer, the GIRK1 N-terminal domain is necessary for a rapid muscarinic-receptor-mediated activation of the channel.

In light of the correlation between the binding of $G_{\alpha\beta\gamma}$ or G_α–GDP to the N-terminal domain of GIRK1 and the enhancement of channel activation by the same region, one possible scenario is that the binding of the trimeric G protein, $G_{\alpha\beta\gamma}$, to the N-terminal domain of GIRK1 may help localizing the G protein in the vicinity of the channel so that the newly released $G_{\beta\gamma}$ could quickly interact with the channel (Slesinger *et al.,* 1995). For the G proteins to also interact with the receptor molecule, the receptors and the channels may have to be segregated in microdomains or compartments, as was suggested to explain the differences in the agonist sensitivity of $I_{K(ACh)}$ and calcium channels activated by the same receptor (Li *et al.,* 1994). This possibility is supported by the analysis of the mobility of activated G proteins in membranes, which revealed that the movement of G proteins is restricted to the vicinity of effectors in the plasma membrane (Graeser and Neubig, 1993; Kwon *et al.,* 1994; Neubig, 1994; Neubig *et al.,* 1994). The mechanism by which specific channels form local compartments with G-protein-linked receptors is currently not known.

Another aspect of $G_{\beta\gamma}$ gating of the GIRK channel is evident in the *weaver* mutant mouse where a mutation in the pore or H5 region of GIRK2 (Patil *et al.,* 1995) disrupts potassium selective permeation (Kofuji *et al.,* 1996; Navarro *et al.,* 1996; Slesinger *et al.,* 1996). It is interesting for this discussion to note that this mutation in the H5 region also affects $G_{\beta\gamma}$-dependent gating of the GIRK2 homomultimer or GIRK1/GIRK2 heteromultimer (for more details, see Chapter 16) (Navarro *et al.,* 1996), although at low levels of channel expression the *weaver* mutant GIRK2 channels still showed G-protein sensitivity (Slesinger *et al.,* 1996). This observation suggests that either the *weaver* mutation not only affects channel selectivity but also causes a dramatic change in protein folding that renders the channel open and insensitive to $G_{\beta\gamma}$, or that the gating of the GIRK channel is somehow linked to its permeation abilities (e.g., the $G_{\beta\gamma}$ inhibitory gate may include part of the selectivity pore structure).

Multiple $G_{\beta\gamma}$ binding domains have also been found in the P/Q-type or α_{1E} neuronal calcium channels, at the cytoplasmic linker between repeat I and II and at the C-terminal cytoplasmic end of the channel (Qin *et al.,* 1997). A chimera of the G-protein-insensitive calcium channel, L-type, and α_{1E}, where the C-terminal $G_{\beta\gamma}$ binding domain was replaced by the C terminus of the L-type, is no longer able to be modulated by G proteins. In contrast, a chimera with the cytoplasmic linker between repeat I and II replaced with that of the L-type channel was still modulated by $G_{\beta\gamma}$. These results suggest that multiple $G_{\beta\gamma}$ contact sites are used for both activation and enhanced binding to effectors. In the case of the calcium channel, $G_{\beta\gamma}$ is probably competing for the binding site of the auxiliary β subunits that promote faster gating kinetics and better functional expression (Pragnell *et al.,* 1994).

The rates of activation and deactivation of GIRK in heart and brain are more rapid than the hydrolysis rate of G_{α}, suggesting an involvement of other mediators for G-protein function. A family of G-protein regulatory proteins (RGS) have been isolated. These proteins display high specificity for certain G proteins and may enhance their GTP exchange (Doupnik *et al.,* 1997; Saitoh *et al.,* 1997) and hydrolysis rates (Berman *et al.,* 1996; Hunt *et al.,* 1996; Watson *et al.,* 1996). Given the functional implication of the N-terminal domain in the rapid activation and deactivation of GIRK1 and the observation of G_α–GDP binding to the N-terminal domain of GIRK1 (Huang *et al.,* 1995), it may be worth examining the possibility that the N-terminal domain of GIRK1 could provide a physical link between RGS proteins, m2 receptor, and G proteins.

Besides the possible involvement of the N-terminal domain in rapid activation and deactivation of the channel, the compartmentalization of channels, G proteins, the RGS, and receptors may also potentially explain how the specificity of activation is achieved in the heart. Although activation of any G-protein-coupled receptor can theoretically liberate $G_{\beta\gamma}$ subunits and therefore activate $I_{K(ACh)}$, only stimulation of $G_{\alpha i}$-coupled but not $G_{\alpha s}$-coupled receptors activates $I_{K(ACh)}$ in heart or brain (Hille, 1992). This specificity of activation cannot be explained by having different $G_{\beta\gamma}$ subunits linked to $G_{\alpha i}$ and $G_{\alpha s}$ because $I_{K(ACh)}$ is activated by all combinations of $G_{\beta\gamma}$ subunits tested, with less than 10-fold difference in sensitivity being reported (Wickman *et al.,* 1994), and a $G_{\alpha s}$-coupled receptor, β2-AR, was shown to activate the channels in *Xenopus* oocytes (Lim *et al.,* 1995). This hypothesis on the involvement of the N-terminal domain in the formation of a molecular complex to ensure specific and rapid activation remains to be tested. One experiment to suggest a mechanism for acquiring signaling specificity was given by Schreibmayer *et al.* (1996), who showed that the activation of a GIRK channel [or IKACh)] by $G_{\beta\gamma}$ can be inhibited by the activated $G_{\alpha i1}$–GTPγS or to a lesser extent $G_{\alpha s}$–GTPγS, but not by $G_{\alpha i2}$ or $G_{\alpha i3}$, at picomolar concentrations (Schreibmayer *et al.,* 1996). According to this evidence, the authors suggested that a new mechanism may be involved in generating signaling specificity by which activation of a receptor that should not activate the channel activates a specific G_α that is able to antagonize the action of its own released $G_{\beta\gamma}$. This hypothesis is rather intriguing but needs to be further tested.

V. $G_{\beta\gamma}$ AND SINGLE-CHANNEL KINETICS

Analysis of the bursting behavior of GIRK channels from bullfrog atrial myocytes by Ivanova-Nikolova and Breitwieser (1997) revealed a mechanism by which $G_{\beta\gamma}$ affects the gating kinetics of a single channel. In the presence of various concentrations of acetylcholine, although opening and closing rates within bursts were altered, no change in single mean burst duration was detected. This suggest that $G_{\beta\gamma}$ binding is permissive for channel bursting but that an intrinsic gating mechanism defines the burst duration. The acetylcholine dose-dependent increase in channel open probability was found by Ivanova-Nikolova and Breitwieser (1997) to be the consequence of a pronounced shift between three channel modes: low, intermediate, and high open channel frequency modes. The relative abundance of these three modes depends on the concentration of acetylcholine (or free $G_{\beta\gamma}$), favoring the longer opening mode with increasing stimulation strength. In all three modes channel openings within a particular mode are composed of three similar open times. These dose-dependent changes in channel modal gating were highly correlated with the acetylcholine dose response of the macroscopic whole cell current. In addition, it was shown that termination of bursting was independent of stimulation strength and had no correlation to any particular channel mode. Thus, Ivanova-Nikolova and Breitwieser (1997) proposed that burst termination is due to channel switching into a $G_{\beta\gamma}$-independent inactivated or desensitized state. This inactivation process is reversible, but no information is available on whether it can be initiated from the open or closed conformation of the channel.

From single-channel recording of the $G_{\beta\gamma}$- or GTPγS-activated GIRK channels in inside-out patches, Luchian *et al.* (1997) were able to demonstrate that the burst duration was shortened by a peptide corresponding to the C-terminal end (T482–D498, IC_{50} of 1.7 μM) but not peptides derived from the N-terminal end or other more upstream parts of the C terminus of GIRK1. This C-terminal peptide did not change intraburst mean open or closed times and did not change the half-maximal effective concentration of $G_{\beta\gamma}$ to activate the channel. The C-terminal peptide-induced reduction in burst duration has a Hill coefficient of 1, suggesting that a single peptide molecule is sufficient to produce this inhibition, and from the shallow voltage dependence of this "blocking" effect it seems that this gating particle does not interact with its receptor deep within the transmembrane electric field (i.e., the receptor is likely to be close to the cytoplasmic mouth of the channel). An open channel blocking mechanism by the C-terminal peptide cannot be excluded at this time, and further detailed analysis is required. However in light of the data presented by Ivanova-Nikolova and Breitwieser (1997), stable channel closures by this C-terminal peptide can also be viewed as a channel intrinsic mechanism to promote inactivation in the presence of $G_{\beta\gamma}$. This interpretation suggests that termination of an active burst may be due, in part, to an interaction of the C-terminal peptide with a receptor on the channel molecule that does not share an interaction site with $G_{\beta\gamma}$.

No information is yet available on the exact location and/or the molecular determinants of the receptor for this gating particle. However, some evidence suggests that the hydrophobic core of the channel plays a role in this process. Chimeric channels between IRK1 and GIRK1 revealed that the hydrophobic core of GIRK1 (M1–H5–M2) is important for determining the brief single-channel bursting behavior, typical of GIRK1, but not important for determining $G_{\beta\gamma}$ sensitivity (Slesinger *et al.*, 1995).

References

Berman, D. M., Wilkie, T. M., and Gilman, A. G. (1996). GAIP and RGS4 are GTPase-activating proteins for the G_i subfamily of G protein α subfamily. *Cell (Cambridge, Mass.)* **86,** 445–452.

Birnbaumer, L., and Brown, A. M. (1987). G protein opening of K^+ channels [letter]. *Nature (London)* **327,** 21–22.

Breitwieser, G. E., and Szabo, G. (1985). Uncoupling of cardiac muscarinic and β-adrenergic receptors from ion channels by a guanine nucleotide analogue. *Nature (London)* **317,** 538–540.

Chan, K. W., Sui, J. L., Vivaudou, M., and Logothetis, D. E. (1996). Control of channel activity through a unique amino acid residue of a G protein-gated inwardly rectifying K^+ channel subunit. *Proc. Natl. Acad. Sci. U.S.A.* **93,** 14193–14198.

Codina, J., Yatani, A., Grenet, D., Brown, A. M., and Birnbaumer, L. (1987). The α subunit of the GTP binding protein G_k opens atrial potassium channels. *Science* **236,** 442–445.

Dascal, N., Schreibmayer, W., Lim, N. F., Wang, W., Chavkin, C., DiMagno, L., Labarca, C., Kieffer, B. L., Gaveriaux-Ruff, C., Trollinger, D., Lester, H. A., and Davidson, N. (1993). Atrial G protein-activated K^+ channel: Expression cloning and molecular properties. *Proc. Natl. Acad. Sci. U.S.A.* **90,** 10235–10239.

Doupnik, C. A., Davidson, N., and Lester, H. A. (1995). The inward rectifier potassium channel family. *Curr. Opin. Neurobiol.* **5,** 268–277.

Doupnik, C. A., Dessauer, C. W., Slepak, V. Z., Gilman, A. G., Davidson, N., and Lester, H. A. (1996). Time resolved kinetics of direct $G_{\beta1\gamma2}$ interactions with the carboxyl terminus of Kir3.4 inward rectifier K^+ channel subunits. *Neuropharmacology* **35,** 923–931.

Doupnik, C. A., Davidson, N., Lester, H. A., and Kofuji, P. (1997). RGS proteins reconstitute the rapid gating kinetics of $G_{\beta\gamma}$-activated inwardly rectifying K^+ channels. *Proc. Natl. Acad. Sci. U.S.A.* **94,** 10461–10466.

Drake, C. T., Bausch, S. B., Milner, T. A., and Chavkin, C. (1997). Girk1 immunoreactivity is present predominantly in dendrites, dendritic spines, and somata in the CA1 region of the hippocampus. *Proc. Natl. Acad. Sci. U.S.A.* **94,** 1007–1012.

Duprat, F., Lesage, F., Guillemare, E., Fink, M., Hugnot, J. P., Bigay, J., Lazdunski, M., Romey, G., and Barhanin, J. (1995). Heterologous multimeric assembly is essential for K^+ channel activity of neuronal and cardiac G-protein-activated inward rectifiers. *Biochem. Biophys. Res. Commun.* **212,** 657–663.

Ferrer, J., Nichols, C. G., Makhina, E. N., Salkoff, L., Bernstein, J., Gerhard, D., Wasson, J., Ramanadham, S., and Permutt, A. (1995). Pancreatic islet cells express a family of inwardly rectifying K^+ channel subunits which interact to form G-protein-activated channels. *J. Biol. Chem.* **270,** 26086–26091.

Glowatzki, E., Fakler, G., Brandle, U., Rexhausen, U., Zenner, H. P., Ruppersberg, J. P., and Fakler, B. (1995). Subunit-dependent assembly of inward-rectifier K^+ channels. *Proc. R. Soc. London B* **261,** 251–261.

Graeser, D., and Neubig, R. R. (1993). Compartmentation of receptors and guanine nucleotide-binding proteins in NG108–15 cells: Lack of cross-talk in agonist binding among the α_2-adrenergic, muscarinic, and opiate receptors. *Mol. Pharmacol.* **43,** 434–443.

Hedin, K. E., Lim, N. F., and Clapham, D. E. (1996). Cloning of a *Xenopus laevis* inwardly rectifying K^+ channel subunit that permits GIRK1 expression of I_{KACh} currents in oocytes. *Neuron* **16,** 423–429.

Hille, B. (1992). G protein-coupled mechanisms and nervous signaling. *Neuron* **9,** 187–195.

Huang, C. L., Slesinger, P. A., Casey, P. J., Jan, Y. N., and Jan, L. Y. (1995). Evidence that direct binding of $G\beta\gamma$ to the GIRK1 G protein-gated inwardly rectifying K^+ channel is important for channel activation. *Neuron* **15,** 1133–1143.

Huang, C. L., Jan, Y. N., and Jan, L. Y. (1997). Binding of the G protein $\beta\gamma$ subunit to multiple regions of G protein-gated inward-rectifying K^+ channels. *FEBS Lett.* **405,** 291–298.

Hunt, T. W., Fields, T., Casey, P. J., and Peralta, E. G. (1996). RGS10 is a selective activator of $G\alpha i$ GTPase activity. *Nature* (*London*) **383,** 175–177.

Inanobe, A., Morishige, K. I., Takahashi, N., Ito, H., Yamada, M., Takumi, T., Nishina, H., Takahashi, K., Kanaho, Y., Katada, T., *et al.* (1995). $G_{\beta\gamma}$ directly binds to the carboxyl terminus of the G protein-gated muscarinic K^+ channel, GIRK1. *Biochem. Biophys. Res. Commun.* **212,** 1022–1028.

Inglese, J., Luttrell, L. M., Iniguez-Lluhi, J. A., Touhara, K., Koch, W. J., and Lefkowitz, R. J. (1994). Functionally active targeting domain of the β-adrenergic receptor kinase: An inhibitor of G $\beta\gamma$-mediated stimulation of type II adenylyl cyclase. *Proc. Natl. Acad. Sci. U.S.A.* **91,** 3637–3641.

Ito, H., Tung, R. T., Sugimoto, T., Kobayashi, I., Takahashi, K., Katada, T., Ui, M., and Kurachi, Y. (1992). On the mechanism of G protein $\beta\gamma$ subunit activation of the muscarinic K^+ channel in guinea pig atrial cell membrane. Comparison with the ATP-sensitive K^+ channel. *J. Gen. Physiol.* **99,** 961–983.

Ivanova-Nikolova, T. T., and Breitwieser, G. E. (1997). Effector contributions to $G_{\beta\gamma}$-mediated signaling as revealed by muscarinic potassium channel gating. *J. Gen. Physiol.* **109,** 245–253.

Karschin, C., Dissmann, E., Stuhmer, W., and Karschin, A. (1996). IRK(1–3) and GIRK(1–4) inwardly rectifying K^+ channel mRNAs are differentially expressed in the adult rat brain. *J. Neurosci.* **16,** 3559–3570.

Kofuji, P., Davidson, N., and Lester, H. A. (1995). Evidence that neuronal G-protein-gated inwardly rectifying K^+ channels are activated by $G_{\beta\gamma}$ subunits and function as heteromultimers. *Proc. Natl. Acad. Sci. U.S.A.* **92,** 6542–6546.

Kofuji, P., Hofer, M., Millen, K. J., Millonig, J. H., Davidson, N., Lester, H. A., and Hatten, M. E. (1996). Functional analysis of the weaver mutant GIRK2 K+ channel and rescue of weaver granule cells. *Neuron* **16,** 941–952.

Krapivinsky, G., Gordon, E. A., Wickman, K., Velimirovic, B., Krapivinsky, L., and Clapham, D. E. (1995a). The G-protein-gated atrial K^+ channel I_{KACh} is a heteromultimer of two inwardly rectifying K^+-channel proteins. *Nature* (*London*) **374,** 135–141.

Krapivinsky, G , Krapivinsky, L., Wickman, K., and Clapham, D. E. (1995b). $G_{\beta\gamma}$ binds directly to the G protein-gated K^+ channel, IKACh. *J. Biol. Chem.* **270,** 29059–29062.

Kubo, Y., Baldwin, T. J., Jan, Y. N., and Jan, L. Y. (1993a). Primary structure and functional expression of a mouse inward rectifier potassium channel. *Nature* (*London*) **362,** 127–133.

Kubo, Y., Reuveny, E., Slesinger, P. A., Jan, Y. N., and Jan, L. Y. (1993b). Primary structure and functional expression of a rat G-protein-coupled muscarinic potassium channel. *Nature* (*London*) **364,** 802–806.

Kunkel, M. T., and Peralta, E. G. (1995). Identification of domains conferring G protein regulation on inward rectifier potassium channels. *Cell* **83,** 443–449.

Kurachi, Y., Nakajima, T., and Sugimoto, T. (1986). On the mechanism of activation of the muscarinic K^+ channels by adenosine in isolated atrial cells: Involvement of GTP-binding proteins. *Pfluegers Arch.* **407,** 264–274.

Kwon, G., Axelrod, D., and Neubig, R. R. (1994). Lateral mobility of tetramethylrhodamine (TMR) labelled G protein α and $\beta\gamma$ subunits in NG 108–15 cells. *Cell. Signalling* **6,** 663–679.

Lesage, F., Duprat, F., Fink, M., Guillemare, E., Coppola, T., Lazdunski, M., and Hugnot, J. P. (1994). Cloning provides evidence for a family of inward rectifier and G-protein coupled K^+ channels in the brain. *FEBS Lett.* **353,** 37–42.

Lesage, F., Guillemare, E., Fink, M., Duprat, F., Heurteaux, C., Fosset, M., Romey, G., Barhanin, J., and Lazdunski, M. (1995). Molecular properties of neuronal G-protein-activated inwardly rectifying K^+ channels. *J. Biol. Chem.* **270,** 28660–28667.

Li, Y., Hanf, R., Otery, A. S., Fischmeister, R., and Szabo, G. (1994). Differential effects of pertussis toxin on the muscarinic regulation of Ca^{2+} and K^+ currents in frog cardiac myocytes. *J. Gen. Physiol.* **104,** 941–959.

Liao, Y. J., Jan, Y. N., and Jan, L. Y. (1996). Heteromultimerization of G-protein-gated inwardly rectifying K^+ channel proteins GIRK1 and GIRK2 and their altered expression in *weaver* brain. *J. Neurosci.* **16,** 7137–7150.

Lim, N. F., Dascal, N., Labarca, C., Davidson, N., and Lester, H. A. (1995). A G protein-gated K channel is activated via β_2-adrenergic receptors and $G_{\beta\gamma}$ subunits in *Xenopus* oocytes. *J. Gen. Physiol.* **105,** 421–439.

Liman, E. R., Tytgat, J., and Hess, P. (1992). Subunit stoichiometry of a mammalian K^+ channel determined by construction of multimeric cDNAs. *Neuron* **9,** 861–871.

Loewi, O. (1921). Über humorale übertragbarkeit der herznervenwirkung. *Pfluegers Arch.* **189,** 239–242.

Logothetis, D. E., Kurachi, Y., Galper, J., Neer, E. J., and Clapham, D. E. (1987). The $\beta\gamma$ subunits of GTP-binding proteins activate the muscarinic K^+ channel in heart. *Nature* (*London*) **325,** 321–326.

Logothetis, D. E., Kim, D. H., Northup, J. K., Neer, E. J., and Clapham, D. E. (1988). Specificity of action of guanine nucleotide-binding regulatory protein subunits on the cardiac muscarinic K^+ channel. *Proc. Natl. Acad. Sci. U.S.A.* **85,** 5814–5818.

Luchian, T., Dascal, N., Dessauer, C., Platzer, D., Davidson, N., Lester, H. A., and Schreibmayer, W. (1997). A C-terminal peptide of the GIRK1 subunit directly blocks the G protein-activated K^+ channel (GIRK) expressed in Xenopus oocytes. *J. Physiol.* (*London*) **505,** 13–22.

Lüscher, C., Jan, L. Y., Stoffer, M., Malenka, R. C., and Nicoll, R. A. (1997). G protein-coupled inwardly rectifying K^+ channels (GIRKs) mediate postsynaptic but not presynaptic transmitter actions in hyppocampal neurons. *Neuron* **19,** 687–695.

MacKinnon, R. (1991). Determination of the subunit stoichiometry of a voltage-activated potassium channel. *Nature* (*London*) **350,** 232–235.

Nair, L. A., Inglese, J., Stoffel, R., Koch, W. J., Lefkowitz, R. J., Kwatra, M. M., and Grant, A. O. (1995). Cardiac muscarinic potassium channel activity is attenuated by inhibitors of $G_{\beta\gamma}$. *Circ. Res.* **76,** 832–838.

Navarro, B., Kennedy, M. E., Velimirovic, B., Bhat, D., Peterson, A. S., and Clapham, D. E. (1996). Nonselective and $G_{\beta\gamma}$-insensitive *weaver* K^+ channels. *Science* **272,** 1950–1953.

Nelson, C. S., Marino, J. L., and Allen, C. N. (1997). Cloning and characterization of Kir3.1 (GIRK1) C-terminal alternative splice variants. *Brain Res. Mol. Brain Res.* **46,** 185–196.

Neubig, R. R. (1994). Membrane organization in G-protein mechanisms. *FASEB J.* **8,** 939–946.

Neubig, R. R., Connolly, M. P., and Remmers, A. E. (1994). Rapid kinetics of G protein subunit association: A rate-limiting conformational change? *FEBS Lett.* **355,** 251–253.

Omori, K., Oishi, K., and Matsuda, H. (1997). Inwardly rectifying potassium channels expressed by gene transfection into the green monkey kidney cell line COS-1. *J. Physiol.* (*London*) **499,** 369–378.

Parton, R. G. (1996). Caveole and caveolins. *Curr. Opin. Cell Biol.* **8,** 542–548.

Patil, N., Cox, D. R., Bhat, D., Faham, M., Myers, R. M., and Peterson, A. S. (1995). A potassium channel mutation in *weaver* mice implicates membrane excitability in granule cell differentiation [see comments]. *Nat. Genet.* **11,** 126–129.

Pfaffinger, P. J., Martin, J. M., Hunter, D. D., Nathanson, N. M., and Hille, B. (1985). GTP-binding proteins couple cardiac muscarinic receptors to a K channel. *Nature (London)* **317,** 536–538.

Ponce, A., Bueno, E., Kentros, C., Vega-Saenz de Miera, E., Chow, A., Hillman, D., Chen, S., Zhu, L., Wu, M. B., Wu, X., Rudy, B., and Thornhill, W. B. (1996). G-protein-gated inward rectifier K^+ channel proteins (GIRK1) are present in the soma and dendrites as well as in nerve terminals of specific neurons in the brain. *J. Neurosci.* **16,** 1990–2001.

Pragnell, M., De Waard, M., Mori, Y., Tanabe, T., Snutch, T. P., and Campbell, K. P. (1994). Calcium channel β-subunit binds to a conserved motif in the I–II cytoplasmic linker of the α_1–subunit [see comments]. *Nature (London)* **368,** 67–70.

Qin, N., Platano, D., Olcese, R., Stefani, E., and Birnbaumer, L. (1997). Direct interaction of $G_{\beta\gamma}$ with a C-terminal $G_{\beta\gamma}$-binding domain of the Ca^{2+} channel α_1 subunit is responsible for channel inhibition by G protein-coupled receptors. *Proc. Natl. Acad. Sci. U.S.A.* **94,** 8866–8871.

Reuveny, E., Slesinger, P. A., Inglese, J., Morales, J. M., Iniguez-Lluhi, J. A., Lefkowitz, R. J., Bourne, H. R., Jan, Y. N., and Jan, L. Y. (1994). Activation of the cloned muscarinic potassium channel by G protein $G_{\beta\gamma}$ subunits. *Nature (London)* **370,** 143–146.

Saitoh, O., Kubo, Y., Miyatani, Y., Asano, T., and Nakata, H. (1997). RGS8 accelerates G-protein mediated modulation of K^+ currents. *Nature (London)* **390,** 525–529.

Schreibmayer, W., Dessauer, C. W., Vorobiov, D., Gilman, A. G., Lester, H. A., Davidson, N., and Dascal, N. (1996). Inhibition of an inwardly rectifying K^+ channel by G-protein α-subunits. *Nature (London)* **380,** 624–627.

Sheng, M., and Kim, E. (1996). Ion channel associated proteins. *Curr. Opin. Neurobiol.* **6,** 602–608.

Silverman, S. K., Lester, H. A., and Dougherty, D. A. (1996). Subunit stoichiometry of a heteromultimeric G protein-coupled inward-rectifier K^+ channel. *J. Biol. Chem.* **271,** 30524–30528.

Slesinger, P. A., Reuveny, E., Jan, Y. N., and Jan, L. Y. (1995). Identification of structural elements involved in G protein gating of the GIRK1 potassium channel. *Neuron* **15,** 1145–1156.

Slesinger, P. A., Patil, N., Liao, Y. J., Jan, Y. N., Jan, L. Y., and Cox, D. R. (1996). Functional effects of the mouse *weaver* mutation on G protein-gated inwardly rectifying K^+ channels. *Neuron* **16,** 321–331.

Spauschus, A., Lentes, K. U., Wischmeyer, E., Dissmann, E., Karschin, C., and Karschin, A. (1996). A G-protein-activated inwardly rectifying K^+ channel (GIRK4) from human hippocampus associates with other GIRK channels. *J. Neurosci.* **16,** 930–938.

Trautwein, W., and Dudel, J. (1958). Zum mechanismus der membranwirkung des acetylcholin an der herzmuskelfaser. *Pfluegers Arch.* **266,** 324–334.

Tucker, S. J., Pessia, M., and Adelman, J. P. (1996). Muscarine-gated K^+ channel: Subunit stoichiometry and structural domains essential for G protein stimulation. *Am. J. Physiol.* **271,** H379–H385.

Velimirovic, B. M., Gordon, E. A., Lim, N. F., Navarro, B., and Clapham, D. E. (1996). A neuronal G protein-gated K channel is a multimer of inward rectifier channel subunits. *FEBS Lett.* **379,** 31–37.

Vivaudou, M., Chan, K. W., Sui, J.-L., Jan, L. Y., Reuveny, E., and Logothetis, D. E. (1997). Probing the G-protein regulation of GIRK1 and GIRK4, the two subunits of the KACh channel, using homomeric mutants. *J. Biol. Chem.* **272,** 31553–31560.

Watson, N., Linder, M. E., Druey, K. M., Kehrl, J. H., and Blumer, K. J. (1996). RGS family members: GTPase activating proteins for heterotrimeric G-protein α-subunit. *Nature* (*London*) **383,** 172–175.

Wickman, K. D., Iniguez-Lluhl, J. A., Davenport, P. A., Taussig, R., Krapivinsky, G. B., Linder, M. E., Gilman, A. G., and Clapham, D. E. (1994). Recombinant G-protein $\beta\gamma$-subunits activate the muscarinic-gated atrial potassium channel. *Nature* (*London*) **368,** 255–257.

Yang, J., Jan, Y. N., and Jan, L. Y. (1995). Determination of the subunit stoichiometry of an inwardly rectifying potassium channel. *Neuron* **15,** 1441–1447.

Yatani, A., Codina, J., Brown, A. M., and Birnbaumer, L. (1987). Direct activation of mammalian atrial muscarinic potassium channels by GTP regulatory protein G_k. *Science* **235,** 207–211.

CHAPTER 18

Effect of Phosphatidylinositol Phosphates on the Gating of G-Protein-Activated K^+ Channels

Jin Liang Sui,[1] Jérôme Petit-Jacques, and Diomedes E. Logothetis
Department of Physiology and Biophysics, Mount Sinai School of Medicine, City University of New York, New York 10029

I. INTRODUCTION

Regulation of protein activity dependent on hydrolysis of cytoplasmic ATP proceeds via protein or lipid kinases. The role of phosphatidylinositol phosphates as membrane-delimited second messengers has begun to receive its overdue attention (for review see Ref. 1). Phosphatidylinositol 4,5-bisphosphate (PIP_2) has been shown to regulate the activity of native and recombinant inwardly rectifying K^+ channels (2–5), the inositol trisphosphate (IP_3) receptor (6), and transporters such as the sodium–calcium exchanger (2). Moreover, anionic phospholipids, and PIP_2 in particular, have been shown to directly bind to proteins as diverse as phospholipases,

[1] Present address: Cambridge Neuroscience, Cambridge, Massachusetts 02139.

1063-5823/99 $30.00

kinases, cytoskeletal proteins, and channel proteins, but the functional importance of such interactions is only now starting to be appreciated.

Muscarinic potassium (K_{ACh}) channels are targets of parasympathetic control in the heart. Acetylcholine released upon vagal stimulation, binds m2 receptors in pacemaker and atrial cells causing an increase in K_{ACh} current and a concomitant decrease in heart rate (7). K_{ACh} provided the first example of a "membrane delimited pathway," where an extracellular signal is transduced sequentially to a final target protein within the plasma membrane, not proceeding through cytoplasmic second messengers (8). $G_{\beta\gamma}$ subunits activate native atrial (K_{ACh}) channels (9) as well as recombinant G-protein-gated inwardly rectifying K^+ (GIRK) channels (10, 11), when applied to inside-out patches. K_{ACh} channels are heterotetramers comprising two each of the component subunits (GIRK1/GIRK4) (10, 12). Binding studies have revealed similar affinities of the native or recombinant channels for $G_{\beta\gamma}$ subunits, confirming the notion of the direct interactions of these proteins (13–18). Intracellular GTP-mediated activation of K_{ACh} (in the presence of external ACh) is also dependent on ATP, as it becomes progressively ineffective (runs down) when internal ATP is removed (5).

Sodium ions have been shown to regulate K_{ACh} channel activity (19). In the absence of G-protein signaling, magnesium-dependent hydrolysis of ATP alters the K_{ACh} residence time in the open state and sensitizes the channel to gating by intracellular Na^+ ions (19). Moreover, channel gating by internal Na^+ does not depend on G protein gating, as $G_{\beta\gamma}$ binding proteins (20), did not interfere with channel activation by Na^+ ions (21).

The work reviewed here addresses the molecular mechanism by which ATP hydrolysis affects K channel gating by the $G_{\beta\gamma}$ subunits or intracellular Na^+ ions. The data support the hypothesis that hydrolysis of ATP generates phosphoinositol phosphates, which in turn regulate K_{ACh} activity directly. These studies reveal a clear functional dependence of G-protein signaling on phosphatidylinositol phospholipids. In light of the dynamic state of phosphoinositol phosphates in plasma membranes, an additional level of regulation may control the function of this channel and its influence on the electrical activity of the heart.

II. MATERIALS AND METHODS

General chemicals were purchased from Sigma (St. Louis, MO), including GTP, ATP, and catalytic subunit of protein kinase A (PKA). Phosphatidylinositol phosphates (PIP) and PIP_2 were from Boehringer Mannheim (Indianapolis, IN). PIP_2 was sonicated on ice for more than 30 min before application. PIP and PIP_2 antibodies were from PerSeptive Biosystems (Framingham, MA). PI-dependent phospholipase C (PI-PLC) was purchased from Boehringer Mannheim and was diluted in Mg^{2+}-free high potassium bath solution with 0.5 mM $MgCl_2$ added (since a high concentration of Mg^{2+} inhibits enzyme activation). PLC-β2 purified protein was provided by Dr. Iyengar (Mount Sinai School of Medicine). The stock of PLC-β2 (0.3 μg/μl) was dissolved in the following (mM): HEPES 10, KCl 100, NaCl 10, EGTA 2, $MgCl_2$ 1, dithiothreitol (DTT) 1, $CaCl_2$ 1.25, pH 7.2. This stock solution was dissolved 60-fold at a final PLC-β2 concentration of 5 μg/ml. Okadaic acid was purchased from RBI (Natick, MA). H-7 was purchased form Calbiochem (San Diego, CA). GIRK4 antibody was purchased from Upstate Biotechnology (Lake Placid, NY). GIRK1 antibodies were provided by Dr. Thornhill (Mount Sinai School of Medicine) and used as previously described (38).

A. Preparation of Chick Atrial Myocytes

The procedure used for isolating cardiac myocytes from chick embrya has been described previously (19). Briefly, atrial tissue was selected from chick embrya from eggs incubated 14–19 days. Atrial tissue (from 4–6 eggs) was incubated for 20–30 min at 37°C in a 15-ml Falcon tube containing 5 ml of Mg^{2+}- and Ca^{2+}-free Hanks' solution supplemented with 1–2% of trypsin-EDTA solution (10×, Gibco). Isolated myocytes were collected by triturating the digested tissue in 5 ml of trypsin-free solution and stored at 4°C for up to 36 hr. The cells were allowed to settle on polylysine-coated coverslips in the recording chamber before recording.

B. Expression of Recombinant Channels in Xenopus Oocytes

Recombinant channel subunits (GIRK1, Genbank Accession No. U39196; GIRK4, Accession No. U39195) were expressed in *Xenopus* oocytes as previously described (22). Briefly, channel subunit coexpression was accomplished by coinjection of equal amounts of each cRNA (~4 ng), with hm2 receptor (~1 ng). cRNA concentrations were estimated from two successive dilutions which were electrophoresed in parallel on formaldehyde gels and compared to known concentrations of RNA marker (GIBCO, Grand Island, NY). Oocytes were isolated and microinjected as previously described (23). All the oocytes were maintained at 18°C, and electrophysiological recordings were performed 2–6 days following injection.

C. Single-Channel Recording and Analysis

Single-channel activity was recorded in the cell-attached or inside-out patch configurations (24, 25) using an Axopatch 200A amplifier (Axon

Instruments, Foster City, CA). All microelectrodes used in the experiments were pulled from WPI-K borosilicate glass (WPI, Sarasota, FL) and gave resistances of 3–8 MΩ. All experiments were performed at room temperature (20–22°C). Single-channel recordings were performed at a membrane potential of −80 mV and in the absence of agonist in the pipette, unless otherwise indicated. Single-channel currents were filtered at 1–2 kHz with a 6-pole low-pass Bessel filter, sampled at 5–10 kHz through the DIGIDATA 1200 interface (Axon Instruments). pCLAMP software (Version 6.01, Axon Instruments) was used for data acquisition.

Xenopus oocytes, were placed in a hypertonic solution (26) for 5–10 min. Shrunk oocytes were first transferred into a V-shaped recording chamber and the vitelline membrane was partially removed, exposing just enough plasma membrane for access with a patch pipette. This procedure increased the success rate of forming gigaseals and reduced loss of the gigaohm seal in long-term cell-attached recordings. The pipette solution contained the following (in m*M*): KCl 96, $CaCl_2$ 1.8, $MgCl_2$ 1, and HEPES 10, pH 7.35. The bath solution contained the following (in m*M*): KCl 96, EGTA 5, and HEPES 10, pH 7.35. In addition, 100 μM gadolinium was routinely added to the pipette solution to suppress native stretch channel activity in the oocyte membrane.

For chick atrial cells the experimental solutions were the same as those used with oocyte recordings with the exceptions of using 140 m*M* KCl (for mammalian cells) instead of 96 m*M* KCI (for *Xenopus* oocytes) and omitting gadolinium since native atrial stretch channels did not present a problem.

Free Mg^{2+} and ATP concentrations were estimated as previously described (27). We generally applied purified G-protein subunits to inside-out patches within 1–2 min following patch excision or withdrawal of MgATP, unless otherwise indicated. Under these conditions activation commenced within 1 min and reached steady-state levels within 2–3 min of application.

Single-channel records were analyzed using pCLAMP software, complemented with our own analysis routine, as described previously (19). Parameters used for single-channel analysis include activity of all the channels in the patch (or the total open probability, NPo), total opening frequency (NFo), and the mean open time (MTo). Results are displayed as averages over 5-sec bins.

III. RESULTS

In our mechanistic studies on the regulation of K_{ACh} activity, we have made extensive use of a model system (expression of recombinant protein in *Xenopus* oocytes) and compared the results to a native system, acutely dissociated chick embryonic atrial cells. Figure 1 shows the biophysical compatibility of the model and native systems, by comparing the channel activity of recombinant GIRK1 and GIRK4 subunits (Fig. 1A) to that of native chick embryonic atrial cells (K_{ACh}; Fig. 1B). One can qualitatively appreciate the similarity in inward rectification properties (lack of channel activity at positive voltages as compared to corresponding negative voltages when the equilibrium potential is chosen to be at 0 mV), single-channel conductance, and kinetics.

Figure 2 plots channel activity (NPo) of GIRK channels, the frequency of channel opening (NFo) and the mean open time (MTo) as a function of time in the experiment. As shown Na^+ ions alone or together with Mg^{2+} ions did not cause channel activation. In contrast, when ATP was added the mean open time, channel opening frequency, and activity all increased with similar kinetics. We have shown previously that hydrolysis of ATP alone can selectively modify the mean open time, with no significant change in channel opening frequency or activity. Following such an ATP-dependent modification, Na^+ ions can stimulate channel opening frequency, which dramatically increases activity (19). Therefore, we interpret the similar

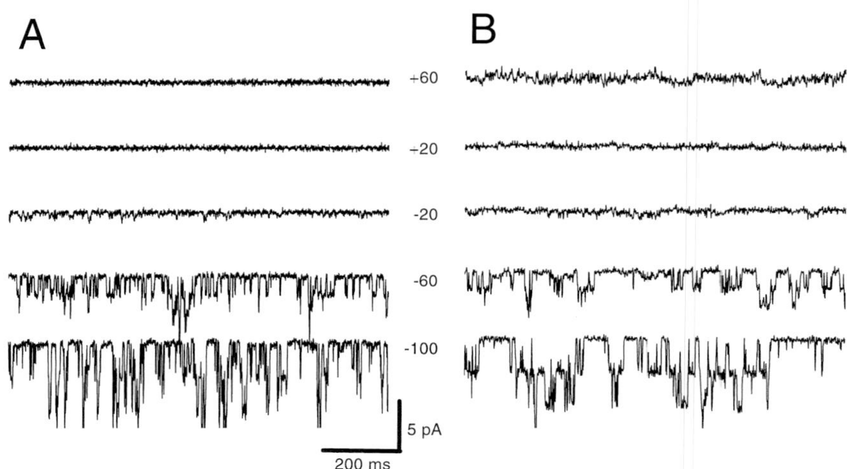

FIGURE 1 Comparison of single-channel records from recombinant versus native atrial G protein-gated K channels. (A) Cell-attached recording from an oocyte expressing recombinant GIRK1/GIRK4 channels, reflecting basal channel activity at voltages between −100 and +60 mV. (B) Cell-attached recording from a chick embryonic atrial cell, reflecting agonist-induced activity (5 μM_{ACh} in the pipette) at voltages between −100 and +60 mV.

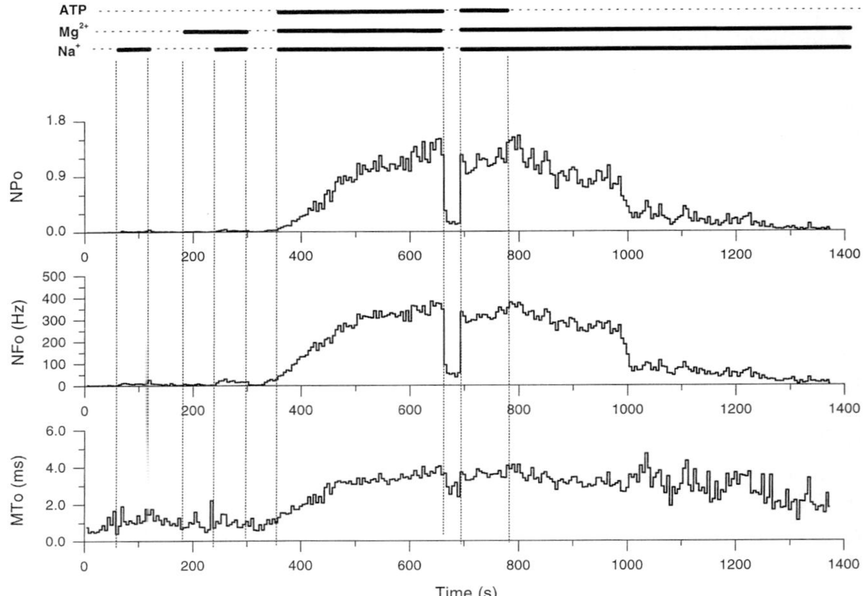

FIGURE 2 Effects of Mg^{2+}, Na^{+}, and ATP on activation of recombinant GIRK channels, as shown by activity (NPo), opening frequency (NFo) of N single channels in the patch, and mean open-time (MTo) plots from an inside-out patch. Perfusion of the patch with Na^{+} ions alone or together with Mg^{2+} failed to stimulate channel activity. In the copresence of ATP, however, Na^{+} and Mg^{2+} ions increased mean open time, opening frequency, and consequently activity with similar kinetics. Reversal of activity, on withdrawal of ATP, proceeded with kinetics similar to those of activation. Applications of Na^{+}, Mg^{2+}, and ATP are indicated by the bars. The membrane potential was clamped at −80 mV in symmetrical K^{+} solutions (96/96 m*M*). Mean values from recordings of consecutive 5-sec bins were plotted in each histogram.

activation kinetics in all three parameters, shown in the experiment of Fig. 2, to reflect open state modification by hydrolysis of ATP with a concurrent stimulation of channel opening frequency and activity by the copresence of Na^{+} ions. Withdrawal of ATP causes a slow reversal of all three parameters with kinetics comparable to those of the activation process.

Figure 3 shows that when GTP is applied at the intracellular side of an atrial inside-out patch, in the presence of ACh in the pipette, it activates K_{ACh} (e.g., Ref. 9). In the prolonged absence of internal ATP, however, the GTP-induced activation is transient, showing rundown characteristics with variable kinetics (5). In chick embryonic atrial patches, brief withdrawal of GTP resulted in an additional rapid decrease in the mean open

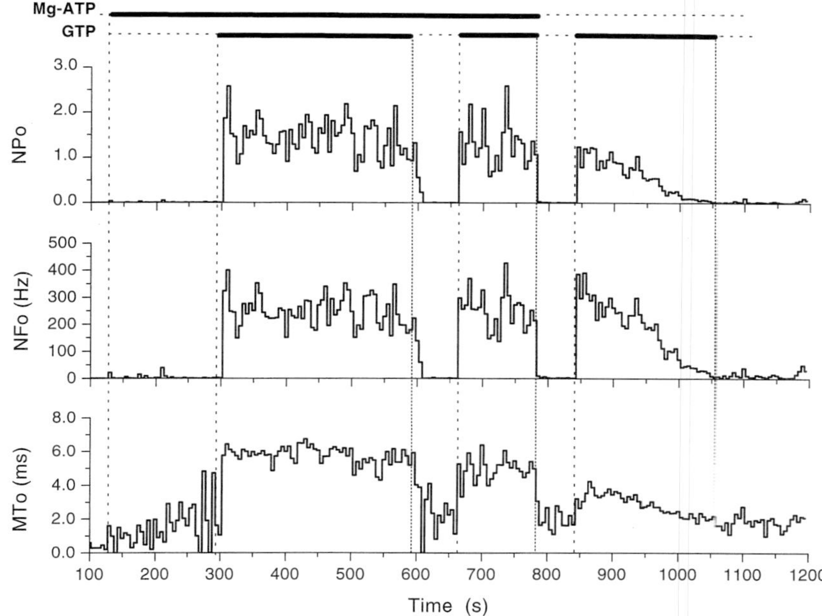

FIGURE 3 MgATP prevents the rundown of G-protein-mediated activation. NPo, NFo, and MTo plots of K_{ACh} channel activity in an inside-out patch from a chick embryonic atrial myocyte show that in the presence of ATP (2 m*M*, Na salt, in the presence of 0.6 m*M* Mg^{2+} in the bath solution), GTP (100 μM) produced sustained activity (10 μM ACh was present in the pipette). Prolonged withdrawal of ATP caused a gradual rundown of the GTP-induced activity. As can be seen at the beginning of the record ATP perfusion alone did not activate the channel. Application of each nucleotide is indicated by the bars.

time, which was fully reversed on application of GTP but not fully reversed in the absence of MgATP. The mechanism for this GTP-dependent effect is not known.

Hydrolysis of ATP had been shown previously to enhance GTP-induced K_{ACh} activity. Mechanisms such as nucleoside diphosphate kinase mediating phosphotransfer to the G protein from ATP or a membrane-bound kinase phosphorylating the channel protein had been proposed (28, 29). At first we explored the possibility that protein phosphorylation underlies the ATP-dependent effects (Figs. 2 and 3) on channel activity. Under our experimental conditions use of kinases, kinase inhibitors, or phosphatase inhibitors failed to affect the ATP dependency of K channel activity (5).

In contrast, PIP_2 mimicked the MgATP sensitization to gating by Na^+ ions of both native atrial and recombinant K channels. Figure 4 compares

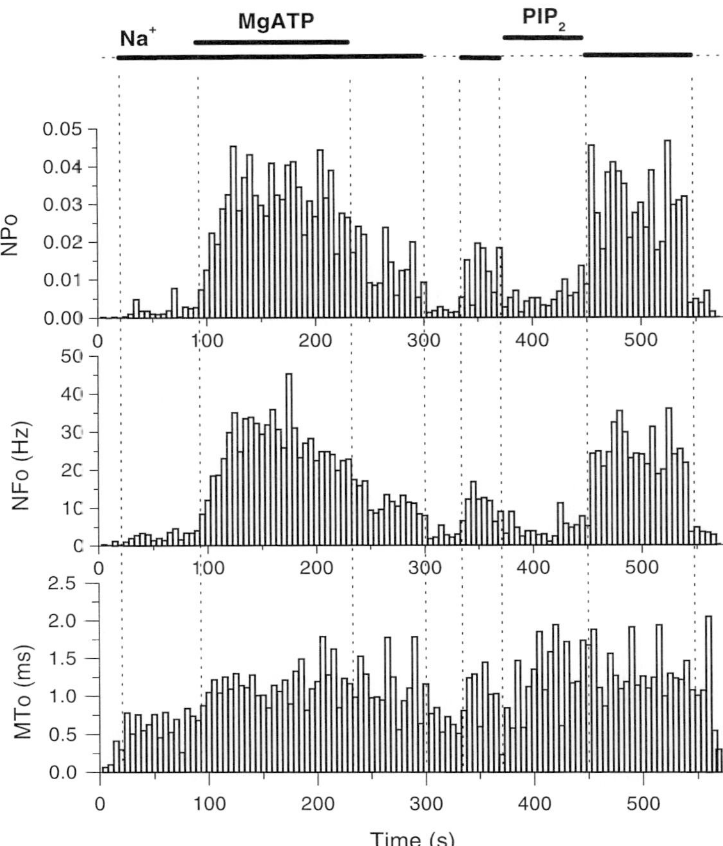

FIGURE 4 PIP_2 mimics the ability of MgATP to sensitize the channel to Na^+ ion gating. NPo, NFo, and MTo plots show channel activity in an inside-out patch from an oocyte expressing GIRK1/GIRK4 subunits. In the absence of MgATP the channels did not respond to Na^+ stimulation (20 m*M*). Addition of MgATP (5 m*M*) modified channels to a longer lived open state and greatly stimulated channel opening frequency and activity. Withdrawal of ATP reversed the effect. PIP_2 (5 μM) produced similar effects on MTo and sensitization to gating by intracellular Na^+ (20 m*M*). PIP_2 application, unlike ATP, had long lasting effects, and therefore it was not necessary to coapply it with Na^+ ions for maximal effects.

the effects of the MgATP exposure with that of PIP_2 on recombinant channels, in the same patch. Application of Na^+ ions to the inside-out patch in a solution lacking MgATP or PIP_2 did not produce appreciable activity or alter the open-time kinetics. Openings in both cases were shorter lived than after treatment with MgATP or PIP_2. The modification to the longer lived open state correlated with the sensitization of the channel to gating by internal Na^+ ions, application of which produced a large increase in activity. The PIP_2 treatment, unlike the corresponding MgATP perfusion, did not require concurrent application with Na^+ ions for maximal effects.

The PIP_2 treatment could also mimic the ATP-dependent persistent activation of K_{ACh} by internal GTP. Figure 5 shows a representative experiment in which 5 μM PIP_2, just like MgATP (Fig. 3), although it did not activate the channel by itself, reversed and prevented rundown of the GTP-induced channel activity (NPo).

We reasoned that block of PIP_2 ought to abolish the MgATP effects on channel activity, if these effects were due to formation of PIP_2 from ATP-mediated phosphorylation of phosphatidyl inositol (PI). Figure 6A shows that treatment of a *Xenopus* oocyte inside-out patch expressing recombinant K_{ACh} with an anti-PIP_2 antibody (PIP_2-Ab) blocked the ability of MgATP and Na^+ ions to sustain channel activity. The effect of the PIP_2-Ab was rapid and robust. An anti-PIP antibody also inhibited K_{ACh} activity but was not as effective as the PIP_2-Ab (data not shown). Similar experi-

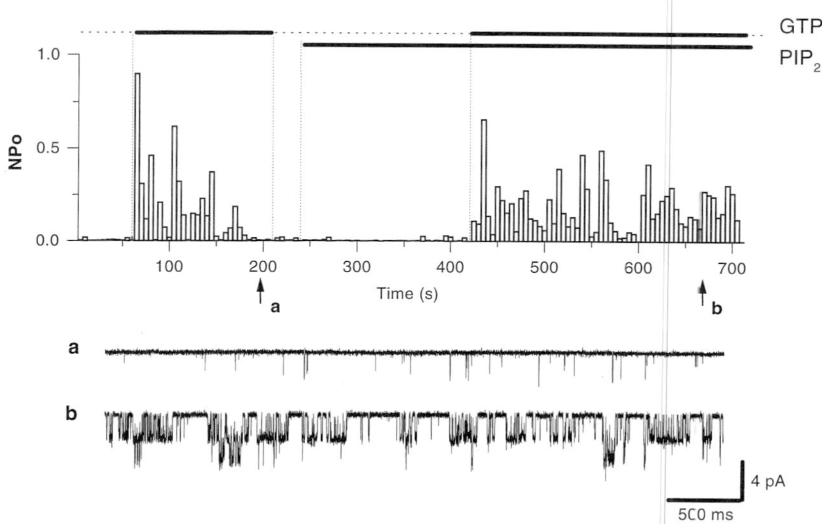

FIGURE 5 PIP_2 mimics MgATP in preventing the rundown of G-protein-mediated activation. PIP_2 restored the ability of GTP to stimulate channel activity and mimicked ATP in its ability to prevent rundown of the GTP-induced channel activity. The solution in the pipette contained 5 μM ACh. Perfusion of the inside-out patch with GTP (100 μM) and PIP_2 (5 μM) in the bath solution is indicated by bars. The arrows labeled a and b show time points during which rundown and restored single-channel activity, respectively, are shown. PIP_2, like ATP, did not stimulate channel activity when perfused alone.

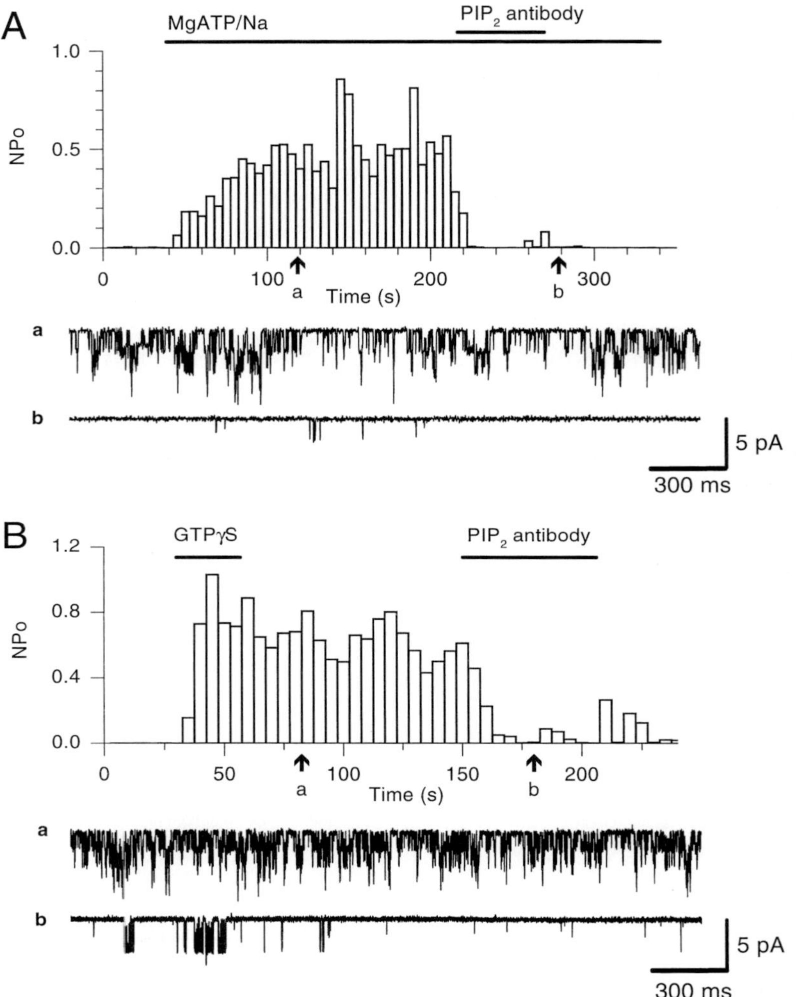

FIGURE 6 Anti-PIP_2 antibody blocks the MgATP/Na or GTPγS activation of K_{ACh} channels. (A) NPo plot from an inside-out patch obtained from a *Xenopus* oocyte expressing GIRK1/GIRK4. K channel activation with MgATP/Na (5/20 m*M*), was blocked by PIP_2-Ab (titer 1:1 or 500-fold dilution from manufacturer's stock), rendering further MgATP/Na perfusion ineffective. The arrows labeled a and b show single-channel activity at time points before and after block by the PIP_2 antibody. (B) Similar plot as in (A) showing GTPγS-mediated, persistent activation (10 μM) and PIP_2-Ab (titer 1:2 or 250-fold dilution from manufacturer's stock) block from an inside-out patch obtained from an atrial cell (5 μM ACh was present in the pipette solution). The arrows labeled a and b show single-channel activity at time points before and after block by the PIP_2 antibody.

ments using antibodies to either the Kv1.1 or the GIRK4 (N-terminal antigen) channels were without effect (data not shown). A GIRK1 (C-terminal antigen) antibody caused a small decrease in channel activity but much less than the consecutive treatment with PIP_2-Ab (data not shown). Similar results were obtained with native K_{ACh} channels from atrial cells (data not shown). The nonhydrolyzable analog of GTP, GTPγS, causes persistent K_{ACh} channel activation (e.g., Ref. 9). PIP_2-Ab treatment of an atrial inside-out patch preactivated by GTPγS blocked K_{ACh} activity in a manner similar to that seen with the MgATP/Na-activated channels but with slower kinetics (Fig. 6B). To rule out that the PIP_2-Ab block was due to a nonprotein impurity, boiled PIP_2-Ab (30 min) was compared in the same patch to unboiled PIP_2-Ab. In all cases boiled PIP_2-Ab was without effect, while unboiled PIP_2-Ab potently blocked channel activity (data not shown). Similar inhibitory actions of the PIP_2-Ab were obtained with recombinant GTPγS-stimulated K_{ACh} channels. PIP_2-Ab was effective also in blocking recombinant K_{ACh} channels previously activated with $G_{\beta\gamma}$ subunits. As with the GTPγS-stimulated K_{ACh} channel activity, Ab-block kinetics were slower than those of the MgATP/Na activity (data not shown).

To further test the hypothesis that PIP_2 plays a central role in the control of channel gating by internal Na^+ ions or G-protein subunits, we tested the effects of inositol-specific phospholipases on recombinant GIRK channels expressed in *Xenopus* oocytes. We first used a phosphatidylinositol-specific PLC which hydrolyzes phosphatidylinositol, thus preventing accumulation of PIP_2, and found that such treatment greatly limited channel sensitization to Na^+ by MgATP (5). $G_{\beta\gamma}$ subunits cause a persistent stimulation of GIRK channel activity. Experiments in which the GIRK channel was activated by $G_{\beta\gamma}$ subunits showed no significant decline in activity in the first 5 min following the end of $G_{\beta\gamma}$ perfusion. In Fig. 7A, channel activation by $G_{\beta\gamma}$ subunits was followed by a 4-min treatment with PLC-β2, which hydrolyzes PIP_2 to IP_3 and diacylglycerol (DAG). PLC-β2-treatment reversed channel activation by $G_{\beta\gamma}$. PLC-β2-treated patches showed on the average a 60% decrease in $G_{\beta\gamma}$-induced activity in the first 2 min of the treatment, whereas in the 2 min prior to the onset of PLC-β2 application, no significant decrease in activity was obtained. Thus, the PLC-β2-mediated inhibition of the $G_{\beta\gamma}$-induced activity was considerably faster than the persistent, unopposed $G_{\beta\gamma}$ stimulation of activity. Again, the PLC-β2 treatment resulted in openings with decreased open times (not shown), consistent with the effect of a reduction in PIP_2 levels. The reduction in MTo of the $G_{\beta\gamma}$-stimulated activity following the PLC-β2 treatment was not as pronounced as that following similar treatment of Na^+ ion-stimulated activity (data not shown). Under conditions that depleted PIP_2, $G_{\beta\gamma}$ treatment of inside-out patches could

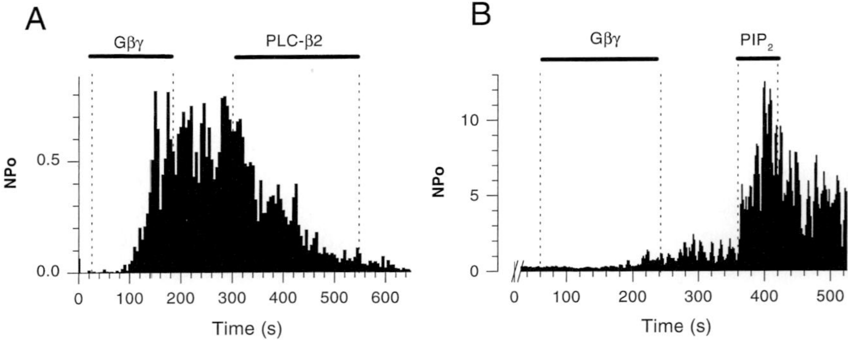

FIGURE 7 Hydrolysis of PIP_2 by PLC-β2 blocks $G_{\beta\gamma}$ stimulation of K_{ACh} channel activity and restoration of PIP_2 rescues $G_{\beta\gamma}$ activation. (A) Single-channel NPo plot, obtained from an inside-out patch of a *Xenopus* oocyte expressing GIRK1/GIRK4. Soon after excision of the patch, $G_{\beta\gamma}$ application (20 nM) caused persistent channel activation. Subsequent treatment of the patch with PLC-β2 (5 μg/ml) inhibited activity. (B) NPo plot as a function of time in the experiment from an inside-out patch of *Xenopus* oocyte expressing GIRK1/GIRK4. The patch was exposed to PLC-β2 (5 μg/ml) for a period greater than 5 min before $G_{\beta\gamma}$ (20 nM) application. PLC-β2 treatment greatly retarded $G_{\beta\gamma}$ effectiveness. Subsequent perfusion with PIP_2 (5 μM) revealed high channel activity, thus rescuing the $G_{\beta\gamma}$ action.

be rendered ineffective. Figure 7B shows such an example together with restoration of PIP_2 and rescue of the $G_{\beta\gamma}$ effect. $G_{\beta\gamma}$ application following pretreatment of an inside-out patch with PLC-β2 for a period greater than 5 min was without effect. Subsequent perfusion of the patch with 5 μM PIP_2 resulted in a significant increase in activity, restoring the effectiveness of $G_{\beta\gamma}$. This result suggested that the G-protein complex remained bound but incapable of gating the channel efficiently, until PIP_2 became available.

We next tested whether the effector molecule was PIP_2 itself or its hydrolysis products IP_3 and DAG. Both the block of PIP_2 accumulation by the PI-PLC (causing a decrease in PIP_2 and its hydrolysis products) as well as hydrolysis of PIP_2 by the PLCβ (causing an increase in the PIP_2 hydrolysis products) had similar inhibitory effects on channel activity. Thus, the similar inhibitory effect seen with the two phospholipases could not be due to PIP_2 hydrolysis products since their action would have opposite effects on the levels of these metabolites. In direct tests, neither IP_3 nor DAG was successful in activating GIRK channels, whereas MgATP/Na in the same patches caused on the average at least 10-fold higher channel activity (data not shown). In addition, these hydrolysis products failed to inhibit activity following stimulation by MgATP/Na (data not shown). Application of arachidonic acid (AA) to inside-out patches had no effect on activation or block of channel activity (data not shown), whereas

MgATP/Na caused on the average a 15-fold increase in activity. Although arachidonic acid has been shown to cause K_{ACh} activation in the cell-attached mode (30, 31), Kurachi and colleagues (30) also found it ineffective when they tested it in inside-out patches. This indicates that the PIP_2 effects were not due to hydrolysis of its metabolite DAG to AA. Thus, these results suggest that PIP_2 acted directly rather than through its hydrolysis products.

Figure 8 shows a model depicting direct interactions of phosphorylated inositol phospholipids with the channel protein and their crucial role on channel activation by $G_{\beta\gamma}$ subunits or Na^+ ions. A two-gate model is postulated. One of the gates is sensitive to phosphorylated inositol phospholipids, the other to the channel-gating molecules ($G_{\beta\gamma}$ or Na^+ ions). In the presence of MgATP, phosphorylated phospholipids are produced but on their own are incapable of gating the channel effectively (state C_1). In the absence of MgATP (and therefore phosphatidylinositol), gating molecules of K_{ACh} (such as $G_{\beta\gamma}$ subunits or Na^+ ions) are also ineffective in activating the channel (state C_2). However, when both phosphorylated phospholipids and gating molecules are present, both gates are open and the channel becomes permeable to K^+ ions (state O).

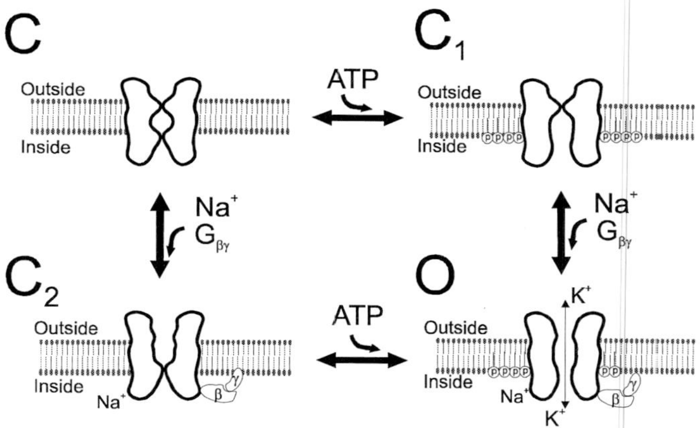

FIGURE 8 A two-gate model of K_{ACh} activation. In state **C**, in the absence of MgATP and Na^+ ions or $G_{\beta\gamma}$ subunits both channel gates are closed. In state $\mathbf{C}_1$, in the presence of MgATP inositol phospholipids are phosphorylated by lipid kinases to produce phosphoinositides. Direct interaction of phosphoinositides with the channel protein results in opening of the bottom gate. However, K^+ ion permeation is limited or absent, since the top gate is closed. In state $\mathbf{C}_2$, in the presence of the channel-gating molecules (Na^+ ions or $G_{\beta\gamma}$ complex), but in the absence of MgATP, the top gate is open but K^+ ion permeation is limited or absent, since the bottom gate is closed. In state **O**, in the presence of intact phosphoinositides and gating molecules (Na^+ ions or $G_{\beta\gamma}$ subunits) both gates are open, resulting in K^+ ion permeation.

IV. DISCUSSION

Evidence for the generation of phosphatidylinositol phosphates by lipid kinases, acting as membrane-delimited second messengers, to regulate ion channel and transporter activity has been presented recently (2–6, 32, reviewed in Ref. 1). K_{ATP} channels, which belong to the same family of inwardly rectifying K^+ channels as K_{ACh}, were shown to be stimulated by anionic phospholipids and PIP_2 in particular. PIP_2 mimicked ATP in preventing channel current rundown or in rescuing activity following rundown (3). Other recombinant inwardly rectifying channels, such as ROMK1 or IRK1, which exhibit MgATP-dependent rundown of their activity, were stimulated by PIP_2 as well (4).

The effects of PIP_2 cannot be explained by an indirect action of its hydrolysis products, as these did not mimic the PIP_2 effects. Evidence for a direct interaction of PIP_2 with inwardly rectifying channels has been provided by Huang and colleagues (4). Binding of glutathione *S*-transferase (GST)-tagged C-terminal segments of ROMK1, IRK1, and GIRK1 with ^3H-PIP_2 liposomes was demonstrated and immunoprecipitation of such GST fusion proteins bound to PIP_2 could be revealed by the PIP_2-Ab. These experiments established the following binding affinities for PIP_2: ROMK1 = IRK1 > GIRK2 = GIRK1/4. In addition, Huang and colleagues compared in giant excised patches the kinetics of PIP_2-Ab block of ATP-stimulated IRK1, ROMK1, GIRK2, and GIRK1/4 activities. In agreement with their biochemical studies they found that the PIP_2-Ab could inhibit the GIRK channels faster than ROMK1 or IRK1, suggesting that the interactions of PIP_2 with the GIRK channels was weaker than those with the other inward rectifiers. Mutagenesis work pointed to positively charged amino acids, such as R188 (4), K187, R217, and K218 (32) as residues of ROMK1 which share strong interactions with PIP_2. The basis for the differences in PIP_2 binding affinities among different inwardly rectifying recombinant channels could lie in differences in such critical positively charged residues, such as K218 which is asparagine in GIRK1, 2, and 4. The validity of such a hypothesis, however, remains to be established.

In the case of the G-protein-gated K^+ channel, we have shown that the inositol phospholipids seem to play a permissive role by regulating the effectiveness of gating molecules such as $G_{\beta\gamma}$ subunits or Na^+ ions, rather than activating the channel themselves. Huang and colleagues suggested that PIP_2 directly activates recombinant GIRK channels (4). However, they conducted their experiments in the presence of 30 m*M* internal Na^+, which is capable of gating these channels. Indeed, as we have shown previously (5, 19, and this chapter) PIP_2 (or hydrolyzable ATP) fails to activate K_{ACh} in the absence of gating molecules such as Na^+ or $G_{\beta\gamma}$ subunits. If our interpretation is correct, then why is PIP_2 capable of activating other recombinant inwardly rectifying K^+ currents? These are considered as background channels which at negative potentials conduct a constant flow of ions. Could it be that nature has critically adjusted the strength of the channel PIP_2 interaction to distinguish between background channels, which by virtue of strong interactions with the lipid could be kept open, versus channels that are gated by a different mechanism, such as K_{ACh}, where the weaker interactions with PIP_2 could serve only a regulatory role? Such hypotheses remain to be tested experimentally.

Huang and colleagues (4) showed that in the presence of $G_{\beta\gamma}$ the block kinetics of channel activity by the PIP_2-Ab become slower, and based on this observation they suggested that $G_{\beta\gamma}$ stabilizes the PIP_2 channel interaction. Our PIP_2-Ab data also support this idea (Ref. 5 and Fig. 6; contrast the PIP_2-Ab block of the MgATP/Na- versus GTPγS-induced channel activities). In addition, following treatment with $G_{\beta\gamma}$ subunits channel activation becomes persistent and less dependent on internal ATP, in contrast to the GTP-induced activation.

G-protein-coupled receptor kinases (GRKs) have been shown to interact directly with anionic phospholipids (33–36). This interaction was shown to be functionally important as it regulated the ability of such kinases to phosphorylate G-protein-coupled receptors and initiate their desensitization. Moreover, $G_{\beta\gamma}$ subunits bind GRKs directly to stimulate receptor phosphorylation, a property that was found to depend on anionic phospholipids, and PIP_2 in particular (33). Thus, the G-protein-gated K^+ channels, just as G-protein-coupled receptor kinases, are lipid-dependent proteins that require phosphatidylinositol phospholipids for regulation of their activity by $G_{\beta\gamma}$ subunits. It is intriguing that other $G_{\beta\gamma}$-sensitive proteins are also PIP_2 binding proteins (e.g., PLCβ isoforms, phosphoinositide 3-kinase). A possible PIP_2 dependence on the $G_{\beta\gamma}$ regulation of the activity of such proteins remains to be examined. However, a PIP_2 requirement for activation of $G_{\beta\gamma}$-insensitive proteins, such as phospholipase D, has been reported as well (37).

Enzymes containing pleckstrin homology (PH) domains have been found in a variety of proteins (38, 39), including protein kinases (e.g., GRKs, PKC), substrates for kinases, regulators of small G proteins, PLC isozymes, and cytoskeletal proteins. Interestingly, this domain has been reported to bind PIP_2 (38, 39), $G_{\beta\gamma}$ subunits (40, 41) and PKC (42). Although specific PIP_2-binding sites have been identified in the cytoskeletal protein α-actinin (43) and in GRKs (34–36), such equivalent sites were not obvious from examination of the GIRK channel primary amino acid sequences.

Many questions remain. What is the physiological relevance of the PIP_2 requirement for channel activity? What lipid kinases and phosphatases are involved? What is the balance of PIP_2 production and hydrolysis, and how are such processes regulated? Is there stereospecificity to the PIP_2 effect? Could the PIP_2 effects be the result of its conversion to PIP_3? Although PIP_3 effects were not tested directly, it is unlikely that this phospholipid plays a major role in the effects described, since it is a trace membrane component and also because the MgATP effect could be reversed by PIP_2-specific proteins (i.e., the antibody and PLC-β2).

Phosphatidylinositol 4,5-bisphosphate is a central molecule in the phosphoinositide cycle, by serving as the precursor of important signaling molecules such as IP_3, DAG, and phosphatidylinositol 3,4,5-trisphosphate. In addition, the results from this study and others demonstrate that PIP_2 can serve as a direct membrane-delimited second messenger molecule itself. Moreover, the novel functional dependence of $G_{\beta\gamma}$ signaling on PIP_2 predicts the possible crosstalk of different membrane-delimited signaling pathways, analogous to that seen with soluble second messengers.

Acknowledgments

We thank Drs. Donald Hilgemann and Chou-Long Huang for discussions, Dr. Massimo Sassaroli for advice regarding handling of lipids, and Dr. Cheng He and Xiaying Wu for advice and technical assistance. We are grateful for the gifts of purified PLC-β2 from Dr. Ravi Iyengar, GIRK1 and Kv1.1 antibodies from Dr. William Thornhill, and purified $G_{\beta\gamma}$ from John Hildebrandt. Finally we thank Drs. Evgeny Kobrinsky, Noelle Langan, Robert Margolskee, Tooraj Mirshahi, Eitan Reuveny, Harel Weinstein, and Hailin Zhang for helpful comments on the manuscript. This work was supported by a fellowship to J.L.S. from the New York Aaron Diamond Foundation and a grant to D.E.L. from the National Institutes of Health (HL59949).

References

1. Hilgemann, D. W. (1997). *Annu. Rev. Physiol.* **59,** 193–220.
2. Hilgemann, D. W., and Ball, R. (1996). *Science* **273,** 956–959.
3. Fan, Z., and Makielski, J. C. (1997). *J. Biol. Chem.* **272,** 5388–5395.
4. Huang, C.-L., Feng, S., and Hilgemann, D. W. (1998). *Nature (London)* **391,** 803–806.
5. Sui, J.-L., Petit-Jacques, J., and Logothetis, D. E. (1998). *Proc. Natl. Acad. Sci. U.S.A.* **95,** 1307–1312.
6. Lupu, V. D., Kaznacheyeva, E. V., and Bezprozvany, I. B. (1998). *Biophys. J.* **74,** A37.
7. Sui, J.-L., Chan, K., Langan, M.-N., Vivaudou, M., and Logothetis, D. E. (1998). *In* "Advances in Second Messenger and Phosphoprotein Research" (D. Armstrong and S. Rossie, eds.), in press.
8. Soejima, M., and Noma, A. (1984). *Pfluegers Arch.* **400,** 424-431.
9. Logothetis, D. E., Kurachi, Y., Galper, J., Neer, E. J., and Clapham, D. E. (1987). *Nature (London)* **325,** 321-326.
10. Krapivinsky, G., Gordon, E. A., Wickman, K., Velimirovic, B., Krapivinsky, L., and Clapham, D. E. (1995). *Nature (London)* **374,** 135-141.
11. Reuveny, E., Slesinger, P. A., Inglese, J., Morales, J. M., Iniguez-Lluhi, J. A., Lefkowitz, R. J., Bourne, H. R., Jan, Y. N., and Jan, L. Y. (1994). *Nature (London)* **370,** 143–146.

12. Silverman, S. K., Lester, H. A., and Dougherty, D. A. (1996). *J. Biol. Chem.* **271,** 30524–30528.
13. Krapivinsky, G., Krapivinsky, L., Wickman, K., and Clapham, D. E. (1995b). *J. Biol. Chem.* **270,** 29059–29062.
14. Kunkel, M. T., and Peralta, E. G. (1995). *Cell* (*Cambridge, Mass.*) **83,** 443–449.
15. Huang, C.-L., Slesinger, P. A., Casey, P. J., Jan, Y. N., and Jan, L. Y. (1995). *Neuron* **15,** 1133–1143.
16. Huang, C.-L., Jan, Y. N., and Jan, L. Y. (1997). *FEBS Lett.,* **405,** 291–298.
17. Inanobe, A., Morishige, K. I., Takahashi, N., Ito, H., Yamada, M., Takumi, T., Nishina, H., Takahashi, K., Kanaho, Y., Katada, T., and Kurachi, Y. (1995). *Biochem. Biophys. Res. Commun.* **212,** 1022–1028.
18. Doupnik, C. A., Dessauer, C. W., Slepak, V. Z., Gilman, A. G., Davidson, N., and Lester, H. A. (1996). *Neuropharmacology* **35,** 923-931.
19. Sui, J. L., Chan, K. W., and Logothetis, D. E. (1996). *J. Gen. Physiol.* **108,** 381–391.
20. Chen, J., DeVivo, M., Dingus, J., Harry, A., Li, J., Sui, J., Carty, D. J., Blank, J. L., Exton, J. H., Stoffel, R. H., Inglese, J., Lefkowitz, R. J., Logothetis, D. E., Hildebrandt, J. D., and Iyengar, R. (1995). *Science* **268,** 1166–1169.
21. Petit-Jacques, J., Sui, J. L., and Logothetis, D. E. (1998). Submitted.
22. Chan, K. W., Langan, M. N., Sui, J. L., Kozak, J. A., Pabon, A., Ladias, J. A. A., and Logothetis, D. E. (1996). *J. Gen. Physiol.* **107,** 381–397.
23. Logothetis, D. E., Movahedi, S., Satler, C., Lindpaintner, K., and Nadal-Ginard, B. (1992). *Neuron* **8,** 531–540.
24. Hamill, O. P., Marty, A., Neher, E., Sakmann, B., and Sigworth, F. J. (1981) *Pfluegers Arch.* **391,** 85–100.
25. Methfessel, C., Witzemann, V., Takahashi, T., Mishina, M., Numa, S., and Sakmann, B. (1986). *Pfluegers Arch.* **407,** 577–588.
26. Stühmer, W. (1992). *In* "Methods in Enzymology" (B. Rudy and L. E Iverson, eds.), Vol. 207, pp. 319–339. Academic Press, San Diego.
27. Vivaudou, M. B., Arnoult, C., and Villaz, M. (1991). *J. Membr. Biol.* **122,** 165–175.
28. Heidbüchel, H., Callewaert, G., Vereecke, J., and Carmeliet, E. (1990). *Pfluegers Arch.* **416,** 213–215.
29. Kim, D. (1991). *J. Physiol.* (*London*) **437,** 133–155.
30. Kurachi, Y., Ito, H., Sugimoto, T., Shimizu, I., Miki, I., and Ui, M. (1989). *Nature* (*London*) **337,** 555–557.
31. Kim, D., Lewis, L. L., Graziadei, L., Neer, E. J., Bar-Sagi, D., and Clapham, D. E. (1989). *Nature* (*London*) **337,** 557–560.
32. Liou, H.-H., Hilgemann, D. W., and Huang, C.-L. (1998). *Biophys. J.* **74,** A116.
33. DebBurman, S. K., Ptasienski, J., Benovic, J. L., and Hosey, M. M. (1996). *J. Biol. Chem.* **271,** 22552–22562.
34. Pitcher, J. A., Touhara, K., Payne, E. S., and Lefkowitz, R. J. (1995) *J. Biol. Chem.* **270,** 11707–11710.
35. DebBurman, S. K., Ptasienski, J., Boetticher, E., Lomansney, J. W., Benovic, J. L., and Hosey, M. M. (1995). *J. Biol. Chem.* **270,** 5742–5747.
36. Pitcher, J. A., Fredericks, Z. L., Stone, W. C., Premont, R. T., Stoffel, R. H., Koch, W. J., and Lefkowitz, R. J. (1996). *J. Biol. Chem.* **271,** 24907–24913.
37. Pertile, P., Liscovitch, M., Chalifa, V., and Cantley, L. C. (1995). *J. Biol. Chem.* **270,** 5130–5135.
38. Harlan, J. E., Hajduk, P. J., Yoon, H. S., and Fesik, S. W. (1994). *Nature* (*London*) **371,** 168–170.

39. Mayer, B. J., Ren, R., Clark, K. L., and Baltimore, D. (1993). *Cell (Cambridge, Mass.)* **73,** 629–630.
40. Touhara, K., Inglese, J., Pitcher, J. A., Shaw, G., and Lefkowitz, R. J. (1994). *J. Biol. Chem.* **269,** 10217–10220.
41. Tsukada, S., Simon, M. I., Witte, O. N., and Katz, A. (1994). *Proc. Natl. Acad. Sci. U.S.A.* **91,** 11256–11260.
42. Yao, L., Kawakami, Y., and Kawakami, T. (1994). *Proc. Natl. Acad. Sci. U.S.A.* **91,** 9175–9179.
43. Fukami, K., Sawada, N., Endo, T., and Takenawa, T. (1996). *J. Biol. Chem.* **271,** 2646–2650.

CHAPTER 19

Functional Analyses of G-Protein Activation of Cardiac K_G Channel

Yukio Hosoya*,† and Yoshihisa Kurachi†,‡
*Department of Nursing, Yamagata School of Health Science, Yamagata, Yamagata 990-22, Japan; †Department of Cell Biology and Signaling, Yamagata University School of Medicine, Yamagata, Yamagata 990-23, Japan; and ‡Department of Pharmacology II, Faculty of Medicine, Osaka University, Osaka 565, Japan

I. INTRODUCTION

In cardiac atrial myocytes, acetylcholine (ACh) binds to m_2-muscarinic receptors and activates the muscarinic $K^+(K_{ACh})$ channels via a pertussis toxin (PTX)-sensitive heterotrimeric G protein (G_K) in a membrane-delimited manner (Pfaffinger *et al.,* 1985; Breitwieser and Szabo, 1985; Kurachi *et al.,* 1986a,b). The $\beta\gamma$ subunits of G_K ($G_{K\beta\gamma}$) directly activate the cardiac K_{ACh} channel (Logothetis *et al.,* 1987; Kurachi *et al.,* 1989; Ito *et al.,* 1992; Yamada *et al.,* 1993; Wickman *et al.,* 1994; Reuveny *et al.,* 1994; Inanobe *et al.,* 1995), which is now supposed to be a heterotetramer of GIRK1/Kir3.1 and GIRK4/Kir 3.4 (Krapivinsky *et al.,* 1995). However, the functional interactions among receptor, G protein, and K_{ACh} channels have not been quantitatively examined.

In this study, the functional model describing the interaction of receptor, G protein, and K_{ACh} channel has been developed in two steps. The first

Current Topics in Membranes, Volume 46

1063-5823/99 $30.00

step is to construct a functional model explaining the $G_{K\beta\gamma}$–K_{ACh} channel interaction, and the second is to incorporate the receptor-mediated G-protein cycle in the model. We previously reported that intracellular GTP activates the K_{ACh} channel in a positive cooperative manner: the Hill coefficient is around 3 (Kurachi *et al.,* 1990; Ito *et al.,* 1991; Yamada *et al.,* 1993). The positive cooperativity may reflect the features in $G_{K\beta\gamma}$–K_{ACh} channel interaction (Ito *et al.,* 1991). Thus, for explaining the positive cooperative effect of GTP on cardiac K_{ACh} channel activation, we have adopted the Monod–Wyman–Changeux (MWC) concerted allosteric model for the $G_{K\beta\gamma}$–K_{ACh} channel interaction (Hosoya *et al.,* 1996). We have further developed the model to include the receptor–G-protein interaction as the second step. To estimate the amount of receptor-activated G protein (i.e., $G_{K\beta\gamma}$), we used the models of either Thomsen *et al.* (1988) or Mackay (1990a,b). The Thomsen model is used to describe the G-protein cycle regulated by a receptor in its low affinity binding state, while the Mackey model is used to describe it in the high affinity binding state. By combining either of these two models with the MWC kinetic model for $G_{K\beta\gamma}$ activation of the K_{ACh} channel, each concentration–response relationship between [GTP] and channel activity in the previous of various concentrations of ACh could be well fitted. It remains, however, to improve the models to account for the fast activation and deactivation of the K_{ACh} channel current on application or washout of ACh, respectively. Development of models is important to further clarify the molecular mechanisms underlying the functional interaction among receptor, G protein, and K_{ACh} channel.

II. MUSCARINIC K^+ CHANNEL ACTIVATION BY G PROTEINS

The cardiac K_{ACh} channels are activated by intracellular GTP (GTP_i) in a concentration-dependent fashion (Fig. 1A). The channel activity was increased by GTP_i in a positive cooperative manner (Kurachi *et al.,* 1990; Ito *et al.,* 1991; Yamada *et al.,* 1993). The $G_{K\beta\gamma}$–K_{ACh} channel interaction was analyzed in the presence of saturating concentrations of ACh (Hosoya *et al.,* 1996). Under these conditions, $G_{\beta\gamma}$ exogenously applied to the internal side of inside-out patch membranes did not further increase the maximum channel activity induced by 10 μM GTPγS. Therefore, in the presence of saturating [ACh], the maximum effect of GTP may be determined by the availability of K_{ACh} channels for $G_{K\beta\gamma}$, but not by the availability of the receptor-activated G proteins for GTP.

Because multiple K_{ACh} channels were usually included in a single inside-out patch membrane from atrial myocytes, the kinetics of the K_{ACh} channel was analyzed using the spectral analysis technique. The power density spectra of current fluctuations at various concentrations of GTP ([GTP])

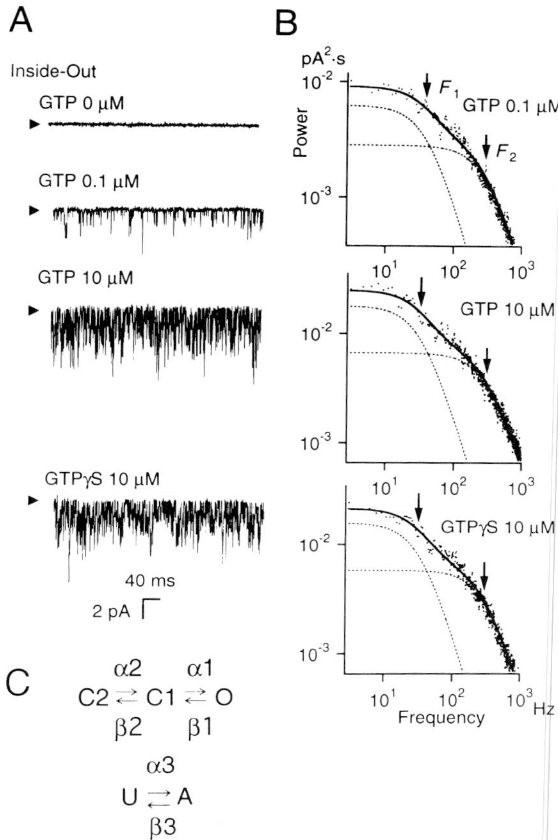

FIGURE 1 Effect of intracellular GTP on the power density spectra of current fluctuations of the K_{ACh} channel. (A) Inside-out patch recording at −60 mV. The pipette solution contained 0.5 μM ACh. Increasing concentrations of GTP were applied to the intracellular side of the patch membrane. GTPγS was finally applied to measure the maximum channel activity of the patch. (B) Power density spectra calculated from the inside-out patch records in (A). Each spectrum could be well-fitted by the sum of two Lorentzian functions. F_1 and F_2 (arrows) indicate corner frequencies of the slow and fast Lorentzian components, respectively. (C) Two independent reactions in K_{ACh} channel kinetics. Fast gating and slow process regulate the channel availability. The slow process is GTP-dependent, while the fast gating kinetics is GTP-independent. U, Unavailable state; A, available state; C2, C1, closed states; O, open state; α1–3, β1–3, rate constants. Reproduced from *The Journal of General Physiology* by copyright permission of The Rockefeller University Press.

were well-fitted by the sum of two Lorentzian functions (Fig. 1B). At increasing $[GTP]_i$, the powers at 0 Hz of the two components increased, but both the ratio of powers at 0 Hz and the corner frequencies of the two components remained almost constant. Thus, the fast gating kinetics of K_{ACh} channels represented by the two Lorentzian components were not affected by the activated G-protein subunits (G_K*, i.e., $G_{K\beta\gamma}$). Because the K_{ACh} channels exhibit one open state (Kurachi *et al.*, 1986a,b; Ito *et al.*, 1991), the fast gating could be described as $C_2 \leftrightarrow C_1 \leftrightarrow O$. In addition to the fast gating, the channels should possess a slow process regulating the functional channel number, because the powers at 0 Hz of two Lorentzian components increased on G-protein activation. This could be described by a slow transition of the channel states, U (unavailable) $\leftrightarrow$ A (available), which is independent of the fast open–close gating transitions. The corresponding component did not appear in the spectral analysis, probably because the transition between two channel states was slow and infrequent compared to the fast gating transitions. Therefore, we could assume the existence of two independent reactions in the K_{ACh} channel kinetics, namely, a fast gating process and a slow one regulating the channel availability (Fig. 1C). The channel is functionally active (available) in the A state and can open when the gate takes the O state. In contrast, in the U state, the channel is functionally inactive (unavailable) and cannot open even when the gate reaches the O state. Only the transition between U and A is GTP-dependent, and $G_{K\beta\gamma}$ shifts the equilibrium to the A state and increases the steady-state probability in A, resulting in an increase of the functional number of K_{ACh} channels. Therefore, the positive cooperative activation of K_{ACh} channels by $G_{K\beta\gamma}$ should be explained in the process between U and A.

We have analyzed the positive cooperative activation of the K_{ACh} channel by G_K* ($G_{K\beta\gamma}$) with the Monod–Wyman–Changeux (MWC) allosteric model (Fig. 2) (Hosoya *et al.*, 1996; Monod *et al.*, 1965). For formation of the model, we assumed that (1) in the presence of saturating [ACh], the concentration of $G_{K\beta\gamma}$ in the membrane linearly increases in the presence of physiological ranges of [GTP]; (2) the K_{ACh} channel consists of a finite number (n) of functionally identical subunits; (3) each of the subunits binds one $G_{K\beta\gamma}$; and (4) the channel is available (i.e., in the A state) when all subunits are in the R state and unavailable (i.e., in the U state) when all in the T state. The relationship between [GTP] and the steady-state availability of the channel was well-fitted with this model when the number of subunits was assumed to be four or more. The values of L, K_T, K_R, and χ^2 obtained by fitting at $n = 1$–6 are listed in Table I. Thus, the MWC allosteric model for the channel state transition (U $\leftrightarrow$ A) can well describe the positive cooperative increase in the channel availability by GTP when $n \geq 4$. The model may indicate that the cardiac K_{ACh} channel can be

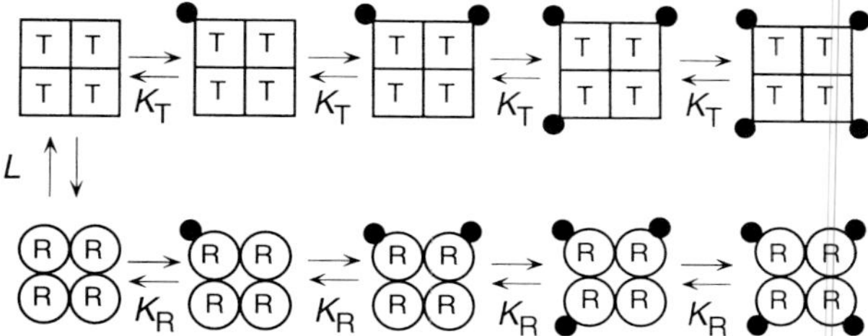

FIGURE 2 Allosteric model of Monod, Wyman, and Changeux. Two different states of the protomers, tense (T) and relaxed (R), are represented by squares and circles, respectively. The relaxed state has a higher affinity for the activated G-protein subunit ($G_{K\beta\gamma}$), which is represented by a small solid circle. In this illustration the K_{ACh} channel is supposed to be an oligomeric protein composed of four functionally identical protomers. Reproduced from *The Journal of General Physiology* by copyright permission of The Rockefeller University Press.

described as a multimer composed of four or more functionally identical subunits, to each of which one $G_{K\beta\gamma}$ might bind. This agrees with the current notion developed by the molecular biological characterization of cardiac K_{ACh} channels that the channel is a heterotetramer of GIRK1 and GIRK4 (Krapivinsky *et al.,* 1995).

III. RECEPTOR–G-PROTEIN INTERACTION

The frequency of K_{ACh} channel openings is enhanced by ACh via G-protein activation. G proteins are membrane-bound proteins that transduce signals from receptors to effectors (Gilman, 1987). These proteins are heterotrimers composed of α, β, and γ subunits. The G-protein activation of the K_{ACh} channel is mediated by $G_{K\beta\gamma}$ (Kurachi, 1995). Therefore, the

TABLE I
Values of Parameters in the Allosteric Model[a]

Number of subunits	L	K_T	K_R	χ^2
1	9.251×10^4	7.203×10^{-7} *M*	5.753×10^{-12} *M*	1.2061
2	1.788×10^6	1.407×10^{-5} *M*	2.240×10^{-10} *M*	0.2602
3	3.985×10^7	1.999×10^{-6} *M*	7.130×10^{-10} *M*	0.2080
4	4.290×10^7	5.780×10^{-7} *M*	2.406×10^{-9} *M*	0.1874
5	3.286×10^8	3.393×10^{-7} *M*	3.128×10^{-9} *M*	0.1862
6	1.028×10^9	2.414×10^{-7} *M*	4.210×10^{-9} *M*	0.1862

[a] See Fig. 2. Reproduced from *The Journal of General Physiology* by copyright permission of The Rockefeller University Press.

dissociation–association kinetics of G-protein subunits must be taken into consideration in the analysis of the receptor-mediated activation of the K_{ACh} channel.

Models for the receptor-mediated cyclic reaction of G_i proteins have been provided by Thomsen *et al.* (1988) and Mackay (1990a,b) (Fig. 3). In the Thomsen model (Fig. 3A), A and R represent agonist (ACh) and m_2-muscarinic receptor, which combine to form the complex, AR. Interaction of agonist and receptor facilitates the formation of a ternary complex between agonist, receptor, and GDP-ligated G protein, designated as AR·G–GDP. The ternary complex AR·G is formed on release of GDP. GTP can bind to the ternary complex and forms AR·G–GTP, which then dissociates into a complex of AR and G-protein subunits ($G_{\alpha\text{-GTP}} + G_{\beta\gamma}$). G_α has intrinsic GTP-ase activity, and, when GTP on G_α is hydrolyzed to GDP, the

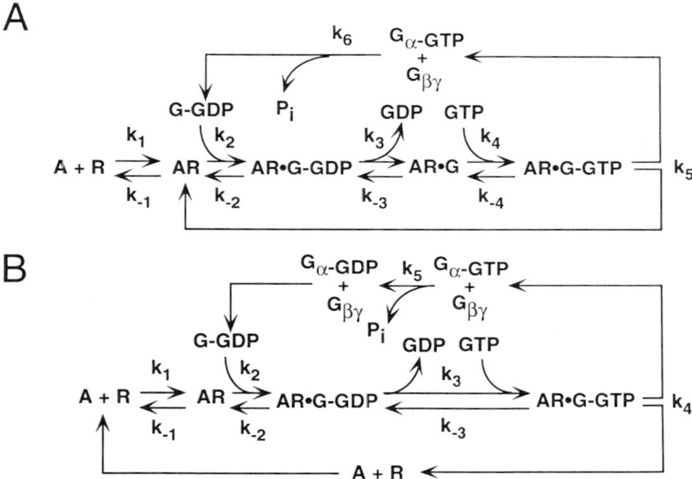

FIGURE 3 The Thomsen and Mackay models for receptor–G-protein interaction. This schematic representation shows the Thomsen (A) and Mackay (B) models, where A is acetylcholine, R is the muscarinic receptor, G is the G protein, G_α is the α subunit of the G protein, and $G_{\beta\gamma}$ is the $\beta\gamma$ subunit of the G protein. The parameters assayed by biochemical techniques in the Thomsen model are as follows: k_1, $5 \times 10^6\ M^{-1}\ \text{sec}^{-1}$; k_{-1}, 0.5 sec^{-1}; k_2, 0.1 sec^{-1}; k_{-2}, 0.1 sec^{-1}; k_3, 0.1 sec^{-1}; k_{-3}, $1 \times 10^{-4}\ \text{sec}^{-1}$; k_4, $1 \times 10^7\ M^{-1}\ \text{sec}^{-1}$; k_{-4}, 0.10 sec^{-1}; k_5, 0.05 sec^{-1}; k_6, 0.10 sec^{-1} (Thomsen *et al.*, 1988). The values of k_4 and k_6 were altered for fitting. The Mackay model is a theoretical one and does not provide any actually measured values of rate constants. Therefore, for the simulation using the Mackay model, we used mostly the same values as for the Thomsen one: k_1, $5 \times 10^6\ M^{-1}\ \text{sec}^{-1}$; k_{-1}, 0.5 sec^{-1}; k_2, 0.1 sec^{-1}; k_{-2}, 0.1 sec^{-1}. The values of k_3 and k_5 in the Mackay model were altered for fitting. In the Mackay model, the k_{-3} value was assumed to be the product of k_{-3} and k_{-4} of the Thomsen model. For k_4, we used the same values as k_5 of the Thomsen model.

GDP-bound form of G_α ($G_{\alpha\text{-GDP}}$) reassociates with $G_{\beta\gamma}$ and forms an inactive ligated form (G–GDP). In the Thomsen model, once the agonist binds to the receptor, G proteins can be activated and deactivated in the continuous binding of agonists to the receptors. Thus, the Thomsen model may well explain the kinetics of G proteins regulated by the low affinity-binding state receptor. On the other hand, the Mackay model assumes that the release of agonist from the membrane receptor is an essential step (Fig. 3B) and may be suited to explain the G-protein kinetics by the receptor in the high affinity binding state. In addition, the Mackay model assumes no accumulation of the AR·G ternary complex. The rate constants of each reaction in the Thomsen model have been provided by *in vitro* measurements, whereas the Mackay model is a theoretical consideration of receptor–G-protein interaction and thus the rate constants of each step have not been actually measured. Therefore, we have mostly adopted the values of the Thomsen model for those in the Mackay model at each corresponding reaction. The values used for the present calculation are listed in the legend of Fig. 3. We have tried to incorporate these two models in the MWC model for $G_{K\beta\gamma}$ activation of the K_{ACh} channel, although there clearly exist limitations in this approach, especially in the case of the Mackay model.

IV. RECEPTOR–G-PROTEIN–MUSCARINIC K^+ CHANNEL INTERACTION

At each [ACh], the channel activity was calculated based on the following assumptions: (1) each K_{ACh} channel is activated by $G_{K\beta\gamma}$ as described by the MWC model with the number of subunits being four (Fig. 2), (2) each $G_{K\beta\gamma}$ is supplied as described by either the Thomsen or the Mackay model, (3) one muscarinic receptor activates one G protein, and (4) the ratio of receptor and G protein is 1:10.

To calculate the channel activity induced by $G_{K\beta\gamma}$, Eq. (2) of Hosoya *et al.*, (1996) was used. In both the Thomsen and Mackay models, in the absence of agonist, there should be no accumulation of $G_{K\beta\gamma}$ in the membrane. Thus, no activation of K_{ACh} channels should be attained, even with very high [GTP]. In the inside-out patches of native cardiac atrial cell membranes, however, significant K_{ACh} channel activity could be induced by high [GTP] in the chloride (Cl^-) internal solution even in the absence of agonist (Ito *et al.,* 1991), probably due to the basal turn-on and slowed turn-off reactions of G_K (Nakajima *et al.,* 1992). To describe the agonist-free basal channel activity detected in the Cl^- internal solution, we also added a parameter of ACh0 to the models. ACh0 was expressed as the [ACh] that generates the [$G_{K\beta\gamma}$] equivalent to the amount produced by the agonist-free basal turn-on reaction. To estimate the concentration of receptor-activated G protein ($G_{K\beta\gamma}$) including the basal turn-on reaction, we substituted ([ACh]+ ACh0) for [ACh] in Eqs. (A12) and (A13) of Thomsen *et al.* (1988) or Eqs. (18)–(20) of Mackay (1990a). ACh0 for the Thomsen model was 9.99×10^{-9} *M*, whereas that for the Mackay model was 9.96×10^{-9} *M*.

To estimate [$G_{K\beta\gamma}$] from [ACh] and [GTP], the modified Eqs. (A12) and (A13) of Thomsen *et al.* (1988) or the modified Eqs. (18)–(20) of Mackay (1990a) were used. For fitting our experimental data using the models, two rate constants (k_4 and k_6 in the Thomsen model, and k_3 and k_5 in the Mackay model; see Fig. 3) were altered and the best-fit values were determined by the least-squares method. Other rate constants were the same as those obtained by Thomsen *et al.* (1988). They are listed in the legend of Fig. 3.

In the Thomsen model, k_4 is the rate constant for GTP-binding reaction, and it determines the affinity of the ternary complex AR·G for GTP. In addition, k_6 is the rate constant for GTPase reaction, and this rate constant determines the speed of deactivation of $G_{K\beta\gamma}$. Because these two rate constants determine the potency and affinity for GTP, we needed to change at least these two rate constants to fit the experimental data.

Figure 4 shows the fitted curves using the Thomsen (Fig. 4A) or Mackay model (Fig. 4B). In the native cardiac K_{ACh} channel, as [ACh] in the pipette was raised, the maximum channel activity (V_{max}) induced by GTP became larger, and the K_d for GTP decreased (see Fig. 4). These features of the relationship between [GTP] and K_{ACh} channel activity could be simulated by incorporating the reaction between receptors and G proteins (the Thomsen and Mackay models) into the MWC allosteric model. The relationships in the presence of 0.01, 0.1, or 1 μM ACh were well-fitted with the Thomsen model (lines in Fig. 4A) but not with the Mackay one. Although the increase of V_{max} and the shift of K_d to the lower [GTP] at higher [ACh] could be also simulated by using the Mackay model, the fit was not good enough for the relations at 0.01 and 0.1 μM ACh. On the other hand, the relation of agonist-free basal channel activity induced by GTP was well-fitted by the Mackay model but not by the Thomsen one (Fig. 4). This result may suggest that some of the qualitative difference in G-protein cycle might exist in the absence and presence of agonist.

Although the concentration–response relationship between [GTP] and channel activity could be well-fitted at various [ACh] by incorporating either the Thomsen or Mackay model in the MWC model for $G_{K\beta\gamma}$ activation of K_{ACh} channel, there exist serious problems in the results. By the best-fitting approach with changing the two parameters (k_4 and k_6 in the Thomsen model and k_3 and k_5 in the Mackay one), the values of these parameters should have been decreased. This resulted in a reasonable accumulation of calculated [$G_{K\beta\gamma}$] in the membrane at various [ACh] and thus in fairly

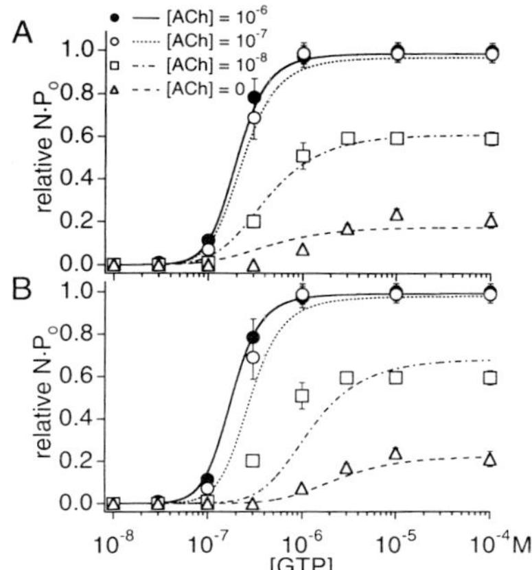

FIGURE 4 Fitting using the Thomsen and Mackay models. In the inside-out patch membranes of guinea pig atrial myocytes, the concentration-dependent effect of GTP on the K_{ACh} channel activity was measured in the absence (triangles) or the presence of 10 n*M* (squares), 100 n*M* (open circles), and 1 μM (filled circles) ACh in the pipette. Symbols and bars represent means ± SEM (n = 6–7). The channel currents were measured at −80 mV. The relative total open probability ($N \cdot P_o$) in the presence of each concentration of GTP and ACh was obtained with reference to the maximum value induced by 10 μM GTPγS in each patch. The lines in the graphs are the fits of the data with the MWC allosteric model combined with the Thomsen (A) or Mackay model (B). For fitting the experimental data, only two rate parameters (k_4 and k_6 in the Thomsen model, and k_3 and k_5 in the Mackay model; see Fig. 3) were changed, and best-fitting values were determined by the least-squares method. On the other hand, other rate constants were kept constant during the fitting, and we used the same values as reported by Thomsen *et al.* (1988). The best-fitting values of each parameter were as follows. In the Thomsen model, the values of k_4, k_6, and ACh0 are $1.28 \times 10^5\ M^{-1}\ sec^{-1}$, $2.11 \times 10^{-2}\ sec^{-1}$, and $9.99 \times 10^{-9}\ M$, respectively. The χ^2 values are 0.00158 (at 1 μM of ACh), 0.0111 (0.1 μM), 0.00490 (0.01 μM), and 0.0166 (0 μM). In the Mackay model, the values of k_3, k_5, and ACh0 are $1.50 \times 10^5\ M^{-1}\ sec^{-1}$, $3.24 \times 10^{-2}\ sec^{-1}$, and $9.96 \times 10^{-9}\ M$, respectively. The χ^2 values are 0.00355 (at 1 μM of ACh), 0.0320 (0.1 μM), 0.0981 (0.01 μM), and 0.00460 (0 μM). In the MWC allosteric model for this fitting including receptor–G-protein interaction, L', C', K_T', and K_R' are 1.926×10^7, 2.075×10^{-6}, 2.031×10^{-5}, and $4.319 \times 10^{-11}\ M$, respectively.

good fittings of the calculated curves to the experimental data. However, the decrease of these rate constants resulted in slowing of both the activation of channel activity on ACh application and the deactivation after washout of the agonist; the half-time for activation of the simulated current on application of 1 μM ACh was ~1 min (with 1 μM GTP in the internal side), and that for deactivation after washout of ACh was ~2.5 min. These values are much larger than those measured in native cardiac K_{ACh} channels or those in heterologously expressed GIRK1 + GIRK4 channels; the time constant for activation time was 0.1–0.3 sec and that for deactivation time was ~0.6–1 sec in native cardiac K_{ACh} channel, whereas in GIRK1 + GIRK4 channels they are 1–2 sec and 10 sec, respectively.

It has been noticed that the time courses of activation and deactivation in the GIRK1 + GIRK4 channels were slower than those measured in native cardiac K_{ACh} channels. However, when the RGS (regulator of G-protein signaling) proteins (Druey *et al.,* 1996; Koelle and Horvitz, 1996; Berman *et al.,* 1996) were coexpressed with the GIRK1 + GIRK4 channel, both activation and deactivation of the channel by ACh were accelerated and became comparable to the values of native cardiac K_{ACh} channels (Doupnik *et al.,* 1997). Because RGS3 and RGS4 are actually expressed in heart, RGS proteins may explain the faster activation and deactivation kinetics of native cardiac K_{ACh} channels than those of heterologously expressed GIRK1 + GIRK4 channels (Szabo and Otero, 1990; Nakajima *et al.,* 1992; Doupnik *et al.,* 1997). Thus, the receptor–G_K cycle interaction in native cardiac cells may be under control of some of RGSs proteins. Because the rate constants measured *in vitro* should be free from the effects of RGS proteins, it is expected that their values, which are subject to the RGS protein effect, are smaller than those *in vivo*. Therefore, for fitting the data of native K_{ACh} channels, it was expected that the values of k_4 and k_6 in the Thomsen model and those of k_3 and k_5 in the Mackay would be increased for the best fit. However, in the actual fitting of this study, we needed to decrease those values (see the legend of Fig. 4 for the values of the rate constants after fitting). One of the possible reasons for this discrepancy might be as follows. The concentration–response relationships between [GTP] and K_{ACh} channel activity were obtained with an internal solution that contained Cl^- as the anion (Ito *et al.,* 1991). Because Cl^- inhibits the GTPase activity of G_α (Gilman, 1987), the relationships in the Cl^- solution might have been shifted to lower [GTP] compared to the *in vivo* relation. Actually, with an internal solution containing sulfate ions (SO_4^{2-}), the relationships shifted to the higher [GTP] and ranged between 1 and 100 μM GTP. No significant basal channel activity was induced in the absence of agonist in the SO_4^{2-} internal solution (Nakajima *et al.,* 1992). Further studies are definitely required to develop the model to explain the fast activation and deactivation of the ACh-induced K_{ACh} current.

Nevertheless, the concentration–response relationships between [GTP] and K_{ACh} channel activity were well-fitted at various [ACh] by incorporating the receptor–G-protein cycle models to the MWC model for $G_{K\beta\gamma}$ activation of K_{ACh} channel. Thus, it may be possible that further insights in the receptor–G protein–K_{ACh} channel interaction will be obtained by analyzing the models. In Fig. 5, we calculated the relationship between [GTP] and $G_{K\beta\gamma}$. Values of $[G_{K\beta\gamma}]/K_R'$ were calculated and plotted against [GTP] at

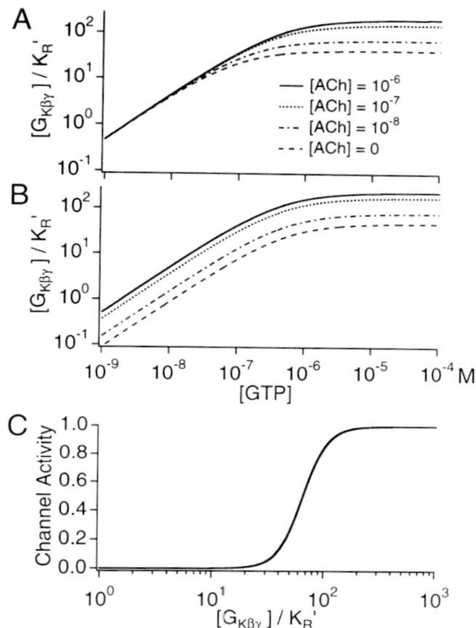

FIGURE 5 (A, B) Relation between [GTP] and receptor-activated G proteins. The value of $[G_{K\beta\gamma}]/K_R'$ was calculated and plotted against [GTP] with each concentration of ACh, by the use of the Thomsen (A) or Mackay (B) model. The parameters used are listed in the legend to Fig. 3. K_R' is a microscopic dissociation constant of a ligand bound to the stereospecific sites in the R (relaxed) state. $[G_{K\beta\gamma}]/K_R'$ $(=\alpha)$ is a normalization for the substrate concentration. (C) Relation between channel activity and receptor-activated G proteins. K_{ACh} channel activity was calculated and plotted against $[G_{K\beta\gamma}]/K_R'$ by the use of the MWC allosteric model with the fitting parameters.

each [ACh], by the use of the Thomsen (Fig. 5A) or Mackay (Fig. 5B) model. K_R' is the microscopic dissociation constant of a ligand bound to the stereospecific sites in the R (relaxed) state of the K_{ACh} channel (see also Fig. 2). $[G_{K\beta\gamma}]/K_R'$ $(=\alpha)$ is a normalization of $G_{K\beta\gamma}$ for the substrate concentration. For fitting the experimental data, two rate constants (k_4 and k_6) were varied in the Thomsen model as described earlier. As the value of k_4 was raised, the curve which represents the relation between [GTP] and $[G_{K\beta\gamma}]/K_R'$ shifted to the left. As the value of k_6 was raised, the curve which represents the relation between [GTP] and $[G_{K\beta\gamma}]/K_R'$ shifted downward. Figure 5C shows the relationship between receptor-activated G proteins and channel activity used for the fitting in Fig. 4. The channel activity was calculated and plotted against $[G_{K\beta\gamma}]/K_R'$ by the use of the MWC allosteric model with the fitting parameters. The dependence of K_{ACh} channel activity on $[G_{K\beta\gamma}]/K_R'$ was sigmoidal. The channel activity was generated at more than 2–3 × 10^1 $[G_{K\beta\gamma}]/K_R'$ and saturated at more than 2 × 10^2 $[G_{K\beta\gamma}]/K_R'$.

In the Thomsen model (Fig. 5A), the $[G_{K\beta\gamma}]/K_R'$ value did not differ significantly up to 10^{-8} *M* GTP among various concentrations of ACh. At ~2 × 10^{-8} *M* GTP, $[G_{K\beta\gamma}]/K_R'$ was ~10^1 in all [ACh]. This value of $[G_{K\beta\gamma}]/K_R'$ is the threshold to generate the channel activity (Fig. 5C). At >10^{-7} *M* GTP, as the [GTP] was increased, the difference in $[G_{K\beta\gamma}]/K_R'$ became gradually bigger in the presence of various [ACh]. As [ACh] was increased from 0 to 10^{-8}, 10^{-7}, and 10^{-6} *M,* the saturating level of the $[G_{K\beta\gamma}]/K_R'$ value became larger, from 4 × 10^1 to 6 × 10^1, 10^2, and 2 × 10^2, respectively. At higher [ACh], $[G_{K\beta\gamma}]/K_R'$ reached its saturating level at higher [GTP] (Fig. 5A). In the presence of 0, 10^{-8}, 10^{-7}, or 10^{-6} ACh, $[G_{K\beta\gamma}]/K_R'$ reached its saturating level at ~10^{-6}, 2 × 10^{-6}, 5 × 10^{-6}, or 10^{-5} *M* GTP, respectively. Thus, at 10^{-6} *M* ACh, $[G_{K\beta\gamma}]/K_R'$ saturated at ~2 × 10^2 at more than 10^{-5} *M* GTP. The $[G_{K\beta\gamma}]/K_R'$ value to induce the maximum channel activity is 2 × 10^2 (Fig. 5C).

Some of the features of the relationships between [GTP] and channel activity (Fig. 4) can be explained on the basis of the above simulation as follows. (1) In the presence of 0 or 0.01 μM ACh, the maximum channel activity (V_{max}) was smaller than that induced by 10 μM GTPγS (Fig. 4). The channel activity against $[G_{K\beta\gamma}]/K_R'$ is not saturated under these conditions in both the Thomsen and Mackay models (Fig. 5), because $[G_{K\beta\gamma}]/K_R'$ saturates at the values below 10^2 even at very high concentrations of GTP. Thus, in the presence of 0 or 0.01 μM ACh, the amount of receptor-activated G protein ($[G_{K\beta\gamma}]/K_R'$) is the determinant of the maximum channel activity. (2) As [ACh] was raised, the threshold [GTP] necessary to induce opening of the channel decreased, and the half-maximum effective concentration of GTP (K_d) decreased (Fig. 4). This is because, as [ACh] is raised, the [GTP] necessary to generate the same amount of $G_{K\beta\gamma}$ decreases (Fig. 5A,B). In the Mackay model (Fig. 5B), on the other hand, the relationship between [GTP] and $[G_{K\beta\gamma}]/K_R'$ shifts to the upward direction in a parallel fashion on increasing [ACh]. The differences in the level of $G_{K\beta\gamma}/K_R'$ among different [ACh] are constant and large in a broad range of [GTP]. This may cause the larger shifts of the relative $N \cdot P_o$–[GTP] relationship by ACh than the actual experimental data (e.g., from 0 to 10^{-8} *M* ACh or from 10^{-8} to 10^{-7} *M* ACh in Fig. 4B). In the Thomsen model, because the difference in $G_{K\beta\gamma}/K_R$ values among different [ACh] values is small at 10^{-8}–10^{-7} *M* GTP, the calculated curves well fit the experimental data (Fig. 4A). The relation of agonist-free background channel activity induced by GTP is, however, well-fitted by the Mackay but not by the Thomsen model (Fig. 4). This may suggest that the property of receptor–G_K interaction in the absence of agonist may better be explained by the Mackay model than the Thomsen one. Thus, some of the differences between the Thomsen and Mackay models may reflect the difference of the receptor–G-protein interactions in the presence and absence of agonist. This might be related either to the low and high affinity binding states of receptor, or to the absence of accumulation of the ternary complex ARG in the Mackay model.

To evaluate the contribution of the high-affinity binding nature of reaction, we simulated the relationship between [GTP] and channel activity using a modified Thomsen's model, where the AR dissociates into A and R on hydrolysis of GTP on G_{α}, but other reactions are the same as original. The obtained fitted curves were nearly identical to those of the original model. Thus, the high and low affinity-binding states may not be mainly responsible for the difference between Thomsen's and Mackay's fittings. The lack of accumulation of the ternary complex AR·G might be involved in the difference of the dependence of $G_{K\beta\gamma}/K_R'$ on intracellular GTP between the two models. Further studies are, however, needed to identify the difference in the receptor–G-protein interaction in the presence and absence of agonist.

V. CONCLUSION

By combining the MWC allosteric model for $G_{K\beta\gamma}$ activation of K_{ACh} channels and either the Thomsen or Mackay model for ACh activation of G proteins, the concentration–response relationships between [GTP] and K_{ACh} channel activity in the presence of various [ACh] could be reasonably simulated. To apply the MWC model to positive cooperative channel activation by G_K*, we assumed multiple functionally identical subunits, to each of which one $G_{K\beta\gamma}$ binds. Then, to incorporate the models for receptor–G protein interactions, we assumed that one receptor activates one G protein. The obtained results suggest that (1) a multiple number of receptors are involved in activation of a K_{ACh} channel and (2) there should exist some qualitative difference in the receptor–G protein interaction in the absence of and in the presence of agonist. Some of the assumptions used in the simulation analyses of the molecular interactions among receptor, G protein, and K_{ACh} channels could be evaluated by the biochemical and molecular biological techniques using heterologous expression of GIRK subunits, receptor, and G-protein subunits. They are as follows. (1) How many $G_{K\beta\gamma}$ bind to one GIRK subunit? (2) How many receptors are involved in activation of one K_{ACh} channel? For example, by making constructs where GIRK subunits and m_2-receptor are connected in tandem, we may be able to examine the stoichiometry between m_2-receptors and K_{ACh} channels. The models presented here have clear limitations to explain the transient behavior of K_{ACh} channel current. Therefore, they should be improved by further studies to explain not only the steady state but also activation and deactivation of K_{ACh} channels on application and wash-out of ACh, respectively. The improved model may provide functional bases to further clarify the molecular mechanism underlying the functional interaction among the agonist, receptor, G protein, and K_G channel.

References

Berman, D.M., Wilkie, T. M., and Gilman, A. G. (1996). GAIP and RGS4 are GTPase-activating proteins for the G_i subfamily of G protein α subunits. *Cell* (*Cambridge, Mass.*) **86,** 445–452.

Breitwieser, G. E., and Szabo, G. (1985). Uncoupling of cardiac muscarinic and β-adrenergic receptors from ion channels by a guanine nucleotide analogue. *Nature* (*London*) **317,** 538-540.

Druey, K. M., Bumer, K. J., Kang, V. H., and Kehrl, J. H. (1996). Inhibition of G-protein-mediated MAP kinase activation by a new mammalian gene family. *Nature* (*London*) **379,** 742–746.

Doupnik, C. A., Davidson, N., Lester, H. A., and Kofuji, P. (1997). RGS proteins reconstitute the rapid gating kinetics of $G\beta\gamma$-activated inwardly rectifying K^+ channels. *Proc. Natl. Acad. Sci. U.S.A.* **94,** 10461–10466.

Gilman, A. G. (1987). G proteins: Transducers of receptor-generated signals. *Annu. Rev. Biochem.* **56,** 615–649.

Hosoya, Y., Yamada, M., Ito, H., and Kurachi, Y. (1996). A functional model for G protein activation of the muscarinic K^+ channel in guinea pig atrial myocytes. Spectral analysis of the effect of GTP on single-channel kinetics. *J. Gen. Physiol.* **108,** 485–495.

Inanobe, A., Morishige, K., Takahashi, N., Ito, H., Yamada, M., Takumi, T., Nishida, H., Takahashi, K., Kanaho, Y., Katada, T., and Kurachi, Y. (1995). $G_{\beta\gamma}$ directly binds to the carboxyl terminus of the G protein-gated muscarinic K^+ channel, GIRK1. *Biochem. Biophys. Res. Commun.* **212,** 1022–1028.

Ito, H., Sugimoto, T., Kobayashi, I., Takahashi, K., Katada, T., Ui, M., and Kurachi, Y. (1991). On the mechanism of basal and agonist-induced activation of the G protein-gated muscarinic K^+ channel in atrial myocytes of guinea-pig heart. *J. Gen. Physiol.* **98,** 517–533.

Ito, H., Tung, R. T., Sugimoto, T., Kobayashi, I., Takahashi, K., Katada, T., Ui, M., and Kurachi, Y. (1992). On the mechanism of G protein $\beta\gamma$ subunit activation of the muscarinic K^+ channel in guinea pig atrial cell membrane. Comparison with the ATP-sensitive K^+ channel. *J. Gen. Physiol.* **99,** 961–983.

Koelle, M. R., and Horvitz, H. R. (1996). EGL-10 regulates G protein signaling in the *C. elegans* nervous system and shares a conserved domain with many mammalian proteins. *Cell* (*Cambridge, Mass.*) **84,** 115–125.

Krapivinsky, G., Gordon, E.A., Wickman, K., Velimirovic, B., Krapivinsky, L., and Clapham, D. E. (1995). The G-protein-gated atrial K^+ channel I_{KACh} is a heteromultimer of two inwardly rectifying K^+ channel protein. *Nature* (*London*) **374,** 135–141.

Kurachi, Y. (1995). G protein regulation of cardiac muscarinic potassium channel. *Am. J. Physiol.* **269,** C821–C830.

Kurachi, Y., Nakajima, T., and Sugimoto, T. (1986a). Acetylcholine activation of K^+ channels in cell-free membrane of atrial cells. *Am. J. Physiol.* **251,** H681–H684.

Kurachi, Y., Nakajima, T., and Sugimoto, T. (1986b). On the mechanism of activation of muscarinic K^+ channels by adenosine in isolated atrial cells: Involvement of GTP-binding proteins. *Pfluegers Arch.* **407,** 264–274.

Kurachi, Y., Ito, H., Sugimoto, T., Katada, T., and Ui, M. (1989). Activation of atrial muscarinic K^+ channels by low concentrations of $\beta\gamma$ subunits of rat brain G protein. *Pfluegers Arch.* **413,** 325–327.
Kurachi, Y., Ito, H., and Sugimoto, T. (1990). Positive cooperativity in activation of the cardiac muscarinic K^+ channel by intracellular GTP. *Pfluegers. Arch.* **416,** 216–218.
Logothetis, D. E., Kurachi, Y., Galper, J., Neer, E. J., and Clapham, D. E. (1987). The $\beta\gamma$ subunits of GTP-binding proteins activate the muscarinic K^+ channel in heart. *Nature (London)* **325,** 321–326.
Mackay, D. (1990a). Interpretation of relative potencies, relative efficacies and apparent affinity constants of agonist drugs estimated from concentration–response curves. *J. Theor. Biol.* **142,** 415–427.
Mackay, D. (1990b). Agonist potency and apparent affinity: Interpretation using classical and steady-tate ternary-complex models. *Trends. Pharmacol. Sci.* **11,** 17–22.
Monod, J., Wyman, J., and Changeux, J.-P. (1965). On the nature of allosteric transitions: A plausible model. *J. Mol. Biol.* **12,** 88–118.
Nakajima, T., Sugimoto, T., and Kurachi, Y. (1992). Effects of anions on the G protein-mediated activation of the muscarinic K^- channel in the cardiac atrial cell membrane: Intracellular chloride inhibition of the GTPase activity of G_K. *J. Gen. Physiol.* **99,** 665–682.
Pfaffinger, P. J., Martin, J. M., Hunter, D. D., Nathanson, N. M., and Hille, B. (1985). GTP-binding proteins couple cardiac muscarinic receptors to a K channel. *Nature (London)* **317,** 536–540.
Reuveny, E., Slesinger, P. A., Inglese, J., Morales, J. M., Iñdiguez-Lluhi, J. A., Lefkowitz, R. J., Bourne, H. R., Jan, Y. N., and Jan, L. Y. (1994). Activation of the cloned muscarinic potassium channel by G protein $\beta\gamma$ subunits. *Nature (London)* **370,** 143–146.
Szabo, G., and Otero, A. S. (1990). G protein mediated regulation of K^+ channels in heart. *Annu. Rev. Physiol.* **52,** 293–305.
Thomsen, W. J., Jacquez, J. A., and Neubig, R. R. (1988). Inhibition of adenylate cyclase is mediated by the high affinity conformation of the α2-adrenergic receptor. *Mol. Pharmacol.* **34,** 814–822.
Wickman, K. D., Iñdiguez-Lluhi, J. A., Davenport, P. A., Taussig, R., Krapivinsky, G. B., Linder, M. E., Gilman, A. G., and Clapham, D. E. (1994). Recombinant G-protein $\beta\gamma$ subunits activate the muscarinic-gated atrial potassium channel. *Nature (London)* **368,** 255–257.
Yamada, M., Jahangir, A., Hosoya, Y., Inanobe, A., Katada, T., and Kurachi, Y. (1993). G_K* and brain $G_{\beta\gamma}$ activate muscarinic K^+ channel through the same mechanism. *J. Biol. Chem.* **268,** 24551–24554.

PART IV

ATP-Sensitive Potassium Channels

CHAPTER 20

Structure and Function of ATP-Sensitive Potassium Channels

Takashi Miki,* Nobuya Inagaki,*[1] Kazuaki Nagashima,* Tohru Gonoi,† and Susumu Seino*

*Division of Molecular Medicine, Center for Biomedical Science,
Chiba University School of Medicine, Chiba 260, Japan; and †Research Center for Pathogenic Fungi and Microbial Toxicoses, Chiba University, Chiba 260, Japan

I. INTRODUCTION

ATP-sensitive K^+ (K_{ATP}) channels are potassium permeable ion channels, regulated by intracellular ATP and ADP concentrations, the ATP/ADP ratio (Noma, 1983). K_{ATP} channels, originally discovered in heart, were later identified in various tissues such as pancreatic β cells, pituitary, skeletal and smooth muscles, and brain (Ashcroft, 1988). An increase in

[1]Present address: Department of Physiology, Akita University School of Medicine, 1-1-1, Hongo, Akita 010, Japan.

Current Topics in Membranes, Volume 46

1063-5823/99 $30.00

the ATP/ADP ratio closes the K_{ATP} channels and a decrease in the ratio opens them, to regulate a variety of cellular functions, such as hormone secretion (Bernardi *et al.,* 1993), excitability of neurons (Ashford *et al.,* 1988) and muscles (Spruce *et al.,* 1987), and cytoprotection (Amoroso *et al.,* 1990; Terzic *et al.,* 1995), by linking the cell's metabolic state to its membrane potential (Ashcroft, 1988).

In pancreatic β cells, K_{ATP} channels play an important role in the regulation of glucose-induced insulin secretion (Cook *et al.,* 1988; Misler *et al.,* 1992). Glucose is transported into the β cells by the glucose transporter GLUT2, and the subsequent metabolism of glucose produces ATP. The increase in the ATP/ADP ratio closes the K_{ATP} channels, which depolarizes the β-cell membrane and leads to the opening of the voltage-dependent calcium channels, which allows calcium influx. The subsequent rise in the intracellular calcium concentration ($[Ca^{2+}]_i$) in the β cell triggers insulin granule exocytosis. Sulfonylureas such as tolbutamide and glibenclamide, widely used in the treatment of non-insulin-dependent diabetes mellitus, stimulate insulin release by closing the K_{ATP} channels directly (Ashcroft and Ashcroft, 1992). Using a radioiodinated derivative of glibenclamide, two types of sulfonylurea receptor have been identified (Schmid-Antomarchi *et al.,* 1987; Gaines *et al.,* 1988). One type, having high affinity for glibenclamide, is found primarily in pancreatic β-cell membrane preparations; the other, having low affinity for glibenclamide, is present primarily in heart, skeletal muscle, and brain. The high affinity receptor was found to be about 140 kDa by photolabeling (Aguilar-Bryan *et al.,* 1990). However, whether the sulfonylurea receptor itself constitutes the K_{ATP} channel or is a regulatory protein in a more complex K_{ATP} channel was not known. Cloning of the sulfonylurea receptors (SUR) and a novel member of the inward rectifying K^+ channel family (Kir6.2) has clarified the molecular composition of K_{ATP} channels.

II. STRUCTURAL AND FUNCTIONAL PROPERTIES OF CLONED K_{ATP} CHANNELS

A. *Molecular Cloning of Kir6.0 Subfamily*

1. Cloning of Kir6.1

Since electrophysiological studies indicate that the K_{ATP} channels have inwardly rectifying K^+ channel properties (Cook and Hale, 1984; Rorsman and Trube, 1985; Misler *et al.,* 1986), we assumed that the K_{ATP} channels would be related structurally to members of the inwardly rectifying K^+ channel (Kir) family. By homology screening of a rat islet cDNA library with GIRK1 as a probe, a novel inward rectifier K^+ channel member, Kir6.1 (uK_{ATP}-1), was identified (Inagaki *et al.,* 1995a). Since Kir6.1 shares only 40 to 45% amino acid identity with known inward rectifier K+ channel members, it represents a new subfamily. Kir6.1 has two putative transmembrane segments with a K^+ ion permeable domain, the H5 region, which is highly conserved among K^+ channels.

2. Cloning of Kir6.2

Northern blot analysis shows that Kir6.1 is ubiquitously expressed in tissues, including pancreatic islets. However, Kir6.1 is not expressed in the various insulin-secreting cell lines such as MIN6, HIT, and RINm5F cells, all of which are known to have K_{ATP} channels. We reasoned there might be another inward rectifier member present in pancreatic β cells. By using Kir6.1 as a probe, we screened a human genomic library and identified the new member, Kir6.2 (called BIR at the time, for β-cell inward rectifier) (Inagaki *et al.,* 1995b). Kir6.2 is a 390-amino acid protein having 71% amino acid identity with Kir6.1.

3. Tissue Distribution of Kir6.2

Northern blot analysis shows that Kir6.2 is expressed at very high levels in pancreatic islets and in the glucose-responsive insulin secreting-cell lines MIN6 and HIT T15, and it is expressed at low to moderate levels in skeletal muscle, heart, and brain, all of which are known to have K_{ATP} channels. Although the molecular structure and the tissue distribution of Kir6.2 strongly suggest a role in the K_{ATP} channels of pancreatic β cells, expression of recombinant Kir6.2 protein in COS-1 cells, HEK239 cells, or *Xenopus leavis* oocytes elicited no K_{ATP} channel currents, indicating that Kir6.2 does not function alone as a K^+ channel. We then coexpressed Kir6.2 and SUR1 in COS-1 cells.

B. Structural and Functional Properties of β-Cell K_{ATP} Channel

1. Molecular Structure of SUR1

A sulfonylurea receptor, called SUR, was cloned from rat insulinoma cell line (RINm5F) cDNA and hamster insulinoma cell line (HIT T15) cDNA libraries by Aguilar-Bryan *et al.* (1995). From HIT T15 cells, they isolated a 140-kDa protein, which was photolabeled by a radioiodinated derivative of glibenclamide, and obtained the peptide sequence that was used to clone SUR (Aguilar-Bryan *et al.,* 1995). This clone has been renamed SUR1, since there is a family of sulfonylurea receptors (Inagaki *et al.,* 1996). SUR1 is a 1582-amino acid protein, having several potential N-linked glycosylation sites, cAMP-dependent protein kinase phosphorylation sites, and protein kinase C-dependent phosphorylation sites. Like the cystic fibrosis transmembrane conductance regulator (CFTR) and the multidrug resistance gene product MDR (P-glycoprotein) and its related protein MRP, SUR1 is a member of the ATP-binding cassette (ABC) transporter superfamily (Higgins, 1992), having two nucleotide binding folds (NBF) in the cytoplasmic side. Each NBF has Walker A and B motifs (Walker *et al.*, 1982), which are important for the functional activity of ABC proteins. Expression of recombinant SUR in COSm6 cells produced a protein with an apparent molecular size of 140 kDa, having high affinity for the glibenclamide derivative. However, the expression of SUR alone or coexpression with Kir1.1a, Kir2.1, or Kir3.4 in *Xenopus leavis* oocytes generated no ATP-sensitive or glibenclamide-sensitive potassium currents (Aguilar-Bryan *et al.*, 1995). Initially, SUR1 was thought to have 13 transmembrane domains, but Tunsnády has recently proposed 17 transmembrane domains by a topology pattern analysis (Tusnády *et al.*, 1997) (Fig. 1).

2. Reconstitution of the β-Cell K_{ATP} Channel from SUR1 and Kir6.2

SUR1 mRNA was expressed at high levels in pancreatic islets and in MIN6, HIT T15, and αTC-6 cells, the same tissue and cell lines in which Kir6.2 mRNA is expressed at high levels, suggesting strongly that SUR1 couples functionally with Kir6.2 in native pancreatic β cells. Accordingly, we coexpressed SUR1 and Kir6.2 in mammalian cell lines (Inagaki *et al.*,

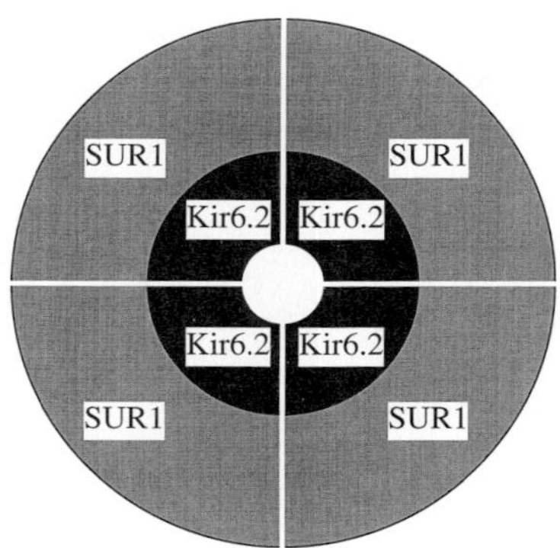

FIGURE 1 K_{ATP} channels are a complex composed of SUR and Kir6.2. The possible membrane topology of SUR1 and Kir6.2 is shown. The membrane topology of SUR is based on Tusnády *et al.* (1997). There are two nucleotide binding folds (NBF1 and NBF2) in SUR. Both NBFs have the Walker A and B consensus motifs (A and B). Kir6.2 possesses two membrane-spanning regions (M1 and M2). The region between the two membrane-spanning regions is H5, which forms the channel pore.

1995b). The COS-1 cells cotransfected with Kir6.2 and SUR1 cDNAs generated distinct K^+ currents. In the presence of 140 m*M* K^+ on both sides of the membrane, the current–voltage relationship exhibited weak inward rectification in the presence of 2 m*M* Mg^{2+} in the intracellular solution. The single-channel conductance was 76.4 ± 1.0 pS at a membrane potential of −60 mV. The reconstituted K^+ currents are inhibited by ATP in a dose-dependent manner with an IC_{50} value of 10 μ*M*. The K^+ currents are also inhibited slightly by 1 m*M* adenosine diphosphate (ADP) or 1 m*M* adenosine monophosphate (AMP) in the presence of 1 m*M* ATP and are almost completely inhibited by 0.1 m*M* AMP-PNP, a nonhydrolyzable ATP analog. The sulfonylurea glibenclamide (0.1 μ*M*), a potent inhibitor of β-cell K_{ATP} channels, blocks the reconstituted K^+ currents. In contrast, diazoxide, a potent opener of β-cell K_{ATP} channels, stimulates the reconstituted K^+ currents in the presence of 2 m*M* Mg^{2+} and 1 m*M* ATP. K^+ currents with the same properties are reconstituted in *Xenopus leavis* oocytes coinjected with Kir6.2 and SUR1 cRNAs. With 90 m*M* K^+ in the external solution, inward currents are increased by diazoxide (100 μ*M*) and decreased by glibenclamide (0.1 μ*M*). K_{ATP} channel currents also were measured as $^{86}Rb^+$ efflux from COSm6 cells coexpressing Kir6.2 and SUR1 in the presence of metabolic inhibitors oligomycin (2.5 mg/ml) and 2 m*M* 2-deoxyglucose. Efflux through the reconstituted channels is inhibited by glibenclamide and activated by diazoxide. Heterologous expression studies in different systems show that K^+ channels reconstituted from Kir6.2 and SUR1 have the features of K_{ATP} channels in native pancreatic β cells. Accordingly, we have concluded that the K_{ATP} channels in pancreatic β cells comprise the two subunits, Kir6.2 and SUR1 (Aguilar-Bryan *et al.*, 1995) (Fig. 2).

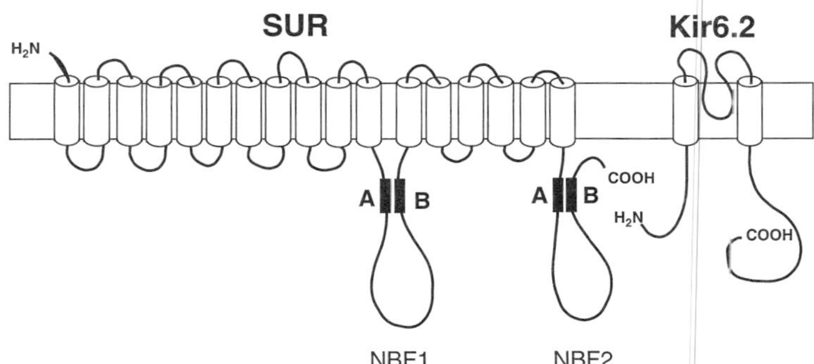

FIGURE 2 Subunit stoichiometry of β-cell K_{ATP} channel. Functional β-cell K_{ATP} channels are heteromultimers, most likely heterooctamers assembled with a tetramer of Kir6.2 and a tetramer of SUR1.

C. Structural and Functional Properties of Cardiac and Skeletal Muscle K_{ATP} Channel

1. Cloning an Isoform of SUR1, SUR2A

Since tissues such as skeletal muscle, heart, and brain, which are known to have K_{ATP} channels, express Kir6.2 but not SUR1, we screened rat brain and rat heart cDNA libraries using hamster SUR1 cDNA as a probe, and isolated cDNA encoding a novel SUR isoform, currently designated SUR2A (Inagaki *et al.*, 1996). SUR2A is a 1545-amino acid protein, having 68% identity with SUR1, with potential N-linked glycosylation sites, cAMP-dependent protein kinase phosphorylation sites, and protein kinase C-dependent phosphorylation sites. Comparison of the amino acid sequences of SUR2A and SUR1 shows that SUR2A has two potential (NBFs) with Walker A and B consensus motifs which are highly conserved between the two proteins (Higgins, 1992).

2. Tissue Distribution of SUR2A

SUR2A mRNA is shown to be expressed at high levels in heart, skeletal muscle, and ovary, at moderate to low levels in pancreatic islets, brain, tongue, testis, and adrenal gland, and at very low levels in stomach, colon, thyroid, and pituitary, by Northern blot analysis.

3. Reconstitution of the Cardiac and Skeletal Muscle K_{ATP} Channel from SUR2A and Kir6.2

The coexpression of SUR2A and Kir6.2 in COS-1 cells reconstituted potassium channels currents (SUR2A/Kir6.2) with inward rectification and unitary channel conductance almost identical to those of β-cell K_{ATP} channels (SUR1/Kir6.2) (Inagaki *et al.*, 1995b). ATP also inhibited the SUR2A/Kir6.2 channel in a dose-dependent manner. However, the SUR2A/Kir6.2 channel is less than one-tenth as sensitive to ATP as the SUR1/Kir6.2 channel. Glibenclamide at 1 μM, a concentration sufficient to block SUR1/Kir6.2 channel currents, only slightly inhibited SUR2/Kir6.2 channels. While the K^+ channel openers cromakalim (30 μM) and pinacidil (30 μM) stimulate channel activity, diazoxide (100 μM) does not activate SUR2/Kir6.2 channels. These properties are similar to those of K_{ATP} channels in native cardiac and skeletal muscle. The electrophysiological and pharmacological properties of cloned K_{ATP} channels in COS-1 cells are summarized in Table I.

Isomoto *et al.* (1996) have reported a variant of SUR2, designated SUR2B. SUR2B is generated by alternative splicing from the SUR2 gene. As a result, SUR2B has 42 amino acid in its C-terminus divergent from SUR2A, but similar to the C terminus of SUR1. Another variant of SUR2,

TABLE I

Electrophysiological Properties of Cloned K_{ATP} Channels Expressed in COS-1 Cells[a]

	β-cell type (SUR1/Kir6.2)	Cardiac and skeletal muscle type (SUR2A/Kir6.2)
Single-channel conductance	76 pS	79 pS
Inward rectification	Weak	Weak
Effect of nucleotides		
ATP	Complete inhibition at 1 mM (IC_{50} 10 μM)	Complete inhibition at 1 mM (IC_{50} 50–100 μM)
ADP	Slight inhibition at 1 mM	Slight inhibition at 1 mM
AMP	Slight inhibition at 1 mM	No inhibition at 1 mM
AMP-PNP	Almost complete inhibition at 0.1 mM	Almost complete inhibition at 1 mM
Effect of channel blocker glibenclamide	Almost complete inhibition at 100 nM	Slight inhibition at 1 μM
Effect of channel openers		
Diazoxide	Stimulation at 100 μM	No stimulation at 100 μM
Cromakalim	n.d.	Stimulation at 30 μM
Pinacidil	n.d.	Stimulation at 30 μM

[a] The data were obtained in the presence of 2 mM Mg^{2+} in the intracellular solution. n.d., Not determined.

SUR2C, has been identified (Chukow *et al.,* 1996; Ashcroft and Gribble, 1998). The coexpression of SUR2B and Kir6.2 elicits K_{ATP} channel currents with properties similar to those of the K_{ATP} channels in smooth muscle (Isomoto *et al.,* 1996). In addition, Yamada *et al.* (1997) have reported that Kir6.1 and SUR2B constitute ATP-insensitive but nucleotide diphosphate-dependent K^+ channels, features similar to those of vascular smooth muscle K^+ channels. These findings suggest that different combinations of SUR isoforms or their variants plus Kir6.2 or Kir6.1 may account, at least in part, for the functional diversity of ATP-sensitve and sulfonylurea-sensitive K^+ channels. Tucker *et al.* (1997) have found that expression of C-terminal truncated Kir6.2 in *Xenopus laevis* oocytes and HEK239 cells generates ATP-sensitive K^+ channel currents in the absence of sulfonylurea receptor, but the precise domain conferring ATP-sensitivity remains to be determined.

III. STOICHIOMETRY OF THE β-CELL K_{ATP} CHANNEL

Clement *et al.* (1997) have shown that SUR1 and Kir6.2 are cophotolabeled by ^{125}I-azidoglibenclamide, when both subunits are coexpressed in

COSm6 cells. Kir6.2 was not photolabeled when expressed alone, indicating that SUR1 and Kir6.2 are physically associated. The association of SUR1 and Kir6.2 was further confirmed by copurification of the glycosylated SUR1 and Kir6.2 on wheat germ agglutinin agarose and by Ni^{2+}–agarose chromatography of histidine-tagged SUR1 plus Kir6.2 complexes. Although inward rectifier K^+ channels function as hetero- or homotetramers (Krapivinsky *et al.*, 1995; Yang *et al.*, 1995; Kofuji *et al.*, 1995; Tinker *et al.*, 1996), the stiochiometry of the K_{ATP} channel was unknown. To investigate the structure of the β-cell K_{ATP} channel, we generated fusion polypeptides with Kir6.2 (α) and SUR1 (β) (Inagaki *et al.*, 1997). Two polypeptides, $\beta\alpha$ polypeptide and $\beta\alpha_2$ polypeptide, consist of a SUR1 (β) subunit and a Kir6.2 (α) subunit and a SUR1 (β) subunit and a dimeric repeat of Kir6.2 (α) subunit, respectively. By analyzing the channel properties in COS-1 cells reconstituted from these polypeptides, the optimal ratio of α and β subunits was determined. The properties of K_{ATP} channels reconstituted by $\beta\alpha$ polypeptide were almost the same as those reconstituted by β and α momomers, suggesting that the coexpression of β and α subunits with a molar ratio of 1 : 1 is sufficient for functional expression of the channel. Coexpression of a monomers with $\beta\alpha$ polypeptide resulted in a decrease in $^{86}Rb^+$ efflux in a dose-dependent manner. In contrast, coexpression of β monomers with $\beta\alpha$ polypeptide did not further increase the $^{86}Rb^+$ efflux, indicating that K_{ATP} channel activity is optimized when SUR1 and Kir6.2 are expressed with a molar ratio of 1 : 1.

Although $^{86}Rb^+$ efflux from COS-1 cells expressing $\beta\alpha_2$ polypeptide was reduced compared to the efflux with β and α monomers, supplementation with coexpression of β monomers restored the efflux, supporting the conclusion. Considering that inward rectifying K^+ channels function as a tetramer, the K_{ATP} channel is likely to function as a heterooctamer, a tetramer of Kir6.2 and a tetramer of SUR1. Similar findings also have been reported by Clement *et al.* (1997).

IV. TRANSGENIC MICE EXPRESSING A DOMINANT-NEGATIVE K_{ATP} CHANNEL

A. *Generation of Transgenic Mice and the Phenotype*

Since the β-cell K_{ATP} channels play an important role in glucose-induced insulin secretion, abnormalities of the SUR1 or Kir6.2 genes could contribute to the development of disorders of glucose homeostasis such as diabetes and hypoglycemia. Genetic mutations of SUR1 or Kir6.2 have been identified in persistent hyperinsulinemic hypoglycemia of infancy (PHHI)

(Thomas *et al.*, 1995; Aguilar-Bryan and Bryan, 1996; Dunne *et al.*, 1997; Nestorowitz *et al.*, 1997), and abnormalities of the β-cell K_{ATP} channel have been implicated in the pathophysiology of PHHI. PHHI is an autosomal recessive, relatively rare, hereditary disease occurring in approximately 1 in 50,000 persons in Western countries, but more frequently in some Arabic communities, where it occurs in 1 in 2500 persons (Thomas, 1997). The disease is characterized by hypoglycemia with inappropriate, excessive insulin secretion in neonates and infants.

To investigate the direct effect of disruption of the β-cell K_{ATP} channel function, we have generated transgenic mice expressing a dominant-negative K_{ATP} channel in pancreatic β cells (Miki *et al.*, 1997). The putative K^+ ion-permeable domain, H5, is highly conserved in K^+ channels, and the motif Gly-Tyr (or Phe)-Gly is thought to be critical for K^+ ion selectivity (Jan and Jan, 1994). Mutant Kir6.2 with a substitution of the first residue of the Gly-Phe-Gly motif with Ser, Kir6.2G132S, shows no K^+ channel activity when coexpressed with SUR1 in COS-1 cells. It also was shown that this mutant Kir6.2, if coexpressed with wild-type Kir6.2 and SUR1 in COS-1 cells, decreases SUR1/Kir6.2 currents by a dominant-negative mechanism, by $^{86}Rb^+$ efflux measurements. We then generated transgenic mice expressing this mutant Kir6.2 specifically in their β cells under the regulation of the human insulin promoter. We have established 13 transgenic lines. Transgenic mice exhibited severe hypoglycemia at birth, and blood glucose levels in most of the neonates were below the sensitivity of the assay [transgenic mice, 37.0 $\pm$ 3.3 (mean $\pm$ SE) mg/dl; control mice, 63.7 $\pm$ 5.3 mg/dl]. The insulin levels remained relatively high, in spite of the hypoglycemia, suggesting a phenotype resembling PHHI in humans. However, blood glucose levels in transgenic mice at 4 weeks became markedly elevated [transgenic mice, 424 $\pm$ 45.1 mg/dl; controls, 185 $\pm$ 15.5 mg/dl] and increased further to 610 $\pm$ 45.5 mg/dl at 6 weeks of age. Glucose-induced insulin secretion was markedly reduced in the transgenic mice.

B. Functional Changes in Pancreatic β Cells of Transgenic Mice

Electrophysiological analysis with the whole-cell configuration showed the normalized K_{ATP} channel conductance of the β cells of transgenic mice to be significantly decreased [transgenic mice (line M4), 0.37 $\pm$ 0.25 nS/pF; control mice, 2.29 $\pm$ 0.27 nS/pF] (Table II). The resting membrane potential of the β cells of transgenic mice (line M4) (-46.7 ± 3.9 mV, $n = 30$) was significantly higher than that of control mice (-63.5 ± 2.0 mV, $n = 22$). Basal intracellular calcium concentrations ($[Ca^{2+}]_i$) also were significantly elevated in the β cells of transgenic mice [transgenic mice

TABLE II
Characteristics of the Transgenic Mice

	Transgenic mice	Control mice
Blood glucose levels at birth	37.0 ± 3.3 mg/dl	63.7 ± 5.3 mg/dl
Blood glucose levels at 4 weeks	424 ± 45 mg/dl	185 ± 15.5 mg/dl
Insulin response to glucose	Impaired	Normal
Normalized K_{ATP} channel conductance	0.37 ± 0.25 nS/pF	2.29 ± 0.27 nS/pF
Resting membrane potential	−46.7 ± 3.9 mV	−63.5 ± 2.0 mV
Basal $[Ca^{2+}]_i$	137.1 ± 15.9 n*M*	90.6 ± 3.5 n*M*

(line M4), 137.1 ± 15.9 n*M;* control mice, 90.6 ± 3.5 n*M*]. These data suggest that K_{ATP} channels are the major determinant of the resting membrane potential and basal $[Ca^{2+}]_i$ in pancreatic β cells, and that the disruption of the channel may lead to chronic depolarization of the membrane and sustained elevations of $[Ca^{2+}]_i$. The characteristics of the transgenic mice are summarized in Table II.

C. Histological Changes in Pancreatic Islets of Transgenic Mice

Histological analysis revealed an abnormal morphological configuration of the islet cells in the transgenic mice. Glucagon-positive α cells, which are present normally in the periphery of pancreatic islets, appear also in the central region of the pancreatic islets of transgenic mice. The morphological abnormalities and the decrease in number of the β cells became more pronounced with age. While the α-cell population in transgenic mice was slightly increased, the β-cell population was significantly decreased by approximately 50%, compared to control mice. Since the decrease in population of β cells might be due to a decrease in cell proliferation or an increase in cell death, the cell proliferation rate was evaluated by 5′-bromodeoxyuridine (5′BrdU) labeling. The number of 5′BrdU-labeled β-cells in transgenic mice was not different from that in control mice (data not shown). To learn if the decrease in β-cell population is due to cell death, the occurrence of apoptosis was investigated by TUNEL (TdT-mediated dUTP nick endolabeling) method. A high frequency of apoptotic β cells was found in the transgenic mice, whereas such cells were found only rarely in controls, suggesting the cause of the decrease in β-cell population in transgenic mice to be acceleration of cell death. Considering these findings together, it is likely that the loss of K_{ATP} channel function causes membrane depolarization and high basal $[Ca^{2+}]_i$ in the β cells, and that the chronic elevation of $[Ca^{2+}]_i$ finally promotes apoptotic cell death, causing hyperglycemia. In addition to their role in the regulation of glucose-induced insulin secretion, K_{ATP} channels may also play an important role in β-cell survival.

V. CONCLUSION

Molecular biological studies have revealed that the K_{ATP} channel is a complex of two subunits: a member of the ATP-binding cassette (ABC) transporter superfamily, SUR, and a member of the inward rectifying K^+ channel family, Kir6.2. While the Kir6.2 subunit forms the K_{ATP} channel pore and determines ion selectivity and the rectification properties of the channel, the SUR subunit confers ATP sensitivity and various pharmacological properties. However, ATP sensitivity might also be endowed by the Kir6.2 subunit, so the precise domain conferring ATP sensitivity is still inconclusive. Clarification of the molecular basis of K_{ATP} channels should provide a better understanding of the relationship of the function of the channel to its structure as well as of its roles in physiological functions and pathophysiological states.

Acknowledgments

Part of the study of transgenic mice was performed in collaboration with Dr. J.-I. Miyazaki (Osaka University) and Dr. T. Iwanaga (Hokkaido University). The studies of our laboratory were supported by Scientific Research Grants from the Ministry of Education, Science and Culture and from the Ministry of Health and Welfare, Japan.

References

Aguilar-Bryan, L., and Bryan, J. (1996). ATP-sensitive potassium channels, sulfonylurea receptors, and persistent hyperinsulinemic hypoglycemia of infancy. *Diabetes Rev.* **4,** 336–346.

Aguilar-Bryan, L., Nelson, D. A., Vu, Q. A., Humphrey, M. B., and Boyd III, A. E. (1990). Photoaffinity labeling and partial purification of the beta cell sulfonylurea receptor using a novel, biologically active glyburide analog. *J. Biol. Chem.* **265,** 8218–8224.

Aguilar-Bryan, L., Nicholas, C. G., Wechsler, S. W., Clement IV, J. P., Boyd III, A. E., Gonzalez, G., Herrera-Sosa, H., Nguy, K., Bryan, J., and Nelson, D. A. (1995). Cloning of the β-cell high-affinity sulfonylurea receptor: A regulator of insulin secretion. *Science* **268,** 423–426.

Amoroso, S., Schmid-Antomarchi, H., Fosset, M., and Lazdunski, M. (1990). Glucose, sulfonylureas, and neurotransmitter release: Role of ATP-sensitive K^+ channels. *Science* **247,** 852–854.

Ashcroft, F. M. (1988). Adenosine 5′-triphosphate-sensitive potassium channels. *Annu. Rev. Neurosci.* **11,** 97–118.

Ashcroft, S. J. H., and Ashcroft, F. M. (1992). The sulfonylurea receptor. *Biochim. Biophys. Acta* **1175,** 45–59.

Ashcroft, F. M., and Gribble, F. M. (1998). Correlating structure and function in ATP-sensitive potassium channels. *Trends Neurosci.* **21,** 288–294.

Ashford, M. L., Sturgess, N. C., Trout, N. J., Gardner, N. J., and Hales, C. N. (1988). Adenosine-5′-triphosphate-sensitive ion channels in neonatal rat cultured central neurons. *Pfluegers Arch.* **412**, 297–304.

Bernardi, H., DeWeille, J. R., Epelbaum, J., Mourre, C., Amoroso, S., Slama, A., Fosset, M., and Lazdunski, M. (1993). ATP-modulated K^+ channels sensitive to antidiabetic sulfonylureas are present in adenohypophysis and are involved in growth hormone release. *Proc. Natl. Acad. Sci. U.S.A.* **90,** 1340–1344.

Chukow, W. A., Simon, M. C., Le Beau, M. M., and Burant, C. F. (1996). Cloning, tissue expression, and chromosomal localization of SUR2, the putative drug-binding subunit of cardiac, skeletal muscle, and vascular K_{ATP} channels. *Diabetes* **45,** 1439–1445.

Clement IV, J. P., Kunjilwar, K., Gozalez, G., Schwanstecher, M., Peanten, U., Aguilar-Bryan, L., and Bryan, J. (1997). Association and stoichiometry of K_{ATP} channel subunits. *Neuron* **18,** 827–838.

Cook, D. L., and Hales, C. H. (1984). Intracellular ATP directly blocks K^+ channels in pancreatic β cells. *Nature (London)* **311,** 271–273.

Cook, D. L., Satin, L. S., Ashford, M. L. J., and Hales, C. N. (1988). ATP-sensitive K^+ channels in pancreatic β-cells. *Diabetes* **37,** 495–498.

Dunne, M. J., Kane, C., Shepherd, R. M., Sanchez, J. A., James R. F. L., Johnson, P. R. V., Aynsley-Green, A., Lu, S., Clement IV, J. P., Lindley, K. J., Seino, S., and Aguilar-Bryan, L. (1997). Familial persistent hyperinsulinemic hypoglycemia of infancy and mutations in the sulfonylurea receptor. *N. Engl. J. Med.* **336,** 703–706.

Gaines, K. L., Hamilton, S., Boyd III, A. E. (1988). Characterization of the sulfonylurea receptor on beta cell membranes. *J. Biol. Chem.* **263,** 2589-2592.

Higgins, C. F. (1992). ABC transporters: From microorganisms to man. *Annu. Rev. Cell Biol.* **8,** 67–113.

Inagaki, N., Tsuura, Y., Namba, N., Masuda, K., Gonoi, T., Horie, M., Seino, Y., Mizuta, M., and Seino, S. (1995a). Cloning and functional characterization of a novel ATP-sensitive potassium channel ubiquitously expressed in rat tissues, including pancreatic islets, pituitary, skeletal muscle, and heart. *J. Biol. Chem.* **270,** 5691–5694.

Inagaki, N., Gogoi, T., Clement IV, J. P., Namba, N., Inazawa, J., Gonzalez, G., Aguilar-Bryan, L., Seino, S., and Bryan, J. (1995b). Reconstitution of I_{KATP}: An inward rectifier subunit plus the sulfonylurea receptor. *Science* **17,** 1166–1170.

Inagaki, N., Gonoi, T., Clement IV, J. P., Wang, C.-Z., Aguilar-Bryan, L., Bryan, J., and Seino, S. (1996). A family of sulfonylurea receptors determines the pharmacological properties of ATP-sensitive K^+ channels. *Neuron* **16,** 1011–1017.

Inagaki, N., Gonoi, T., and Seino, S. (1997). Subunit stoichiometry of the pancreatic β-cell K^+ channel. *FEBS Lett.* **409,** 232–236.

Isomoto, S., Kondo, C., Yamada, M., Matsumoto, S., Higashiguchi, O., Hoyio, Y., Matsuzawa, Y., and Kurachi, Y. (1996). A novel sulfonylurea receptor forms with BIR (Kir6.2) a smooth muscle type ATP-sensitive K^+ channels. *J. Biol. Chem.* **271,** 24321–24324.

Jan, L. Y., and Jan, Y. N. (1994). Potassium channels and their evolving gates. *Nature (London)* **371,** 119–122.

Kofuji, P., Davidson, N., and Lester, H. A. (1995). Evidence that neuronal G-protein-gated inwardly rectifying K^+ channels are activated by $G\beta\gamma$ subunits and function as heteromultimers. *Proc. Natl. Acad. Sci. U.S.A.* **92,** 6542–6546.

Krapivinsky, G., Gordon, E. A., Wickman, K., Velimirovic, B., Krapivinsky, L., and Clapham, D. E. (1995). The G-protein gated atrial K^+ channel IKACh is a heteromultimer of two inwardly rectifying K^+ channel proteins. *Nature (London)* **374,** 135–141.

Miki, T., Tashiro, F., Iwanaga, T., Nagashima, K., Yoshitomi, H., Aihara, H., Nitta, Y., Gonoi, T., Inagaki, N., Miyazaki, J.-I., and Seino, S. (1997). Abnormalities of pancreatic islets by targeted expression of a dominant-negative K_{ATP} channel. Proc. *Natl. Acad. Sci. U.S.A.* **94,** 11969–11973.

Misler, S., Falke, L. C., Gillis, K., and McDaniel, M. L. (1986). A metabolite-regulated potassium channel in rat pancreatic β-cells. *Proc. Natl. Acad. Sci. U.S.A.* **83,** 7119–7123.

Misler, S., Barnett, D. W., Gillis, K. D., and Pressel, D. M. (1992). Electrophysiology of stimulus–secretion coupling in human β-cells. *Diabetes* **41,** 1211–1228.

Nestorowitz, A., Inagaki, N., Gonoi, T., Schoor, K. P., Wilson, B. A., Glaser, E., Landau, H., Stanley, C. A., Thorenton, P. S., Seino, S., and Permutt, A. (1997). A nonsense mutation in the inward rectifier potassium channel gene, Kir6.2, is associated with familial hyperinsulinism. *Diabetes* **46,** 1743–1748.

Noma, A. (1983). ATP-regulated K^+ channels in cardiac muscle. *Nature (London)* **305,** 147–148.

Rorsman, P., and Trube, G. (1985). Glucose dependent K^+-channels in pancreatic β-cells are regulated by intracellular ATP. *Pfluegers Arch.* **405,** 305–309.

Schmid-Antomarchi, H., De Weille, J., Fosset, M., and Lazdunski, M. (1987). The receptor for antidiabetic sulfonylurea controls the activity of the ATP-modulated K^+ channel in insulin-secreting cells. *J. Biol. Chem.* **262,** 15840–15844.

Spruce, A. E., Standen, N. B., and Stanfield, P. R. (1987). Voltage-dependent ATP-sensitive potassium channels of skeletal muscle membrane. *J. Physiol. (London)* **382,** 213–236.

Terzic, A., Jahangir, A., and Kurachi, Y. (1995). Cardiac ATP-sensitive K^+ channels: Regulation by intracellular nucleotides and K^+ channel-opening drugs. *Am. J. Physiol.* **269,** C525–C545.

Thomas, P. M., Cote, G. J., Wohllk, N., Haddad, B., Mathew, P. M., Rabl, W., Aguilar-Bryan, L., and Bryan, J. (1995). Mutations in the sulfonylurea receptor gene in familial persistent hyperinsulinemic hypoglycemia of infancy. *Science* **268,** 426–429.

Thomas, P. M. (1997). Molecular basis for familial hyperinsulinemic hypoglycemia of infancy. *Curr. Opin. Endocrinol. Diabetes* **4,** 272–276.

Tinker, A., Jan, Y. N., and Jan, L. Y. (1996). Regions responsible for the assembly of inwardly rectifying potassium channels. *Cell* (Cambridge, Mass.) **87,** 857–868.

Tucker, S. J., Gribble, F. M., Zhao, C., Trapp, S., and Ashcroft, F. M. (1997). Truncation of Kir6.2 produces ATP-sensitive K^+ channels in the absence of the sulfonylurea receptor. *Nature (London)* **387,** 179–183.

Tusnády, G. E., Bakos, É. Váradi, A., and Sarkadi, B. (1997). Membrane topology distinguishes a subfamily of the ATP-binding cassette (ABC) transporters. *FEBS Lett.* **402,** 1–3.

Walker, J. E., Saraste, M., Runswick, M. J., and Gay, N. J. (1982). Distantly related sequences in the α- and β-subunits of ATP synthase, myosin kinases and other ATP-requiring enzymes and a common nucleotide binding fold. *EMBO J.* **8,** 945–951.

Yamada, M., Isomoto, S., Matsumoto, S., Kondo, C., Horio, Y., and Kurachi, Y. (1997). Sulfonylurea receptor 2B and Kir6.1 form a sulfonylurea-sensitive but ATP-insensitive K^+ channel. *J. Physiol. (London)* **499,** 715–720.

Yang, J., Jan, Y. N., and Jan, L. Y. (1995). Determination of the subunit stoichiometry of an inwardly rectifying potassium channel. *Neuron* **15,** 1441–1447.

CHAPTER 21

Molecular Structure and Function of Cardiovascular ATP-Sensitive Potassium Channels

Mitsuhiko Yamada, Eisaku Satoh, Chikako Kondo, Vez P. Repunte, and Yoshihisa Kurachi
Department of Pharmacology II, Faculty of Medicine, Osaka University, Osaka 565, Japan

I. INTRODUCTION

The ATP-sensitive K^+ (K_{ATP}) channel is a weakly inwardly rectifying K^+ channel which is inhibited by intracellular ATP (ATP_i) and activated by intracellular nucleoside diphosphates (NDP_i). Thus, it provides a link between cellular metabolism and excitability. This channel was first identified in cardiac myocytes by Noma (1983) and subsequently by other groups in pancreatic β cells and skeletal muscle (for review, see Ashcroft, 1988;

1063-5823/99 $30.00

Terzic *et al.*, 1995). It has been believed that activation of cardiac and skeletal muscle K_{ATP} channels attenuates muscle contraction during metabolic inhibition or oxygen shortage, whereas suppression of the channel in pancreatic β cells is responsible for the glucose-induced secretion of insulin. The K_{ATP} channels also exhibit characteristic pharmacological properties: they are selectively inhibited by antidiabetic sulfonylurea derivatives (SUs), such as glibenclamide and tolbutamide, and activated by a certain class of vasorelaxants, such as pinacidil, levcromakalim, and nicorandil, which are collectively termed K^+ channel openers (KCOs) (Table I).

The pancreatic, cardiac, and skeletal muscle K_{ATP} channels all exhibit a single-channel conductance of ~70–90 pS under the symmetrical 150 m*M* K^+ conditions (Table I; Ashcroft, 1988; Terzic *et al.*, 1995). Their responses to intracellular nucleotides or to pharmacological agents are, however, diverse (Table 1). The pancreatic and skeletal muscle K_{ATP} channels are more sensitive to Mg^{2+}-free than Mg^{2+}-bound ATP_i, whereas the cardiac K_{ATP} channel is equally sensitive to both forms of ATP_i. The pancreatic K_{ATP} channel is inhibited by lower concentrations of tolbutamide, while submillimolar concentrations of this agent are needed to inhibit the cardiac K_{ATP} channel. The pancreatic K_{ATP} channel is activated by diazoxide but not by pinacidil, and vice versa for cardiac and skeletal muscle K_{ATP} channels. Thus, KCOs exhibit clear tissue specificity. These K_{ATP} channels, therefore, may be similar but distinct members of the same family of K^+ channels. This class of K_{ATP} channels has often been referred to as "classical-type" K_{ATP} channels.

The fact that various KCOs such as pinacidil and diazoxide potently induce vasorelaxation in an SU-sensitive manner indicates that vascular smooth muscle may also possess some types of K_{ATP} channels (Edwards and Weston, 1993). Indeed, many distinct types of K^+ channels have been reported as K_{ATP} channels in vascular smooth muscle cells (Table I) (Quayle *et al.*, 1997). However, despite the similarity in pharmacology, these channels exhibit single-channel characteristics and nucleotide regulation distinct from those of the classic K_{ATP} channels. For example, the most commonly observed vascular K_{ATP} channel, which is often called the "small-conductance" K_{ATP} channel or the "nucleoside diphosphate-dependent" K^+ (K_{NDP}) channel (Kajioka *et al.*, 1991; Beech *et al.*, 1993a,b; Kamouchi and Kitamura, 1994; Zhang and Bolton, 1995, 1996), exhibits less than half of the single-channel conductance of the classic K_{ATP} channels (Table I). The classic K_{ATP} channels open spontaneously when ATP_i is removed from the internal surface of the membrane, while the K_{NDP} channel requires NDP_i to open. Furthermore, the K_{NDP} channel is reported to be activated rather than inhibited by ATP_i. Such striking differences between classic K_{ATP} channels and vascular K_{NDP} channels have given some confusion as to the identity of the so-called K_{ATP} channels in vascular smooth muscle cells.

In this chapter, we review progress in the molecular dissection of cardiovascular K_{ATP} channels and show that the apparent differences between cardiac K_{ATP} and vascular K_{NDP} channels can be explained in terms of different combinations of subunits with similar molecular structures.

II. MOLECULAR STRUCTURE OF CARDIOVASCULAR K_{ATP} CHANNELS

It has been shown that the K_{ATP} channels are heteromers composed of sulfonylurea receptors (SURs) and Kir6.0 subunits (Fig. 1) (Inagaki *et al.*, 1995b, 1996; Sakura *et al.*, 1995; Isomoto *et al.*, 1996; Yamada *et al.*, 1997, Aguilar-Bryan *et al.*, 1998). Three distinct SURs (SUR1, SUR2A, and SUR2B) have been functionally characterized in detail. SUR1, SUR2A, and SUR2B are believed to be included in K_{ATP} channels of pancreatic β cells, cardiac or skeletal muscle myocytes, and smooth muscle cells, respectively.

Aguilar-Bryan *et al.* (1995) for the first time cloned a high-affinity sulfonylurea receptor (SUR) from rat insulinoma (RINm5F) and hamster insulin-secreting tumor cell (HIT T15) cDNA libraries. The SUR is a protein composed of 1581 amino acids and possesses two putative nucleotide-binding folds (NBFs) each of which has the Walker A and Walker B consensus motifs (Walker *et al.*, 1982), indicating that it is a member of the ATP-binding cassette (ABC) superfamily. The SUR is likely to possess 13 putative transmembrane domains with the amino and carboxyl termini in the extracellular and intracellular spaces, respectively. The first NBF is located in the cytoplasmic loop between the ninth and tenth transmembrane domains, whereas the second is on the intracellular carboxyl terminus (for details see Chapter 20). The heterologously expressed SUR formed a high-affinity SU binding site but did not show K_{ATP} channel activity by itself. However, Thomas *et al.* (1995) identified two separate SUR gene mutations in the patients with familial persistent hyperinsulinemic hypoglycemia of infancy (PHHI), strongly indicating a crucial role for SUR in regulation of insulin secretion. Inagaki *et al.* (1995b) and Sakura *et al.* (1995) for the first time found that SUR forms the pancreatic type of the classic K_{ATP} channel when coexpressed with an inwardly rectifying K^+ channel (Kir) subunit, BIR (Kir6.2) (Fig. 1). Kir6.2 was cloned from a human genomic library by homology screening by using uK_{ATP}-1 (Kir6.1) as a probe. It is composed of 390 amino acids (Inagaki *et al.*, 1995a, 1996). Consistent with the functional study, both SUR1 and Kir6.2 mRNAs are abundantly ex-

TABLE I

Biophysical and Pharmacological Properties of the Native and Cloned K^{ATP} Channels

Channel	Single-channel conductance[a]	Effect of ATP_i: (IC_{50})[b]	Effect of ATP_i: Hill coefficient[b]	Glibenclamide (IC_{50})	Tolbutamide (IC_{50})	Pinacidil (EC_{50})	Diazoxide (EC_{50})	References[c]
Native K_{ATP} channels								
Pancreatic β cells	~50–90 pS	Mg^{2+} (−)[d]: ~5 μM	~1	~5–30 nM	~5–20 μM	>500 μM	~20–100 μM	1
		Mg^{2+} (+)[e]: ~15–45 μM	~1					
Cardiac myocytes	~70–90 pS	Mg^{2+} (−) and (+): ~20–100 μM	~2–3	~5 nM	~400 μM	~10–30 μM	No effect, or inhibitory at 500 μM	1–4
Skeletal muscle cells	~55–75 pS	Mg^{2+} (−): ~10–20 μM	~1.5–2	~10–200 nM	~50 μM	~100 μM	No effect	5–10
		Mg^{2+} (+): ~200 μM	~1.5–2					
Smooth muscle cells	~30 pS[f]	Mg^{2+} (−): ~30–200 μM[g]		~25nM	~350 μM[h]	~0.5 μM[h]	~40 μM[h]	11–15
		Mg^{2+} (+): stimulatory at mM or inhibitory at ~1 mM[g]						
	~100–250 pS[i]							
SUR/Kir channels								
SUR1/Kir6.2	~75 pS	Mg^{2+} (−): ~10 μM		~2–10 nM	~5–30 μM	No effect a 1 mM	~50 μM	16–19
		Mg^{2+} (+): ~10–300 μM	~1					
SUR2A/Kir6.2	~80 pS	Mg^{2+} (−): ~150 μM	~2	~150 nM	~120 μM	~10 μM	No effect at 200 μM	20, 21
		Mg^{2+} (+): ~100–150 μM	~2					
SUR2B/Kir6.2	~80 pS	Mg^{2+} (−): ~70 μM	~2	>70% inhibition at 1 μM	>70% inhibition at 500 μM	~1 μM	Stimulatory at 200 μM	22
		Mg^{2+} (+): ~300 μM	~1.5					

SUR2B/Kir6.1	~33 pS	Mg^{2+} (−): no effect[j] Mg^{2+} (+): stimulatory at m*M*, inhibitory at 10 m*M*[k]		>90% inhibition at 3 μ*M*		Stimulatory at ≥1 μ*M*[l]	Stimulatory at 200 μ*M*[l]	23, 24

[a] The values under the symmetrical 150 m*M* K^+ or similar conditions.

[b] Only applicable when ATP inhibits the channels.

[c] References: 1, Ashcroft and Ashcroft, 1990; 2, Hamada *et al.*, 1990; 3, Findlay, 1992; 4, Faivre and Findlay, 1989; 5, Weik and Neumcke, 1989, 1990; 6, Vivaudou *et al.*, 1991; 7, Benton and Haylett, 1992; 8, Allard and Lazdunski, 1993; 9, McKillen *et al.*, 1994; 10, Allard *et al.*, 1995; 11, Kajioka *et al.*, 1991; 12, Beech *et al.*, 1993a,b; 13, Kamouchi and Kitamura, 1994; 14, Zhang and Bolton, 1995, 1996; 15, Quayle *et al.*, 1995, 1997; 16, Inagaki *et al.*, 1995b; 17, Sakura *et al.*, 1995; 18, Nichols *et al.*, 1996; 19, Gribble *et al.*, 1997a; 20, Inagaki *et al.*, 1996; 21, Okuyama *et al.*, 1998; 22, Isomoto *et al.*, 1996; Yamada *et al.*, 1997; 24, Satoh *et al.*, 1998.

[d] In the absence of intracellular Mg^{2+}.

[e] In the presence of intracellular Mg^{2+}, whose concentration differs among studies but is usually several hundred μ*M* to several m*M* in total.

[f] The small-conductance ATP-sensitive K^+ channel or the nucleoside diphosphate-dependent K^+ channel.

[g] There is a controversy over the effect of ATP. Some groups reported only stimulatory effects, while others showed both stimulatory and inhibitory effects.

[h] These data were obtained from smooth muscle cells of the mesenteric artery in which the nucleoside diphosphate-dependent K^+ channels are known to exist. Because the experiments were done in the whole-cell configuration (Quayle *et al.*, 1995), however, the single-channel conductance of the channels responding to these agents was unknown.

[i] So-called large-conductance ATP-sensitive K^+ channels in smooth muscle cells, which are so heterogeneous that no representative values are given to these channels in this table.

[j] This means no stimulatory effect. An inhibitory effect could not be examined because this channel does not possess spontaneous opening and because neither nucleoside diphosphates nor K^+ channel openers activated the channel in the absence of Mg^{2+}.

[k] Effect of ATP on its own in the absence of spontaneous activity, nucleoside diphosphates, or K^+ channel openers.

[l] These effects of K^+ channel openers are completely dependent on intracellular Mg^{2+} and nucleoside di- or tri phosphates.

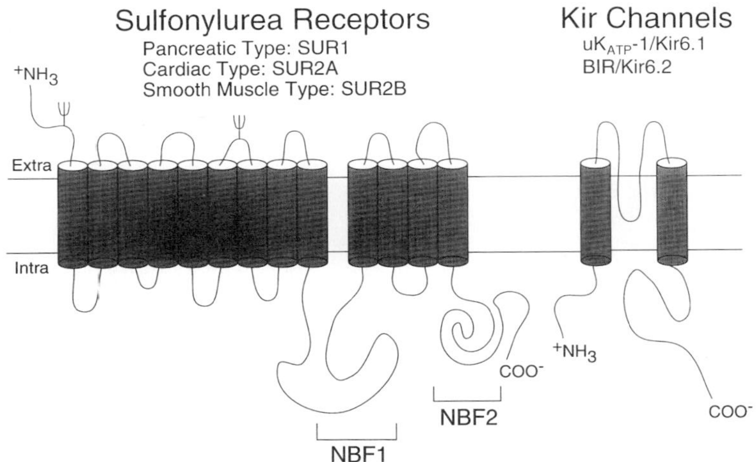

FIGURE 1 Molecular structure of ATP-sensitive K^+ channels. Pancreatic, cardiac, and skeletal muscle ATP-sensitive K^+ (K_{ATP}) channels are composed of two distinct subunits: a sulfonylurea receptor (SUR) and a K^+ (Kir) channel subunit, Kir6.2 (Inagaki *et al.,* 1995a, 1996; Sakura *et al.,* 1995; Isomoto, 1996; Aguilar-Bryan *et al.,* 1997). On the other hand, the complex of SUR2B and Kir6.1 corresponds to the small-conductance K_{ATP} or nucleoside diphosphate-dependent K^+ (K_{NDP}) channel in vascular smooth muscle (Yamada *et al.,* 1997).

pressed in pancreatic β cells (Table II) (Inagaki *et al.,* 1995b, 1996). Clement *et al.* (1997) proposed that the pancreatic K_{ATP} channel is composed of four SUR/Kir6.2 complexes.

The SUR which forms the cardiac type K_{ATP} channel was cloned by Inagaki *et al.* (1996) from a rat brain cDNA library. This novel SUR is composed of 1545 (or 1546) amino acids whose sequence is 68% identical with that of the SUR cloned by Aguilar-Bryan *et al.* (1995) (Inagaki *et al.,* 1996; Isomoto *et al.,* 1996). Thus, it was named SUR2, and accordingly the SUR cloned by Aguilar-Bryan *et al.* (1995) is now referred to as SUR1. The deduced amino acid sequence indicates that SUR2 also possesses thirteen transmembrane domains and two potential NBFs. With Kir6.2, SUR2 forms the classic K_{ATP} channel whose characteristics are similar to those of the cardiac K_{ATP} channel (Inagaki *et al.,* 1996; Okuyama *et al.,* 1998). In fact, both mRNAs for SUR2 and Kir6.2 are expressed in heart (Table II) (Inagaki *et al.,* 1995b, 1996).

The third isoform of SUR was isolated from a mouse cardiac cDNA library (Isomoto *et al.,* 1996) (Fig. 2A). This SUR is composed of 1546 amino acids whose sequence is 67 and 97% identical with those of rat SUR1 and SUR2, respectively (Aguilar-Bryan *et al.,* 1995; Inagaki *et al.,* 1996). This SUR and SUR2 are divergent only in the last 42-amino acid stretch

TABLE II
Distribution of mRNA for SUR and Kir6[a]

Location	SUR1	SUR2A	SUR2B	Kir6.1	Kir6.2
Brain	±	+	+	+	+
Heart	±	+	+	+	+
Lung	−	−	+	+	−
Liver	−	−	+	+	−
Pancreas	++	−	+	+	++
Spleen	NE	−	+	NE	NE
Kidney	−	−	+	+	−
Stomach	−	−	+	+	−
Small intestine	−	−	+	+	−
Colon	−	−	+	+	−
Adrenal	−	NE	NE	+	−
Testis	−	NE	NE	+	−
Ovary	−	−	+	+	−
Uterus	NE	−	+	NE	NE
Urinary bladder	NE	+	+	NE	NE
Skeletal muscle	−	+	+	+	+

[a] Northern analysis was used to measure the expression of mRNA for SUR1 (Inagaki *et al.*, 1995b), Kir6.1 (Inagaki *et al.*, 1995a), and Kir6.2 (Inagaki *et al.*, 1996). Reverse transcription polymerase chain reaction was used to measure the expression of mRNA for SUR2A and SUR2B (Isomoto *et al.*, 1996). −, undetectable; ±, moderately expressed; +, expressed; ++, strongly expressed; NE, not examined.

in the carboxyl-terminal tail (Fig. 2B), indicating that they are splicing variants derived from the same gene. We designated the novel SUR SUR2B and the SUR2 originally cloned by Inagaki *et al.* (1996) SUR2A (Isomoto *et al.*, 1996). Interestingly, the sequence of the last 42 amino acids of SUR2B exhibits 74 and 33% identity with those of the corresponding regions of rat SUR1 and mouse SUR2A, respectively (Fig. 2C). SUR2B mRNA is ubiquitously expressed throughout various organs (Table II). When coexpressed with Kir6.2, SUR2B forms a classic K_{ATP} channel whose nucleotide regulation and pharmacological properties are significantly different from those of the SUR2A/Kir6.2 channel. As discussed in Section V, SUR2B seems to form a certain type of vascular K_{ATP} channel by coupling with Kir6.1 (Isomoto *et al.*, 1996; Yamada *et al.*, 1997). Kir6.1 is composed of 424 amino acids whose sequence is 71% identical with that of Kir6.2 (Inagaki *et al.*, 1995a). Like the mRNA for SUR2B, the Kir6.1 mRNA is ubiquitously expressed in various organs (Table II).

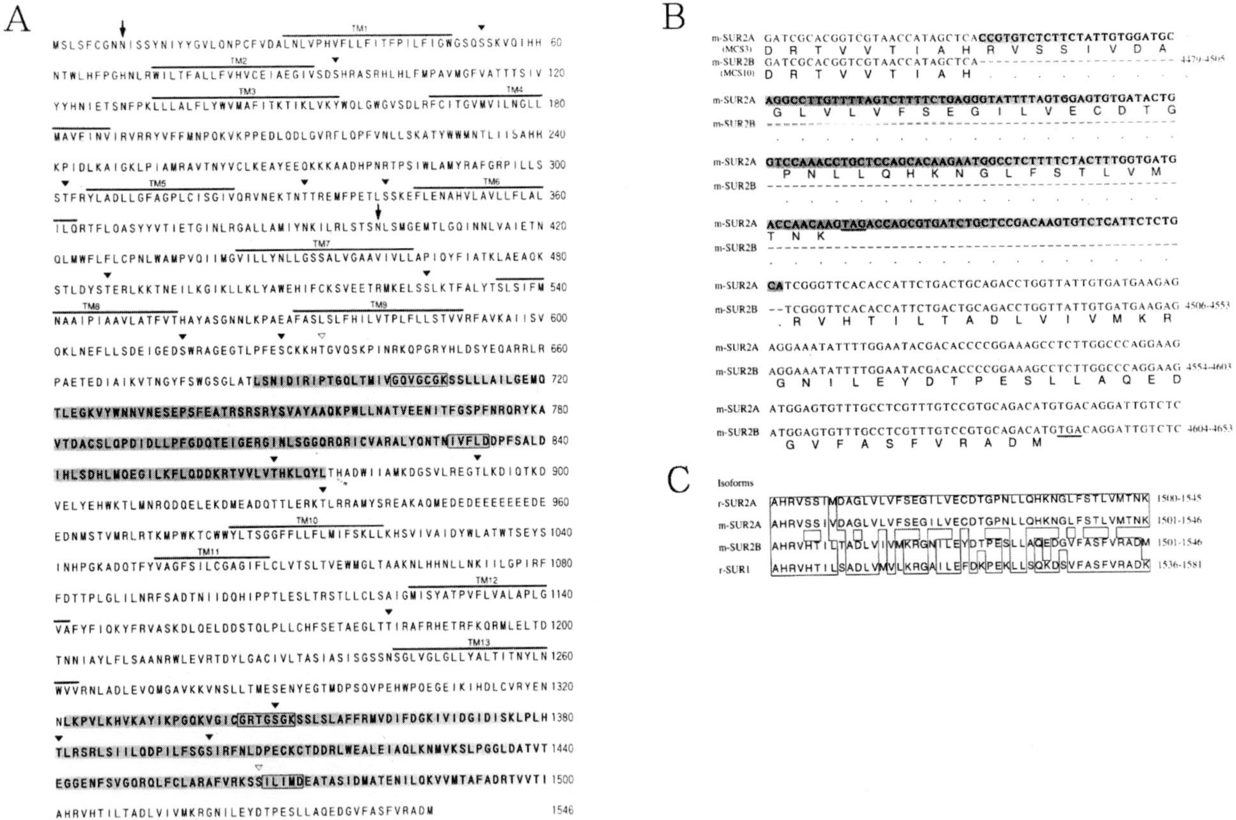

A

```
MSLSFCGNNISSYNIYYGVLQNPCFVDALNLVPHVFLLFITFPILFIGWGSQSSKVQIHH 60
NTWLHFPGHNLRWILTFALLFVHVCEIAEGIVSDSHRASRHLHLFMPAVMGFVATTTSIV 120
YYHNIETSNFPKLLLALFLYWVMAFITKTIKLVKYWQLGWGVSDLRFCITGVMVILNGLL 180
MAVFINVIRVRRYVFFMNPQKVKPPEDLQDLGVRFLQPFVNLLSKATYWWMNTLIISAHH 240
KPIDLKAIGKLPIAMRAVTNYVCLKEAYEEQKKKAADHPNRTPSIWLAMYRAFGRPILLS 300
STFRYLADLLGFAGPLCISGIVQRVNEKTNTTREMFPETLSSKEFLENAHVLAVLLFLAL 360
ILQRTFLQASYYVTIETGINLRGALLAMIYNKILRLSTSNLSMGEMTLGQINNLVAIETN 420
QLMWFLFLCPNLWAMPVQIIMGVILLYNLLGSSALVGAAVIVLLAPIQYFIATKLAEAQK 480
STLDYSTERLKKTNEILKGIKLLKLYAWEHIFCKSVEETRMKELSSLKTFALYTSLSIFM 540
NAAIPIAAVLATFVTHAYASGNNLKPAEAFASLSLFHILVTPLFLLSTVVRFAVKAIISV 600
QKLNEFLLSDEIGEDSWRAGEGTLPFESCKKHTGVQSKPINRKQPGRYHLDSYEQARRLR 660
PAETEDIAIKVTNGYFSWGSGLATLSNIDIRIPTGQLTMIVGQVGCGKSSLLLAILGEMQ 720
TLEGKVYWNNVNESEPSFEATRSRSRYSVAYAAQKPWLLNATVEENITFGSPFNRQRYKA 780
VTDACSLQPDIDLLPFGDQTEIGERGINLSGGQRQRICVARALYQNTNIVFLDDPFSALD 840
IHLSDHLMQEGILKFLQDDKRTVVLVTHKLQYLTHADWIIAMKDGSVLREGTLKDIQTKD 900
VELYEHWKTLMNRQDQELEKDMEADQTTLERKTLRRAMYSREAKAQMEDEDEEEEEEEDE 960
EDNMSTVMRLRTKMPWKTCWWYLTSGGFFLLFLMIFSKLLKHSVIVAIDYWLATWTSEYS 1040
INHPGKADQTFYVAGFSILCGAGIFLCLVTSLTVEWMGLTAAKNLHHNLLNKIILGPIRF 1080
FDTTPLGLILNRFSADTNIIDQHIPPTLESLTRSTLLCLSAIGMISYATPVFLVALAPLG 1140
VAFYFIQKYFRVASKDLQELDDSTQLPLLCHFSETAEGLTTIRAFRHETRFKQRMLELTD 1200
TNNIAYLFLSAANRWLEVRTDYLGACIVLTASIASISGSSNSGLVGLGLLYALTITNYLN 1260
WVVRNLADLEVQMGAVKKVNSLLTMESENYEGTMDPSQVPEHWPQEGEIKIHDLCVRYEN 1320
NLKPVLKHVKAYIKPGQKVGICGRTGSGKSSLSLAFFRMVDIFDGKIVIDGIDISKLPLH 1380
TLRSRLSIILQDPILFSGSIRFNLDPECKCTDDRLWEALEIAQLKNMVKSLPGGLDATVT 1440
EGGENFSVGQRQLFCLARAFVRKSSILIMDEATASIDMATENILQKVVMTAFADRTVVTI 1500
AHRVHTILTADLVIVMKRGNILEYDTPESLLAQEDGVFASFVRADM               1546
```

(Labels above the sequence: TM1–TM13)

B

```
m-SUR2A   GATCGCACGGTCGTAACCATAGCTCACCGTGTCTCTTCTATTGTGGATGC
(MCS3)    D  R  T  V  V  T  I  A  H  R  V  S  S  I  V  D  A
m-SUR2B   GATCGCACGGTCGTAACCATAGCTCA------------------------ 4479-4505
(MCS10)   D  R  T  V  V  T  I  A  H

m-SUR2A   AGGCCTTGTTTTAGTCTTTTCTGAGGGTATTTTAGTGGAGTGTGATACTG
           G  L  V  L  V  F  S  E  G  I  L  V  E  C  D  T  G
m-SUR2B   --------------------------------------------------

m-SUR2A   GTCCAAACCTGCTCCAGCACAAGAATGGCCTCTTTTCTACTTTGGTGATG
            P  N  L  L  Q  H  K  N  G  L  F  S  T  L  V  M
m-SUR2B   --------------------------------------------------

m-SUR2A   ACCAACAAGTAGACCAGCGTGATCTGCTCCGACAAGTGTCTCATTCTCTG
          T  N  K
m-SUR2B   --------------------------------------------------

m-SUR2A   CATCGGGTTCACACCATTCTGACTGCAGACCTGGTTATTGTGATGAAGAG
m-SUR2B   --TCGGGTTCACACCATTCTGACTGCAGACCTGGTTATTGTGATGAAGAG 4506-4553
           . R  V  H  T  I  L  T  A  D  L  V  I  V  M  K  R

m-SUR2A   AGGAAATATTTTGGAATACGACACCCCGGAAAGCCTCTTGGCCCAGGAAG
m-SUR2B   AGGAAATATTTTGGAATACGACACCCCGGAAAGCCTCTTGGCCCAGGAAG 4554-4603
           G  N  I  L  E  Y  D  T  P  E  S  L  L  A  Q  E  D

m-SUR2A   ATGGAGTGTTTGCCTCGTTTGTCCGTGCAGACATGTGACAGGATTGTCTC
m-SUR2B   ATGGAGTGTTTGCCTCGTTTGTCCGTGCAGACATGTGACAGGATTGTCTC 4604-4653
            G  V  F  A  S  F  V  R  A  D  M
```

C

Isoforms

```
r-SUR2A   AHRVSSIMDAGLVLVFSEGILVECDTGPNLLQHKNGLFSTLVMTNK 1500-1545
m-SUR2A   AHRVSSIVDAGLVLVFSEGILVECDTGPNLLQHKNGLFSTLVMTNK 1501-1546
m-SUR2B   AHRVHTILTADLVIVMKRGNILEYDTPESLLAQEDGVFASFVRADM 1501-1546
r-SUR1    AHRVHTILSADLVMVLKRGAILEFDKPEKLLSQKDSVFASFVRADK 1536-1581
```

III. EXPRESSION OF RECOMBINANT K_{ATP} CHANNELS

Various heterologous expression systems have been successfully utilized to express functional SUR/Kir6.0 channels. The examples are COS-1, COSm6, and HEK293T cells (Inagaki *et al.*, 1995b, 1996; Sakura *et al.*, 1995; Isomoto *et al.*, 1996) and *Xenopus* oocytes (Gribble *et al.*, 1997a,b). We have used HEK293T cells because they can efficiently express functional K_{ATP} channels when cotransfected with cDNAs encoding SUR and Kir6.0 in different combinations. The density of the expressed K_{ATP} channels in the plasma membrane of HEK293T cells is usually so high that cell-attached and inside-out recordings with the patch-clamp method often result in measurement of macroscopic currents. All of the data shown in the following were obtained under such conditions. In the cell-attached and inside-out configurations, the patch pipette contained the following (in mM): 140 KCl, 1 $CaCl_2$, 1 $MgCl_2$, and 5 HEPES–KOH (pH 7.4); the bath was perfused with an internal solution containing the following (in mM): 140 KCl, 5 EGTA, 2 $MgCl_2$, and 5 HEPES–KOH (pH 7.3). The free Mg^{2+} concentration in this solution was calculated to be ~1.4 mM. When nucleotides were added to the solution in the inside-out patch experiments, $MgCl_2$ was added to the solution to keep the free Mg^{2+} concentration at ~1.4 mM. To remove intracellular Mg^{2+} from the solution, no $MgCl_2$ was added while EGTA was replaced with equimolar EDTA. In the whole-cell configuration, pipettes were filled with the internal solution containing 1–3 mM of total ATP and ~1.4 mM of free Mg^{2+}, whereas the bath was perfused with a solution containing (in mM) 136.5 NaCl, 5.4 KCl, 1.8 $CaCl_2$, 0.53 $MgCl_2$, 5.5 glucose, and HEPES–NaOH (pH 7.4).

In the following sections, we detail the functional features of the cardiac K_{ATP} (SUR2A/Kir6.2) and vascular K_{NDP} (SUR2B/Kir6.1) channels. We describe the features of each channel in comparison with those of the

FIGURE 2 Amino acid and nucleotide sequences of mouse SUR2B. (A) Deduced amino acid sequence of mouse SUR2B. Putative transmembrane regions (TM1–TM13) are overlined; putative NBFs are shaded; the Walker A and B consensus sequences are boxed; possible phosphorylation sites for the cyclic AMP-dependent protein kinase and the protein kinase C are indicated by inverted filled and open triangles, respectively and two potential N-linked glycosylation sites are indicated by arrows. (B) Alignment of the nucleotide sequences at the 3′ ends and the deduced amino acid sequences at the C-terminal ends of mouse SUR2A (m-SUR2A) and mouse SUR2B (m-SUR2B). Termination codons are underlined. The nucleotide sequence of m-SUR2A possesses an insertion of 176 bp, which is shaded, between positions 4505 and 4506 of m-SUR2B. (C) Comparison of amino acid sequences in the C termini of rat SUR2A (r-SUR2A), m-SUR2A, m-SUR2B, and rat SUR1 (r-SUR1). Identical residues are boxed. Adapted from Isomoto *et al.* (1996) with permission from The American Society for Biochemistry and Molecular Biology.

SUR2B/Kir6.2 channel. Differences between SUR2A/Kir6.2 and SUR2B/Kir6.2 channels indicate the functional task of SUR in K_{ATP} channels, whereas comparison between SUR2B/Kir6.1 and SUR2B/Kir6.2 channels delineate the functionality of Kir subunits.

IV. COMPARISON BETWEEN THE SUR2A/Kir6.2 AND SUR2B/Kir6.2 CHANNELS

A. *Channel Activity in the Cell-Attached and Inside-out Configurations*

Figure 3Aa shows a cell-attached recording obtained from a HEK293T cell cotransfected with SUR2A and Kir6.2 (Okuyama *et al.*, 1998). Pinacidil added to the bath evoked K^+ channel currents. The K^+ channel currents appeared in bursts with a unitary current amplitude of ~4.8 pA at −60 mV in the presence of ~145 m*M* extracellular K^+ (K^+_o). These channel currents were almost completely inhibited by glibenclamide applied to the bath. On the other hand, diazoxide (300 μ*M*), an effective activator of the SUR1/Kir6.2 channel (Inagaki *et al.*, 1995b), did not induce any channel activity (not shown but see Table I). On patch excision (IO in Fig. 3Ab), spontaneous openings of the channels appeared immediately and then gradually decreased in the inside-out patch membrane (rundown). Application of ATP_i inhibited the channel activity in a reversible manner. These data indicate that the SUR2A/Kir6.2 channel has the characteristics of the classic K_{ATP} channel as previously indicated (Inagaki *et al.*, 1996).

On the other hand, coexpression of SUR2B and Kir6.2 resulted in formation of K^+ channels that were activated by both diazoxide and pinacidil in the cell-attached configuration (Fig. 3Ba and Table I) (Isomoto *et al.*, 1996). These channel currents were also inhibited by tolbutamide and glibenclamide. Patch excision caused strong spontaneous opening, which was inhibited by ATP_i. This channel activity ran down spontaneously (not shown) or in response to intracellular $CaCl_2$ and was restored by treatment of the patch with $MgATP_i$ (Fig. 3Bb). The rundown channel was also activated by intracellular UDP. These data indicate that the SUR2B/Kir6.2 channel also exhibits the characteristic features of the classic K_{ATP} channels. Therefore, the SUR2A/Kir6.2 and SUR2B/Kir6.2 channels have more or less similar characteristics except for the response to KCOs.

Figure 3Ac shows the single-channel recordings of the SUR2A/Kir6.2 channel at different membrane potentials (Okuyama *et al.*, 1998). The channel opened in long-lasting bursts at various membrane potentials. The channel currents reversed in direction at ~0 mV, close to the K^+ equilibrium potential (E_K) under the experimental condition. The single-channel

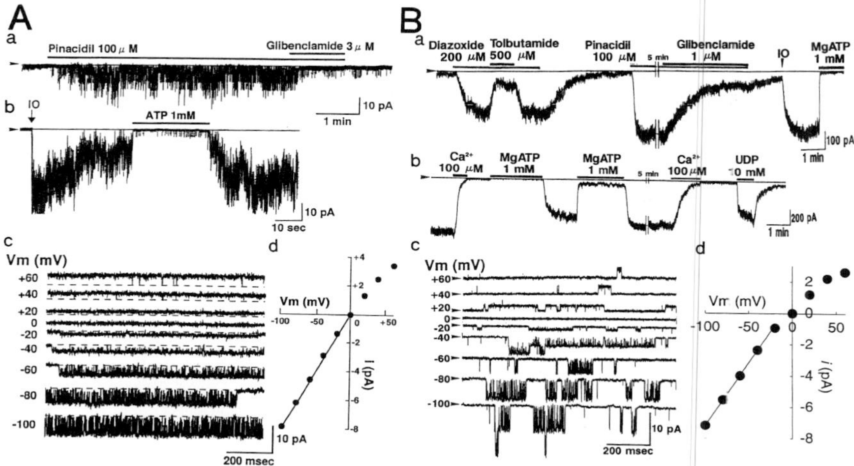

FIGURE 3 Pharmacological and single-channel properties of the SUR2A/Kir6.2 and SUR2B/Kir6.2 channels. In this and the following figures, the membrane potential (V_m) was held at -60 mV unless otherwise indicated. The protocols of bath perfusion are indicated above each trace. Arrowheads or thin straight or broken lines in current traces indicate the zero current level. (A) The SUR2A/Kir6.2 channel. (a) Cell-attached recording. (b) Recording before and after patch excision at the timing indicated by IO. (c) Single-channel recordings at different V_m in an inside-out patch membrane. (d) Single-channel current–voltage (I–V) relationship obtained from the data shown in Fig. 3Ac. The line is the fit of the data at negative V_m and has a slope of ~80.2 pS. Adapted from *Pfluegers Archiv,* The effects of nucleotides and potassium channel openers on the SUR2A/Kir6.2 complex K^+ channel expressed in a mammalian cell line, HEK293T cells, Y. Okuyama, M. Yamada, C. Kondo, E. Satoh, S. Isomoto, T. Shindo, Y. Horio, M. Kitakaze, M. Hori, and Y. Kurachi, Volume 435, pp 595–603, Figures 1 and 2, 1998, copyright Springer-Verlag. (B) The SUR2B/Kir6.2 channel. (a) Recording before and after patch excision. (b) Recording in the inside-out configuration. (c) Single-channel recordings at different V_m in an inside-out patch membrane. (d) Single-channel I–V relationship obtained from the data shown in Fig. 3Bc. The line is the fit of the data at V_m between -100 and -20 mV and has a slope of ~80.3 pS. Adapted from Isomoto *et al.* (1996) with permission from The American Society for Biochemistry and Molecular Biology.

current–voltage relationship showed a weak inward rectification in the presence of Mg^{2+}_i (Fig. 3Ad). The unitary conductance of the channel was ~80 pS at negative membrane potentials. Other single-channel characteristics, including open and closed times in bursts and the voltage dependency of open probability, were also very similar between SUR2A/Kir6.2 and native cardiac or skeletal muscle K_{ATP} channels (Trube and Hescheler, 1984; Spruce *et al.,* 1987). The single-channel recordings of SUR2B/Kir6.2 channels at different membrane potentials are shown in Fig. 3Bc (Isomoto *et al.,* 1996). These channel currents also appeared in bursts and reversed in direction at E_K. As is the case for the SUR2A/Kir6.2 channel, the channel exhibited weak inward rectification and the single-channel conductance of ~80 pS at negative membrane potentials (Fig. 3Bd). Thus, the conductance properties of unitary currents are practically identical between SUR2A/Kir6.2 and SUR2B/Kir6.2 channels. The SUR1/Kir6.2 channel is also known to exhibit a similar but slightly smaller single-channel conductance (Table I) and a similar extent of inward rectification. Therefore, the properties of the pore of the SUR/Kir6.2 channels are likely to be determined by the Kir6.2 subunit.

Comparison of Fig. 3Ac and Fig. 3Bc, however, suggests that the burst behavior is somehow different between the SUR2A/Kir6.2 and SUR2B/Kir6.2 channels. Inagaki *et al.* (1996) showed that the SUR2A/Kir6.2 channel exhibits a longer burst duration than the SUR1/Kir6.2 channel. Tucker *et al.* (1997) found that a Kir6.2 whose last 26 amino acids at the carboxyl terminus were deleted (Kir6.2ΔC26) can be functionally expressed in the absence of SUR. Kir6.2ΔC26 exhibits distinct single-channel kinetics with short openings not clearly grouped into bursts. When SUR2A or SUR2B was coexpressed with Kir6.2ΔC26, the channel opened in clear bursts (M. Yamada, E. Satoh, C. Kondo, V. P. Repunte, and Y. Kurachi, 1997). Therefore, SUR may modulate the burst behavior of classic K_{ATP} channels.

B. Inhibition by Intracellular ATP

One of the hallmarks of the classic K_{ATP} channels is the inhibition of channel activity by micromolar concentrations of ATP_i (Noma, 1983; Ashcroft, 1988; Terzic *et al.*, 1995). Figure 4A shows the response of the SUR2A/Kir6.2 channel to ATP_i (Okuyama *et al.*, 1998). Both in the presence and in the absence of $Mg^{2+}{}_i$, ATP_i fully inhibited the SUR2A/Kir6.2 channel activity in a range of concentrations between 30 μM and 1 mM (Fig. 4Aa). ATP_i inhibited the spontaneous channel activity with the half-maximum inhibitory concentration (IC_{50}) of 70–280 μM (172 ± 47 μM, $n = 5$) in the presence of $Mg^{2+}{}_i$ (Fig. 4Ab) and 60–290 μM (148 ± 34 μM, $n = 6$) in the absence of $Mg^{2+}{}_i$ (Fig. 4Ac). The Hill coefficient was ~1.8 both in the presence and in the absence of $Mg^{2+}{}_i$ ($n = 6$ for each). Thus, neither the IC_{50} value nor the Hill coefficient was significantly affected by $Mg^{2+}{}_i$, as is the case for the native cardiac K_{ATP} channel (Table I). On the other hand, pancreatic and skeletal muscle K_{ATP} channels are reported to be more sensitive to ATP_i in the absence than in the presence of $Mg^{2+}{}_i$ (Table I).

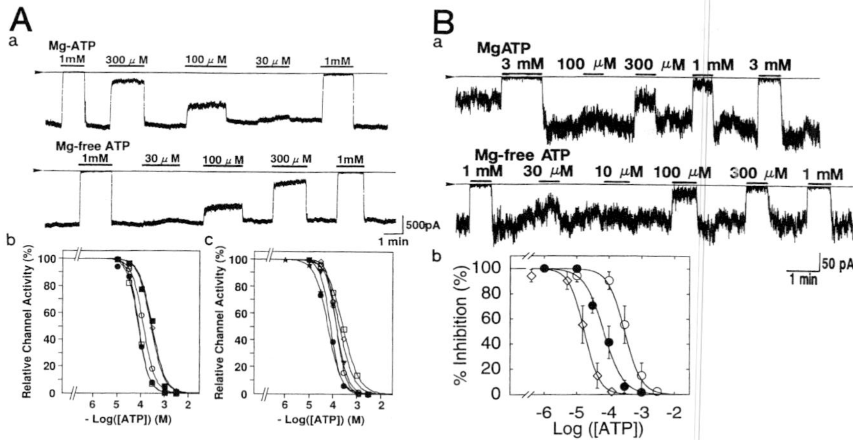

FIGURE 4 Inhibition of the SUR2A/Kir6.2 and SUR2B/Kir6.2 channels by intracellular ATP. (A) The SUR2A/Kir6.2 channel. (a) Effect of intracellular ATP in inside-out patch membranes in the presence (top trace) or the absence (bottom trace) of ~1.4 m*M* intracellular free Mg^{2+}. (b and c) Relationship between total ATP concentrations and channel activity in the presence (b) and the absence (c) of Mg^{2+}. The data are expressed as a percentage of the value obtained in the absence of ATP. Different symbols indicate the data obtained from different patches. Lines are the fit of each set of the data with the Hill equation. Adapted from *Pfluegers Archiv,* The effects of nucleotides and potassium channel openers on the SUR2A/Kir6.2 complex K^+ channel expressed in a mammalian cell line, HEK293T cells, Y. Okuyama, M. Yamada, C. Kondo, E. Satoh, S. Isomoto, T. Shindo, Y. Horio, M. Kitakaze, M. Hori, and Y. Kurachi, Volume 435, pp 595–603, Figure 4, 1998, copyright Springer-Verlag. (B) The SUR2B/Kir6.2 channel. (a) Effect of ATP in inside-out patch membranes in the presence (top trace) or the absence (bottom trace) of ~1.4 m*M* free Mg^{2+}. (b) Relationship between ATP concentrations and channel activity in the presence (open circles) and the absence (filled circles) of Mg^{2+}. Symbols and bars indicate the mean ± SEM ($n = 3$ for each point). Open and filled circles are the plot against the total ATP concentrations, whereas open diamonds are the plot of the data obtained in the presence of Mg^{2+} against the calculated Mg^{2+}-free ATP concentrations. Lines are the best fit of each set of data by the Hill equation. Adapted from Isomoto *et al.* (1996) with permission from The American Society of Biochemistry and Molecular Biology.

The response to ATP_i of the SUR2B/Kir6.2 channel was significantly different from that of the SUR2A/Kir6.2 channel (Fig. 4B) (Isomoto *et al.*, 1996). In the presence of $Mg^{2+}{}_i$, ATP_i concentration-dependently inhibited the spontaneous channel activity in a range of concentrations between 0.1 and 3 m*M* (Fig. 4Ba, top trace) with an IC_{50} of 300 μ*M* (Fig. 4Bb, open circles). In the absence of $Mg^{2+}{}_i$, the inhibitory effect of ATP_i occurred between 10 μ*M* and 1 m*M* (Fig. 4Ba, bottom trace), and the IC_{50} value of ATP_i was 67.9 μ*M* (Fig. 4Bb, closed circles). The Hill coefficient was 1.85 independently of $Mg^{2+}{}_i$. Open diamonds in Fig. 4Bb plot channel activity in the presence of $Mg^{2+}{}_i$ and ATP_i against the calculated concentration of Mg^{2+}-free ATP_i. This plot was located significantly leftward of the concentration–response curve experimentally obtained under the Mg^{2+}-free condition (closed circles). Therefore, the SUR2B/Kir6.2 channel is more sensitive to Mg^{2+}-free than Mg^{2+}-bound ATP_i.

Kir6.2ΔC26 expressed in the absence of SURs is inhibited by ATP_i (Tucker *et al.,* 1997). A charge-neutralization mutation on lysine 185 of Kir6.2ΔC26 reduces the ATP_i sensitivity of the Kir6.2ΔC26 channel by ~40 times. The ATP_i sensitivity of the Kir6.2ΔC26 channel was increased by 5–8 times by coexpression of SUR1. Thus, it is likely that the primary inhibitory ATP_i binding site resides in Kir6.2, while SUR1 increases the ATP_i sensitivity of Kir6.2. Koster *et al.* (1998) showed that the complex of SUR1 and Kir6.2 whose N-terminal 30 amino acids were deleted (Kir6.2ΔN30) exhibited ~10 times lower ATP_i sensitivity than the SUR1/Kir6.2 channel. Interestingly, the SUR1/Kir6.2ΔN30 channel was also less sensitive to intracellular ADP (ADP_i) and tolbutamide than the SUR1/Kir6.2 channel. It is known that not only tolbutamide but ADP_i interact with K_{ATP} channels through SUR (see Section IVC). Therefore, Kir6.2 might interact with SUR1 through its N terminus, and the low ATP_i sensitivity of the SUR1/Kir6.2ΔN30 channel might be due to impaired coupling between SUR1 and Kir6.2.

It is unknown how SUR1 increases the ATP_i sensitivity of Kir6.2. Gribble *et al.* (1997b) found that the ATP_i sensitivity of the SUR1/Kir6.2 channel was not modified by mutations on either or both of the two conserved lysine residues in the Walker A motifs in the first or the second NBF of SUR1 (K719A and K1384M, respectively). On the other hand, Ueda *et al.* (1997) demonstrated that SUR1 possesses two distinct ATP_i binding sites with high and low affinities. The high-affinity binding site was saturated with 10 μM ATP_i in the absence of $Mg^{2+}{}_i$. Substitution of the conserved lysine residue in the Walker A motif (K719R and K719M) or the aspartate residue in the Walker B motif (D854N) in the first NBF all abolished the high-affinity ATP_i binding, while the corresponding mutations in the second NBF did not cause any significant effect. Because Ueda *et al.* (1997) and Gribble *et al.* (1997b) used different mutations (K719R K719M or D854N versus K719A, respectively), it is not clear whether the ATP_i binding found by Ueda *et al.* (1997) underlies the sensitization of Kir6.2 to ATP_i by SUR1.

No corresponding studies have been done in SUR2s. However, ATP_i inhibits the SUR2A/Kir6.2 and the Kir6.2ΔC26 channels with similar potencies (~100 μM) (Tucker *et al.,* 1997). Thus, SUR2A may not substantially enhance the ATP_i sensitivity of Kir6.2. However, the native cardiac K_{ATP} channel has been reported to be ~3 to 10 times more sensitive to ATP_i than the SUR2A/Kir6.2 channel (Table I). Thus, some unidentified factors in cardiac myocytes might sensitize the SUR2A/Kir6.2 channel to ATP_i *in vivo*. Actually, various factors including intracellular polyvalent cations and actin polymerization have been suggested to affect the ATP_i sensitivity of the cardiac K_{ATP} channel (Deutsch *et al.*, 1994; Terzic and Kurachi, 1996).

The SUR2A/Kir6.2 channel was equally sensitive to Mg^{2+}-free and Mg^{2+}-bound ATP_i (Fig. 4A), whereas the SUR2B/Kir6.2 channel was more sensitive to Mg^{2+}-free than Mg^{2+}-bound ATP_i. This difference should be ascribed to the difference in amino acid sequence of the C terminus between SUR2A and SUR2B (Isomoto *et al.*, 1996). As stated Section II, the sequence of the last 42 amino acids in the C terminus of SUR2B is more similar to that of the corresponding part of SUR1 than SUR2A, and the SUR1/Kir6.2 channel is more sensitive to Mg^{2+}-free than Mg^{2+}-bound ATP_i (Nichols *et al.*, 1996). Therefore, the last 42 amino acids in the C terminus of SURs may be involved in discrimination between Mg^{2+}-bound and Mg^{2+}-free ATP_i by K_{ATP} channels. This part is very close to the second NBF in primary structure (Fig. 2), and the second NBF is known to play a crucial role in NDP_i-induced activation of K_{ATP} channels (Nichols *et al.*, 1996; Gribble *et al.*, 1997b). Ueda *et al.* (1997) found in SUR1 that the binding of ADP_i to the second NBF potently antagonized the ATP_i binding to the first NBF of the same protein in the presence of $Mg^{2+}{}_i$. Therefore, the C-terminal tail of SURs might serve to regulate either ATP_i hydrolysis on or ADP_i binding to the second NBF in the presence of $Mg^{2+}{}_i$.

ATP_i inhibited both SUR2A/Kir6.2 and SUR2B/Kir6.2 channels, with the Hill coefficient being significantly larger than unity (~1.8). No cooperativity was, however, detected with the SUR1/Kir6.2 channel nor the Kir6.2DC26 channel (Gribble *et al.*, 1997a,b; Tucker *et al.*, 1997). Therefore, SUR2s but not SUR1 may function to create the cooperative interaction between ATP_i and K_{ATP} channels through an unknown molecular mechanism.

C. Response to Intracellular Nucleoside Diphosphates

Intracellular NDPs such as UDP exhibit distinct effects on the cardiac K_{ATP} channel before and after rundown (Terzic *et al*, 1994, 1995). UDP antagonizes the inhibitory effect of ATP_i before rundown. After rundown, UDP restores the channel activity without attenuating the ATP_i sensitivity of the channels (Tung and Kurachi, 1991). The SUR2A/Kir6.2 channel well-mimicked such a dualistic response of the cardiac K_{ATP} channel to intracellular NDPs (Fig. 5) (Okuyama *et al.*, 1998).

As shown in Fig. 5Aa, ATP_i exhibited a weaker inhibitory effect on spontaneous channel activity of the SUR2A/Kir6.2 channel in the presence than in the absence of UDP. Removal of UDP almost completely restored the ATP_i sensitivity of the channels. Thus, UDP antagonized the ATP_i-mediated inhibition of channel activity before rundown.

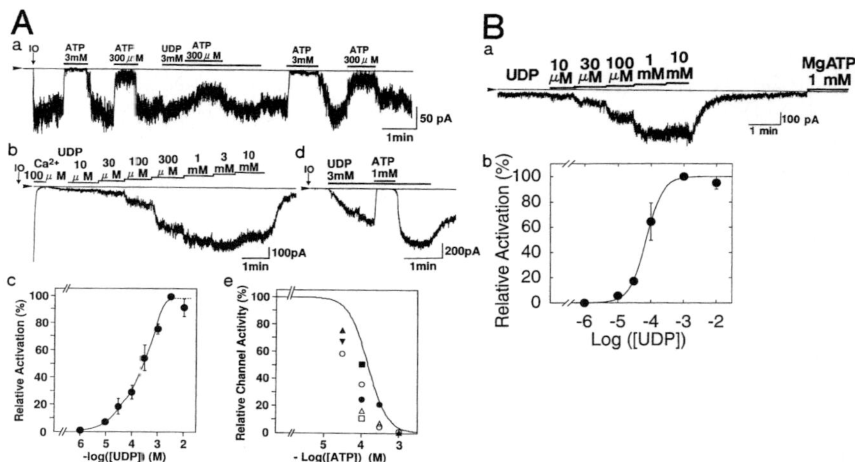

FIGURE 5 Effect of nucleoside diphosphates on the SUR2A/Kir6.2 and SUR2B/Kir6.2 channels. (A) The SUR2A/Kir6.2 channel. (a, b, and d) Inside-out patch recordings. (c) Relationship between UDP concentrations and channel activity after rundown. Channel activity is expressed as a percentage of the maximum activity induced by 3 m*M* UDP. Symbols and bars indicate the mean ± SEM (n = 3–12 for each point). The line is the best fit of the data with the Hill equation. (e) The concentration-dependent inhibitory effect of ATP on channel activity induced by 3 m*M* UDP after rundown in the presence of Mg^{2+} (symbols). Different symbols indicate data from different patches. The line indicates the averaged inhibitory effect of ATP on the spontaneous channel activity in the absence of UDP (Fig. 4Ab). Adapted from *Pfluegers Archiv,* The effects of nucleotides and potassium channel openers on the SUR2A/Kir6.2 complex K^+ channel expressed in a mammalian cell line, HEK293T cells, Y. Okuyama, M. Yamada, C. Kondo, E. Satoh, S. Isomoto, T. Shindo, Y. Horio, M. Kitakaze, M. Hori, and Y. Kurachi, Volume 435, pp 595–603, Figure 6, 1998, copyright Springer-Verlag. (B) The SUR2B/Kir6.2 channel. (a) Inside-out patch recording. (b) Relationship between UDP concentrations and channel activity after rundown. Channel activity is expressed as a percentage of the maximum activity induced by 1 m*M* UDP. Symbols and bars indicate the mean ± SD (n = 3 for each point). The line is the best fit of the data with the Hill equation. Adapted from Isomoto *et al.* (1996) with permission from the American Society for Biochemistry and Molecular Biology.

After rundown, UDP activated the channel in a concentration-dependent manner (Fig. 5Ab) with a half-maximum effective concentration (EC_{50}) of 240 μM (Fig. 5Ac). This effect was completely dependent on $Mg^{2+}{}_i$ (not shown). ATP_i inhibited the channel activity induced by UDP (Fig. 5Ad) in a concentration-dependent manner (symbols in Fig. 5Ae) as potently as inhibiting spontaneous activity in the absence of UDP (line in Fig. 5Ae, which is the average of the data shown in Fig. 4Ab). Thus, the SUR2A/Kir6.2 and native cardiac K_{ATP} channels respond to NDP_i in a very similar way (Terzic *et al.,* 1994, 1995).

UDP activated the postrundown SUR2B/Kir6.2 channel in a concentration-dependent manner with an EC_{50} of 71.7 μM and a Hill coefficient of 1.74 (Fig. 5B). This response was also dependent on $Mg^{2+}{}_i$ (not shown). UDP also antagonized the inhibitory effect of ATP_i on SUR2B/Kir6.2 channel before but not after rundown as is the case for the SUR2A/Kir6.2 channel (M. Yamada, S. Isomoto, and Y. Kurachi, unpublished observation, 1996). Overall, the responses to NDP_i were very similar between SUR2A/Kir6.2 and SUR2B/Kir6.2 channels.

Nichols *et al.* (1996) found that a human PHHI mutation (G1479R) in the second NBF of SUR1 abolished the antagonizing effect of ADP_i on the ATP_i-induced inhibition of the SUR1/Kir6.2 channel. Tucker *et al.* (1997) demonstrated that $MgADP_i$ inhibited the Kir6.2ΔC26 channel but activated the SUR1/Kir6.2ΔC26 channel. Gribble *et al.* (1997b) showed that either the K719A or K1384M mutation of rat SUR1 abolished the stimulatory effects of ADP_i on the partially rundown SUR1/Kir6.2 channel both in the presence and in the absence of ATP_i. Furthermore, hydrolysis-resistant $\alpha\beta$-methylene-ADP failed to activate the SUR1/Kir6.2 channel. Thus, it is likely that hydrolysis or binding of $MgADP_i$ (and probably also the other intracellular NDPs) at the NBFs of SUR1 may be critically involved in the activating effect of the nucleotides.

Although no corresponding studies have been done in SUR2s, a similar mechanism could underlie the activating effect of NDP_i on the SUR2/Kir6.2 channel. The molecular mechanism responsible for the dualistic responses of the SUR2/Kir6.2 channels to NDP_i is, however, unknown. As discussed in Section IV,D, both spontaneous opening and rundown of K_{ATP} channels seem to be the phenomena associated with the Kir6.2 but not the SUR subunit. The dualistic behavior of SUR2/Kir6.2 channels may indicate that the interaction between SUR2 and Kir6.2 is modulated by the functional states of the Kir6.2 subunit. Koster *et al.* (1998) indicated that Kir6.2 might interact with SUR through its N terminus. Therefore, the interaction between SUR and Kir6.2 might be modulated by a conformational change of the N terminus of Kir6.2 which occurs depending on the functional states of Kir6.2.

D. Rundown

Rundown of channel activity in inside-out patch membranes is not necessarily a phenomenon specifically associated with K_{ATP} channels but can be seen in other Kir channels as well. Nevertheless, many investigators have been interested in the mechanism underlying the rundown of K_{ATP} channels because K_{ATP} channels run down more prominently than the other Kir channels (Ashcroft, 1988; Terzic *et al.,* 1995). The K_{ATP} channels can be reactivated after rundown with ATP_i in the presence but not in the absence of $Mg^{2+}{}_i$ (Findlay and Dunne, 1986; Ohno-Shosaku *et al.,* 1987; Takano *et al.,* 1990). Nonhydrolyzable ATP analogs cannot mimic this effect of ATP_i even in the presence of $Mg^{2+}{}_i$ (Ohno-Shosaku *et al.,* 1987; Takano *et al.,* 1990). Therefore, rundown/refreshment of K_{ATP} channel activity might be crucially related to hydrolysis of ATP_i or phosphorylation/dephosphorylation of K_{ATP} channels (Ashcroft, 1988; Terzic *et al.,* 1995).

Kir6.2ΔC26 channels also exhibit rundown and can be reactivated with $MgATP_i$ (Tucker *et al.,* 1997), indicating that the rundown/refreshment is primarily associated with a certain functional alteration of Kir6.2. Hilgeman and Ball (1996) showed that phosphatidylinositol bisphosphate (PIP_2) added to the intracellular side of the membrane could restore K_{ATP} channel activity after rundown. Huang *et al.* (1998) showed that various Kir channels, including Kir1.1, Kir2.1, and Kir3.2 homomeric and Kir3.1/Kir3.4 heteromeric channels, could be reactivated by PIP_2 after rundown. Furthermore, the ATP_i-mediated restoration of activity was inhibited by antibodies against PIP_2. Thus, PIP_2 and its generation by ATP-dependent lipid kinases appear to be critically involved in spontaneous Kir channel activity. The dualistic behavior of K_{ATP} channels to intracellular NDPs (Terzic *et al.,* 1994, 1995) may have to be reexamined with respect to PIP_2.

E. Pharmacological Properties

Potassium channel openers such as pinacidil, levcromakalim, and nicorandil but not diazoxide have been known to activate K_{ATP} channels in cardiac myocytes (Terzic *et al.,* 1995). On the other hand, the pancreatic K_{ATP} channel is activated by diazoxide but not by pinacidil or by nicorandil (Table I). Nicorandil caused moderate activation of the SUR2A/Kir6.2 channel only at 1 m*M* (Fig. 6A,B) (Okuyama *et al.,* 1998). Diazoxide did not significantly affect the membrane current up to 200 μM. Pinacidil (100 μM) induced a large current, which was inhibited by glibenclamide in a concentration-dependent manner with an IC_{50} of 160 ± 11 n*M* (n = 6) and by tolbutamide with an IC_{50} of 120 ± 16 μM (n = 6) (not shown) (Okuyama *et al.,* 1998). The concentration-dependent effect of KCOs on the SUR2A/Kir6.2 channel is summarized in Fig. 6C. Diazoxide was practically ineffective. Nicorandil (1 m*M*) induced only ~20% of the current evoked by 100 μM pinacidil. The EC_{50} and the maximum effective concentrations of pinacidil were 10 and 100 μM, respectively. Pinacidil showed some inhibitory effects at >100 μM. Consistent with a previous report (Inagaki *et al.,* 1996), these effects of KCOs and SUs on the SUR2A/Kir6.2 channel

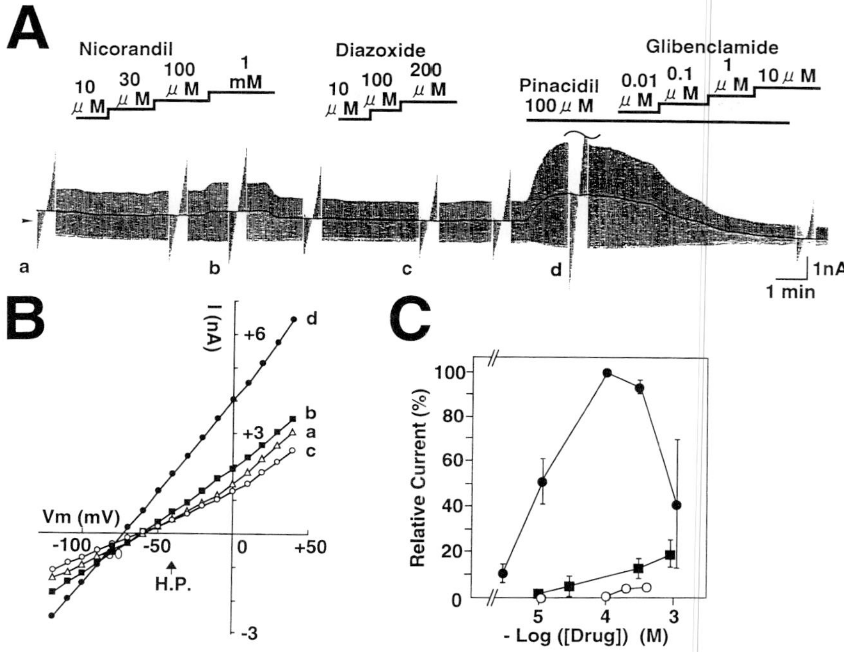

FIGURE 6 Effect of K^+ channel openers on the SUR2A/Kir6.2 channel in the whole-cell configuration. (A) The whole-cell recording of the SUR2A/Kir6.2 channel current in the presence of K^+ channel openers (KCOs) and glibenclamide. A pair of 500-msec voltage steps to 0 and −80 mV was continuously applied to the membrane from the holding V_m of −40 mV every 3 sec. Intermittently, 500-msec pulses to potentials ranging between −120 and +40 mV were applied in 10-mV increments to analyze the I–V relationship. (B) Steady-state I–V relationships measured under each condition indicated by a–d in A. (C) Concentration-dependent effect on the SUR2A/Kir6.2 channel of pinacidil (filled circles), nicorandil (squares), and diazoxide (open circles). The channel current was measured at 0 mV and was expressed as a percentage of that induced by 100 μM pinacidil in the same cells. Symbols and bars indicate the mean ± SEM (n = 3–6 for each point). Reproduced from *Pfluegers Archiv,* The effects of nucleotides and potassium channel openers on the SUR2A/Kir6.2 complex K^+ channel expressed in a mammalian cell line, HEK293T cells, Y. Okuyama, M. Yamada, C. Kondo, E. Satoh, S. Isomoto, T. Shindo, Y. Horio, M. Kitakaze, M. Hori, and Y. Kurachi, Volume 435, pp 595–603, Figure 3, 1998, copyright Springer-Verlag.

are the same as those reported in native cardiac and skeletal muscle K_{ATP} channels but different from those of pancreatic or smooth muscle K_{ATP} channels (Ashcroft and Ashcroft, 1990; Weston and Edwards, 1992; Edwards and Weston, 1993; Terzic *et al.,* 1995).

As has been shown in Fig. 3Ba, the SUR2B/Kir6.2 channel responded to both diazoxide and pinacidil. In the whole-cell configuration, 100 μM

nicorandil also activated the SUR2B/Kir6.2 channel to the same extent as 100 μM pinacidil (not shown). The EC_{50} value of nicorandil for the SUR2B/Kir6.2 channels was ~1 μM. These differences in the response to KCOs between SUR2A/Kir6.2 and SUR2B/Kir6.2 channels may indicate that the diverging C-terminal region of SUR2s might be a part of the receptor site for diazoxide and nicorandil but not for pinacidil. Alternatively, the C-terminal region may indirectly modulate the effect of diazoxide and nicorandil by affecting the interaction between nucleotides and these KCOs on SUR2/Kir6.2 channels. Further studies are necessary to elucidate the molecular mechanism by which the C-terminal region determines the sensitivity of K_{ATP} channels to various KCOs.

V. COMPARISON BETWEEN SUR2B/Kir6.2 AND SUR2B/Kir6.1 CHANNELS

The SUR2B/Kir6.2 channel exhibits a response to KCOs similar to that of K_{ATP} channels in vascular smooth muscle (Table I). However, KCOs are known to activate many different types of K^+ channels in vascular smooth muscle cells (Quayle *et al.,* 1997). These K^+ channels are often classified into two categories: large and small conductances (Table I) (Quayle *et al.,* 1997). The former group is so heterogeneous that common features shared by these K^+ channels can be hardly defined. Some of them, however, closely resemble the large-conductance Ca^{2+}-activated K^+ channels (Quayle *et al.,* 1997). It is the small-conductance K_{ATP} channels which have been consistently identified by different groups of investigators as the primary target of KCOs in vascular smooth muscle cells (Table I) (Kajioka *et al.,* 1991; Weston and Edwards, 1992; Beech *et al.,* 1993a,b; Kamouchi and Kitamura, 1994; Zhang and Bolton, 1995, 1996). This channel possesses the single-channel conductance of 30–35 pS in the presence of 150 mM K^+_o, lacks spontaneous openings in the absence of ATP_i, and opens in response to ATP_i or NDP_i in the presence of Mg^{2+}_i (Table I). This channel is sensitive to KCOs and SUs as are the classic K_{ATP} channels (Table I). It was proposed to designate this channel a K_{NDP} channel in order to distinguish it from the classic K_{ATP} channels and emphasize the importance of NDP_i for physiological regulation of the channel activity (Beech *et al.,* 1993a).

Studies indicated that pharmacological properties of classic K_{ATP} channels are determined by SURs whereas the single-channel conductance is determined by Kir6.2 (Inagaki *et al.,* 1996). We, therefore, hypothesized that the K_{NDP} channel might be a complex of SUR and a Kir subunit other than Kir6.2. Among more than 10 Kir subunits cloned so far, Kir6.1 exhibits the highest amino acid sequence identity (71%) with Kir6.2 (Inagaki *et al.*, 1995a). Kir6.1 is ubiquitously expressed in different tissues, as is SUR2B, whereas SUR1 and SUR2A are expressed only in limited tissues including the pancreas, heart, and brain (Table II). We, therefore, examined the properties of K^+ channels formed in HEK293T cells cotransfected with Kir6.1 and SUR2B (Yamada *et al.*, 1997).

A. Electrophysiological Properties

In HEK293T cells cotransfected with SUR2B and Kir6.1, pinacidil applied to the bath induced opening of a channel in bursts in the cell-attached configuration (Fig. 7A) (Yamada *et al.*, 1997). These channel currents reversed in direction at ~0 mV (i.e., ~E_K) (Fig. 7B) and showed a quasi-linear single-channel current–voltage relationship (Fig. 7C). When the patch was excised in the continuous presence of pinacidil, the burst duration significantly increased (Fig. 7A).

When the patch was excised in the absence of KCO, however, no spontaneous opening was elicited (Fig. 8A). Application of UDP (Fig. 8A,B) or GDP (Fig. 8C) to the internal side of inside-out patch membranes induced

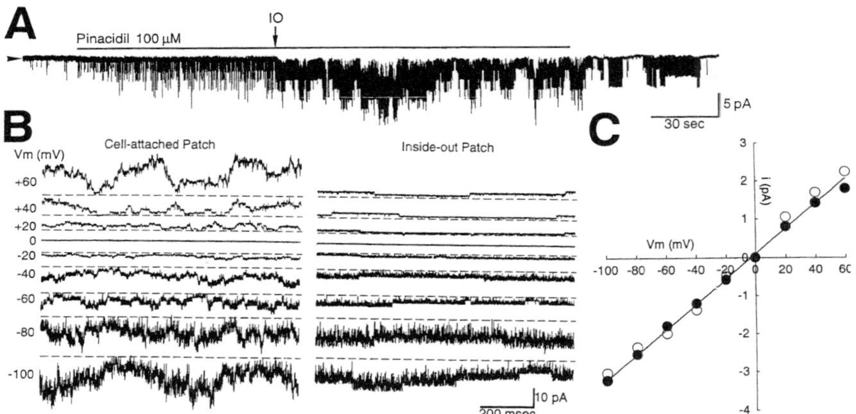

FIGURE 7 Single-channel characteristics of the SUR2B/Kir6.1 channel in the cell-attached and inside-out configurations. (A) Single-channel recording of the SUR2B/Kir6.1 channels before and after patch excision in the presence of 100 μM pinacidil in the bath. (B) Pinacidil-induced single-channel currents in the cell-attached and inside-out patch membranes at different V_m. (C) Single-channel I–V relationship in the cell-attached (open circles) and inside-out patches (filled circles) shown in B. The regression line has a slope of 32.9 pS. Reproduced from Yamada *et al.* (1997) with permission from The Physiological Society.

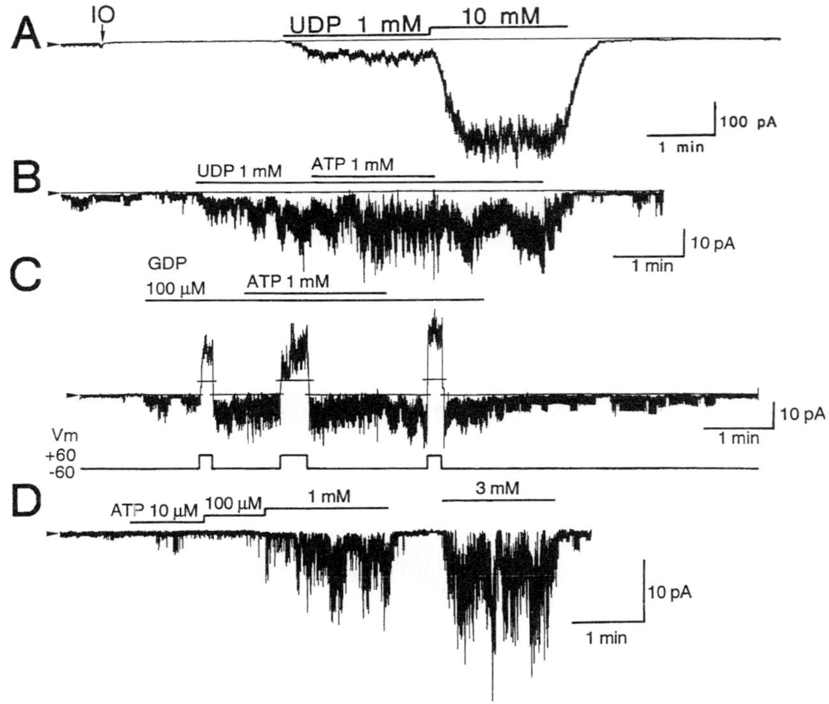

FIGURE 8 Response of the SUR2B/Kir6.1 channel to intracellular nucleoside di- and triphosphates in the inside-out configuration. (A) Recording before and after patch excision. (B–D) Recordings in the inside-out configuration. In C, V_m was changed as indicated by the voltage protocol shown below the current trace. Adapted from Yamada *et al.* (1997) with permission from The Physiological Society.

opening of the same channel that was activated by pinacidil in the cell-attached configuration (Fig. 7A). This channel activity was observed only in the presence of $Mg^{2+}{}_i$ (not shown). Surprisingly, ATP_i (1 m*M*) did not markedly inhibit the NDP_i-induced channel activity (Fig. 8B,C). Even at 10 m*M,* ATP_i inhibited the UDP-induced channel activity by less than ~60%, although 10 μ*M* glibenclamide completely inhibited the channel activity (not shown). Furthermore, ATP_i by itself activated the channel at millimolar concentrations (Fig. 8D). ATP_i induced lower channel activity at 10 than 3 m*M,* indicating that ATP_i may have an inhibitory effect at relatively high concentrations (not shown).

Thus, K_{ATP} and K_{NDP} channels are similar in pharmacological properties but different in single-channel conductance and in nucleotide regulation, probably because they share SUR but differ in Kir subtype. In other words, the small difference in the primary structure between Kir6.1 and Kir6.2 may account for the big difference in the characteristics between K_{ATP} and K_{NDP} channels. We are currently analyzing the molecular mechanism responsible for the functional differences between SUR/Kir6.1 and SUR/Kir6.2 channels by constructing chimeras of Kir6.1 and Kir6.2. Our preliminary results indicate that (1) the M1–M2 linker region including the H5 portion determines single-channel conductance and (2) the proximal N terminus of Kir6.2 is mandatory for spontaneous opening while further inclusion of the proximal C terminus of Kir6.2 into chimeras enhances spontaneous activity after patch excision (Kondo *et al.*, 1998). It is, however, not clear how these two cytosolic parts of Kir6.2 but not Kir6.1 generate spontaneous opening.

We deal with the difference in the response to ATP_i between SUR2B/Kir6.2 and SUR2B/Kir6.1 channels in Section V,B and show that ATP_i has both activating and inhibitory effects on the SUR2B/Kir6.1 channel by using pinacidil as a pharmacological tool to separate the two effective concentration ranges of ATP_i.

B. Pharmacological Properties

Although pinacidil strongly activated the SUR2B/Kir6.1 channel in the cell-attached and whole-cell configurations, it failed to do so in the inside-out configuration (Fig. 9Aa) (Yamada *et al.*, 1997). When ATP_i was further applied to the patch with pinacidil, the SUR2B/Kir6.1 channel was effectively activated. When the channel was first activated by ATP_i, the channel activity was strongly enhanced by the subsequent application of pinacidil (Fig. 9Ab). Figure 9Ac summarizes the relationship between concentrations of ATP_i and SUR2B/Kir6.1 channel activity in the presence and in the absence of 100 μM pinacidil. It is clear from these data that ATP_i and pinacidil synergistically activated the channel.

Note that the concentration–response curve in the presence of pinacidil has a bell shape, indicating that ATP_i activated the channel at lower concentrations but inhibited it at higher concentrations. This was always the case when different concentrations (10–300 μM) of pinacidil were used with ATP_i (not shown) (Satoh *et al.*, 1998). As the concentration of pinacidil was increased, the ascending limb of the bell-shaped concentration–response curves progressively shifted leftward while the slope of the descending limb was constant (see Fig. 11). Thus, pinacidil activated the SUR2B/Kir6.1 channel by specifically increasing the potency of ATP_i in activating the channel.

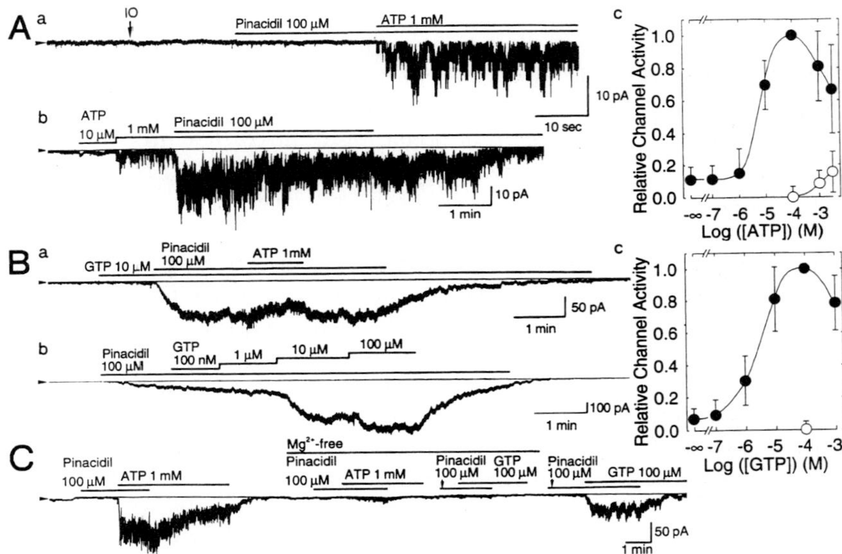

FIGURE 9 Interaction between pinacidil and nucleotide triphosphates on the SUR2B/Kir6.1 channel in inside-out patch membranes. (A) Interaction between pinacidil and ATP. (a) Recording before and after patch excision. (b) Inside-out patch recording. (c) Relationship between ATP concentrations and channel activity in the absence (open circles) and the presence (filled circles) of 100 μM pinacidil. The channel activity was normalized to the maximum in the presence of 100 μM pinacidil plus 100 μM ATP. Symbols and bars indicate the mean $\pm$ SEM (n = 3–6 for each point). The lines were drawn by eye. (B) Interaction between pinacidil and GTP. (a and b) Recordings in the inside-out configuration. (c) Relationship between GTP concentrations and channel activity in the absence (open circles) and the presence (filled circles) of 100 μM pinacidil. Channel activity was normalized to the maximum in the presence of 100 μM pinacidil plus 100 μM GTP. Symbols and bars indicate the mean $\pm$ SEM (n = 3–6 for each point). The line was drawn by eye. (C) Inside-out patch recording. Adapted from Yamada *et al.* (1997) with permission from The Physiological Society.

The IC_{50} of ATP_i for the SUR2B/Kir6.1 channel estimated from the fitting of the constant descending limb was ~300 μM in the presence of $Mg^{2+}{}_i$ (Satoh *et al.,* 1998). This value is not significantly different from that for the spontaneously opening SUR2B/Kir6.2 channel (Isomoto *et al.,* 1996). However, we postulated that ATP_i may inhibit the SUR2B/Kir6.1 and SUR2B/Kir6.2 channels through different molecular mechanisms for the following two reasons. (1) GTP also synergistically activated the SUR2B/Kir6.1 channel with pinacidil (Fig. 9Ba,b). The relationship between GTP concentrations and channel activity in the presence of pinacidil was also in a bell shape (Fig. 9Bc). The IC_{50} value of GTP estimated from these data was ~300 μM. Thus, GTP and ATP were equipotent in inhibiting the channel, which is not the case for classic K_{ATP} channels (Ashcroft, 1988). (2) ADP and GDP also exhibited bell-shaped concentration–response curves in the presence and in the absence of different concentrations of pinacidil (not shown) (Satoh *et al.,* 1998). UTP and UDP exhibited only stimulatory effects. Therefore, the SUR2B/Kir6.1 channel may be inhibited by purine but not pyrimidine nucleotides. Such nucleotide selectivity is also different from that of classic K_{ATP} channels (Ashcroft, 1988).

As described in Section IV,B, lysine at amino acid 185 (K185) and the first 30-amino acid stretch of the N terminus of Kir6.2 are thought to crucially determine the ATP_i sensitivity of the SUR1/Kir6.2 channel (Tucker *et al.,* 1997; Koster *et al.,* 1998). Kir6.1 bears arginine at the site corresponding to K185 of Kir6.2 while possessing an amino acid sequence ~50% identical to that of Kir6.2 within the N-terminal region (Inagaki *et al.,* 1995a). It is unknown whether these differences are responsible for the different nucleotide selectivity between the SUR2B/Kir6.2 and SUR2B/Kir6.1 channels.

ATP, GTP, UTP, ADP, GDP, and UDP caused the stimulatory effect even in the absence of pinacidil in a $Mg^{2+}{}_i$-dependent manner (Fig. 9C and data not shown). None of nucleoside monophosphates nor nucleosides activated the channel even in the presence of pinacidil (Satoh *et al.,* 1998). These data indicate that hydrolysis or binding of the Mg^{2+}-bound form of nucleoside di- and triphosphates might be responsible for activation of the SUR2B/Kir6.1 channel.

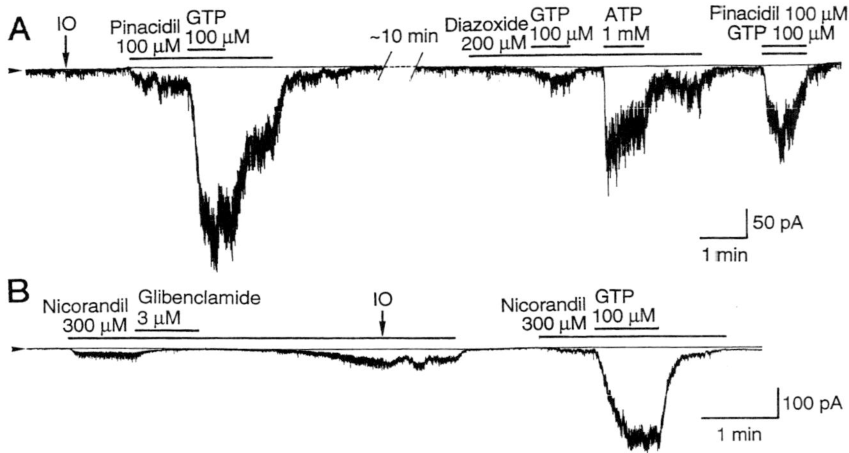

FIGURE 10 Interaction between nucleoside triphosphates and diazoxide or nicorandil on the SUR2B/Kir6.1 channel. Recordings before and after patch excision are shown.

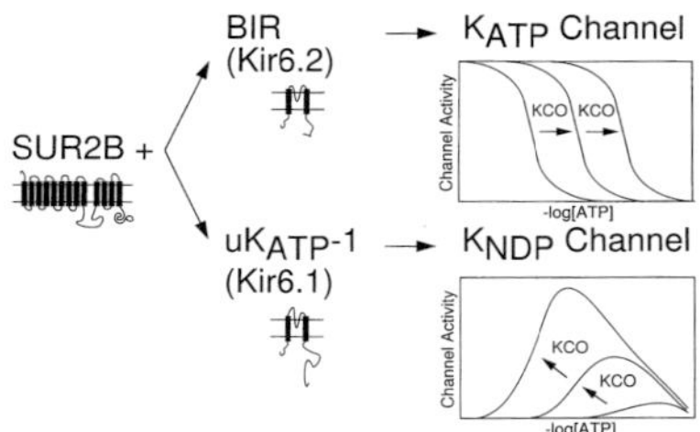

FIGURE 11 Schematic representation of the effect of K^+ channel openers on the ATP-sensitive and nucleoside diphosphate-dependent K^+ channels. SUR2B forms K_{ATP} channels with Kir6.2 and forms nucleoside diphosphate-dependent K^+ (K_{NDP}) channels with Kir6.1. Two graphs schematically indicate the effect of K^+ channel openers (KCO) on the concentration-dependent effect of ATP on each type of channel.

Finally, KCOs other than pinacidil also activated the SUR2B/Kir6.1 channel by synergistically interacting with intracellular nucleotides. Figure 10 shows such examples. Diazoxide required ATP or GTP to activate the channel in the inside-out configuration (Fig. 10A). The effect of nicorandil was also strongly enhanced by GTP (Fig. 10B) or ATP (1 m*M*) (not shown but see Fig. 2 in Yamada *et al.,* 1997).

Pinacidil as well as most KCOs activate classic K_{ATP} channels mainly by antagonizing the inhibitory effect of ATP_i (Fig. 11) (Terzic *et al.,* 1995). Therefore, the mode of action of KCOs is clearly different between SUR2B/Kir6.1 and classic K_{ATP} channels, indicating that the interaction between SUR and Kir subunits may differ between SUR/Kir6.1 and SUR/Kir6.2 channels. Such a difference could be utilized to develop novel KCOs which specifically act on K_{NDP} channels in vascular smooth muscle but not on the classic K_{ATP} channels in other tissues.

VI. CONCLUSIONS

Molecular dissection of Kir channels and SURs has identified molecular structures of K_{ATP} channels in the cardiovascular system. Further understanding at the molecular level of the K_{ATP} channels in the cardiovascular system may enable us to clarify the roles of these channels in cardiovascular physiology and pathophysiology, which may allow further development of strategies and pharmacological agents to treat various cardiovascular diseases.

References

Aguilar-Bryan, L., Nichols, C. G., Wechsler, S. W., Clement IV, J. P., Boyd III, A. E., González, G., Herrera-Sosa, H., Nguy, K., Bryan, J., and Nelson, D. A. (1995). Cloning of the β cell high-affinity sulfonylurea receptor: A regulator of insulin secretion. *Science* **268,** 423–426.

Aguilar-Bryan, L., Clement IV, J. P., Gonzalez, G., Kunjilwar, K., Babenko, A., and Bryan, J. (1998). Toward understanding the assembly and structure of K_{ATP} channels. *Physiol. Rev.* **78,** 227–245.

Allard, B., and Lazdunski, M. (1993). Pharmacological properties of ATP-sensitive K^+ channels in mammalian skeletal muscle cells. *Eur. J. Pharmacol.* **236,** 419–426.

Allard, B., Lazdunski, M., and Rougier, O. (1995). Activation of ATP-dependent K^+ channels by metabolic poisoning in adult mouse skeletal muscle: Role of intracellular Mg^{2+} and pH. *J. Physiol. (London)* **485,** 283–296.

Ashcroft, F. M. (1988). Adenosine 5′-triphosphate-sensitive potassium channels. *Annu. Rev. Neurosci.* **11,** 97–118.

Ashcroft, S. J. H., and Ashcroft, F. M. (1990). Properties and functions of ATP-sensitive K-channels. *Cell. Signalling* **2,** 197–214.

Beech, D. J., Zhang, H., Nakao, K., and Bolton, T. B. (1993a). K channel activation by nucleotide diphosphates and its inhibition by glibenclamide in vascular smooth muscle cells. *Br. J. Pharmacol.* **110,** 573–582.

Beech, D. L., Zhang, H., Nakao, K., and Bolton, T. B. (1993b). Single channel and whole-cell K-currents evoked by levcromakalim in smooth muscle cells from the rabbit portal vein. *Br. J. Pharmacol.* **110,** 583–590.

Benton, D. C., and Haylett, D. G. (1992). Effects of cromakalim on the membrane potassium permeability of frog skeletal muscle *in vitro. Br. J. Pharmacol.* **107,** 152–155.

Clement IV, J. P., Kunjilwar, K., Gonzalez, G., Schwanstecher, M., Panten, U., Aguilar-Bryan, L., and Bryan, J. (1997). Association and stoichiometry of K_{ATP} channel subunits. *Neuron* **18,** 827–838.

Deutsch, N., Matsuoka, S., and Weiss, J. N. (1994). Surface charge and properties of cardiac ATP-sensitive K^+ channels. *J. Gen. Physiol.* **104,** 773–800.

Edwards, G., and Weston, A. H. (1993). The pharmacology of ATP-sensitive potassium channels. *Annu. Rev. Pharmacol. Toxicol.* **33,** 597–637.

Faivre, J.-F., and Findlay, I. (1989). Effects of tolbutamide, glibenclamide and diazoxide upon action potentials recorded from rat ventricular muscle. *Biochim. Biophys. Acta* **984,** 1–5.

Findlay, I. (1992). Inhibition of ATP-sensitive K^+ channels in cardiac muscle by the sulphonylurea drug glibenclamide. *J. Pharmacol. Exp. Ther.* **261,** 540–545.

Findlay, I., and Dunne, M. J. (1986). ATP maintains ATP-inhibited K^+ channels in an operational state. *Pfluegers Arch.* **407,** 238–240.

Gribble, F. M., Ashfield, R., Åmmälä, C., and Ashcroft, F. M. (1997a). Properties of cloned ATP-sensitive K^+ currents expressed in *Xenopus* oocytes. *J. Physiol. (London)* **498,** 87–98.

Gribble, F. M., Tucker, S. J., and Ashcroft, F. M. (1997b). The essential role of the Walker A motifs of SUR1 in K-ATP channel activation by Mg-ADP and diazoxide. *EMBO J.* **16,** 1145–1152.

Hamada, E., Takikawa, R., Ito, H., Iguchi, M., Terano, A., Sugimoto, T., and Kurachi, Y. (1990). Glibenclamide specifically blocks ATP-sensitive K^+ channel currents in atrial myocytes of guinea pig heart. *Jpn. J. Pharmacol.* **54,** 473–477.

Hilgeman, D. W., and Ball, R. (1996). Regulation of cardiac Na^+, Ca^{2+} exchanger and K_{ATP} potassium channels by PIP_2. *Science* **273,** 956–959.

Huang, C.-L., Feng, S., and Hilgeman, D. W. (1998). Direct activation of inward rectifier potassium channels by PIP_2 and its stabilization by $G_{\beta\gamma}$. *Nature (London)* **391,** 803–806.

Inagaki, N., Gonoi, T., Clement IV, J. P., Namba, N., Inazawa, J., Gonzalez, G., Aguilar-Bryan, L., Seino, S., and Bryan, J. (1995a). Reconstitution of I_{KATP}: An inward rectifier subunit plus the sulfonylurea receptor. *Science* **270,** 1164–1170.

Inagaki, N., Tsuura, Y., Namba, N., Masuda, K., Gonoi, T., Horie, M., Seino, Y., Mizuta, M., and Seino, S. (1995b). Cloning and functional characterization of a novel ATP-sensitive potassium channel ubiquitously expressed in rat tissues, including pancreatic islets, pituitary, skeletal muscle, and heart. *J. Biol. Chem.* **270,** 5691–5694.

Inagaki, N., Gonoi, T., Clement IV, J. P., Wang, C.-Z., Aguilar-Bryan, L., Bryan, J., and Seino, S. (1996). A family of sulfonylurea receptors determines the pharmacological properties of ATP-sensitive K^+ channels. *Neuron* **16,** 1011–1017.

Isomoto, S., Kondo, C., Yamada, M., Matsumoto, S., Higashiguchi, O., Horio, Y., Matsuzawa, Y., and Kurachi, Y. (1996). A novel sulfonylurea receptor forms with BIR(Kir6.2) a smooth muscle type ATP-sensitive K^+ channel. *J. Biol. Chem.* **271,** 24321–24324.

Kajioka, S., Kitamura, K., and Kuriyama, H. (1991). Guanosine diphosphate activates an adenosine 5′-triphosphate-sensitive K^+ channel in the rabbit portal vein. *J. Physiol.* (*London*) **444,** 397–418.

Kamouchi, M., and Kitamura, K. (1994). Regulation of ATP-sensitive K^+ channels by ATP and nucleotide diphosphate in rabbit portal vein. *Am. J. Physiol.* **266,** H1687–H1698.

Kondo, C., Yamada, M., Satoh, E., Okuyama, Y., Isomoto, S., Repunte, C. V., Horio, Y., and Kurachi, Y. (1998). The structure–function of pore forming subunit of K_{ATP} channel. *Biophys. J.* **74,** A116.

Koster, J. C., Shyng, S.-L., Sha, Q., and Nichols, C. G. (1998). Involvement of the N-terminus of Kir6.2 in regulating ATP-sensitivity of K_{ATP} channels. *Biophys. J.* **74,** A230.

McKillen, H.-C., Davies, N. W., Stanfield, P. R., and Standen, N. B. (1994). The effect of intracellular anions on ATP-dependent potassium channels of rat skeletal muscle. *J. Physiol.* (*London*) **479,** 341–351.

Nichols, C. G., Shyng, S.-L., Nestorowicz, A., Glaser, B., Clement IV, J. P., Gonzalez, G., Aguilar-Bryan, L., Permutt, M. A., and Bryan, J. (1996). Adenosine diphosphate as an intracellular regulator of insulin secretion. *Science* **272,** 1785–1787.

Noma, A. (1983). ATP-regulated K^+ channels in cardiac muscle. *Nature* (*London*) **305,** 147–148.

Ohno-Shosaku, T., Zünkler, B. J., and Trube, G. (1987). Dual effect of ATP on K^+ currents of mouse pancreatic β-cells. *Pfluegers Arch.* **408,** 133–138.

Okuyama, Y., Yamada, M., Kondo, C., Satoh, E., Isomoto, S., Shindo, T., Horio, Y., Kitakaze, M., Hori, M., and Kurachi, Y. (1998). The effects of nucleotides and potassium channel openers on the SUR2A/Kir6.2 complex K^+ channel expressed in a mammalian cell line, HEK293T cells. *Pfluegers Arch.* **435,** 595–603.

Quayle, J. M., Bonev, A. D., Brayden, J. E., and Nelson, M. T. (1995). Pharmacology of ATP-sensitive K^+ currents in smooth muscle cells from rabbit mesenteric artery. *Am. J. Physiol.* **269,** C1112–C1118.

Quayle, J. M., Nelson, M. T., and Standen, N. B. (1997). ATP-sensitive and inwardly rectifying potassium channels in smooth muscle. *Physiol. Rev.* **77,** 1165–1232.

Sakura, H., Ämmälä, C., Smith, P. A., Gribble, F. M., and Ashcroft, F. M. (1995). Cloning and functional expression of the cDNA encoding a novel ATP-sensitive potassium channel subunit expressed in pancreatic β-cells, brain, heart and skeletal muscle. *FEBS Lett.* **377,** 338–344.

Satoh, E., Yamada, M., Kondo, C., Okuyama, Y., Isomoto, S., Horio, Y., and Kurachi, Y. (1998). Kir subunits determine the molecular mode of pinacidil-activation of SUR/Kir6.0 complex potassium channels. *J. Physiol.* (*London*) **511,** 633–674.

Spruce, A. E., Standen, N. B., and Stanfield, P. R. (1987). Studies of the unitary properties of adenosine-5′-triphosphate-regulated potassium channels of frog skeletal muscle. *J. Physiol.* (*London*) **382,** 213–236.

Takano, M., Qin, D., and Noma, A. (1990). ATP-dependent decay and recovery of K^+ channels in guinea-pig cardiac myocytes. *Am. J. Physiol.* **258,** H45–H50.

Terzic, A., and Kurachi, Y. (1996). Actin microfilament disrupters enhance K_{ATP} channel opening in patches from guinea-pig cardiomyocytes. *J. Physiol. (London)* **492,** 395–404.

Terzic, A., Findlay, I., Hosoya, Y., and Kurachi, Y. (1994). Dualistic behavior of ATP-sensitive K^+ channels toward intracellular nucleoside diphosphates. *Neuron* **12,** 1–20.

Terzic, A., Jahangir, A., and Kurachi, Y. (1995). Cardiac ATP-sensitive K^+ channels: Regulation by intracellular nucleotides and K^+ channel-opening drugs. *Am. J. Physiol.* **269,** C525–C545.

Thomas, P. M., Cote, G. J., Wohllk, N., Haddad, B., Mathew, P. M., Rabl, W., Aguilar-Bryan, L., Gagel, R. F., and Bryan, J. (1995). Mutations in the sulfonylurea receptor gene in familial persistent hyperinsulinemic hypoglycemia of infancy. *Science* **268,** 426–428.

Trube, G., and Hescheler, J. (1984). Inward-rectifying channels in isolated patches of the heart cell membrane: ATP-dependence and comparison with cell-attached patches. *Pfluegers Arch.* **401,** 178–184.

Tucker, S. J., Gribble, F. M., Zhao, C., Trapp, S., and Ashcroft, F. M. (1997). Truncation of Kir6.2 produces ATP-sensitive K^+ channels in the absence of the sulphonylurea receptor. *Nature (London)* **387,** 179–183.

Tung, R., and Kurachi, Y. (1991) On the mechanism of nucleotide diphosphate activation of the ATP-sensitive K^+ channel in ventricular cell of guinea-pig. *J. Physiol. (London)* **437,** 239–256.

Ueda, K., Inagaki, N., and Seino, S. (1997). MgADP antagonism to Mg^{2+}-independent ATP binding of the sulfonylurea receptor SUR1. *J. Biol. Chem.* **272,** 22983–22986.

Vivaudou, M. B., Arnoult, C., and Villaz, M. (1991). Skeletal muscle ATP-sensitive K^+ channels recorded from sarcolemmal blebs of split fibers: Inhibition is reduced by magnesium and ADP. *J. Membr. Biol.* **122,** 165–175.

Walker, J. E., Saraste M., Runswick, M. J., and Gay, N. J. (1982). Distantly related sequences in the α and β subunits of ATP synthase, myosin, kinase and other ATP-requiring enzymes and common nucleotide binding fold. *EMBO J.* **1,** 945–951.

Weik, R., and Neumcke, B. (1989). ATP-sensitive potassium channels in adult mouse skeletal muscle: Characterization of the ATP-binding site. *J. Membr. Biol.* **110,** 217–226.

Weik, R., and Neumcke, B. (1990). Effects of potassium channel openers on single potassium channels in mouse skeletal muscle. *Naunyn-Schmiedeberg's Arch. Pharmacol.* **342,** 258–263.

Weston, A. H., and Edwards, G. (1992). Recent progress in potassium channel opener pharmacology. *Biochem. Pharmacol.* **43,** 47–54.

Yamada, M., Isomoto, S., Matsumoto, S., Kondo, C., Shindo, T., Horio, Y., and Kurachi, Y. (1997). Sulphonylurea receptor 2B and Kir6.1 form a sulphonylurea-sensitive but ATP-insensitive K^+ channel. *J. Physiol. (London)* **499,** 715–720.

Zhang, H., and Bolton, T. B. (1995). Activation by intracellular GDP, metabolic inhibition and pinacidil of a glibenclamide-sensitive K-channel in smooth muscle cells of rat mesenteric artery. *Br. J. Pharmacol.* **114,** 662–672.

Zhang, H.-L., and Bolton, T. B. (1996). Two types of ATP-sensitive potassium channels in rat portal vein smooth muscle cells. *Br. J. Pharmacol.* **118,** 105–114.

CHAPTER 22

Role of ATP-Sensitive Potassium Channels in Ischemia/Reperfusion-Induced Ventricular Arrhythmias

Makoto Arita and Sakuji Shigematsu
Department of Physiology, Oita Medical University, Hasama, Oita 879-5593, Japan

I. INTRODUCTION

An interesting phenomenon called "ischemic preconditioning" has been reported on by Murry *et al.* (1986), and is defined as a brief episode(s) of ischemia resulting in increased tolerance of the myocardium to subsequent severe ischemic insults. However, this protective effect is lost when the interval between the initial brief ischemia and the subsequent severe ischemia becomes longer than 1 hr (Murry *et al.*, 1991). Gross and Auchampach (1992) have suggested that activation of the ATP-sensitive K^+ channels (K_{ATP} channels) (Noma, 1983; Nakamura *et al.*, 1989) contributes to this phenomenon. We provided some evidence that ischemic preconditioning is indeed attributable to the persistent activation of K_{ATP} channels during reperfusion periods (Shigematsu *et al.*, 1995). Regarding this, we demonstrate here that the activation of K_{ATP} channels, elicited by preceding

1063-5823/99 $30.00

ischemia, persists for some time after reperfusion, and that this, keeping K_{ATP} channels open, plays a pivotal role in postischemic contractile recovery.

The electrophysiological effects of K_{ATP} channel modulation during ischemia, and its implication for arrhythmogenesis, have been extensively reviewed by Wilde and Janse (1994). However, little is known about how the modulation of K_{ATP} channel activity during the reperfusion period affects the occurrence of postischemic ventricular arrhythmias. Furthermore, studies by our laboratory (Wu *et al.,* 1992) and others (Horie *et al.,* 1997) have revealed that some antiarrhythmic drugs affect cardiac as well as pancreatic K_{ATP} channels. In cardiac muscle, activation of K_{ATP} channels shortens the duration of the action potential (APD) (and therefore the effective refractory period), and in pancreatic β cells it inhibits the secretion of insulin. Thus, it is conceivable that drugs which modulate pancreatic K_{ATP} channel (such as glibenclamide and cromakalim) should affect the incidence and severity of arrhythmias during the ischemia–reperfusion and that antiarrhythmic drugs capable of modulating cardiac K_{ATP} channels would affect insulin secretion from pancreatic β cells.

In this chapter, we discuss here the pros and cons of the persistent activation of cardiac K_{ATP} channels during reperfusion on reperfusion induced-arrhythmias and on myocardial stunning, with the use of various antiarrhythmic and antidiabetic drugs that may or may not affect cardiac K_{ATP} channels.

II. MATERIALS AND METHODS

In the present study, we used two different methods. First, we measured intracellular action potential and contractile tension from coronary-perfused right ventricular free-wall muscles of guinea pigs, stimulated at 3 Hz. The preparation was perfused at a constant flow rate (1 ml/g tissue/min) with an oxygenated Tyrode's solution (with or without drugs). Global ischemia (10 or 20 min) was introduced by shutting off an electromagnetic value installed close to the orifice of the aorta, followed by 60 min of reperfusion. During the procedure, the action potentials and contraction were recorded simultaneously. The precise methods have been described previously (Shigematsu *et al.,* 1995).

Second, we measured action potentials and K_{ATP} current from single ventricular myocytes harvested from guinea pigs using a whole cell recording mode of patch clamp, stimulated at 1 Hz. Anoxia (simulated ischemia) was introduced by switching the oxygenated Tyrode's solution in the bath to a glucose-free anoxic solution ($PO_2 < 0.5$ torr), using a specially designed airtight clamber (Shigematsu and Arita, 1997). The pipette solution consisted of 150 m*M* KCl, 10 m*M* HEPES, and 20 or 0.01 m*M* EGTA (the pH was adjusted to 7.2 with KOH).

III. THE PERSISTENT ACTIVATION OF K_{ATP} CHANNELS

Figure 1A,B (open circles) shows the time course of changes in the action potential duration measured at 90% repolarization (APD_{90}) before, during, and after the introduction of global ischemia for 10 min, in the coronary-perfused right ventricular preparations of guinea pigs. The APD_{90} decreased after the onset of ischemia, and reached 59% of preischemic values in 10 min (i.e., from 157 ± 5 msec to 92 ± 5 msec). On reperfusion, the APD_{90} was mostly restored for the initial 10 min, whereas it remained slightly shortened even until ~60 min after reperfusion.

The shortening of the APD suggests the activation of the outward K_{ATP} current. To determine the role of K_{ATP} channels on the shortening of the APD during ischemia and reperfusion, we examined the effect of glibenclamide, a potent blocker of K_{ATP} channels, on these action potential changes. The application of glibenclamide (10μM) was started either 20 min before the introduction of ischemia (preischemia group, Fig. 1A) or from the onset of reperfusion (after-reperfusion group, Fig. 1B). Under normal (preischemic) conditions, glibenclamide had no effect on the electrical activity (APD_{90}, the resting potential, or the action potential amplitude) or mechanical activity (the resting tension or the developed tension). However, in the presence of glibenclamide (preischemia group), the extent of APD shortening caused by ischemia was markedly attenuated, and the recovery of APD_{90} after reperfusion was faster (Fig. 1A). The APD_{90} shortened only 28.1 ± 2.8% during 10 min of ischemia in the treated preparations, versus 41.3 ± 3.1% in untreated preparations ($p < 0.05$). The APD_{90} was restored up to 98.6 ± 2.6% of the preischemic value within 5 min of reperfusion in treated preparations, versus 90.5 ± 2.7% in untreated preparations ($p < 0.05$). The APD difference, between glibenclamide-treated and untreated preparations, is depicted at the bottom of Fig. 1A as a possible measure of the time course of the intensity of K_{ATP} channel activation.

To gain further insight into the contribution of K_{ATP} channels to the persistent shortening of the APD in the reperfusion phase, we applied glibenclamide starting from the onset of reperfusion (Fig. 1B). In this experiment, the time course of APD shortening during ischemia was naturally identical to that of untreated preparations; however, during reperfusion, the APD_{90} returned to preischemic values more quickly than in the

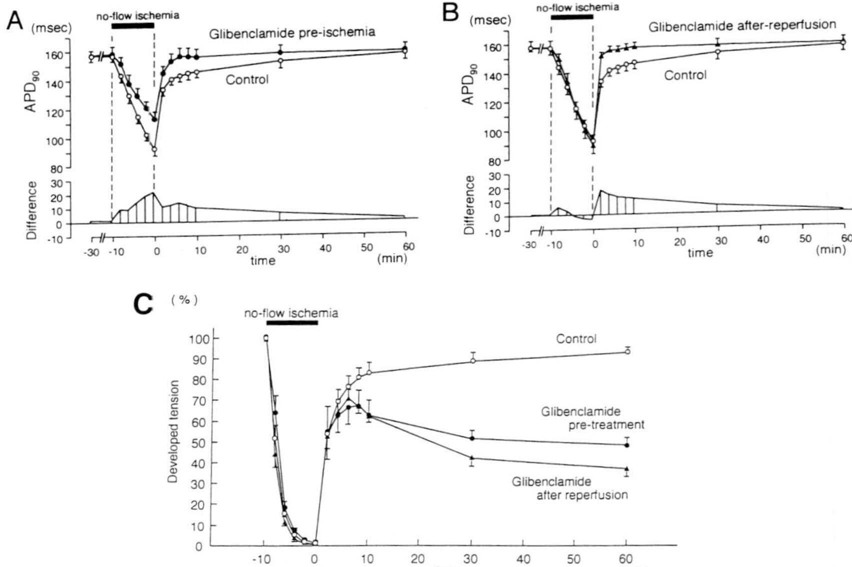

FIGURE 1 Effect of glibenclamide on APD_{90} during 10 min of no-flow ischemia and 60 min of reperfusion in coronary-perfused right ventricular muscle preparations from guinea pigs. (A) Serial changes of APD_{90} in the absence (○, $n = 7$) and presence (●, $n = 5$) of glibenclamide (10 μM) applied from the preischemic period. (B) Comparison of the changes of APD_{90} in the absence (○, same as A) and presence (▲, $n = 5$) of glibenclamide (10 μM); glibenclamide was applied immediately after the initiation of reperfusion. In both A and B, the bottom section shows the difference in the APDs shown above, as a handy measure of the activation of K_{ATP} channels. (C) Changes of contractile tension seen in the absence (○) and in the presence of glibenclamide (10 μM), from the preischemic period (●) or after reperfusion (▲). Note that K_{ATP} channels were activated during ischemia and remained activated for as long as ~60 min in the reperfusion phase. Adapted from Shigematsu *et al.* (1995, Figs. 2 and 4) with permission of the American Heart Association.

untreated preparations, as was the case for Fig. 1A. The APD difference during reperfusion (Fig. 1B, bottom) was similar to that seen in Fig. 1A, suggesting that the intensity and time course of K_{ATP} channel activation (during reperfusion) are practically the same between the experiments. These observations lend support to the notion that an outward K^+ current through K_{ATP} channels, activated during the 10 min of ischemia, remains activated in the early phase of reperfusion for as long as 30–60 min, irrespective of the channel condition during ischemia (unblocked or blocked), thereby accounting for maintenance of APD shortening in this phase of reperfusion.

IV. PATHOPHYSIOLOGICAL ROLES OF K_{ATP} CHANNEL ACTIVATION

A. *Facilitation of Contraction Recovery during Reperfusion via Sustained Activation of K_{ATP} Channels*

The application of glibenclamide attenuated the shortening of the APD not only during ischemia but also in the early phase of reperfusion. Therefore, this drug is expected to affect the contractile function as well. Figure 1C summarizes the changes of contractile tension before, during, and after 10 min of no-flow ischemia with and without glibenclamide. In untreated preparations (control), the developed tension declined rapidly after the introduction of ischemia, and was lost with 10 min; this occurred without significant changes in the resting tension (not illustrated). After reperfusion, the developed tension gradually recovered, reaching as large as 92.0 ± 6.4% of preischemic values in 60 min. The resting tension was slightly, but significantly, elevated on reperfusion and eventually returned to the preischemic level (not illustrated).

In the presence of glibenclamide from the preischemic period (Fig. 1C, filled circles) the developed tension decreased rapidly during ischemia, as was the case for untreated preparations, with significant elevation of the resting tension that persisted for the entire 60 min of reperfusion (not illustrated). The developed tension was restored quickly on reperfusion and reached a peak value within 7 to 8 min of reperfusion. However, it must be noted that the developed tension gradually declined hereafter in the reperfusion phase. The net recovery of developed tension estimated at 30 and 60 min of reperfusion was significantly less in the glibenclamide-pretreated preparations than in the untreated preparations.

When glibenclamide was applied from the onset of reperfusion (Fig. 1C, triangles), a rapid recovery of the developed tension was noted also, albeit it was again followed by a subsequent gradual decrease of contraction, a finding identical to that seen in the pretreatment group (cf. Fig. 1C, filled circles). The developed tensions during reperfusion were significantly lower than those of untreated preparations (when measured at 10, 30, and 60 min of reperfusion) but were not significantly different from those of glibenclamide-pretreated preparations.

These observations suggest that the intentional blockage of K_{ATP} channels, which otherwise remain active in the reperfusion phase, suppressed contraction in this phase. We then asked whether an enforced activation of K_{ATP} channels by a K_{ATP} channel opener promotes the recovery of contraction during reperfusion. Therefore, we examined the effect of cromakalim, a K_{ATP} channel opener, with a protocol entailing 20 min of no-flow ischemia followed by 60 min of reperfusion. We used 20 min of ischemia (instead of 10 min) in this series of experiments, because the 20 min of ischemia produced, without exception, a sustained, significant contractile depression (myocardial stunning, see below) after 60 min of reperfusion.

The alterations of the APD_{90} and the tension developed during 20 min of no-flow ischemia and the 60 min of reperfusion in the presence or absence of cromakalim are illustrated in Fig. 2. In untreated preparations, ischemia markedly shortened the APD_{90}, and in 2 of 4 preparations the electrical excitability was lost within 17 min (Fig. 2A). In these preparations, action potentials could not be elicited, even when the intensity of stimulation

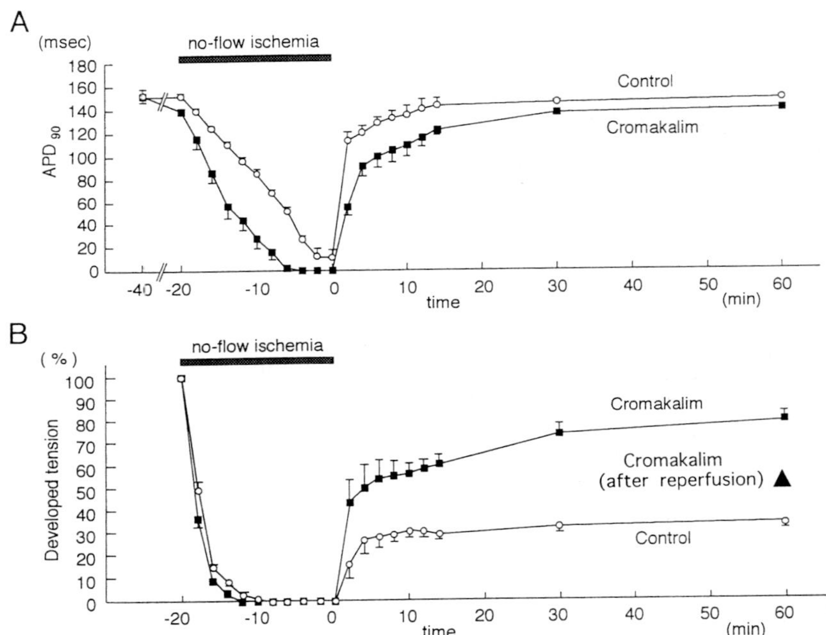

FIGURE 2 Effects of cromakalim on APD_{90} and contractile tension, during 20 min of no-flow ischemia and 60 min of reperfusion. (A) Time course of changes in APD_{90} with (■, $n = 5$) and without (○, $n = 4$) cromakalim. Cromakalim (2 μM) facilitated the shortening of APD_{90} during ischemia as well as during reperfusion. (B) Effects of cromakalim (2 μM) on the developed tension (expressed as percent of preischemic values). In the untreated preparation (○, $n = 4$), the developed tension markedly decreased during ischemia and the reperfusion led to only a limited restoration of tension. In the presence of cromakalim (■, $n = 5$), the rate of decline of developed tension during ischemia was slightly faster (albeit not significant) and the recovery of contraction after reperfusion was markedly enhanced ($p < 0.05$, versus untreated group). A large triangle ($n = 4$) indicates the developed tension where cromakalim was introduced just after initiation of reperfusion. Adapted from Shigematsu *et al.* (1995, Figs. 5 and 6) with permission of the American Heart Association.

was increased 10-fold. This electrical inexcitability was probably not due to a decrease in the resting potential (i.e., by means of inactivation of the inward Na^+ current), because the resting potential remained relatively large (-69.5 ± 0.2 mv, $n = 4$), even at the end of 20 min of ischemia. Consequently, this electrical inexcitability is attributed to a large increase in the outward current mediated by K_{ATP} channels. The ischemia significantly depolarized the resting potential, and the amplitude of the action potential declined progressively. On reperfusion, all electrical parameters (APD_{90}, action potential amplitude, and resting potential) were restored to the preischemic condition within 60 min, whereas the developed tension remained markedly depressed (only ~34% of the preischemic contraction) even at 60 min of reperfusion (Fig. 2B, circles), with the resting tension also remaining significantly elevated (not shown). These impaired contractile parameters did not return to preischemic levels, even when the reperfusion period was extended to 3 hr ($n = 2$). This indicates that the 20 min of no-flow ischemia led to "ischemic myocardial stunning," which is known as a sustained postischemic depression of the contractile function.

Under control (preischemic) conditions, cromakalim (2 μM) slightly but significantly shortened the APD_{90} (by $8.8 \pm 0.7\%$) with no apparent effects on action potential amplitude, resting potential, or contractile parameters. During ischemia and reperfusion, however, this agent markedly modified the action potentials and contractility. Cromakalim shortened the APD_{90} more quickly and severely during ischemia, with eventual electrical inexcitability in all preparations tested ($n = 5$), and at much earlier times (~15 min) than seen in untreated preparations. The shortening of APD_{90} in the reperfusion phase was also greater, and lasted longer. However, the most outstanding finding was that the recovery of contraction during reperfusion was clearly greater in the presence of the drug (Fig. 2B, squares); at 60 min after reperfusion, the developed tension was restored up to $79.3 \pm 4.1\%$ of the preischemic value, versus only $34.0 \pm 3.2\%$ in untreated preparations ($p < 0.05$). The elevation of resting tensions during ischemia was not affected by the presence of cromakalim, although the reperfusion-induced increase in resting tension was significantly ameliorated in the treated preparations.

Because the presence of cromakalim from the preischemic period remarkably improved contraction recovery, we examined the effect of cromakalim given exclusively after the onset of reperfusion. The application of cromakalim enhanced the shortening of the APD during reperfusion, that is, the APD_{90} was 112.0 ± 2.4 msec at 10 min and 136.0 ± 1.8 msec at 30 min of reperfusion ($n = 5$, $p < 0.05$ versus untreated preparations). In contrast, the developed tension measured at 60 min of reperfusion was significantly greater ($48.7 \pm 3.2\%$ of preischemic values, $n = 7$; $p < 0.05$;

see the large triangle put at 60 min of reperfusion, Fig. 2B) than that measured in untreated preparations (Fig. 2B, circles).

From these observations, we conclude that the pharmacological blockade of K_{ATP} channels during reperfusion impairs the recovery of contraction, whereas pharmacologically enforced opening of the K_{ATP} channels facilitates the recovery of contraction during the reperfusion phase.

B. Dual Effects of K_{ATP} Channel Activation on Reperfusion-Induced Arrhythmias

1. Inhibitory Effects of K_{ATP} Channel Activation on Triggered Arrhythmias

As shown in Fig. 1C (triangles), the blockade of K_{ATP} channels, via the introduction of glibenclamide after the onset of reperfusion, markedly depressed the recovery of contraction thereafter. Such a finding prompted us to test the blockage of K_{ATP} channels that was started some time after beginning of reperfusion on the later contraction recovery. Figure 3 illustrates an experiment in which glibenclamide was introduced not immediately but 15 min after the initiation of reperfusion. This procedure led to a prompt lengthening of the APD_{90} (by ~10 msec) (Fig. 3A) and a (transient) increase in the developed tension, albeit this increase was soon followed by an obvious interruption of the smooth recovery of contractile tension (Fig. 3B) that otherwise would be seen during this phase of reperfusion (cf. Fig. 1C, control). Similar tests were repeated with glibenclamide applied at different time intervals after beginning of reperfusion (i.e., 15, 30, and 60 min), and the amplitude of the transient increase in developed tension (mean percent increase from the preglibenclamide value) was 82, 23, and 0%, respectively. This result implies that the earlier the blockage of K_{ATP} channels, the greater the transient increase in developed tension. This was because the transient lengthening of the APD_{90} (by glibenclamide) was much greater when glibenclamide was applied at earlier phases of reperfusion.

Therefore, the persistent APD shortening seen during reperfusion was mostly attributable to the residual activation of K_{ATP} channels during this phase. Such persistent APD shortening, although relatively small compared with the APD shortening during ischemia, is nonetheless important for the recovery of the contractile function. In theory, shorter APDs lead to a reduction of the time for Ca^{2+} influx via voltage-gated Ca^{2+} channels, and they increase the time during which the Na^{+}–Ca^{2+} exchanger may operate to extrude Ca_{2+} in exchange for the influx of Na^{+} (Kimura *et al.,* 1987; Sheu *et al.,* 1986). Resulting decreases in the transsarcolemmal Ca^{2+} influx would

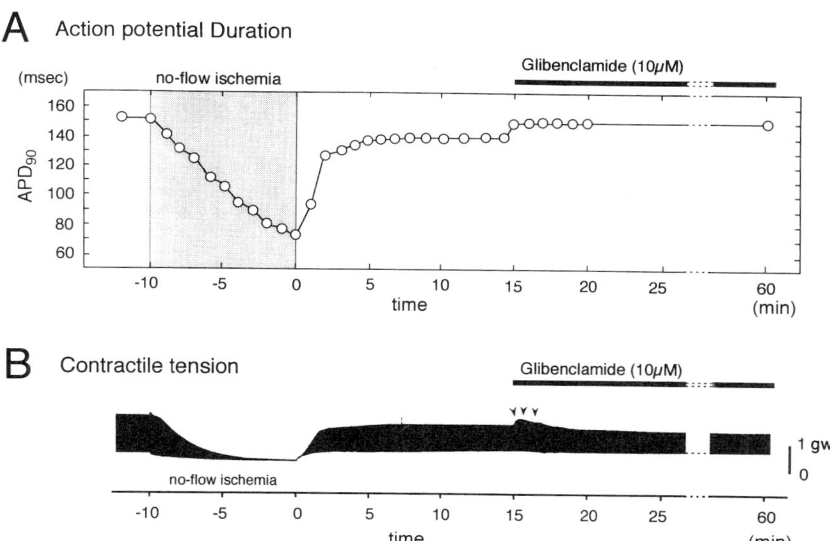

FIGURE 3 Effects of glibenclamide (applied from 15 min after the beginning of reperfusion to the end of observation) on the APD_{90} (A) and contractile tension (B). Note the transient increase of developed tension (arrows) seen immediately after the introduction of the drug (B), which was caused by a ~10-msec prolongation of APD_{90} (A). The transient increase of contraction led to the secondary decline of contraction in the later phase of reperfusion (B).

help maintain the intracellular Ca^{2+} concentration, $[Ca^{+2}]_i$, at physiological levels when other Ca^{2+} extrusion mechanisms, such as the Ca^{2+} pump, were impaired due to decreased $[ATP]_i$ (Krause and Gess, 1985). Accordingly, in the early reperfusion phase, when the Ca^{2+} extrusion mechanisms have not yet been fully restored, the application of glibenclamide (i.e., a sudden prolongation of APD) might have increased the $[Ca^{2+}]_i$. The rise in $[Ca^{2+}]_i$ (Ca^{2+} overload) might have impaired the subsequent recovery of contractility (Kusuoka *et al.,* 1987). Such a notion is supported by the finding that glibenclamide produced a rapid but only transient recovery of contraction at the very initial phase of reperfusion (Fig. 1C, filled circles and triangles; Fig. 3B), thereby leading to a subsequent severe decline of the contraction throughout the later phase of the reperfusion period.

Our speculation that a sudden prolongation of the APD, caused by the blockage of K_{ATP} channels in a relatively early phase of reperfusion, may lead to intracellular Ca^{2+} overload is supported by experiments using single ventricular cells exposed to simulated ischmia (Shigematsu and Arita, 1997).

Figure 4A shows action potentials recorded from a single guinea pig ventricular myocyte (stimulated at 1 Hz), subjected to 22 min of simulated ischemia (glucose-free anoxia). Reoxygenation quickly lengthened the APD, due to prompt blockage of K_{ATP} channels, and led to the evolution of delayed afterdepolarization (DAD), as indicated by arrows (Fig. 4B); the development of triggered activity eventually followed (Fig. 4C). Such was electrophysiological evidence of intracellular Ca^{2+} overload as examined in the following experiments (Fig. 4D–G). When the rate of stimulation was altered stepwise from 0.2 to 2 Hz, relatively lower stimulation rates (0.2 and 0.5 Hz) suppressed the evolution of DAD with no sign of triggered activity (Fig. 4D,E), whereas the higher stimulation rates (1 and 2 Hz) enhanced DAD and provoked triggered arrhythmias (Fig. 4F,G), perhaps due to further increases in $[Ca^{2+}]_i$. These findings suggest that persistent activation of K_{ATP} channels, seen during reperfusion, plays a central role in preventing the heart from developing triggered arrhythmias, even in coronary-perfused heart preparations. In accordance with this, we observed only sporadic ventricular premature contractions (VPCs) during reperfusion in 4 of 7 coronary-perfused ventricular preparations tested. Short runs

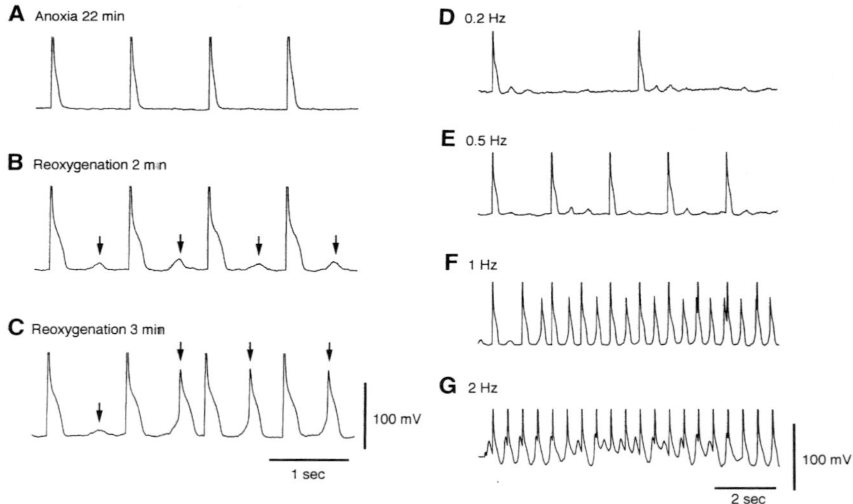

FIGURE 4 Delayed afterdepolarizations and triggered activity (arrows) induced by a prompt and marked prolongation of the APD, due to the blockage of K_{ATP} channels caused by reoxygenation, after 22 min of glucose-free anoxia, in a guinea pig ventricular myocyte stimulated at 1 Hz (A–C). After recording the trace in C, the rate of stimulation was changed stepwise from 0.2 to 2 Hz, as indicated. Note that the faster the rate of stimulation, the more frequent the incidence of triggered arrhythmias (D–G).

of ventricular tachycardias (VTs) were seen in only 1 of 7 preparations tested. In the presence of K_{ATP} channel blockade (by glibenclamide), however, VPCs evolved in all 5 preparations tested, and short runs of VTs evolved in 2 of 5 preparations. Thus blockage of K_{ATP} channels during reperfusion increased the incidence of ventricular tachyarrhythmias arising from DAD. However, it must be noted that in different animal models (i.e., rat, rabbit, pig, and dog) of acute myocardial ischemia, the incidence of ventricular fibrillation (Vf) was decreased by sulfonylurea derivatives during first 30–60 min of ischemia (Wilde and Janse, 1994), probably due to the drug-induced lengthening of the ventricular APD or the prolongation of the effective refractory period.

2. Facilitating Effects of K_{ATP} Channel Activation on Reentrant Arrhythmias

Introduction of cromakalim causes improvements in postischemic mechanical recovery and ameliorates myocardial stunning (Fig. 2). Others have postulated that some K_{ATP} channel openers have prevented irreversible cell injury during ischemia–reperfusion (Cole *et al.*, 1991; Kempsford and Hawgood, 1989; McCullough *et al.*, 1991). Cromakalim increases coronary flow by means of its vasodilating effect (Cook and Quast, 1990). However, in our study, the rate of coronary perfusion was constant throughout the experiments. The protective effect of cromakalim could be attributable, at least in part, to a reduction of energy consumption due to decreased contractility (Nichols *et al.*, 1991). Such an energy-sparing effect may facilitate contractile recovery during reperfusion. In our experiments, however, this mechanism did not appear to play a primary role, because the rate of decline in contractions seen during ischemia, in the presence of cromakalim, did not differ significantly from that found in the absence of this drug (Fig. 2B). Accordingly, we attribute the protective effect of cromakalim mostly to the attenuation of $[Ca^{2+}]_i$ during ischemia and reperfusion.

Cromakalim improved contractile recovery even when it was applied from the onset of reperfusion. Such an improvement is thought to be associated with an enhanced shortening of the APD during reperfusion. However, in the presence of a very short APD, VTs and Vf are more likely to occur during reperfusion, because the reentry mechanism is facilitated due to the markedly shortened effective refractory period. Figure 5 shows such an example of reperfusion-induced arrhythmias occurring in the presence of cromakalim ($5 \mu M$). In the presence of the drug (from the preischemic period), the reperfusion introduced after 15 min of no-flow ischemia readily produced sustained VT (not shown), followed by Vf lasting longer than 5 min (Fig. 5D,E); at this moment glibenclamide ($10\ \mu M$) was given

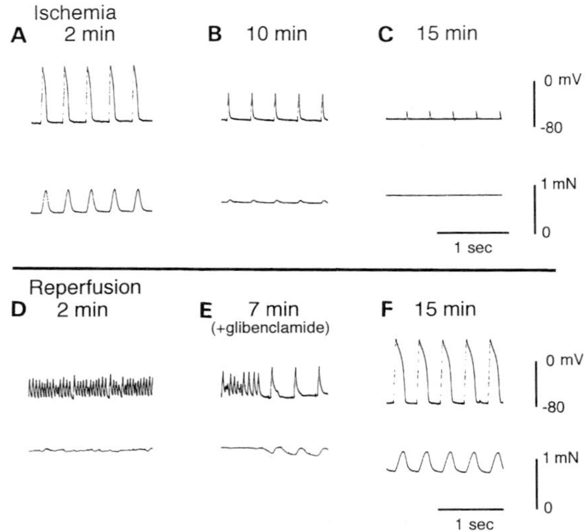

FIGURE 5 Development of ventricular tachycardia and fibrillation during reperfusion after 15 min of ischemia. The records were taken with cromakalim (5 μM) present throughout the experiment. Note that the fibrillation was terminated immediately after the application of glibenclamide (i.e., after the lengthening of the APD), suggesting that the mechanism of arrhythmia could be reentry. Adapted from Shigematsu *et al.* (1995, Fig. 7) with permission of the American Heart Association.

(in the continued presence of cromakalim) to examine the contribution of the K_{ATP} current to this tachyarrhythmia. With the introduction of glibenclamide, Vf promptly terminated after a marked prolongation of the APD, indicating the prolongation of the effective refractory period (Fig. 5E). In our experiments, Vf that continued for more than 3 min never terminated spontaneously. Glibenclamide might have abolished Vf by lengthening the APD or the effective refractory period by blocking K_{ATP} currents.

In non-drug-treated preparations, reperfusion after 20 min of no-flow ischemia (Fig. 2, circles) resulted in more frequent and severe arrhythmias than those with reperfusion after 10 min of ischemia (Fig. 1, open circles). Namely, VPCs and/or short runs of VT were observed in all 7 preparations tested, though sustained VT (duration 108 sec) followed by Vf was seen in only one preparation. The presence of cromakalim (from the preischmic period) did not increase the incidence of VPCs during reperfusion; however, the procedure tended to increase the incidence of VTs and Vf (in 20 and 40% of 10 preparations tested, respectively). Applying cromakalim from the beginning of reperfusion also produced VTs in 29% and Vf in 57% of 7 preparations tested. These findings suggest that the presence of cromakalim, that is, the drug-induced extra opening of K_{ATP} channels, tends to increase the risk of VT and Vf during reperfusion; the mechanism may be attributable to the enhanced probality of reentry due to the marked shortening of the effective refractory period. Similar trends have been noted in rabbit and canine hearts, although K_{ATP} channel openers may also act as an antiarrhythmic agent whenever the arrhythmia was caused by enhanced automaticity or triggered activity (Wilde and Janse, 1994).

V. DIVERGENT EFFECTS OF ANTIARRHYTHMIC DRUGS ON K_{ATP} CHANNELS AND THEIR IMPLICATIONS

The K_{ATP} channel plays an essential role in shortening APD in ischemic myocardium (Fig. 1) and in the secretion of insulin from pancreatic β cells (Nichols *et al.,* 1991; Horie *et al.,* 1997). In the 1980s, when disopyramide was first introduced for clinical use for the treatment of arrhythmias, there appeared several reports to document that hypoglycemia was a possible side effect of this agent (Goldberg *et al.,* 1980). The same side effect was also reported with cibenzoline, introduced several years later (Hilleman *et al.,* 1987; Gachot *et al.,* 1988). These drugs are Class Ia antiarrhythmic drugs (according to the modified Vaugham Williams classification), prevent the hypoxia-induced shortening of the APD in cardiac tissues (Millar and Vaugham Williams, 1982), and allegedly cause hypoglycemia in some patients. These reports prompted us to study the effect of various antiarrhythmic drugs, especially Class I drugs, on cardiac K_{ATP} channels (Wu *et al.,* 1992; Sato *et al.,* 1993; Wang *et al.,* 1995).

Table I summarizes the effects of Class I and III drugs investigated so far on the K_{ATP} currents of guinea pig ventricular myocytes. It is interesting that most Class Ia drugs (cibenzoline, disopyramide, procainamide) blocked K_{ATP} channels at concentractions lower than twice the maximum therapeutic concentrations (Wu *et al.,* 1992). Class Ic drugs did not affect K_{ATP} currents, at least at concentrations lower than twice the maximum therapeutic concentration, although much higher concentrations of flecainide (10 μM) (but not of pilsicainide) blocked the current only when the current was measured as an outwardly directed current (voltage-dependent block; Wang *et al.,* 1995).

To our surprise, a Class Ib drug, mexiletine, did not block but actually activated K_{ATP} channel currents, provided that the intracellular side of the membrane patch was perfused with solution containing a nucleoside diphosphate like UDP or ADP (Sato *et al.,* 1995); mexiletine had a character tantamount to that of nicorandil (Shen *et al.,* 1991), the only approved

TABLE I
Effects of Antiarrhythmics on $I_{K,ATP}$[a]

Drugs	Concentration (μM)	Effects	Conditions
Class Ia			
Cibenzoline	10	Block	
Disopyramide	30	Block	
Procainamide	100	Block	
Class Ib			
Mexiletine	30, 100	Activation	In presence of NDP
Class Ic			
Flecainide	2	No effect	
	10	Block	Outward current only
Pilsicainide	50	No effect	
Class III			
E4031	10	No effect	

[a] The effects of antiarrhythmic drugs on cardiac K_{ATP} channels were measured using either a whole-cell voltage clamp or an inside out patch voltage clamp method, in guinea pig ventricular myocytes. The concentrations tested were up to twice the therapeutic concentration, except for 100 μM mexiletine and 10 μM flecainide.

K_{ATP} channel opener prescribed to patients. E4031 is classified as a Class III drug and is regarded as an authentic blocker of the delayed rectifier K^+ channel; it had no effect on K_{ATP} channels.

From Class I drugs, which are known to block cardiac Na^+ channels, we chose a representative from each subgroups, Ia, Ib, and Ic, to compare their effects on the recovery of contractile tension during reperfusion in the coronary-perfused ventricular preparations. From Class Ia drugs, we chose cibenzoline, because this agent blocked the K_{ATP} current most effectively among all Class I drugs tested (Wu *et al.,* 1992). Selected from Class Ib was mexiletine, which activated K_{ATP} channels, especially under conditions of ischemia, where intracellular levels of nucleoside diphosphates (ADP and/or UDP) may be elevated (Sato *et al.,* 1995). From Class Ic, pilsicainide was chosen because the agent was regarded as a pure Na^+ channel blocker without effect on other ion channels, including the K_{ATP} channel in heart (Inomata *et al.,* 1987; Wu *et al.,* 1992). The concentration of each drug was adjusted to approximately twice the therapeutic concentration (Rothbart and Saksena, 1986; Zipes and Troup, 1987).

Figure 6A compares the degree of recovery of developed tension (expressed as the percentage of the respective preischemic value), evaluated at 60 min of reperfusion afer 10 min of ischemia, in the absence (untreated) and presence of various antiarrhythmic drugs, along with glibenclamide

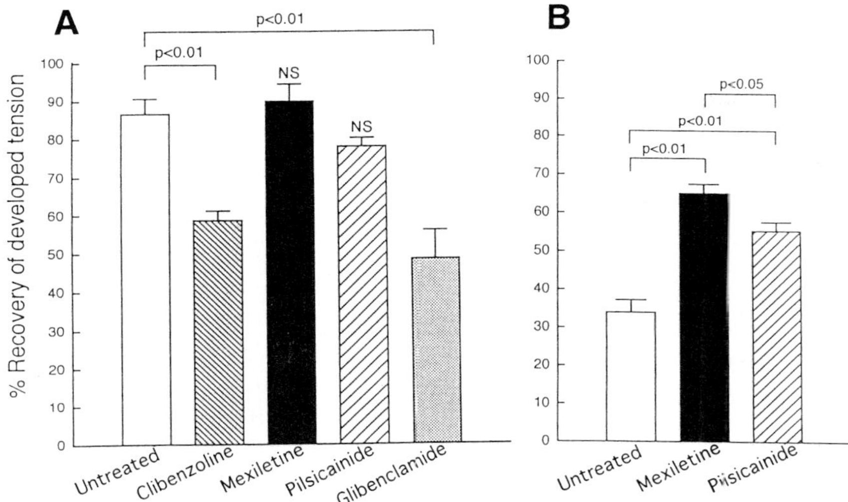

FIGURE 6 Comparison of the recovery of contraction measured at 60 min of reperfusion after 10 min of ischemia (A) or 20 min of ischemia (B), in the absence (untreated) and presence of various Class I antiarrhythmic drugs (5 μM cibenzoline, 10 μM mexiletine, 10 μM pilsicainide) and glibenclamide (10 μM). Recovery was expressed as the percentage of respective preischemic values. The number of observations was 4–7 in each column.

(used as a reference drugs). The drugs were present from the preischemic period to the end of the observation. It must be noted that in the presence of cibenzoline (5 μM), which blocked K_{ATP} channels, the recovery of contraction during reperfusion was significantly depressed, as was the case for glibenclamide (10 μM) (see also Fig. 1C). In contrast, in the presence of pilsicainide (10 μM), which has no effect on K_{ATP} channels, there was no significant difference in contraction recovery as compared to that seen in untreated preparations. In the presence of mexiletine (10 μM), used to activate K_{ATP} channels, the contraction recovery tended to be maximal though it did not reach statistical significance (as compared to the recovery of untreated preparations). We then repeated the same experiments (using mexiletine and pilsicainide) with a much longer ischemic period of 20 min (instead of 10 min) followed by 60 min of reperfusion; the results are summarized in Fig. 6B. As mentioned earlier (Fig. 2), 20 min of ischemia (followed by 60 min of reperfusion) resulted in a severe depression of contractile tension, reaching as low as 34% of preischemic values (Fig. 6B, left bar; see also Fig. 2B). In the presence of mexiletine (10 μM), however, the contraction recovery was raised significantly to 64% of preischemic levels (Fig. 6B, middle bar), suggesting that additional openings of the

K_{ATP} channels caused by mexiletine promoted the recovery of contraction. However, it must be noted here that significant facilitation of contractile recovery (up to 55% of control, Fig. 6B, right bar) was also recognized in the presence of pilsicainide (10 μM), a pure Na^+ channel blocker, even though the recovery did not reach the level attained by mexiletine. Such findings suggest that the activation of K_{ATP} channels, and the blockage of Na^+ channels during ischemia and reperfusion, are both beneficial for the recovery of contraction after ischemic insults.

Class Ic drugs (flecainide and pilsicainide) were reported to improve contractile function after global ischemia in isolated rat hearts (Liu *et al.*, 1993), a finding squaring with our observation. We found that pilsicainide improved contraction recovery without affecting the APD during ischemia. Flecainide at concentrations close to therapeutic levels (~2 μM) did not block K_{ATP} channels. Therefore, the ameliorating effects of flecainide and pilsicainide on contraction recovery should be attributed to their blocking effects on Na^+ currents. It is worth noting that mexiletine facilitated the recovery of contraction significantly more than did pilsicainide. The beneficial effects of mexilentine on contraction recovery may be ascribed to the extra shortening of APD during ischemia–reperfusion, caused by the drug-induced extra activation of K_{ATP} channels (in addition to the blockage of Na^+ channels).

Currently, the Cardiac Arrhythmia Suppression Trial (CAST) has raised a question about whether the suppression of ventricular arrhythmias by class I antiarrhythmic drugs protects patients from eventual arrhythmogenic death (Echt *et al.*, 1991). Surprisingly the trial showed an increase in mortality of patients receiving some class I antiarrhythmic drugs, probably due to their proarrhythmic effects. However, another possible explanation could be deduced, at least in part, from our present observtion: that is, the "decreased myocardial contractility" might occur because of the drug-induced blockage of K_{ATP} channels during ischemia–reperfusion, channels which would otherwise be kept in an open state at least during the early phase (~60 min) of reperfusion.

We conclude that the persistent activation of K_{ATP} channels during the early reperfusion phase is essential for a smooth and full recovery of contractile function, as well as for maintenance of electrical stability, in heart that has been subjected to ischemia.

References

Cole, W. C., McPherson, C. D., and Sontag, D. (1991). ATP-regulated K^+ channels protect the myocardium against ischemia/reperfusion damage. *Cir. Res.* **69,** 571–581.

Cook, N. S., and Quast, U. (1990). Potassium channels: Structure, classification, function and therapeutic potential. *In* "Potassium Channel Pharmacology" (N. S. Cook, ed.), pp. 181–255. Ellis Harwood, Chester, U.K.

Echt, D. S., Liebson, R. R., Mitchell, L. B., Peters, R. W., Obias-Manno, D., Barker, A. H., Arensberg, D., Barker, A., Friedman, L., Greene, H. L. Huther, M. L., and Richardson, D. W. (1991). Mortality and morbidity in patients receiving encainide, flecainide, or placebo. The Cardiac Arrhythmia Suppression Trials. *N. Engl. J. Med.* **324,** 781–788.

Gachot, B. A., Bazier, M., Cherrier, J. F., and Daubeze, J. (1988). Cibenzoline and hypoglycemia. *Lancet* **2,** 280.

Goldberg, I. J., Brown, L. K., and Rayfeld, E. J. (1980). Disopyramide (Norpace®) increased hypoglycemia. *Am. J. Med.* **69,** 463–466.

Gross, G. J., and Auchampach, J. A. (1992). Blockade of ATP-sensitive potassium channels prevents myocardial preconditioning in dogs. *Circ. Res.* **70,** 223–233.

Hillemann, D. E., Mohiuddin, S. M., and Ahmed, I. S. (1987). Cibenzoline-induced hypoglycemia. *Drug. Intell. Clin. Pharm.* **21,** 38–89.

Horie, M., Ishida-Taka, A., Ai, T., Nishimoto, T., Tsuura, Y., Ishida, H., Seino, Y., and Sasayama, S. (1997). Insulin secretion and its modulation by antiarrhythmic and sulfonylurea drugs. *Cardiovasc.* Res. **34,** 69–72.

Inomata, N., Ishihara, T., and Akaike, N. (1987). SUN1165: A new antiarrhythmic Na current blocker in ventricular myocytes of guinea-pig. *Comp. Biochem. Physiol.* **87C,** 237–243.

Kempsford, R. D., and Hawgood, B. J. (1989). Assessment of the antiarrhythmic activity of nicorandil during myocardial ischemia and reperfusion. *Eur. J. Pharmacol.* **163,** 61–68.

Kimura, J., Miyamae, S., and Noma, A. (1987). Identification of sodium–calcium exchange current in single ventricular cells of guinea-pig. *J. Physiol. (London)* **384,** 199–222.

Krause, S. M., and Gess, M. L. (1985). Characterization of cardiac sarcoplasmic reticulum dysfunction during short-term normothermic global ischemia. *Circ. Res.* **55,** 176–184.

Kusuoka, H., Porterfield, J. K., Weisman, H. F., Weistfeld, M. L., and Marban, E. (1987). Pathophysiology and pathogenesis of stunned myocardium. Depressed Ca^{2+} activation as a consequence of reperfusion-induced cellular calcium overload in ferret hearts. *J. Clin. Invest.* **79,** 950–961.

Liu, J.-X., Tanonaka, K., Ohtsuka, Y., Sasaki, Y., and Takeo, S. (1993). Improvement of ischemia/reperfusion-induced contractile dysfunction of perfused hearts by class Ic antiarrhythmic drugs. *J. Pharmacol. Exp. Ther.* **266,** 1247–1254.

McCullough, J. R., Normandin, D. E., Conder, M. L., Sleph, P. G., Dzwonczyk, S., and Grover, G. J., (1991). Specific block of the anti-ischemic actions of cromakalim by sodium-5-hydroxy-decanoate. *Circ. Res.* **69,** 949–958.

Millar, J. S., and Vaugham Williams, E. M. (1982). Effects on rabbit nodal, atria, ventricular and Purkinje cell potentials of a new antiarrhythmic drug, cibenzoline, which protects against action potential shortening in hypoxia. *Br. J. Pharmacol.* **75,** 469–478.

Murry, C. E., Jennings, R. B., and Reimer, K. A. (1986). Preconditioning with ischemia: A delay of lethal cell injury in ischemic myocardium. *Circulation* **74,** 1124–1136.

Murry, C. E., Richard, V. J., Jennings, R. B., and Reimer, K. A. (1991). Myocardial protection is lost before contractile function recovers from ischemic preconditioning. *Am. J. Physiol.* **260,** H796–H804.

Nakamura, S., Kiyosue, T., and Arita, M. (1989). Glucose reverses 2,4-dinitrophenol induced changes in action potentials and membrane currents of guinea pig ventricular cells via enhanced glycolysis. *Cardiovasc. Res.* **23,** 286–294.

Nichols, C. G., Ripoll, C., and Lederer, W. J. (1991). ATP-sensitive potassium channel modulation of the guinea pig ventricular action potential and contraction. *Cir. Res.* **68,** 280–287.

Noma, A. (1983). ATP-regulated K^+ channels in cardiac muscle. *Nature* (*London*) **305,** 147–148.

Rothbart, S. T., and Saksena, S. (1986). Clinical electrophysiology efficacy and safety of chronic oral cibenzoline therapy in refractory ventricular tachycardia. *Am. J. Cardiol.* **57,** 941–946.

Sato, T., Wu, B., Nakamura, S., Kiyosue, T., and Arita, M. (1993). Cibenzoline inhibits diazoxide and 2,4-dinitrophenol-activated ATP-sensitive K^+ channels in guinea-pig ventricular cells. *Br. J. Pharmacol.* **108,** 549–556.

Sato, T., Shigematsu, S., and Arita, M. (1995). Mexiletine-induced shortening of the action potential duration of ventricular muscles by activation of ATP-sensiive K^+ channels. *Br. J. Pharmacol.* **115,** 381–382.

Shen, W. K., Tung, P. T., Machulda, M. M., and Kurachi, M. (1991). Essential role of nucleotide diphosphate in nicrorandil-mediated activation of cardiac ATP-sensitive K^+ channel: A comparison with pinacidil and lemakalim. *Circ. Res.* **69,** 1152–1158.

Sheu, S., Sharma, V. K., and Uglesity, A. (1986). Na^+–Ca^{2+} exchange contributes to increase of cytosolic Ca^{2+} concentration during depolarization in heart muscle. *Am. J. Physiol.* **250,** C651–C656.

Shigematsu, S., and Arita, M. (1997). Anoxia-induced activation of ATP-sensitive K^+ channels in guinea pig ventricular cells and its modulation by glycolysis. *Cardiovasc. Res.* **35,** 273–282.

Shigematsu, S., Sato, T., Abe, T., Saikawa, T., Sakata, T., and Arita, M. (1995). Pharmacological evidence for the persistent activation of ATP-sensitive K^+ channels in early phase of reperfusion and its protective role against myocardial stunning. *Circulation* **92,** 2266–2275.

Wang, D. W., Sato, T., and Arita, M. (1995). Voltage dependent inhibition of ATP-sensitive potassium channels by flecainide in guinea pig ventricular cells. *Cardiovasc. Res.* **29,** 520–525.

Wilde, A. A. M., and Janse, M. J. (1994). Electrophysiological effects of ATP sensitive potassium channel modulation: Implications for arrhythmogensis. *Cardiovasc. Res.* **28,** 16–24.

Wu, B., Sato, T., Kiyosue, T., and Arita, M. (1992). Blockade of 2,4-dinitrophenol induced ATP sensitive potassium current in guinea pig ventricular myocytes by class I antiarrhythmic drugs, *Cardiovasc. Res.* 26, 1095–1101.

Zipes, D. P., and Troup, P. J. (1987). New antiarrhythmic agents. *Am. J. Cardiol.* **41,** 1005–1024.

CHAPTER 23

Role of K_{ATP} Channels in Cardioprotection

Masafumi Kitakaze*, Yasuhiko Sakata*, Kazuhisa Kodama†, Tsunehiko Kuzuya*, and Masatsugu Hori*

*The First Department of Medicine, Osaka University Medical School, Suita, Japan; and †The Cardiovascular Division, Osaka Police Hospital, Osaka, Japan

I. INTRODUCTION

Adenosine is known to be cardioprotective because (1) adenosine affects coronary smooth muscle, endothelial cells, cardiomyocytes, platelets, leukocytes, and sympathetic nerves, which may contribute to the cardiovascular protection against ischemia and reperfusion injury, and (2) adenosine is involved in the infarct size-limiting effect and the attenuation of myocardial stunning. Adenosine can open ATP-sensitive K^+ channels (K_{ATP} channels), which is one of the major factors in the adenosine-induced cardioprotection. Here, we discuss the role of the opening of K_{ATP} channels in the cardioprotection from the viewpoint of adenosine receptor activation.

II. ADENOSINE IN ISCHEMIC HEARTS

When the heart becomes ischemic due to insufficient supply of myocardial oxygen and nutrition, ATP in cardiomyocytes is rapidly degraded to ADP,

1063-5823/99 $30.00

which cannot be resynthesized to ATP because of the lack of the supply of creatinine phosphate from the mitochondria. Under such conditions, ADP is converted to AMP, from which adenosine is massively produced (Berne, 1980; Hori and Kitakaze, 1991). Adenosine is also produced in endothelial cells of the coronary arteries (Newman *et al.,* 1988). Although the total mass of endothelial cells is about 2% in the heart, the capacity of adenosine synthesis in an endothelial cells is 50 times higher than in a myocardial cell (Forrester and Williams, 1977).

In the nonischemic condition, the basal level of adenosine is mainly attributed to *S*-adenosylhomocysteine by *S*-adenosylhomocysteine hydrolase (Cshrader *et al.,* 1981); however, during ischemia, the pathway of adenosine synthesis is switched to the AMP-adenosine pathway via 5′-nucleotidase. This is proved by the fact that the substrate for *S*-adenosylhomocysteine hydrolase is only 1.5 times increased, but the substrate for 5′-nucleotidase is 50–100 times increased. Therefore, in the ischemic condition, adenosine may be produced by mainly 5′-nucleotidase, which is located on the cell membrane (ecto-5′-nucleotidase) and in the cytoplasm (cytosolic 5′-nucleotidase) (Newby *et al.,* 1987). The adenosine produced is rapidly taken up by the surrounding tissues (i.e., myocytes, endothelial cells, and erythrocytes) and drained into the coronary vein, in which adenosine is degraded to inosine and hypoxanthine via adenosine deaminase, and finally to uric acid (Rovetto, 1985). Some of adenosine is resynthesized to AMP which is, in turn, degraded to IMP via AMP deaminase (Fig. 1). The adenosine produced may act at its own receptors.

Adenosine receptors are classified into A_1–A_3 receptors. The activation of adenosine A_1 receptors coupled with G_i proteins attenuates the synthesis of cAMP when β-adrenoceptors are stimulated, and attenuates Ca^{2+} influx through voltage-dependent Ca^{2+} channels and Na^+/Ca^{2+} exchange (Brechler *et al.,* 1990). Furthermore, A_1 receptors are coupled with K_{ATP} channels in myocardial cells: Adenosine A_1 receptor stimulation opens the K_{ATP} channels, leading to a decrease in Ca^{2+} influx through the shortening of the action potential duration (Kirsch *et al.,* 1990). Furthermore, the opening of K_{ATP} channels may attenuate the release of norepinephrine from sympathetic nerve terminals. These effects attenuate Ca^{2+} influx and preserve metabolic function to attenuate the utilization of the energy during ischemia. Activation of adenosine A_1 receptors of neutrophils enhances chemotaxis (Engler and Covell, 1987). On the other hand, A_2 receptors are coupled with G_s proteins, and activation of the receptors may enhance the synthesis of cAMP; activation of the A_2 receptors exerts coronary vasodilation (Hori *et al.,* 1988; Hori *et al.,* 1989), inhibits aggregation of platelets, and inhibits free radical formation by neutrophils (Agarwal, 1987; Kitakaze *et al.,* 1991a).

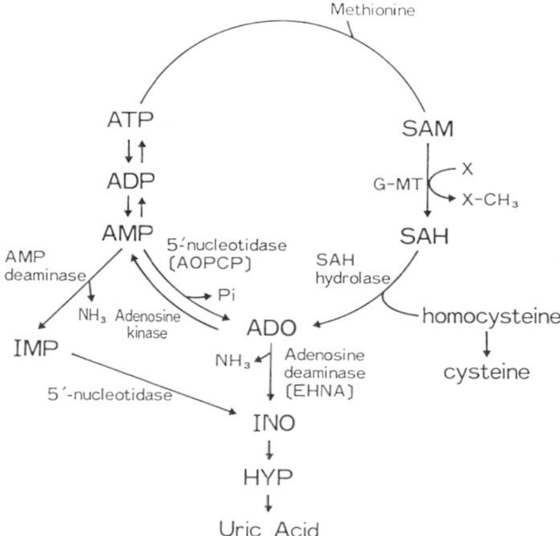

FIGURE 1 Adenosine metabolism. SAM, *S*-Adenosylmethionine; SAH, *S*-adenosylhomocysteine; ADO, adenosine; AMP, adenosine 5′-monophosphate; IMP, inosine 5′-monophosphate; INO, inosine; HYP, hypoxanthine; AOPCP, α,β-methylene adenosine 5′-diphosphate; EHNA, erythro-9-(2-hydroxy-3-nonyl)adenosine. Modified from Schrader, J. (1983). Metabolism of adenosine and site of production in the heart. *In* "Regulatory Function of Adenosine" (Berne, R. M., Rall, T. W., and Rubio, eds.), pp. 133–156. Martinus Nijhoff, Boston.

III. ACTIONS OF ADENOSINE AND K_{ATP} CHANNELS IN CORONARY VASODILATION

Adenosine is quickly released during myocardial ischemia and serves as a potent coronary vasodilator through A_2 receptors; the adenosine concentration in the coronary vein is increased within 30 sec of ischemia and dilates coronary resistance vessels. Indeed, reactive hyperemia after 10–20 sec of occlusion of the coronary artery is partially attenuated by treatment with aminophylline, an inhibitor of adenosine receptors (Schutz *et al.,* 1977; Merrill *et al.,* 1978; Kitakaze *et al.,* 1987, 1995; Morioka *et al.,* 1995). Interestingly, α_1-adrenoceptor activity regulates the release of adenosine from ischemic myocardium (Fig. 2) (Kitakaze *et al.,* 1987), and α_2-adrenoceptor activity regulates the adenosine-induced coronary vasodilation (Fig. 3; Hori *et al.,* 1988). It was reported that K_{ATP} channels are also involved in coronary vasodilation during ischemia (Komaru *et al.,* 1991). K_{ATP} channels are distributed in vascular smooth muscles and other tissues, such as skeletal muscle, β-islet cells in the pancreas, and nerve cells (Cook and Hales, 1984; Ashford *et al.,* 1988). When ATP is decreased, this channel opens, allowing increases in K^+ efflux and rendering the myocardial cells hyperpolarized. Accordingly, the action potential is shortened and Ca^{2+} influx through

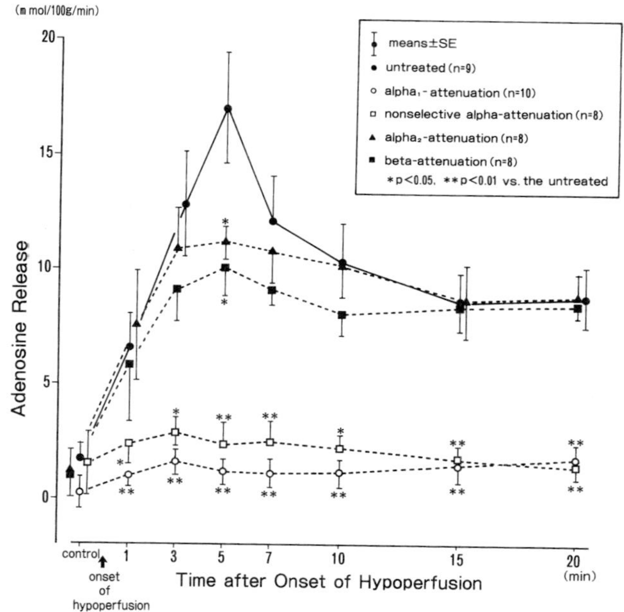

FIGURE 2 Adenosine release during coronary hypoperfusion during α_1-adrenoceptor blockade [prazosin 4 μg/kg/min via intracoronary infusion (ic)], α_2-receptor blockade (yohimbine 9 μg/kg/min ic + propranolol 0.3 mg/kg ic), nonselective α-receptor blockade (phentolamine 9 μg/kg/min ic + propranolol 0.3 mg/kg ic), and β-adrenoceptor blockade (propranolol 0.3 mg/kg ic). Administration of prazosin and phentolamine markedly reduced the release of adenosine during coronary hypoperfusion. Values are means, bars = SEM. $*p < 0.05$, $**p < 0.01$ versus untreated. From Kitakaze *et al.* (1987) with permission of the American Heart Association.

voltage-dependent Ca^{2+} channels is decreased. This is the underlying mechanism of vasodilation by opening of K_{ATP} channels.

It is considered that the K_{ATP} channels have an ATP binding site and when the ATP level is high enough, K_{ATP} channels are closed. If the ATP concentration is decreased, the ATP binding site may not be occupied with ATP, and thus, the open probability of the K^+ channels is increased. However, it is known that the K_i of ATP for the opening of K_{ATP} channels is 20–100 μM, which is much lower than the concentration observed during brief ischemia (Nichols and Lederer, 1991). Thus, a decrease in ATP concentration during a brief period of ischemia alone could not cause the opening of K_{ATP} channels. However, Aversano *et al.* (1991) reported that decreases in coronary vascular resistance during ischemia are significantly attenuated by glibenclamide, indicating that K_{ATP} channels are opened during ischemia. It is reported that adenosine opens K_{ATP} channels, as adenosine-induced coronary vasodilation is attenuated by treatment with glibenclamide (Lederer and Nichols, 1989). In our preliminary study, we observed that K_{ATP} channel openers, both nicorandil and cromakalim, enhance the adenosine-induced coronary vasodilation; a small dose of these agents (which did not change the basal coronary blood flow) further increased the coronary hyperemic flow induced by adenosine. These results indicate that both decreases in ATP content and release of adenosine during ischemia may contribute to coronary vasodilation. It is of note that acidosis during ischemia may augment these effects because a decrease in pH could enhance

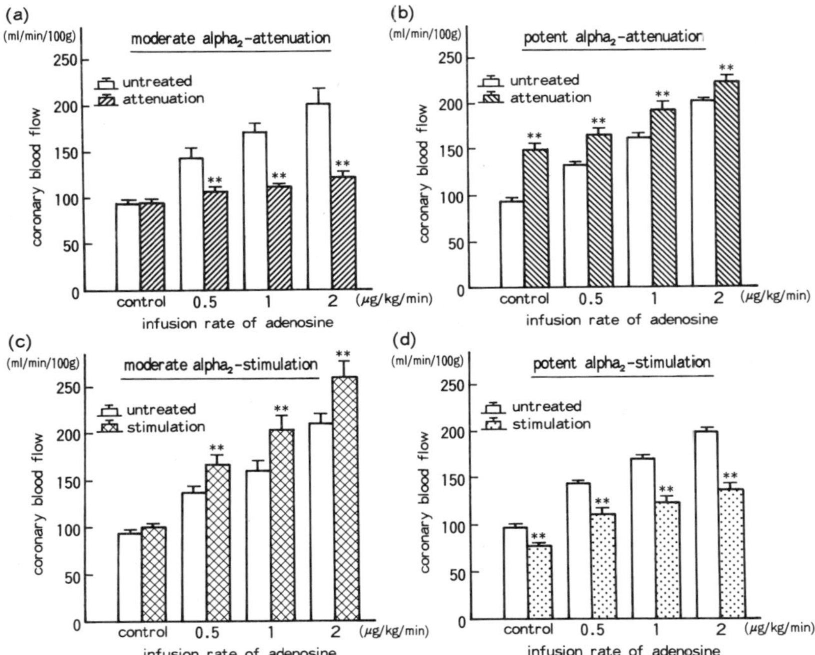

FIGURE 3 Hyperemic coronary blood flow during intracoronary infusion of three doses of adenosine in moderate (a) or potent α_2-adrenoceptor attenuation (b), and moderate (c) or potent α_2-adrenoceptor stimulation (d). Moderate α_2 stimulation enhances adenosine-induced coronary vasodilation (c), and moderate α_2 blockade attenuates the flow response to adenosine (a). Open columns are controls (propranolol 0.3 mg/kg ic); hatched columns in A, propranolol 0.3 mg/kg ic + yohimbine 9 μg/kg/min ic; hatched columns in b, propranolol 0.3 mg/kg ic + yohimbine 20 μg/kg/min ic; cross-hatched columns in c, propranolol 0.3 mg/kg ic + norepinephrine 0.03 μg/kg/min ic + prazosine 6 μg/kg/min ic; shaded columns in d, propranolol 0.3 mg/kg ic + norepinephrine 0.3 μg/kg/min ic + prazosin 6 μg/kg/min ic. Values are means, bars = SEM. $*p < 0.05$, $**p < 0.01$ versus control (propranolol) From Hori *et al.* (1988) with permission of the American Physiological Society.

the opening of K_{ATP} channels and the sensitivity of vascular smooth muscle to adenosine (Lederer and Nichols, 1989).

IV. EXPERIMENTAL EVIDENCE FOR K_{ATP} CHANNEL-MEDIATED CARDIOPROTECTION

When coronary perfusion is restored after a brief period of ischemia before irreversible cellular injury occurs, contractile function remains depressed for a long time, a phenomenon known as myocardial stunning (Braunwald and Kloner, 1982). As the underlying mechanisms, several lines of evidence suggest that Ca^{2+} overload due to an excessive Ca^{2+} influx on reperfusion is one of the major causes of stunning (Kusuoka *et al.,* 1987). ATP depletion and microcirculatory disturbance are also involved in the pathogenesis of stunning. ^{23}Na-NMR studies showed that Na^+ influx is increased via Na^+/H^+ exchange to extrude intracellular H^+ during ischemia (Pike *et al.,* 1990). Since H^+ inhibits Na^+/Ca^{2+} exchange during ischemia, Ca^{2+} overload during myocardial ischemia becomes moderate. However, on reperfusion, myocardial cellular acidosis is rapidly recovered, which may decrease the inhibition of Na^+/Ca^{2+} exchange and increase the intracellular Ca^{2+} levels. Since the Ca^{2+} pump activity is ATP-dependent and is depressed during ischemia, Ca^{2+} efflux is insufficient to avoid Ca^{2+} overload.

It is reported that adenosine could attenuate myocardial stunning (Takeo *et al.,* 1988; Kitakaze *et al.,* 1991b), and exogenous adenosine improved functional recovery after 15 min of coronary occlusion in dogs. 8-Phenyltheophylline attenuated the contractile function after reperfusion, indicating that both endogenous and exogenous adenosine could attenuate the reperfusion injury after brief ischemia. *N*^{6}-Cyclohexyladenosine (an A_1 receptor agonist) also improved contractile function after reperfusion, indicating that activation of adenosine A_1 receptors mediates myocardial stunning. We also observed that CGS21680C (an A_2 receptor agonist) attenuates the stunning to a similar extent as *N*^{6}-cyclohexyladenosine. These results indicate that inhibition of Ca^{2+} influx and free radical formation by A_1 receptor stimulation may be largely involved in the pathogenesis of myocardial stunning, and inhibition of platelet and neutrophil activation by A_2 receptors may also be involved.

Yao and Gross (1993) reported that glibenclamide antagonizes A_1 receptor-mediated cardioprotection in stunned myocardium in dogs. They made repetitive coronary occlusions in dogs and reperfused for 2 hr. Administration of glibenclamide (0.1 μg/kg), which did not alter the basal myocardial contractility, completely abolished the beneficial effect of cyclopentyladenosine, the A_1 receptor agonist. Thus, the improvement of stunning by

A_1 receptors may be partially mediated via opening of K_{ATP} channels. Grover *et al.* (1992) reported that cromakalim significantly improved reperfusion function in isolated rat hearts, and reduced the infarct size in dogs subjected to 90 min of coronary occlusion and 5 hr of reperfusion. It is also reported that nicorandil improves functional recovery in rat heart (Mitani *et al.*, 1991). Taken together, the results suggest that opening of K_{ATP} channels may contribute to protection of the heart from the ischemia–reperfusion injury in concert with endogenous adenosine.

Much attention has been paid to the pathophysiology of ischemic preconditioning, a phenomenon that preceding brief ischemia renders the heart resistant to sustained ischemia followed by reperfusion. This is because the infarct size-limiting effect is the most potent among various interventions previously reported (Murry *et al.*, 1986). Liu *et al.* (1991) have shown that endogenous adenosine mediates ischemia preconditioning via adenosine A_1 receptors. We observed that adenosine release in the coronary vein is significantly increased in heart subjected to ischemic preconditioning, namely, four episodes of 5 min of ischemia (Kitakaze *et al.*, 1993). Therefore, we hypothesized that augmented production of adenosine during preconditioning and sustained ischemia may contribute to the cardioprotective effects, probably through attenuation of tissue acidosis and Ca^{2+} overload after reperfusion. To clarify the mechanism of increase in adenosine production after ischemic preconditioning, we measured 5′-nucleotidase activity, which mainly regulates the rate of adenosine synthesis during ischemia. Both ecto- and cytosolic 5′-nucleotidase activities were increased after ischemic preconditioning (Nichols and Lederer, 1991). Blockade of ecto-5′-nucleotidase by AMP-CP attenuated the adenosine release and prevented the infarct size-limiting effect of preconditioning (Kitakaze *et al.*, 1994a). These results strongly suggest that ischemic preconditioning augments adenosine production via activation of ecto-5′-nucleotidase. Our results are in accordance with previous reports in which acetylcholine, bradykinin, and norepinephrine that activate protein kinase C could mimic the infarct size-limiting effect of ischemic preconditioning; we have demonstrated that activation of protein kinase C through stimulation of α_1-adrenoceptors mimicked the infarct size-limiting effect even without ischemic preconditioning (Kitakaze *et al.*, 1994b). From our and other studies we can hypothesize that any intervention which activates protein kinase C could increase adenosine production via 5′-nucleotidase, which may protect the heart from the ischemic injury.

How, then, does adenosine affect the metabolic function of ischemic hearts? The physiological effects of cardiac A_1 receptors are attenuation of cAMP-dependent contraction and Ca^{2+} influx (Kirsch *et al.*, 1990; Cribier *et al.*, 1992). Both actions may attenuate energy consumption of the heart and myocardial injury due to Ca^{2+} overload. It is also reported that A_1 receptors are coupled with G_i proteins, which may modulate K_{ATP} channels in atrial muscle (Kirsch *et al.*, 1990). Coupling of A_1 receptors with K_{ATP} channels in ventricular muscle cells has not been characterized yet. Opening of K channels may be involved in ischemic preconditioning because K_{ATP} channel openers (e.g., cromakalim and nicorandil) mimic the infarct size-limiting effect (Fig. 4), and the administration of glibenclamide abolished the beneficial effect of ischemic preconditioning (Fig. 5). Our preliminary observations suggest that both cromakalim and nicorandil activate 5′-nucleotidase and increase adenosine release, and that the infarct size-limiting effect due to the K_{ATP} channel openers are lost if 5′-nucleotidase is inhibited by AMP-CP. Moreover, it is of interest that reduction of infarct size by either cromakalim or nicorandil is prevented by 8-phenyltheophylline, indicating that the opening of K_{ATP} channels per se may cause augmentation of adenosine production. Thus, the activation of adenosine receptors and the opening of K_{ATP} channels may exert a synergistic action in an autocrine manner, although precise mechanisms are to be determined.

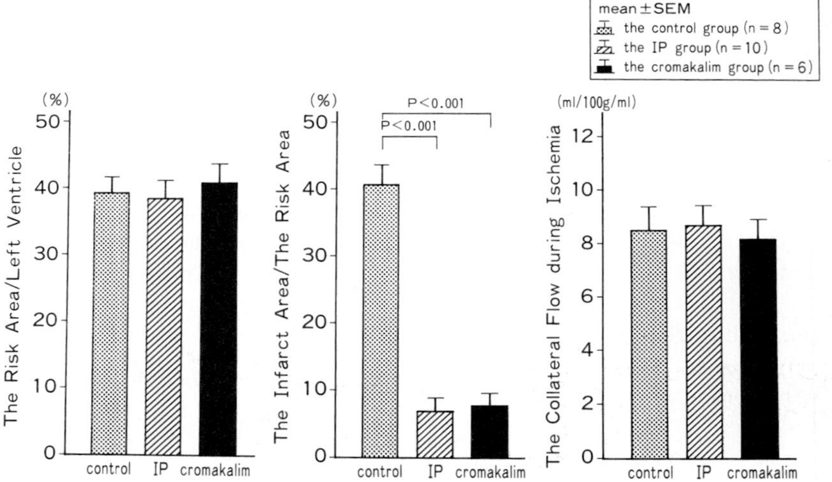

FIGURE 4 Effects of cromakalim on infarct size. Cromakalim attenuated infarct size following 90 min of ischemia and 6 hr of reperfusion to the same extent seen in ischemic preconditioning (IP group).

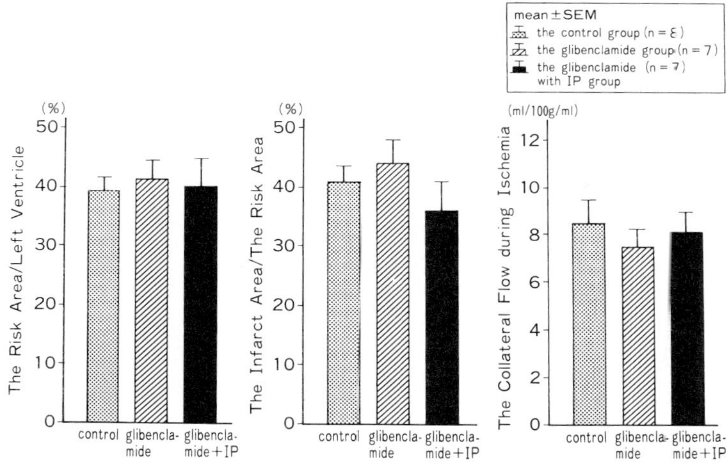

FIGURE 5 Effects of glibenclamide on the infarct size-limiting effect of ischemic preconditioning (IP). Glibenclamide attenuated the reduction of infarct size following 90 min of ischemia and 6 hr of reperfsuion.

V. CLINICAL EVIDENCE FOR K_{ATP} CHANNEL-MEDIATED CARDIOPROTECTION

Evidence of ischemic preconditioning has accumulated in clinical settings (Deutsch *et al.*, 1990; Cribier *et al.*, 1992; Sakata *et al.*, 1998) as well as in the animal experiments (Murry *et al.*, 1986). Although the important role of the interaction between activation of adenosine receptors and opening of K_{ATP} channels in ischemic preconditioning has not been fully investigated in human myocardium, recent studies suggested those factors are involved in the mechanism of ischemic adaptation in humans. Leesar *et al.* (1997) reported that the degrees of chest pain and electrocardiographic changes during repeated coronary occlusions at the time of percutaneous transluminal coronary angioplasty were significantly less in adenosine-treated patients than in control patients, suggesting that adenosine can precondition human myocardium against ischemia. Tomai *et al.* (1994) suggested the involvement of opening of K_{ATP} channels in ischemic preconditioning in clinical settings. In their study, oral administration of 10 mg glibenclamide, a selective K_{ATP} channel blocker, 60 min before coronary angioplasty entirely abolished the ischemic preconditioning effect, revealed by less anginal chest pain and electrocardiographic changes during the second balloon inflation compared with the first, seen in patients without glibenclamide treatment.

Thus, considering these lines of evidence and the fact that adenosine is closely associated with activation of K_{ATP} channels of myocardium, the linkage between adenosine and K_{ATP} channels may be a possible explanation for the mechanism of ischemic preconditioning of human myocardium.

Many physicians and clinical investigators have a great interest in the roles of K_{ATP} channels in clinical settings. Several reports have shown that chronic administration of sulfonylureas, which block K_{ATP} channels, resulted in higher mortality from cardiovascular events and the poorest outcome at the time of acute myocardial infarction in diabetic patients (University Group Diabetes Program VIII, 1982; Rytter *et al.,* 1985; Hofmann, 1993). In contrast, we have reported that nicorandil, a hybrid of a K_{ATP} channel opener and a nicotinamide nitrate, restored myocardial blood flow to the reperfused myocardium and improved functional recovery in patients with acute myocardial infarction (Sakata *et al.,* 1997a,b) (Fig. 6), suggesting the clinical implication of K_{ATP} channel openers. Saito *et al.* (1995) also suggested that nicorandil reduced myocardial ischemia during coronary angioplasty. Thus K_{ATP} channel openers may be one of the most desired drugs in the field of cardioprotection. Further investigations with larger sample sizes are required for the clinical use of K_{ATP} channel openers in the near future.

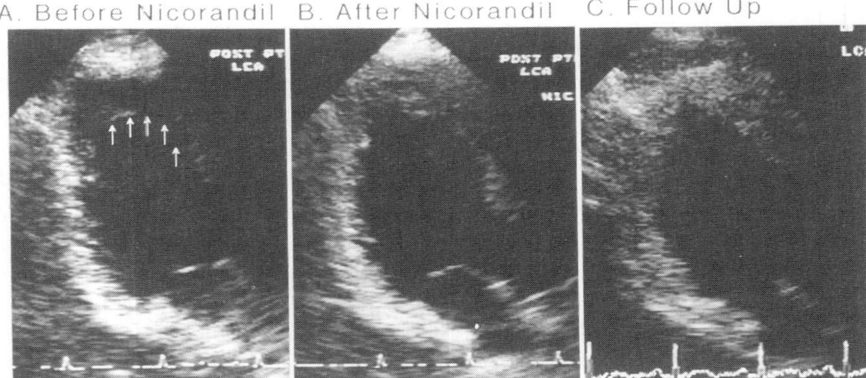

FIGURE 6 Myocardial contrast echocardiographic images (apical long-axis view) of a patient with acute myocardial infarction who underwent successful coronary recanalization for the left anterior descending coronary artery. Contrast medium was injected into the left coronary artery. Before intracoronary administration of nicorandil (A), a broad area of contrast defect, defined as the "no-reflow" area, was found in the ventricular septum and the cardiac apex (arrows). Shortly after nicorandil administration (B), contrast enhancement was observed in the area showing the no-reflow phenomenon before nicorandil administration, indicating that intracoronary nicorandil provided homogeneous myocardial perfusion. From Sakata *et al.* (1997) with permission of the Japanese Circulation Journal.

VI. SUMMARY

Adenosine is quickly released during ischemia and exerts potent coronary vasodilation to maintain coronary blood flow through A_2 receptors. During ischemia, K_{ATP} channels are also opened, contributing to coronary vasodilation. We have showed that adenosine-induced coronary vasodilation is partly mediated by K_{ATP} channels. Since K_{ATP} channel openers augment the adenosine-induced coronary vasodilation, adenosine and K_{ATP} channels may be coupled. Reperfusion injury (e.g., stunning and myocardial necrosis due to sustained ischemia and reperfusion) is significantly attenuated by adenosine and K_{ATP} channels. Thus, the activation of adenosine receptors and the opening of K_{ATP} channels may have a beneficial role in minimizing ischemia and reperfusion injury.

Acknowledgment

This work was supported by a Grant-in-Aid for Scientific Research (No. 05454272 and 05670617) from the Ministry of Education, Science, and Culture, Japan.

References

Agarwal, K. C. (1987). Adenosine and platelet function. *In* "Role of Adenosine in Cerebral Metabolism and Blood Flow" (V. Stefanovich, Okayuz, and I. Baklouti, eds.), pp. 107–124. VNU Science Press, Utrecht, The Netherlands.

Ashford, M. L. J., Sturgess, N. C., and Trout, N. J. (1988). Adenosine-5′-triphosphate-sensitive ion channels in neonatal rat cultures central neurons. *Pfluegers. Arch.* **412,** 297–304.

Aversano, T., Ouyang, P., and Silverman, H. (1991). Blockade of the ATP-sensitive potassium channel modulate reactive hyperemia in the canine coronary circulation. *Circ. Res.* **69,** 618–622.

Berne, R. M. (1980). The role of adenosine in the regulation of coronary blood flow. *Circ. Res.* **47,** 807–813.

Braunwald, E., and Kloner, R. A. (1982). The stunned myocardium prolonged, postischemic ventricular dysfunction. *Circulation* **66,** 1146–1148.

Brechler, V., Pavoine, C., Lotersztajn, S., Garbarz, E., and Pecker, F. (1990). Activation of Na^+/Ca^{2+} exchange by adenosine in ewe heart sarcolemma is mediated by a pertussis toxin-sensitive G protein. *J. Biol. Chem.,* **265,** 16851–16855.

Cook, D. L., and Hales, C. N. (1984). Intracellular ATP directly blocks K^+ channels in pancreatic β cells, *Nature* (*London*) **311,** 271–273.

Cribier, A., Korsatz, L., Koning, R., Rath, P., Gamra, H., Stix, G., Merchant, S., Chan, C., and Letac, B. (1992). Improved myocardial ischemic response and enhanced collateral circulation with long repetitive coronary occlusion during angioplasty: A prospective study. *J. Am. Coll. Cardiol.* **20,** 578–586.

Cshrader, J., Schutz W., and Bardenheuer, H. (1981). Role of *S*-adenosylhomocysteine hydrolase in adenosine metabolism in mammalian heart. *Biochem. J.* **196(1),** 65–70.

Deutsch, E., Berger, M., Kussmaul, W. G., Hirshfeld, J. W., Herrmann, H. C., and Laskey, W. K. (1990). Adaptation to ischemia during percutaneous transluminal coronary angioplasty: Clinical, hemodynamic, and metabolic features. *Circulation* **82,** 2044–2251.

Engler, R., and Covell, J. W. (1987). Consequences of activation and adenosine-mediated inhibition of granulocytes during myocardial ischemia. *Fed. Proc.* **46,** 2407–2412.

Forrester, T., and Williams, C. A. (1977). Release of adenosine triphosphate from isolated adult heart cells in response to hypoxia. *J. Physiol. (London)* **268,** 371–390.

Grover, G. J., Sleph, P. G., and Dzwonczyk, S. (1992). Role of myocardial ATP-sensitive potassium channels in mediating preconditioning in the dog heart and their possible interaction with adenosine A1-receptors. *Circulation* **86,** 1310–1316.

Hofmann, O. L. H. (1993). Potassium channel blockade and acute myocardial infarction: Implications for management of non-insulin requiring diabetic patient. *Eur. Heart J.* **14,** 1585–1589.

Hori, M., and Kitakaze, M. (1991). Adenosine, the heart, and coronary circulation. *Hypertension (Dallas)* **18,** 565–574.

Hori, M., Kitakaze, M., Tamai, J., Koretsune, Y., Iwakura, K., Kagiya, T., Kitabatake, A., Inoue, M., and Kamada, T. (1988). α_2-Adrenoceptor activity exerts dual coronary blood flow in canine artery. *Am. J. Physiol.* **255,** H250–H260.

Hori, M., Kitakaze, M., Tamai, J., Iwakura, K., Kitabatake, A., Inoue, M., and Kamada, T. (1989). α_2-Adrenoceptor stimulation can augment coronary vasodilation maximally induced by adenosine in dogs. *Am. J. Physiol.* **257,** H132–H140.

Kirsch, G. E., Codina, J., Birnbaumer, L., and Brown, A. M. (1990). Coupling of ATP-sensitive K^+ channels to A_1 receptors by G proteins in rat ventricular myocytes. *Am. J. Physiol.* **259,** H820–H826.

Kitakaze, M., Hori, M., and Tamai, J. (1987). Alpha$_1$-adrenoceptor activity regulates release of adenosine from the ischemic myocardium in dogs. *Circ. Res.* **60,** 631–639.

Kitakaze, M., Hori, M., Sato, H., Takashima, S., Inoue, M., Kitabatake, A., and Kamada, T. (1991a). Endogenous adenosine inhibits platelet aggregation during myocardial ischemia in dogs. *Circ. Res.* **69,** 1402–1408.

Kitakaze, M., Hori, M., and Sato, H. (1991b). Beneficial effects of alpha$_1$-adrenoceptor activity on myocardial stunning in dogs. *Circ. Res.* **68,** 1322–1339.

Kitakaze, M., Hori, M., Takashima, S., Sato, H., Inoue, M., and Kamada, T. (1993). Ischemic preconditioning increases adenosine release and 5′-nucleotidase activity during myocardial ischemia and reperfusion in dogs. Implications for myocardial salvage. *Circulation* **87,** 208–215.

Kitakaze, M., Hori, M., Morioka, T., Minamino, T., Inoue, M., and Kamada, T. (1994a). The infarct size limiting effect of ischemic preconditioning is blunted by inhibition of 5′-nucleotidase activity and attenuation of adenosine release. *Circulation* **89,** 1237–1246.

Kitakaze, M., Hori, M., Morioka, T., Minamino, T., Takashima, S., Sato, H., Shinozaki, Y., Chujo, M., Mori, H., Inoue, M., and Kamada, T. (1994). Alpha$_1$-adrenoceptor activation mediates the infarct size-limiting effect of ischemic preconditioning through augmentation of 5′-nucleotidase activity, *J. Clin. Invest.* **93,** 2197–2205.

Kitakaze, M., Hori, M, Morioka, T., Minamino, T., Takashima, S., Okazaki, Y., Node, K., Komamura, K., Iwakura, K., Inoue, M., and Takenobu K. (1995). α_1-Adrenoceptor activation increases ectosolic 5′-nucleotidase activity and adenosine release in rat cardiomyocytes by activating protein kinase C. *Circulation* **91,** 2226–2234.

Komaru, T., Lamping, K. G., and Eastham, C. L. (1991). Role of ATP-sensitive potassium channels in coronary microvascular autoregulatory responses. *Circ. Res.* **69,** 1146–1151.

Kusuoka, H., Porterfield, J. K., Weisman, H. F., Weisfeldt, M. L., and Marban, E. (1987). Pathophysiology and pathogenesis of stunned myocardium. Depressed Ca^{2+} activation of contraction as a consequence of reperfusion-induced cellular calcium overload in ferret hearts. *J. Clin. Invest.* **79,** 950–961.

Leesar, M. A, Stoddard, M., Ahmed, M., Broadbent, J., Bolli, R. (1997). Preconditioning of human myocardium with adenosine during coronary angioplasty. *Circulation* **95,** 2500–2507.

Lederer, W. J., Nichols, C. G. (1989). Nucleotide modulation of the activity of the rat heart ATP-sensitive K^+ channels in isolated membrane patches. *J. Physiol.* (*London*) **419,** 193–211.

Liu, G. S., Thornton, J., Van Winkle, D. M., Stanley, A. W. H., Olsson, R. A., and Downey, J.M. (1991). Protection against infarction afforded by preconditioning is mediated by A1-adenosine receptors in rabbit heart. *Circulation* **84,** 350–356.

Merrill, G. F., Haddy, F. J., and Dabney, J. M. (1978). Adenosine, theophylline and perfusate pH in the isolated perfused guinea pig heart. *Circ. Res.* **42,** 225–229.

Mitani, A., Kinoshita, K., Fukamachi, K., Sakamoto, M., Kurisu, K., Tsuruhara, Y., Fukumura, F., Nakashima, A., and Tokunaga, K. (1991). Effects of glibenclamide and nicorandil on perfused rat heart. *Am. J. Physiol.* **261,** H1864–H1871.

Morioka, T., Kitakaze, M., Minamino, T., Takashima, T., Node, K., Sato, H., Shinozaki Y., Chujo, M., Mori, H., Inoue, M., Hori, M., and Kamada, T. (1995). Evidence for impaired coronary autoregulation during reperfusion following brief period of ischemia in dogs. *Am. J. Physiol.* **269,** H1237–H1245.

Murry, C. E., Jennings, R. B., and Reimer, K. A. (1986). Preconditioning with ischemia: A delay of lethal cell injury in ischemic myocardium. *Circulation* **74,** 1124–1136.

Newby, A. C., Worku, Y., Meghji, P. (1987). *In* "Topics and Perspectives in Adenosine Research. Critical Evaluation of the Role of Ecto- and Cytosolic 5′-Nucleotidase in Adenosine Formation" (E. Gerlach, and B. F. Becker, eds.), pp. 155–168. Springer-Verlag, Berlin.

Newman, W. H., Becker, B. F., Heier, M., Nees, S., and Gerlach, E. (1988). Endothelium-mediated coronary activation: Studies in isolated guinea pig heart. *Pfluegers Arch.* **413,** 1–7.

Nichols, C. G., and Lederer, W. J. (1991). Adenosine triphosphate-sensitive potassium channels in cardiovascular system. *Am. J. Physiol.* **261,** H11675–H1686.

Pike, M., Kitakaze, M., and Marban, E. (1990). ^{23}Na NMR measurements of intracellular sodium in intact perfused ferret heart during ischemia and reperfusion. *Am. J. Physiol.* **259,** H1767–H1773.

Rovetto, M. J. (1985). Myocardial nucleotidase transport. *Annu. Rev. Physiol.* **47,** 605–616.

Rytter, L., Troelsen, S., and Beck-Nielsen, H. (1985). Prevalence and mortality of acute myocardial infarction in patients with diabetes. *Diabetes Care* **8,** 230–234.

Saito, S., Mizumura, T., Takayama, T., Honye, J., Fukui, T., Kamata, T., Moriuchi, M., Hibiya, K., Tamura, Y., Ozawa, Y., Kanmatsuse, K., Ozawa, K., Ishihata, F., Nakakimura, H., Sakai, K. (1995).Antiischemic effects of nicorandil during coronary angioplasty in humans. *Cardiovasc Drugs Ther.* **9**(Suppl 2), 257–263.

Sakata, Y., Kodama, K., Ishikura, F., Komamura, K., Hasegawa, S., Sakata, Y., and Hirayama, A. (1997a). Disappearance of the "no-reflow" phenomenon after adjunctive intracoronary administration of nicorandil in a patient with acute myocardial infarction. *Jpn. Circ. J.* **61,** 455–458.

Sakata, Y., Kodama, K., Komamura, K., Lim, Y. J., Ishikura, F., Hirayama, A., Kitakaze, M., Masuyama, T., and Hori, M. (1997b). Salutary effect of adjunctive intracoronary nicorandil administration on restoration of myocardial blood flow and functional improvement in patients with acute myocardial infarction. *Am. Heart J.* **133,** 616–621.

Sakata, Y., Kodama, K., Kitakaze, M., Masuyama, T., Hirayama, A., Lim, Y. J., Ishikura, F., Sakai, A., Adachi, T., and Hori, M. (1997). Different mechanisms of ischemic adaptation to repeated coronary occlusion in patients with and without recruitable collaterals. *J. Am. Coll. Cardiol.* **30,** 1679–1686.

Schutz, W., Zimpfer, M., and Raberger, G. (1977). Effect of aminophylline on coronary reactive hyperemia following brief and long occlusion periods. *Cardiovasc. Res.* **11,** 507–511.

Sparks, H. V., and Bardenheuer, H. (1986). Regulation of adenosine formation in the heart. *Circ. Res.* **58,** 193–201.

Takeo, S., Tanonaka, K., Miyake, K., and Imago, M. (1988). Adenine nucleotide metabolites are beneficial for recovery of cardiac contractile force after hypoxia. *J. Mol. Cell. Cardiol.* **20,** 187–199.

Tomai, F., Crea, F., Gaspardone, A., Versaci, F., De Paulis, R., Penta de Peppo, A., Chiariello, L., and Gioffre, P. A. (1994). Ischemic preconditioning during coronary angioplasty is prevented by glibenclamide, a selective ATP-sensitive K^+ channel blocker. *Circulation* **90,** 700–705.

University Group Diabetes Program, VIII. (1982). Evaluation of insulin therapy: Final report. *Diabetes* **31,** 1–81.

Yao, Z., and Gross, G. J. (1993). Glibenclamide antagonizes adenosine A1 receptor-mediated cardioprotection in stunned canine myocardium. *Circulation* **88,** 235–244.

CHAPTER 24

Subcellular Heterogeneity of Energy Metabolism and K_{ATP} Current Oscillation in Cardiac Myocytes

Brian O'Rourke, Dmitry N. Romashko, and Eduardo Marbán
Section of Molecular and Cellular Cardiology, Department of Medicine, The Johns Hopkins University, Baltimore, Maryland 21205

I. INTRODUCTION

Modern molecular techniques have given biologists the opportunity to deconstruct even the most complicated systems of the cell into component proteins that can be studied with unprecedented biophysical detail. Complete understanding of the pieces of a puzzle, however, does not guarantee that the complete picture can be reconstructed. Thus, it is vitally important to study physiological systems in intact cells. It is also becoming increasingly clear that examination of the steady-state behavior of the cell is not enough; one must take into consideration how the system changes with time and how

1063-5823/99 $30.00

these changes are distributed throughout the system. A notable example is inositol trisphosphate-mediated Ca^{2+} signalling in nonexcitable cells, where frequency encoding rather than amplitude modulation is the dominant mechanism for regulating cellular activity (Rooney *et al.* 1989; Thomas *et al.* 1996; woods *et al.* 1986). Other important examples of periodic phenomena include spontaneous sarcoplasmic reticular Ca^{2+} oscillations (Kort *et al.*, 1985; Wier *et al.*, 1983), pacemaker activity in cardiac cells and neurons, and the bursting pattern of insulin secretion in pancreatic β cells (Corkey *et al.*, 1988; Henquin, 1990). In the latter, oscillations in energy metabolism play a prominent role the response (Nilsson *et al.*, 1996; Pralong *et al.*, 1990). In all of these cases, the spatial distribution of the oscillatory signal is important.

A well-studied biochemical system that has spurred theoretical consideration of periodic phenomena in biological systems is the oscillation of energy metabolism in yeast (Duysens and Amesz, 1957; Hommes, 1964) or in cell-free cytoplasmic extracts (Chance *et al.*, 1964). This pioneering work demonstrated that the complex metabolic control inherent in a biochemical pathway like glycolysis, while enabling rapid responsiveness and fine-tuning of the system to environmental factors, also introduces instability which may lead to periodic fluctuations in the flow of metabolites (Frenkel, 1968a). Remarkably, such oscillations can be initiated by changes in the substrate supply to the pathway that are not far from the normal operating range of the system (Boiteux *et al.*, 1975; Hess and Boiteux, 1968).

We have previously found that oscillations in energy metabolism also occur in cardiac myocytes deprived of exogenous substrates (O'Rourke *et al.*, 1994). The oscillations could be detected by following changes in the fluorescence of cellular NADH. Periods of suppressed energy metabolism were indicated by oxidation of NADH and were closely associated with activation of ATP-sensitive K^+ currents (K_{ATP} currents) in the cardiac sarcolemma. A dependence on the external glucose concentration suggested that glycolytic oscillations, known to occur in cardiac cytoplasmic extracts (Frenkel, 1968b), could be involved in the phenomenon. In addition to the oscillations in K_{ATP} current and excitation–contraction coupling, it was clear that mitochondrial function also oscillated under these conditions, since the NADH signal arises primarily from the mitochondrial matrix (Eng *et al.*, 1989). We have explored the link between mitochondrial function and sarcolemmal K_{ATP} channel activation using mitochondrial flavoproteins as indicators of the mitochondrial redox state (Romashko *et al.*, 1998). Confocal fluorescence imaging has revealed a surprising degree of subcellular spatial and temporal heterogeneity of mitochondrial redox and membrane potential during oscillation. The results have provided fundamental insights into the organization and synchronization of the mitochondrial network in the intact cardiac cell and have stimulated exploration into how metabolic signals are transmitted in the myocardium.

II. METHODS

A. Experimental Conditions and Confocal Imaging

All experiments were carried out at room temperature on adult guinea pig ventricular myocytes prepared by enzymatic dispersion. Cells were typically stored in a cardioplegic high potassium solution containing (in m*M*) 120 potassium glutamate, 25 KCl, 1 $MgCl_2$, 10 HEPES, 0.1 EGTA, and 10 glucose, or, in control experiments, a modified Tyrode's solution containing (in m*M*) 140 NaCl, 5 KCl, 1 $MgCl_2$, 10 HEPES, 10 glucose, and 0.2 $CaCl_2$. Experimental recordings began following a short equilibration period (5–10 min) in glucose-free Tyrode's solution (with 2 m*M* $CaCl_2$) in an open superfusion chamber. Fluorescence images were collected on a Nikon Diaphot-300 inverted microscope with a PCM-2000 confocal attachment with Argon or He–Ne laser excitation. Images were analyzed off-line on a Silicon Graphics O2 workstation (Silicon Graphics, Mountain View, CA) using the Tcl language interface to the SGI ImageVision library or on a computer with the shareware program ImageTool (University of Texas at San Antonio Health Sciences Center).

B. Indices of Mitochondrial Metabolism

1. NADH

Although electrophysiological techniques are available to measure changes in sarcolemmal ion channel activity and numerous fluorescent indicators have been developed to detect changes in intracellular ion concentrations (e.g., Ca^{2+}, H^+), fewer indicators are available for examining the metabolic state of isolated single cells. Fortunately, the cardiac myocyte has provided us with built-in fluorescent indicators of redox potential, namely, NADH and the flavin-containing proteins of the mitochondria. NADH is the major component of autofluorescence when cells are excited with ultraviolet light of ~360 nm wavelength (Chance *et al.*, 1965), and its emission peak is the range of 450–500 nm (Chance and Williams, 1955). Oscillations induced by substrate withdrawal are characterized by a close correlation between K_{ATP} current transients and decreases of the NADH fluorescence indicative of net oxidation of the cellular NAD^+/NADH pool

(O'Rourke *et al.,* 1994) (Fig. 1). Two pieces of evidence indicate that the measured redox response obtained from whole-cell fluorescence recordings originates from the mitochondrial space. First, it is known that a relatively small fraction of the total NADH fluorescence is from the cytoplasmic compartment (Eng *et al.,* 1989). Second, the observed phase relation between the NADH redox state and cytoplasmic ATP/ADP was the opposite of that predicted for a simple glycolytic oscillator. In cell-free extracts, the NADH signal reflects the activity of glyceraldehyde 3-phosphate dehydrogenase (GAPDH) in the glycolytic pathway, which produces NADH in the substrate level phosphorylation of glyceraldehyde 3-phosphate. With no mitochondria present, production of ADP resulting from an increase in phosphofructokinase activity would stimulate NADH production by GAPDH; thus, an increase in ADP would correlate with net reduction of cytoplasmic NAD^+/NADH (Frenkel, 1968c). During metabolic oscillation in the intact cell, however, the phase of decreased energy availability (indicated by K_{ATP} channel activation) corresponded to NAD^+/NADH oxidation. This phase relation could be obtained if ADP generated in the cyto-

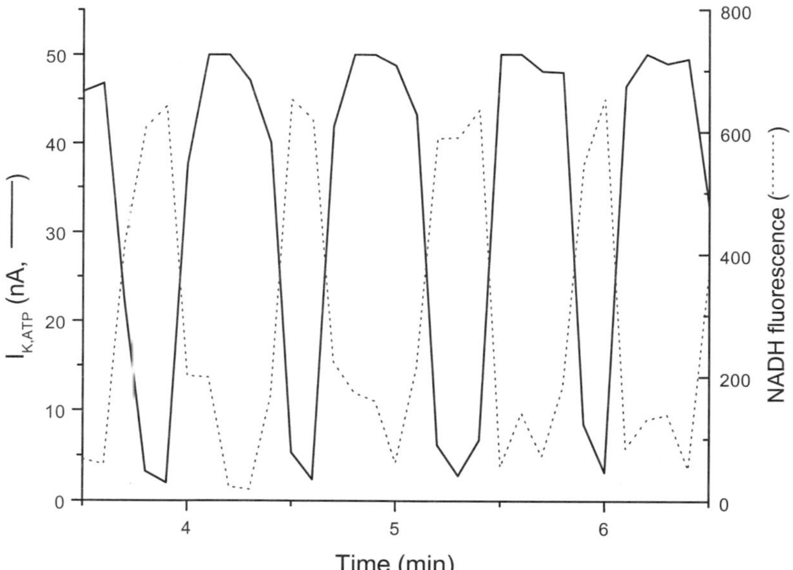

FIGURE 1 Phase relation between whole-cell K_{ATP} current oscillations (solid line) and NADH fluorescence (dotted line) in substrate-deprived guinea pig cardiomyocytes. The phase of K_{ATP} channel opening corresponded with transient oxidation of cellular NADH, indicated by transient drops in autofluorescence (excitation wavelength 365 nm; emission wavelength 450 nm).

plasm stimulated mitochondrial respiration or if mitochondrial energy production transiently dropped to cause the decrease in cytoplasmic ATP/ADP ratio. Therefore, we sought a direct confirmation that the mitochondrial redox potential was oscillating.

2. Flavoproteins

We utilized another component of cellular autofluorescence to index mitochondrial matrix redox potential directly. At a longer excitation wavelength ~480 nm, the autofluorescence emission of the myocyte at ~520 nm originates from flavin-bound enzymes in the mitochondrial matrix (Chance *et al.,* 1967). These include pyruvate dehydrogenase (PDH), α-ketoglutarate dehydrogenase (αKDA), succinate dehydrogenase (SDH), and the electron transfer flavoprotein of the β-oxidation pathway (Kunz, 1986; Kunz and Kunz, 1985). The multienzyme complexes PDH and αKDH, which employ covalently bound flavin adenine dinucleotide (FAD) in the dihydrolipoyl dehydrogenase component of the reaction, contribute the most to the fluorescence signal at these wavelengths. Since the reaction is linked to the oxidation state of matrix NADH, their contribution to the total fluorescence can be determined by maximally oxidizing the mitochondrial NAD^+/NADH pool in the absence and presence of rotenone, which selectively inhibits NADH oxidation by the electron transport chain. This is illustrated in Fig. 2, which shows that the oxidizing effect of the mitochondrial uncoupler 2,4-dinitrophenol (DNP) on the flavoprotein fluorescence (oxidation of the flavoproteins increases the fluorescence) is nearly eliminated by treatment with rotenone. Approximately 80% of the fluorescence was attributable to NADH-linked flavoproteins.

In addition to the precise localization of the fluorescence signal to mitochondrial enzymes, the use of flavoprotein fluorescence afforded the opportunity to use confocal fluorescence imaging with standard fluorescein filter sets to examine the spatial distribution of mitochondrial redox potential in the isolated cardiomyocytes. Figure 3A shows flavoprotein fluorescence images of a guinea pig cardiomyocyte in (a) control solution, (b) cyanide-containing solution (to maximally reduce the mitochondria), and (c) DNP (to maximally oxidize the mitochondria). The tight packing of mitochondria along the myofibrils and absence of mitochondria at the sarcomeric Z-lines is evident in the fully oxidized state (Fig. 3Ac). Figure 3B shows the mean flavoprotein fluorescence over the course of the experiment and indicates the time frame over which the images in Fig. 3A were averaged.

3. Tetramethylrhodamine Methyl Ester

A third technique for determining the metabolic state of the cell employs tetramethylrhodamine ethyl ester (TMRE). TMRE is a positively charged

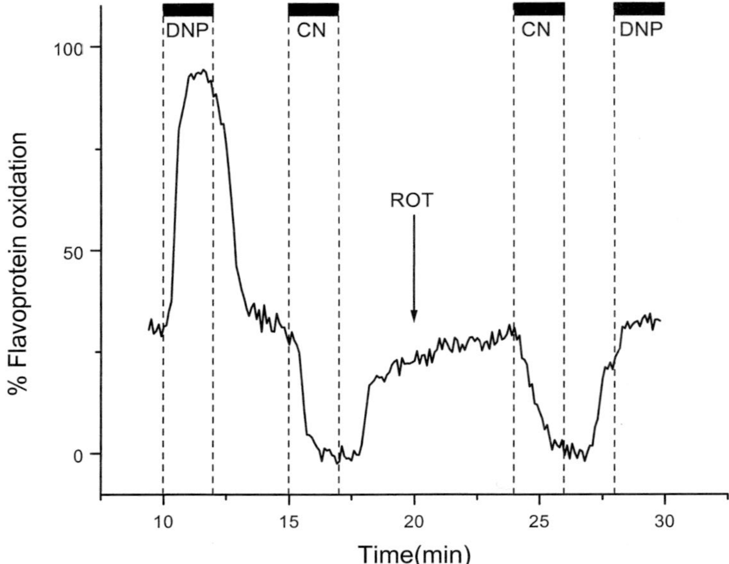

FIGURE 2 Sensitivity of mitochondrial flavoprotein redox potential to retenone. Flavoproteins were fully oxidized by 2,4-dinitrophenol (DNP; 200 μM) and fully reduced by cyanide (CN; 4 mM) prior to treatment with the NADH dehydrogenase inhibitor rotenone (ROT). After rotenone treatment, the oxidizing effect of DNP was largely eliminated.

lipophilic compound whose distribution within cellular compartments is determined by the membrane potential; thus, it readily accumulates in the mitochondrial matrix (Ehrenberg *et al.*, 1988; Loew *et al.*, 1993). Depolarization of the mitochondrial membrane potential results in redistribution of the TMRE fluorescence from the mitochondrial matrix to the cytoplasm. In our experiments, 100 nM TMRE was added to the external solution and allowed to equilibrate for 20 min before images of TMRE fluorescence (543 nm excitation; 605 ± 16 nm emission) were collected.

FIGURE 3 Confocal fluorescence imaging of mitochondrial flavoprotein redox state. (A) Fluorescence images of an isolated guinea pig ventricular myocyte in (a) control medium (modified Tyrode's containing 10 mM glucose), (b) 4 mM cyanide, and (c) 200 μM 2,4-dinitrophenol. Pixel brightness increases with flavoprotein oxidation. (B) Time course of mean cellular autofluorescence change for the experiment depicted in (A), where a, b, and c denote the time frames used to generate the average images in (A). Fluorescence was excited with the 488-nm line of an argon laser and emission was collected at 520 nm. Adapted from Romashko *et al.* (1998).

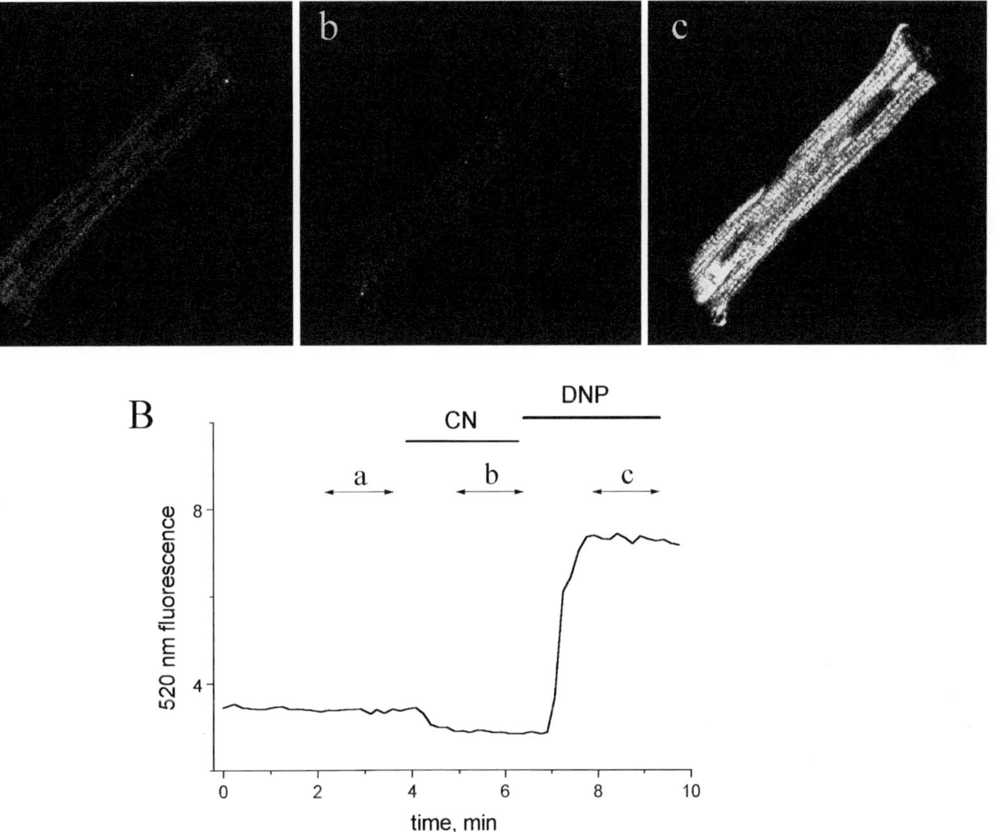
A
a
b
c
B
CN
DNP
a
b
c
520 nm fluorescence
8
4
0
2
4
6
8
10
time, min

III. RESULTS

A. *Complex Mitochondrial Redox Oscillations*

As suspected from the studies employing NADH as the metabolic index, periodic oxidation of the mitochondrial flavoproteins correlated with the activation of sarcolemmal K_{ATP} currents in cardiomyocytes studied under substrate-free conditions. This confirmed that oscillations in mitochondrial redox potential were an integral part of the oscillatory phenomenon. Often, the pattern of oscillation of the whole-cell signal could be quite complex (Fig. 4), suggesting either that there was a single oscillator capable of switching between different activity states or, more likely, that multiple oscillatory foci were present within a single myocyte. We were thus motivated to study the spatial patterns of redox change within isolated cells.

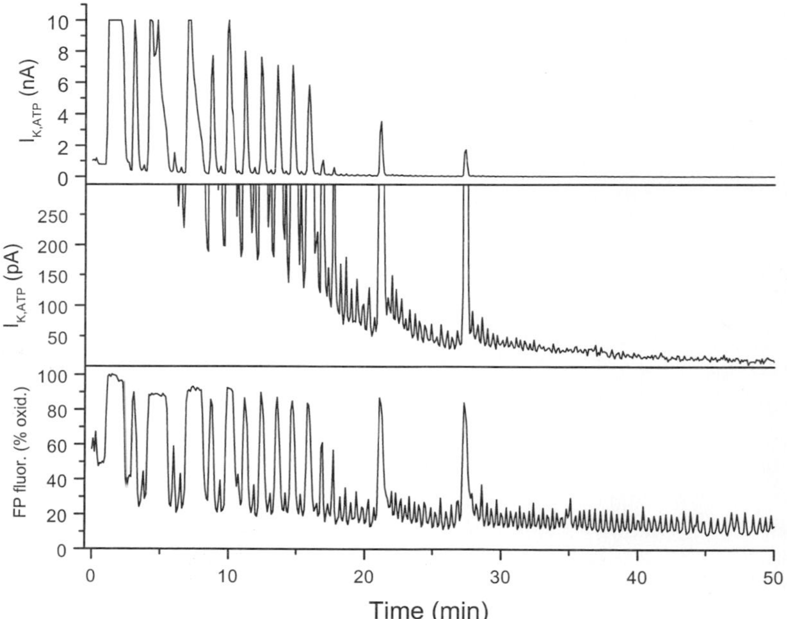

FIGURE 4 Complex oscillations of whole-cell fluorescence and K_{ATP} current in a single cardiomyocyte. As with the NADH measurements shown in Fig. 1, K_{ATP} current activation corresponded with the oxidation phase of mitochondrial flavoproteins. The top graph shows the very large K_{ATP} current peaks activated by metabolic oscillation, whereas the middle graph is scaled up to illustrate the small oscillations in membrane current corresponding to the small redox transients seen in the flavoprotein redox plot at bottom.

B. Cellular and Subcellular Heterogeneity

Confocal fluorescence imaging of mitochondrial flavoproteins revealed that (1) redox transients can occur independently in myocytes that are physically separated from each other within a microscopic field, (2) redox transients can be localized to small regions (mitochondrial clusters) within cells while the remainder of the cell stays reduced, and (3) mitochondrial redox waves can propagate within cardiac myocytes (Figs. 5–7). These results suggest that each cardiomyocyte has an intrinsic capability for metabolic oscillation that may be triggered under metabolic stress. Furthermore, the trigger does not involve an external factor (other than substrate withdrawal) or messenger, since the responses of different cells were not entrained. The most important finding was that the metabolic state of an individual mitochondrion or a cluster of mitochondria could differ markedly from its neighbors within a single cardiomyocyte. This is exemplified by the higher magnification image in Fig. 6A, showing redox transitions occurring in three different regions of a myocyte. Highly localized flavoprotein oxidations occur in two of the three regions (see time course plots in Fig. 6B), but the region between the clusters remains reduced throughout the experiment. Metabolic foci such as these could explain the very small oscillations in whole-cell fluorescence associated with small oscillations in K_{ATP} current shown in Fig. 4.

C. Intracellular and Intercellular Propagated Waves

Propagation of an intracellular redox wave would indicate that a signal generated by a cluster of mitochondria undergoing a redox transient can spread in a regenerative manner to other mitochondria in the cell. Such propagation is evident in the image triptych of Fig. 7. The rate of propagation of this wave was approximately 10 μm/sec, a value identical to the estimated rate of diffusion of a molecule such as ADP in the cytoplasm (10 μm/sec). This finding raises questions about how metabolic signals are transmitted within cells. Is a diffusible messenger involved? Is the coordinated response of a cluster of mitochondria due to interconnections between their matrices? Regarding the question of diffusible messengers, in some experiments we observed that a redox wave could propagate not only within a cell, but also between cells connected by gap junctions. Figure 8 shows an example of this behavior. Over the time course of this experiment, we observed both localized redox oscillations and propagated waves that crossed the intercellular junction in either direction. Since gap junctions behave somewhat like molecular sieves with a pore size of ~1000 MW

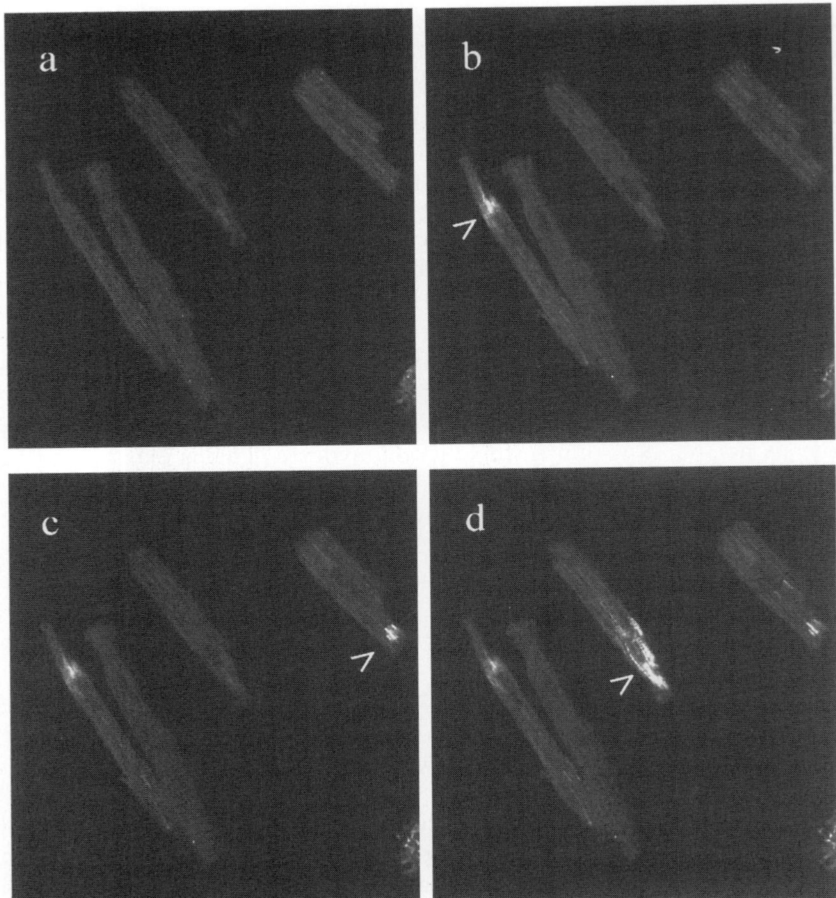

FIGURE 5 Heterogeneity of mitochondrial redox transitions in guinea pig cardiomyocytes. In myocytes bathed in substrate-free modified Tyrode's solution, redox transitions in different cells in the microscopic field occurred asynchronously and involved either small or larger regions of the cell (and sometimes the entire cell; not shown). Flavoprotein fluorescence was imaged as in Fig. 3. Arrows indicate the appearance of newly oxidized regions of the cells. The oxidations in b, c, and d occurred 4.4, 7.4, and 12.3 min, respectively, from the start of the image acquisition series.

(Veenstra, 1996), the results support the idea that a small cytoplasmic messenger could be responsible for the propagation of redox waves. Although this messenger has not been identified, the size and functional effects of ADP would fit the bill, especially since we have previously shown that the photolytic release of ADP can initiate metabolic oscillations in quiescent myocytes (O'Rourke *et al.*, 1994).

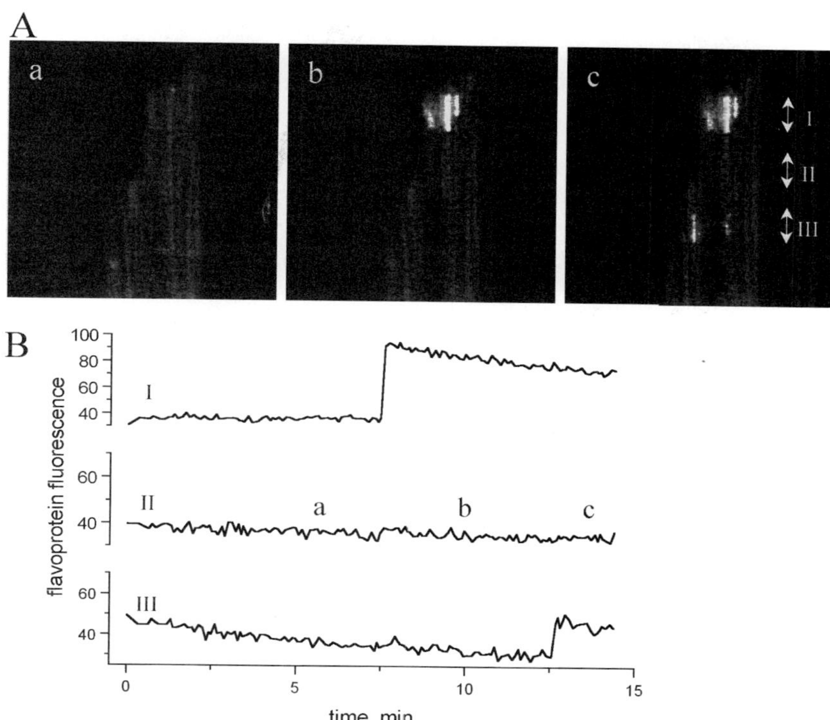

FIGURE 6 Redox transitions in mitochondrial clusters. (A) Mitochondrial flavoprotein oxidations were often localized to small clusters or chains of mitochondria spanning several sarcomeres, as seen in b and c. The region between clusters (region II) remained reduced throughout the experiment. (B) Time course of redox change in regions I, II, and III during the experiment from which the images in (A) were taken (at time points a, b, and c). Adapted from Romashko *et al.* (1998).

D. Clusters of Mitochondria

Although redox wave propagation suggested that mitochondrial responses could be triggered by diffusible messengers, the organization of subcellular oscillatory foci into mitochondrial clusters which respond as a unit provides evidence for intrinsic communication between mitochondria. Using TMRE, we tested whether the spatial patterns of mitochondrial oxidation were associated with the localized collapse of mitochondrial membrane potential. Myocytes that were predominantly reduced displayed intense TMRE loading of the mitochondrial matrix space (Fig. 9a) compatible with the known tight packing of mitochondria along myofibrils. The TMRE

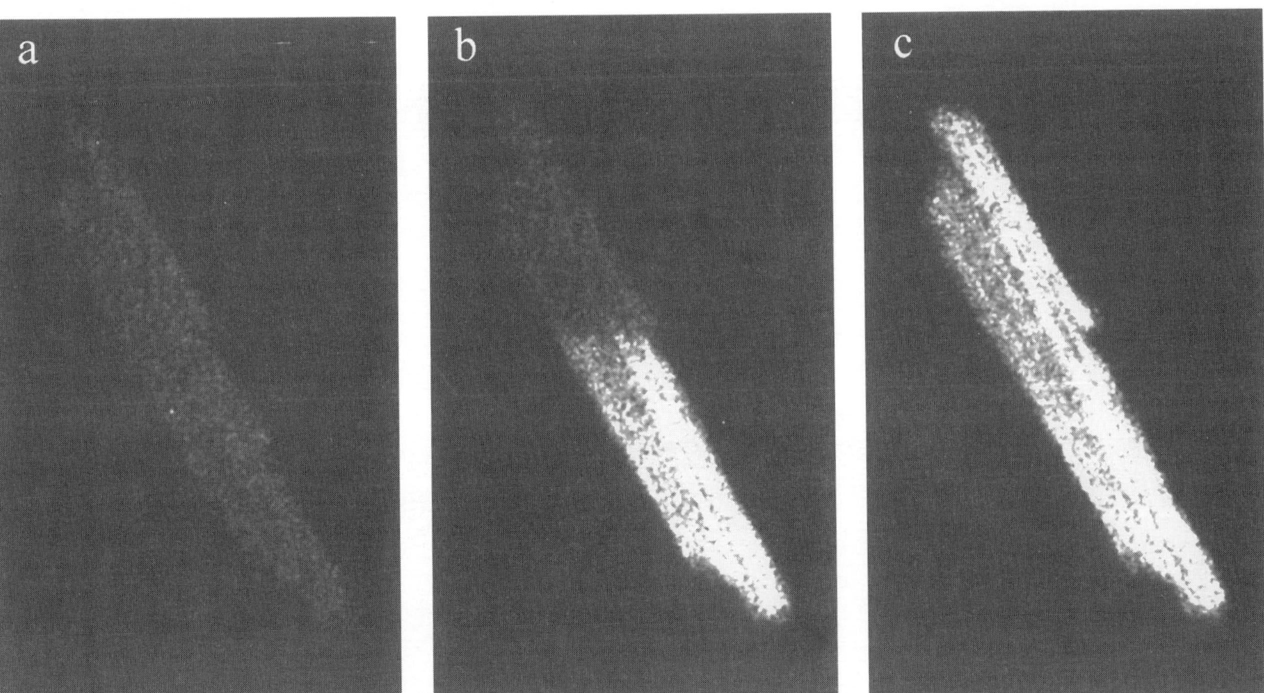

FIGURE 7 Propagating redox wave in a single cardiomyocyte. Images in a, b., and c were taken 6 sec apart and show a mitochondrial redox wave propagating at a rate of ~10 μm/sec.

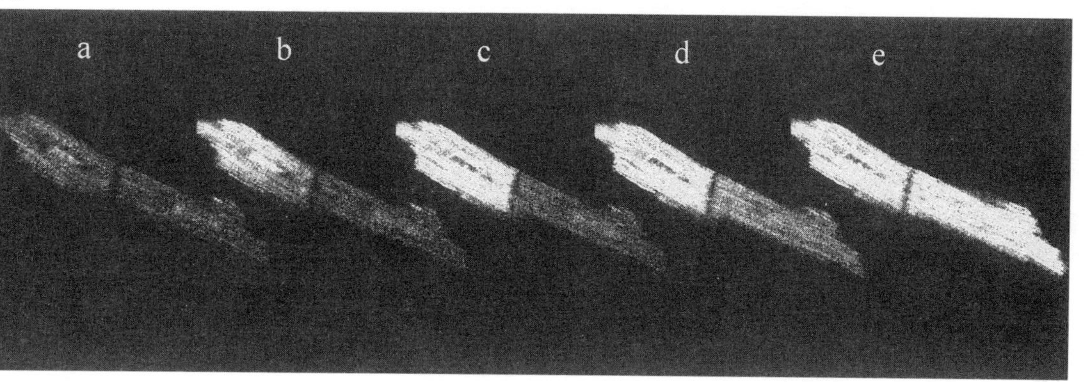

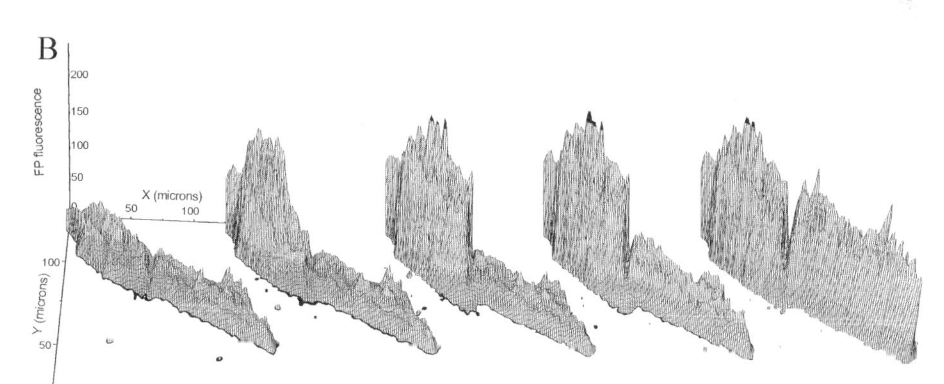

FIGURE 8 Redox wave propagation between two myocytes connected by an intercalated disc. (A) An intracellular redox wave initiated in the top left-hand cell propagated to the cell–cell junction (evident in image c) and then crossed into the second cell (d and e). (B) Fluorescence intensity maps from the images shown in (A). Adapted from Romashko *et al.* (1998).

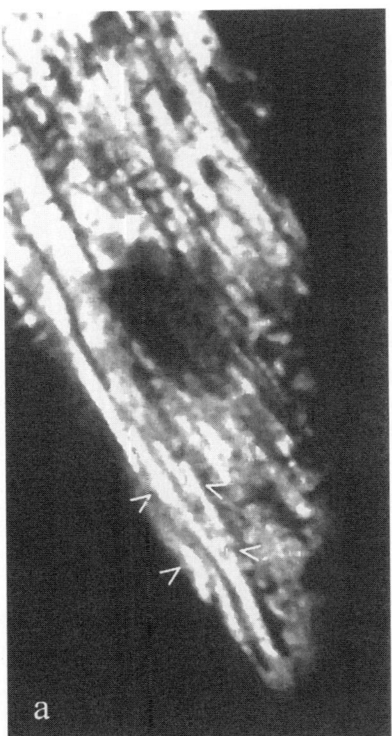

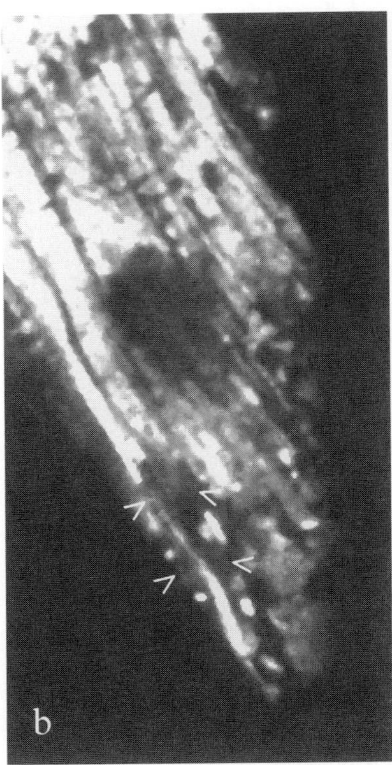

FIGURE 9 Depolarization of membrane potential in a mitochondrial cluster. In cardiomyocytes loaded with the mitochondrial membrane potential probe tetramethylrhodamine ethyl ester (TMRE), polarized mitochondria were readily visualized as bright fluorescent spots or chains (fluorescence excitation wavelength 547 nm; emission >605 nm). Transient oxidation of the mitochondrial matrix (indexed by flavoprotein fluorescence, not shown in these images) was associated with the collapse of mitochondrial membrane potential in mitochondrial clusters, as is evident when images a and b are compared. Arrowheads indicate the sudden loss of TMRE fluorescence in a mitochondrial chain. Note that nearby mitochondria remained polarized during this transition.

distribution also correlated well with the flavoprotein fluorescence distribution, as expected for signals arising from the matrix space. Most striking was the observation that mitochondrial flavoprotein redox transients were accompanied by transients in the TMRE fluorescence signal (Fig. 9b), confirming that clusters of mitochondria were becoming depolarized in parallel with the redox transitions. The rapidity of the change in membrane potential, along with its strict limitation to a defined region of the cell, argues for a structural connection between mitochondria in the cluster.

E. Consideration of Mechanisms

We tested whether the redox transitions involved two known mechanisms that influence the mitochondrial redox state: mitochondrial permeability transitions and Ca^{2+} regulation of dehydrogenase activity. To determine if mitochondrial permeability transition (MPT) pores contributed to the oscillatory behavior, we tested whether cyclosporin A (CsA), a known blocker of MPT, could interrupt the redox oscillations. As shown in Fig. 10, CsA did not block the metabolic transients or oscillations. Similarly, a negative finding was obtained for a mechanism involving Ca^{2+}. Although the propagating redox waves resembled traveling Ca^{2+} waves resulting from spontaneous SR Ca^{2+} release in Ca^{2+}-overloaded heart cells (Capogrossi *et al.*, 1986) and Ca^{2+} is known to modulate mitochondrial dehydrogenase activity (Denton and McCormack, 1990; Moreno-Sanchez and Hansford, 1988), several pieces of evidence argue against Ca^{2+} as the mediator of the redox waves. First, no sarcomere shortening or contractions of the cells were observed during the redox transients; second, the redox oscillations could be routinely observed even in the presence of intracellular Ca^{2+} buffer (1–5 m*M* EGTA); third, there was no correlation between resting

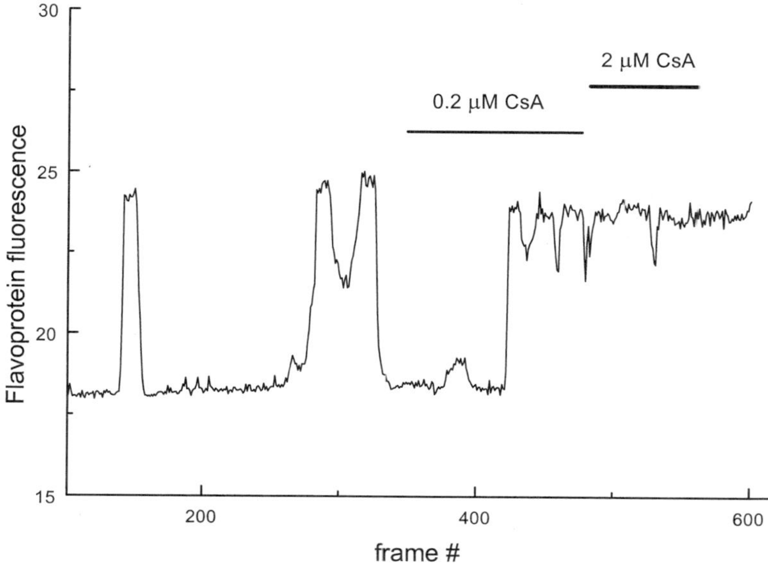

FIGURE 10 Lack of effect of cyclosporin A (CsA) on mitochondrial redox oscillations in cardiomyocytes. Oscillations were not inhibited by a low dose (0.2 μM) or a high dose (2 μM) of CsA, indicating a lack of involvement of mitochondrial permeability transition preos.

Ca^{2+} and K_{ATP} current oscillations (O'Rourke *et al.,* 1994); and, finally, ruthenium red had no effect on the oscillations, ruling out Ca^{2+} entry through the mitochondrial Ca^{2+} uniporter.

IV. DISCUSSION

Our previous work, which demonstrated the phenomenon of metabolic oscillation coupled to altered K_{ATP} current activation and excitation–contraction coupling (O'Rourke *et al.,* 1994), indicated that a mild metabolic perturbation, substrate withdrawal, can push cardiac energy metabolism toward instability. Because factors that influenced glycolytic flux appeared to modulate the periodicity and amplitude of the oscillations, a model centered on glycolytic oscillation was offered as a possible mechanism. This explanation was convenient, since a wealth of experimental and theoretical work was available establishing glycolytic oscillation as a prototypical periodic biochemical phenomenon (Berridge and Rapp, 1979; Boiteux *et al.,* 1977; Hess and Boiteux, 1971), even in cardiac cell-free extracts (Frenkel, 1968a–c). The present finding, that rapid transitions in mitochondrial redox and membrane potential can occur in highly localized regions of the cardiac cell, suggests that the mechanism of oscillation may be more complex than first thought. Instead of a cytoplasmic oscillator driving changes in mitochondrial function, we must also consider the possibility that the mitochondria are the source of the oscillations, with glucose metabolism simply biasing the energy state of the cell toward or away from instability. The interconnected and branched metabolic pathways in the cytoplasm and the mitochondria contain a myriad of potential control sites and feedback loops to consider as possible sources of oscillation, making the identification of cause and effect difficult, if not impossible. Despite this difficulty, the oscillations are an important tool for examining which regulatory components are intimately involved, since their amplitude, phase, and periodicity reflect the overall control properties of the system. In this regard, there is currently much debate about what actually regulates mitochondrial energy metabolism (reviewed in Brown, 1992; Heineman and Balaban, 1990). Leading hypotheses of respiratory control include regulation by ATP utilization (via ADP levels or the phosphorylation potential), substrate availability, $NAD^+/NADH$ redox balance, and mitochondrial Ca^{2+}. In truth, it appears that control may be distributed among several sites in the pathway and may shift depending on the energy state of the cell. As yet, we do not know if the oscillations we observe will provide insight into the normal physiological control of metabolism or whether the controls are altered by metabolic stress.

By discovering a situation in which synchronization of mitochondrial function is lost, we can explore what factors play a role in coordination of the mitochondrial network. Is the network synchronized by continuity of the matrix space via intermitochondrial connections (Amchenkova *et al.*, 1988)? Are diffusible cytoplasmic or mitochondrial messengers involved? How are redox waves propagated? Can we develop models of metabolic propagation within the myocardial syncytium analogous to those of electrical propagation? Many of these questions can be addressed by examining the phenomena we have described. Ultimately, the goal is to understand metabolism from the viewpoint of a distributed network of functional units, both at the subcellular and multicellular level, and to examine how metabolism is linked to electrical excitability. This approach will be crucial to understanding how metabolic perturbation leads to electrical and mechanical dysfunction in the heart.

Acknowledgments

This work was supported by National Institutes of Health Grants R01HL54598 (B.O'R., D.R.) and R37HL36957 (E.M.)

References

Amchenkova, A. A., Bakeeva, L. E., Chentsov, Y. S., Skulachev, V. P., and Zorov, D. B. (1988). Coupling membranes as energy-transmitting cables. I. Filamentous mitochondria in fibroblasts and mitochondrial clusters in cardiomyocytes. *J. Cell Biol.* **107,** 481–495.

Berridge, M. J., and Rapp, P. E. (1979). A comparative survey of the function, mechanism and control of cellular oscillators. *J. Exp. Biol.* **81,** 217–279.

Boiteux, A., Goldbeter, A., and Hess, B. (1975). Control of oscillating glycolysis of yeast by stochastic, periodic, and steady source of substrate: A model and experimental study. *Proc. Natl. Acad. Sci. U.S.A.* **72,** 3829–3833.

Boiteux, A., Hess, B., and Plesser, T. (1977). Oscillatory phenomena in biological systems. *FEBS Lett.* **75,** 1–4.

Brown, G. C. (1992). Control of respiration and ATP synthesis in mammalian mitochondria and cells. *Biochem. J.* **284,** 1–13.

Capogrossi, M. C., Suarez-Isla, B. A., and Lakatta, E. G. (1986). The interaction of electrically stimulated twitches and spontaneous contractile waves in single cardiac myocytes. *J. Gen. Physiol.* **88,** 615–633.

Chance, B., and Williams, G. R. (1955). A method for the localization of sites for oxidative phosphorylation. *Nature (London)* **176,** 250–254.

Chance, B., Schoener, B., and Elsaesser, S. (1964). Control of the waveform of oscillations of the reduced pyridine nucleotide level in a cell-free extract. *Proc. Natl. Acad. Sci. U.S.A.* **52,** 337–341.

Chance, B., Williamson, J. R., Jamieson, D., and Schoener, B. (1965). Properties and kinetics of reduced pyridine nucleotide fluorescence of the isolated and in vivo rat heart. *Biochem. Z.* **341,** 357–377.

Chance, B., Ernster, L., Garland, P. B., Lee, C. P., Light, P. A., Ohnishi, T., Ragan, C. I., and Wong, D. (1967). Flavoproteins of the mitochondrial respiratory chain. *Proc. Natl. Acad. Sci. U.S.A.* **57,** 1498–1505.

Corkey, B. E., Deeney, J. T., Glennon, M. C., Matschinsky, F. M., and Prentki, M. (1988). Regulation of steady-state free Ca^{2+} levels by the ATP/ADP ratio and orthophosphate in permeabilized RINm5F insulinoma cells. *J. Biol. Chem.* **263,** 4247–4253.

Denton, R. M., and McCormack, J. G. (1990). Ca^{2+} as a second messenger within mitochondria of the heart and other tissues. *Annu. Rev. Physiol.* **52,** 451–466.

Duysens, L. N. M., and Amesz, J. (1957). Fluorescence spectrophotometry of reduced phosphopyridine nucleotide in intact cells in the near ultraviolet and visible region. *Biochim. Biophys. Acta* **24,** 19–26.

Ehrenberg, B., Montana, V., Wei, M. D., Wuskell, J. P., and Loew, L. M. (1988). Membrane potential can be determined in individual cells from the nernstian distribution of cationic dyes. *Biophys. J.* **53,** 785–794.

Eng, J., Lynch, R. M., and Balaban, R. S. (1989). Nicotinamide adenine dinucleotide fluorescence spectroscopy and imaging of isolated cardiac myocytes. *Biochem. J.* **55,** 621–630.

Frenkel, R. (1968a). Control of reduced diphosphopyridine nucleotide oscillations in beef heart extracts. I. Effects of modifiers of phosphofructokinase activity. *Arch. Biochem. Biophys.* **125,** 151–156.

Frenkel, R. (1968b). Control of reduced diphosphopyridine nucleotide oscillations in beef heart extracts. II. Oscillations of glycolytic intermediates and adenine nucleotides. *Arch. Biochem. Biophys.* **125,** 157–165.

Frenkel, R. (1968c). Control of reduced diphosphopyridine nucleotide oscillations in beef heart extracts. III. Purification and kinetics of beef heart phosphofructokinase. *Arch. Biochem. Biophys.* **125,** 166–174.

Heineman, F. W., and Balaban, R. S. (1990). Control of mitochondrial respiration in the heart in vivo. *Annu. Rev. Physiol.* **52,** 523–542.

Henquin, J.-C. (1990). Glucose-induced electrical activity in β-cells: Feedback control of ATP-sensitive K^+ channels by Ca^{2+}. *Diabetes* **39,** 1457–1460.

Hess, B., and Boiteux, A. (1968). Mechanism of glycolytic oscillation in yeast. I. Aerobic and anaerobic growth conditions for obtaining glycolytic oscillation. *Hoppe-Seyler's Z. Physiol. Chem.* **349,** 1567–1574.

Hess, B., and Boiteux, A. (1971). Oscillatory phenomena in biochemistry. *Annu. Rev. Biochem.* **40,** 237–258.

Hommes, F. A. (1964). Oscillatory reductions of pyridine nucleotides during anaerobic glycolysis in brewers' yeast. *Arch. Biochem. Biophys.* **108,** 36–46.

Kort, A. A., Lakatta, E. G., Marban, E., Stern, M. D., and Wier, W. G. (1985). Fluctuations in intracellular calcium concentration and their effect on tonic tension in canine cardiac Purkinje fibres. *J. Physiol.* (*London*) **367,** 291–308.

Kunz, W. S. (1986). Spectral properties of fluorescent flavoproteins of isolated rat liver mitochondria. *FEBS Lett.* **195,** (1–2), 92–96.

Kunz, W. S., and Kunz, W. (1985). Contribution of different enzymes to flavoprotein fluorescence of isolated rat liver mitochondria. *Biochim. Biophys. Acta* **841**(3), 237–246.

Loew, L. M., Tuft, R. A., Carrington, W., and Fay, F. S. (1993). Imaging in five dimensions: Time-dependent membrane potentials in individual mitochondria. *Biophys. J.* **65,** 2396–2407.

Moreno-Sanchez, R., and Hansford, R. G. (1988). Relation between cytosolic free calcium and respiratory rates in cardiac myocytes. *Am. J. Physiol.* **255,** H347–H357.

Nilsson, T., Schultz, V., Berggren, P. O., Corkey, B. E., and Tornheim, K. (1996). Temporal patterns of changes in ATP/ADP ratio, glucose 6-phosphate and cytoplasmic free Ca^{2+} in glucose-stimulated pancreatic beta-cells. *Biochem. J.* **314,** 91–94.

O'Rourke, B., Ramza, B. M., and Marban, E. (1994). Oscillations of membrane current and excitability driven by metabolic oscillations in heart cells. *Science* **265,** 962–966.

Pralong, W.-F., Bartley, C., and Wollheim, C. B. (1990). Single islet β-cell stimulation by nutrients: Relationship between pyridine nucleotides, cytosolic Ca^{2+} and secretion. *EMBO J.* **9,** 53–60.

Romashko, D. N., Marban, E., and O'Rourke, B. (1998). Subcellular metabolic transients and mitochondrial redox waves in heart cells. *Proc. Natl. Acad. Sci. U.S.A.* **95,** 1618–1623.

Rooney, T. A., Sass, E. J., and Thomas, A. P. (1989). Characterization of cytosolic calcium oscillations induced by phenylephrine and vasopressin in single fura-2-loaded hepatocytes. *J. Biol. Chem.* **264,** 17131–17141.

Thomas, A. P., Bird, G. S., Hajnoczky, G., Robb-Gaspers, L. D., and Putney, J. W., Jr. (1996). Spatial and temporal aspects of cellular calcium signaling. *FASEB J.* **10,** 1505–1517.

Veenstra, R. D. (1996). Size and selectivity of gap junction channels formed from different connexins. *J. Bioenergetics Biomembr.* **28,** 327–337.

Wier, W. G., Kort, A. A., Stern, M. D., Lakatta, E. G., and Marban, E. (1983). Cellular calcium fluctuations in mammalian heart: Direct evidence from noise analysis of aequorin signals in Purkinje fibers. *Proc. Natl. Acad. Sci. U.S.A.* **80,** 7367–7371.

Woods, N. M., Cuthbertson, K. S., and Cobbold, P. H. (1986). Repetitive transient rises in cytoplasmic free calcium in hormone-stimulated hepatocytes. *Nature (London)* **319,** 600–602.

PART V

Other Inwardly Rectifying Potassium Channels

CHAPTER 25

Glial Inwardly Rectifying Potassium Channels

Yoshiyuki Horio and Yoshihisa Kurachi
Department of Pharmacology II, Faculty of Medicine, Osaka University, Osaka 565, Japan

I. SPATIAL BUFFERING OF K^+ IONS IN GLIAL CELLS

Neural tissues consist of two major classes of cells, namely, neurons and glial cells. Glial cells surround both the cell bodies and processes of neurons and fill up the interneuronal spaces. The importance of glial cells for brain function is suggested by the increase in their number during evolution; glial cells constitute 25, 65, and 90% of the brain cells in *Drosophila,* rodents, and humans, respectively (Pfrieger and Barres, 1995).

One of the most characteristic properties of glial cells is their high K^+ conductance. In the central nervous system, including the retina, neural excitation causes an increase of extracellular K^+ ions ($[K^+]_o$) especially at synaptic sites. For example, the $[K^+]_o$ at the retinal inner plexiform layer,

1063-5823/99 $30.00

which contains abundant synapses, increases prominently after a flush of light (Oakley and Green, 1976). If the elevated $[K^+]_o$ were uncorrected, it would result in loss of synaptic transmission by depolarizing the presynaptic axonal termini. Kuffler and co-workers (1966) reported that glial cells possess a deep membrane potential and high permeability to K^+ ions. Using optic nerves of mud puppy (*Necturus*) and frog, Orkand *et al.* (1966) found that nerve impulses cause a slow depolarization of glial cells. They interpreted this phenomenon as indicating that K^+ ions released from neurons were aspirated by glial cells, which resulted in depolarization of the glial cells. On the basis of this observation, they proposed that glial cells have a function to draw in K^+ ions at the sites where $[K^+]_o$ is high, transport them within the cells, and extrude the ions at sites where $[K^+]_o$ is low (Fig. 1). They called this function of glial cells "K^+-spatial buffering" (Orkand *et al.,* 1966). Inwardly rectifying K^+ (Kir) channels on the glial cell membrane have been supposed to be responsible for this regulatory function, on the basis of their high expression in glial cells and their biophysical properties (Brew *et al.,* 1986). Although the Kir channels in various glial cells have been examined electrophysiologically, molecular dissection of Kir channels has enabled us to study the glial Kir channels at the molecular level, as discussed in this chapter.

II. Kir CHANNELS IN RETINAL MÜLLER GLIAL CELLS

A. Expression of Kir Channels in Müller Cells

Müller cells are the principal glial cells in the retina. They belong to the so-called radial glial cells; each Müller cell contacts both the outer and

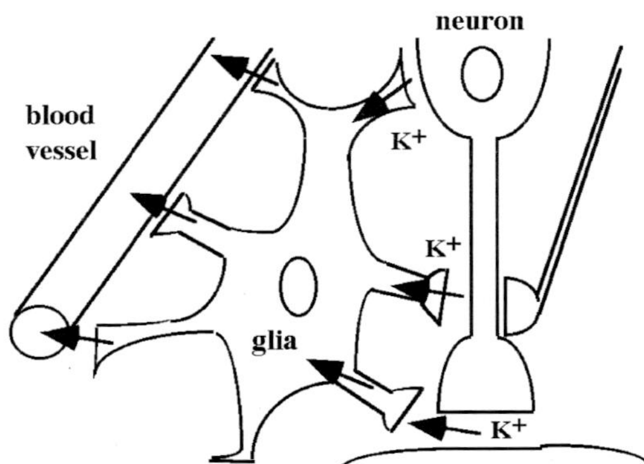

FIGURE 1 Potassium ion spatial buffering function of a glial cell. The glial cell draws in K^+ ions released from neurons, transports them within the cell, and extrudes the ions at a site (e.g, blood vessel) where $[K^+]_o$ is low.

inner surfaces of the retina and thus bridges the whole thickness of the neuroretina. Müller cells have been widely used to elucidate the properties and mechanism of the K^+-spatial buffering action of glial cells because they have several advantages over brain glial cells: (1) The retina has an organized structure of six layers composed of various neural elements, such as neuronal cell bodies, nerve fibers, and synaptic regions. Müller cells surround and face all of these neuronal elements. Thus, the dynamics of K^+ movement in the retina could be characterized more easily than in the brain. (2) Müller cells are larger in size and more simple in shape than brain glial cells. They can be isolated and identified easily, and thus can be subject to electrophysiological and other studies. Newman *et al.* (1984) actually showed dissociated amphibian Müller cells to possess the capability of aspirating extracellular K^+ ions from their distal ends and secreting the cations from their proximal endfeet into the vitreous body. They termed this phenomenon the K^+-siphoning action of Müller cells.

Patch-clamp studies demonstrated expression of a Kir channel in salamander Müller cells. In amphibia, more than 90% of the K^+ conductance exists in the endfoot of Müller cells, due to the higher density of Kir channels in this region (Brew *et al.,* 1986). This characteristic distribution of Kir channels is supposed to be crucial for the K^+-siphoning action in the retina of this species (Newman *et al.,* 1984). However, the Kir channel on salamander Müller cells has not yet been examined at the molecular level.

For mammalian Müller cells, the situation is different from the salamander cells. Predominant expression of Kir channels was also found in mammalian Müller cells, but the Kir channel activity was not accumulated at the endfoot (Nilius and Reichenbach, 1988; Newman, 1987). Although Nilius and Reichenbach (1988) reported three kinds of Kir channels differentially distributed on the rabbit Müller cell membrane, recent studies have identified only a single population of Kir channels in rabbit, monkey, and human retinal Müller cells (Ishii *et al.,* 1997; Kusaka and Puro, 1997; Tada *et al.,* 1998). The mammalian Müller cell Kir channel possesses a unitary conductance of ~25 pS with 150 m*M* $[K^+]_o$ and exhibits intermediate inward-rectification. No other types of Kir channels have been identified in mammalian Müller cells so far.

More than 10 cDNA clones encoding various Kir channels have been isolated from cDNA libraries of brain, kidney, and other tissues (Isomoto *et al.,* 1997; Bond *et al.,* 1994). Among them, only Kir4.1/K_{AB}-2/BIR10 mRNA was found to be predominantly expressed in glial cells of the brain (Takumi *et al.,* 1995) (Fig. 2). Using rabbit and rat Müller cells, Ishii *et al.* (1997) demonstrated that only a single population of Kir channels is diffusely expressed on the Müller cell membrane from the distal end to proximal endfoot. The Kir channels exhibited electrophysiological properties identical to those of the cloned Kir4.1 channel expressed in a mammalian

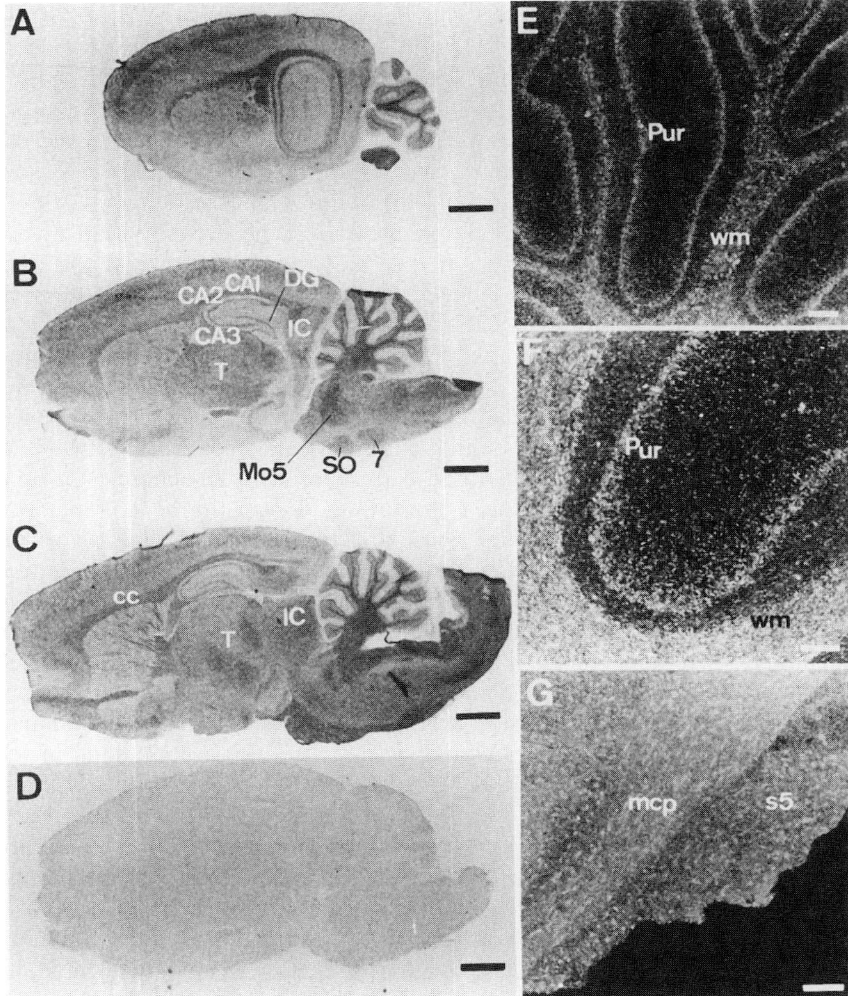

FIGURE 2 Kir4.1 mRNA expression in the brain. (A–D) X-Ray film autoradiographs illustrating the distribution of Kir4.1 mRNA in sagittal sections of rat brain. Sections were hybridized with antisense probe (A–C) or sense probe (D) and then exposed to X-ray film. Strong expression of the mRNA was detected in cerebellum, and moderate expression was found in corpus callosum (cc), hippocampus (CA1–3 and DG), thalamus (T), inferior colliculus (IC), and brain stem. No hybridization signal was found in D. (E–G) Dark-field photomicrographs of sagittal sections of cerebellum (E and F) and coronal section of brain stem (G). Kir4.1 mRNA was detected in Bergmann glial cells of the Purkinje cell layer (Pur) and white matter (wm) of cerebellum (E) and also in sensory root trigeminal ganglion (s5) and middle cerebellar peduncle (mcp) (G). CA1–3, fields CA1-3 of Ammon's horn; DG, dentate gyrus; Mo5, motor trigeminal nucleus; SO, superior olive; 7, facial nucleus. Scale bars: (A–D) 2 mm; (E) 200 μm; (F and G) 100 μm.

cell line, HEK293T cells (Ishii *et al.,* 1997; Tada *et al.,* 1998). Kusaka and Puro (1997) demonstrated predominant expression of 20 pS Kir channels in 100 m*M* $[K^+]_o$ in human and monkey Müller cells, the conductance and kinetic properties of which are quite similar to those of Kir4.1. Ishii *et al.* (1997) further identified this Kir channel as Kir4.1 with molecular biological and immunohistochemical techniques. Immunoelectron microscopic examination of rat Müller cells demonstrated high expression of Kir4.1 on the cell membrane adjacent not only to the vitreous body but also to photoreceptor cells, pericytes, and endothelial cells of capillaries (Ishii *et al.,* 1997). These results strongly suggest that the K^+-spatial buffering mechanism in retinal Müller cells is performed by Kir4.1. Because Kir4.1 shows an intermediate inwardly rectifying property, Kir4.1 can extrude K^+ ions more effectively than the strong inwardly rectifying K^+ channels such as Kir2.x (Ishii *et al.,* 1997). This property of Kir4.1 suggests that the Kir channels participate in extrusion as well as in intrusion of K^+ ions from Müller cells for the spatial buffering action.

B. Localization of Kir Channels in Müller Cells

To intrude and extrude K^+ ions effectively, Müller cells seem to localize the Kir4.1 channels in the membrane adjacent to synaptic clefts, blood vessels, and the vitreous body as described above. Figure 3 shows the immunoreactivity of Kir4.1 in rat Müller cells. The channel immunoreactivity distributed in a clustered manner from the distal end to the endfoot of each Müller cell (Horio *et al.,* 1997; Ishii *et al.,* 1997). This clustered distribution may be important for Kir4.1 to be localized at the proper sites for intrusion and extrusion of K^+ ions for the spatial buffering action.

How, then, do Müller cells cluster Kir4.1 and localize the clusters of Kir4.1 in appropriate positions on their cell membrane? The primary amino acid sequence of Kir4.1 indicates that the carboxyl-terminal region of this channel subunit ends with the sequence of Ser-Asn-Val (Takumi *et al.,* 1995). An anchoring protein of the PSD-95/SAP90 family was shown to bind and cluster NMDA receptors and voltage-gated K^+ channels, which have the sequence Ser-X-Val (where X is any amino acid residue) in their carboxyl-terminal ends (Kornau *et al.,* 1995; Kim *et al.,* 1995). Furthermore, in *Drosophila,* synaptic clustering of *Shaker* K^+ channels was abolished by mutations of *dlg,* a gene encoding one of the PSD-95/SAP90 family (Tejedor *et al.,* 1997). Therefore, PSD-95/SAP90 family proteins might control the cluster formation and localization of Kir4.1 on the Müller cell membrane. Actually, Horio *et al.* (1997) found that SAP97, a mammalian homolog of *dlg,* is expressed in Müller cells and distributes on the cell membrane in a

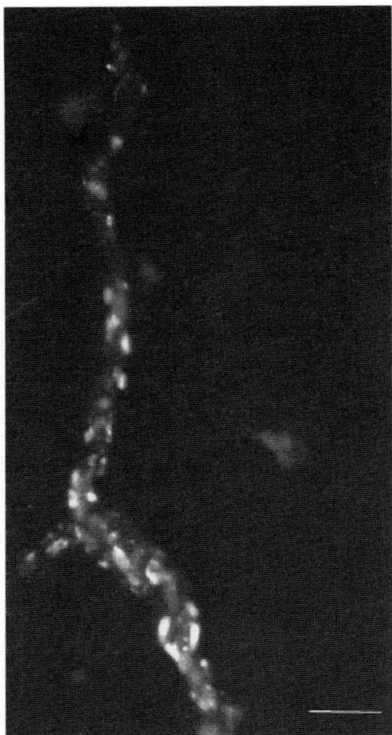

FIGURE 3 Expression of Kir4.1 on the membrane of Müller cells. Isolated rat Müller cells were stained with anti-Kir4.1 antibody and then with FITC-conjugated anti-rabbit IgG. Kir4.1 immunoreactivity (white) was distributed in a clustered manner on the membrane. Scale bar: 10 μm.

similar clustered manner to Kir4.1. Furthermore, in HEK293T cells, the clustering of Kir4.1 on the cell membrane by PSD-95/SAP90 and SAP97 proteins could be reconstituted: When Kir4.1 was expressed alone in HEK293T cells, the channel immunoreactivity was distributed diffusely. In contrast, when Kir4.1 was coexpressed with PSD-95/SAP90 or SAP97, prominent clustering of Kir4.1 on the cell membrane occurred. Kir4.1 was coimmunoprecipitated with PSD-95/SAP90 in the coexpressed cells. The interaction between Kir4.1 and PSD-95/SAP90 disappeared when the Ser-Asn-Val motif of Kir4.1 was deleted. Because SAP97 was reported to bind protein 4.1, which interacts with actin (Leu *et al.,* 1994), it is possible that cytoskeletons may also be involved in the control of clusters of Kir4.1 channels through SAP97. In addition to clustering, it was also noticed that cotransfection of SAP97 causes an increase of the whole-cell Kir4.1 current magnitude by threefold (Horio *et al.,* 1997). Because this current increase may be due to an increase in functional channel number, SAP97 might also regulate the translocation of Kir4.1 to the Müller cell membrane.

Immunoelectron microscopic examination suggested that Kir4.1 channels on Müller cells are localized in the vicinity of neurons and blood vessels (Ishii *et al.*, 1997). This localization of the cluster of channels may facilitate the efficacy of the glial cell K^+-buffering action. Some unidentified factors derived from neurons or vessels may control the localization of the clusters of Kir4.1 channel–SAP97 complexes on the Müller cell membrane. Ishii *et al.* (1997) showed that laminin is necessary for the clustered expression of Kir4.1 on the membrane of cultured Müller cells. Therefore, extracellular matrix, which is provided by neurons, pericytes, and/or endothelial cells, might be one of the factors regulating the distribution of Kir4.1–SAP97 clusters on the Müller cell membrane. Obviously further experiments are needed to clarify the mechanism underlying the control of the proper distribution of Kir4.1–SAP97 clusters on the Müller cell membrane.

III. Kir CHANNELS IN SCHWANN CELLS

Schwann cells wrap themselves around axons to provide electrical insulation in the form of a myelin sheath in peripheral nerve fibers. In Schwann cells, Wilson and Chiu (1990) reported two types of K^+ currents, namely, delayed rectifying outward currents and inwardly rectifying currents. These K^+ currents did not distribute homogeneously on the membrane of Schwann cells. Wilson and Chiu (1990) recorded a 40-fold higher density of K^+ currents in the node of Ranvier than in the cell body. Mi *et al.* (1996) demonstrated that Kir2.1/IRK1 and Kir2.3/IRK3 are localized at the microvilli of the nodes of Ranvier using a specific antibody which recognizes both channels. Because the microvilli have a large surface area, the localization of Kir2.1 and Kir2.3 on the microvilli of nodal space suggests involvement of these Kir channels in the K^+-spatial buffering action by Schwann cells. Because Kir2.1 and Kir2.3 have a strong inward rectification property, these channels might be mainly responsible for intrusion of K^+ ions into Schwann cells.

Mi *et al.* (1995) also demonstrated that Kv1.5, a member of the *Shaker*-like family of voltage-dependent K^+ channels, is localized on the outer surface of Schwann cells in the vicinity of nodes of Ranvier. Kv1.5 channels are located at some distance from the axonal membranes. They require a depolarization to be opened, and, at depolarized potentials, K^+ efflux would be favored. Kv1.5 may thus extrude K^+ ions from Schwann cells.

IV. Kir CHANNELS IN OTHER GLIAL CELLS

A. Ependymoglial Cells

Ependymoglial cells line the internal cavities of the brain and spinal cord. Their epithelial arrangement reflects the origin of the central nervous system from an epithelial tube. Kubo *et al.* (1996) cloned a Kir channel, sWIRK, from salmon brain. The amino acid sequence of sWIRK exhibited 70% identity with that of Kir4.1. Furthermore, sWIRK possesses a Walker type-A ATP-binding domain and the sequence Ser-Asn-Val in its carboxyl-terminal end. The rectification property of sWIRK is intermediate. These characters of sWIRK are very similar to Kir4.1. *In situ* hybridization of salmon brain showed that the sWIRK mRNA is expressed in the ependymoglial cells on the surface of the ventricles as well as in the small perineuronal glialike cells in midbrain and medulla. Although it is not known whether sWIRK is a salmon homolog of mammalian Kir4.1, it might participate in intrusion and extrusion of K^+ ions of ependymoglial cells in salmon brain.

B. Astrocytes

Astrocytes in the brain have a function to take up K^+ ions from extracellular spaces. In studies using brain slices, extracellular cesium caused epileptiform activity characterized by synchronous burst discharges of pyramidal cells and prevented maintenance of long-term depression (LTD) in the CA1 hippocampal region (Schwartzkroin and Prince, 1980; Schwartzkroin and Wyler, 1980). Cesium enhanced the amplitude and duration of the stimulation-evoked increase in $[K^+]_o$. Recordings from astrocytes in hippocampal slices revealed that cesium mainly inhibits Kir channels of the glial cells (Janigro *et al.,* 1997). Thus, the K^+ uptake by astrocytes through Kir channels may be necessary for maintaining long-term depression in the CA1 hippocampal region.

Brain astrocytes are classified into two types: type-1 and type-2. Type-1 astrocytes are polygonal glial cells of the brain. Eighty-five percent of cultured type-1 astrocytes isolated from rat optic nerves possess Kir channels, whereas a delayed rectifying K^+ current is present in 100% of the cells (Barres *et al.,* 1990a). The Kir current is sensitive to extracellular barium ions, but not to charybdotoxin, a blocker of some calcium-dependent K^+ channels. Almost no process remained in acutely isolated type-1 astrocytes. Barres *et al.* (1990a) developed a method to isolate process-bearing astrocytes using a nitrocellulose filter. The type-1 astrocytes thus obtained, which possess processes, express Kir channels at a much higher density than do the acutely isolated cells. These results suggest that Kir channels are predominantly expressed at the fine processes of type-1 astrocytes (Barres *et al.,* 1990a). The molecular properties of the Kir channels in type-1 astrocytes have not yet been examined.

Type-2 astrocytes and oligodendrocytes are descended from a common bipotential glial precursor cell, the O-2A progenitor (Raff *et al.,* 1983, 1984). It was reported that 100% of O-2A progenitor cells in culture possess Kir channels, although the Kir current is not recorded in acutely isolated O-2A progenitors (Barres *et al.,* 1990b). Because acutely isolated O-2A progenitor cells lost their fine processes during isolation, the Kir channels might be localized in the processes of O-2A progenitors.

Type-2 astrocytes are stellate cells with radially distributed fine processes and can be distinguished from type-1 cells by morphology and by their surface antigens. Type-2 astrocytes developed from O-2A progenitor cells of rat optic nerves possess much larger Kir currents than the O-2A cells themselves (Barres *et al.,* 1990b). Because the processes of type-2 astrocytes are suggested to be closely apposed to the nodes of Ranvier (ffrench-Constant and Raff, 1986; Miller *et al.,* 1989), the Kir channels of fine processes in type-2 astrocytes might participate in the K^+-spatial buffering action of the cells.

C. Oligodendrocytes

Oligodendrocytes can be differentiated from O-2A progenitor cells *in vitro* (Raff *et al.,* 1983). The current through Kir channels was recorded in cultured oligodendrocytes derived from O-2A progenitors of optic nerves (Barres *et al.,* 1990b), but the magnitude of Kir currents in these cells was smaller than that of type-2 astrocytes. *In situ* hybridization using rat brain slices with Kir4.1 cDNA as a probe suggests that Kir4.1 is expressed in oligodendrocytelike cells of the cerebellum, middle cerebellar peduncle, and corpus callosum (Fig. 1) (Takumi *et al.,* 1995). The Kir current recorded in oligodendrocytes may flow through Kir4.1.

Karschin *et al.* (1994) found that serotonin, somatostatin, and muscarinic acetylcholine receptors are expressed in rat brain oligodendrocytes. Activation of these receptors resulted in inhibition of constitutively active Kir channels within 1 sec after the addition of ligands. The inhibition was pertussis toxin-sensitive. By contrast, in heart and brain, similar pathways activate G-protein-gated Kir channels. Thus, the transmitter-mediated blockade of Kir channels appears to be specific in oligodendrocytes. At present, the molecular mechanism underlying the inhibition of constitutively active Kir channels by these neurotransmitters is unclear.

D. Microglias

Microglias participate in the removal of degenerated materials, immune responses within the central nervous system, production of interleukin-1, and initiation of subsequent growth factor cascades. They are thought to be the resident macrophages of the central nervous system (Perry and Gordon, 1988). Cultured rat microglias possess Kir currents similar to those found in peritoneal macrophages (Kettenmann *et al.,* 1990). In the majority of cultured microglial cells, Kir currents often disappeared within the first several minutes of whole-cell recording (Korotzer and Cotman, 1992). This phenomenon is thought to reflect loss of soluble factors necessary for the channel to open (McCloskey and Cahalan, 1990). Phagocytosis by macrophages was reported to be accompanied by their prolonged hyperpolarization (Kouri *et al.,* 1980), which was due to an increase of K^+ channel activity (Ince *et al.,* 1988). This suggests the possibility that Kir channels may also have some functional roles in the phagocytosis of microglias. Further studies are needed to elucidate the functional roles of Kir channels in microglias.

E. Satellite Cells

Satellite cells surround ganglion cells of the cochlea in the inner ear. Satellite cells wrap a ganglion cell with several layers of myelinlike sheaths and separate the ganglion cells from one another. Hibino *et al.* (1997) found expression of Kir4.1 in satellite cells using an anti-Kir4.1 antibody. Kir4.1 was localized on the myelinlike sheaths of the satellite cell membrane adjacent to the ganglion cell (Hibino *et al.,* submitted for publication). Kir4.1 in the satellite cells may participate in the regulation of $[K^+]_o$ for ganglion neurons.

V. DEVELOPMENTAL REGULATION OF GLIAL Kir CHANNELS

Kir currents were not recorded from type-1 astrocytes 2 and 3 days after birth. By 5, 8, and 10 days after birth, 40, 50, and 100%, respectively, of the cells expressed Kir current (Barres *et al.,* 1990a). The Kir current of cultured type-1 astrocytes was increased when cocultured with neurons (Barres *et al.,* 1990a). Nonmyelinating Schwann cells of the cervical sympathetic trunk also possess Kir channels (Konishi, 1994). The density of the Kir current in these cells increased 1–4 weeks after birth and then stayed constant for up to 12 weeks. The mean current density of the Kir current in cells 12 weeks after birth was about twice that in cells 1 week after birth. The block of nerve conduction in the cervical sympathetic trunk by tetrodotoxin for 5 days resulted in a significant decrease in both magnitude and density of Kir currents in these Schwann cells. Kir4.1 is found to be expressed in retinal Müller cells and satellite cells of the inner ear (Horio *et al.,* 1997; Ishii *et al.,* 1997; Hibino *et al.,* 1997). The expression of Kir4.1 in these cells was not detected 1 to 5 days after birth. Weak immunoreactivity was detected 8 days after birth, and the expression of Kir4.1 then increased during the following days and reached an adult level at 14 days after birth (Hibino *et al.,* 1997; Kusaka *et al.,* submitted for publication). These results suggest that the expression of Kir channels of glial cells is developmentally controlled and is regulated by some factors from neurons. These factors may be released by neural excitation. At present, there is no evidence that K^+ ions themselves increase the expression of glial Kir channels.

Human Müller cells also express Kir channels (Kusaka and Puro, 1997). Francke *et al.* (1997) found that the Kir current of Müller cells from patients suffering from retinal detachment, secondary glaucoma, melanoma, and perforating injury is lost or severely reduced. On the other hand, the Müller cells from the eyes of patients with corneal ulcers show a Kir current similar to that found in healthy retinas. Because the Na^+ current in the Müller cells from patients with these retinal diseases was larger than that of healthy Müller cells, the disappearance of the Kir current seems to be specific to the pathological processes in Müller cells. Electron microscopic examination showed a dramatic change of the intermediate cytoskeleton in reactive Müller cells after retinal detachment (Fisher *et al.,* 1993). Thus, the environment around Müller cells affects the structure of glial cells and changes the expression of Kir4.1.

The Kir current of cultured glial cells from frog optic nerve diminished after 3 days in culture and disappeared after 5 days (Philippi *et al.,* 1996). Similarly, when isolated Müller cells from rat retina were cultured on poly-D-lysine-coated disks, the expression of Kir4.1 promptly disappeared (Ishii *et al.,* 1997). Addition of insulin to the culture medium induced the expression of Kir4.1 mRNA, but no functional Kir4.1 was recorded. Immunocytochemical staining using anti-Kir4.1 antibody showed that Kir4.1 seems to exist in the cytosol of these Müller cells under this condition. However, isolated Müller cells cultured on laminin-coated disks in the presence of insulin expressed functional Kir4.1 channels on their membrane. These results suggest that insulin promotes the expression of Kir4.1 mRNA and that the laminin signal promotes the translocation of Kir4.1 from the cytosol to the membrane (Fig. 4; see color plate) (Ishii *et al.,* 1997). Such dynamic regulation of expression and translocation of Kir4.1 might underlie the disappearance and reappearance of Kir currents in retinal diseases and after cure, respectively. Further studies are needed to clarify the mechanism of the regulation of glial Kir channels.

VI. SUMMARY OF GLIAL INWARDLY RECTIFYING POTASSIUM CHANNELS

Glial cells surround both cell bodies and processes of neurons. Neural excitation causes an increase of extracellular K^+ ($[K^+]_o$) especially at synaptic sites. If the elevated $[K^+]_o$ were uncorrected, it would result in loss of synaptic transmission by depolarizing the neurons. Glial cells are supposed to aspirate K^+ ions released from neurons and then transport and extrude them to the sites where $[K^+]_o$ is low. This function of glial cells has been called K^+-spatial buffering. Glial cells possess a high K^+ conductance flowing through inwardly rectifying K^+ (Kir) channels, which is considered to be responsible for their K^+-spatial buffering action. The expression and function of glial Kir channels have been extensively studied using retinal Müller glial cells, Schwann cells, astrocytes, and oligodendrocytes. Expression of Kir4.1/K_{AB}-2 was identified in brain and retinal glial cells, whereas those of Kir2.1/IRK1 and Kir2.3/IRK3 were demonstrated in Schwann cells. Our studies on the distribution and expression of Kir4.1 in retinal glial Müller cells indicate that glial Kir channels may be dynamically controlled by various humoral factors, extracellular, matrix, and intracellular proteins.

References

Barres, B. A., Koroshetz, W. J., Chun, L. L. Y., and Corey, D. P. (1990a). Ion channel expression by white matter glia: The type-1 astrocyte. *Neuron* **5,** 527–544.

Barres, B. A., Koroshetz, W. J., Swartz, K. J., Chun, L. L. Y., and Corey, D. P. (1990b). Ion channel expression by white matter glia: The O-2A glial progenitor cell. *Neuron* **4,** 507–524.

Bond, C. T., Pessia, M., Xia, X. M., Lagrutta, A., Kavanauph, M. P., and Adelman, J. P. (1994). Cloning and expression of a family of inward rectifier potassium channels. *Recept. Channels* **2,** 183–191.

Brew, H., Gray, P. T. A., Mobbs, P., and Attwell, D. (1986). Endfeet of retinal glial cells have higher densities of ion channels that mediate K^+ buffering. *Nature* (*London*) **324,** 466–468.

ffrench-Constant, C., and Raff, M. C. (1986). The oligodendrocyte-type-2 astrocyte cell lineage is specialized for myelination. *Nature* **323,** 335–338.

Fisher, S. K., Lewis, G. P., Guerin, C. J., Anderson, D. H., and Matsumoto, B. (1993). 3-Dimensional changes in the intermediate cytoskeleton of reactive Müller cells after retinal detachment. *Invest. Ophthalmol. Visual Sci* **34,** 1039.

Francke, M., Pannicke, T., Biedermann, B., Faude, F., Wiedemann, P., Reichenbach, A., and Reichelt, W. (1997). Loss of inwardly rectifying potassium currents by human retinal glial cells in diseases of the eye. *Glia* **20,** 210–218.

Hibino, H., Horio, Y., Inanobe, A., Doi, K., Ito, M., Yamada, M., Gotow, T., Uchiyama, Y., Kawamura, M., Kubo, T., and Kurachi, Y. (1997). An ATP-dependent inwardly rectifying potassium channel, K_{AB}-2 (Kir4.1), in cochlear stria vascularis of inner ear: Its specific

subcellular localization and correlation with the formation of endocochlear potential. *J. Neurosci.* **17,** 4711–4721.
Hibino, H., Horio, Y., Fujita, A., Inanobe, A., Doi, K., Gotow, T., Uchiyama, Y., Kubo, T., and Kurachi, Y. (1998). Expression of an inwardly rectifying K^+ channel, Kir4.1, in the satellite cells of rat cochlear ganglions. Submitted for publication.
Horio, Y., Hibino, H., Inanobe, A., Yamada, M., Ishii, M., Tada, Y., Satoh, E., Hata, Y., Takai, Y., and Kurachi, Y. (1997). Clustering and enhanced activity of an inwardly rectifying potassium channel, Kir4.1, by an anchoring protein, PSD-95/SAP90. *J. Biol. Chem.* **272,** 12885–12888.
Ince, C., Coremans, J. M. C. C., Ypey, D. L., Leijh, P. C. J., Verveen, A. A., and van Furth, R. (1988). Phagocytosis by human macrophages is accompanied by changes in ionic channel currents. *J. Cell Biol.* **106,** 1873–1878.
Ishii, M., Horio, Y., Tada, Y., Hibino, H., Inanobe, A., Ito, M., Yamada, M., Gotow, T., Uchiyama, Y., and Kurachi, Y. (1997). Expression and clustered distribution of an inwardly rectifying potassium channel, K_{AB}-2/Kir4.1, on mammalian retinal Müller cell membrane: Their regulation by insulin and laminin signals. *J. Neurosci.* **17,** 7725–7735.
Isomoto, S., Kondo, C., and Kurachi, Y. (1997). Inwardly rectifying potassium channels: Their molecular heterogeneity and function. *Jpn. J. Physiol.* **47,** 11–39.
Janigro, D., Gasparini, S., D'Ambrosio, R., McKhann II, G., and DiFrancesco, D. (1997). Reduction of K^+ uptake in glia prevents long-term depression maintenance and causes epileptiform activity. *J. Neurosci.* **17,** 2813–2824.
Karschin, A., Wischmeyer, E., Davidson, N., and Lester, H.A. (1994). Fast inhibition of inwardly rectifying K^+ channels by multiple neurotransmitter receptors in oligodendroglia. *Eur. J. Neurosci.* **6,** 1756–1764.
Kettenmann, H., Hoppe, D., Gottmann, K., Banati, R., and Kreutzberg, G. (1990). Cultured microglial cells have a distinct pattern of membrane channels different from peritoneal macrophages. *J. Neurosci. Res.* **26,** 278–287.
Kim, E., Niethammer, M., Rothschild, A., Jan, Y. N., and Sheng, M. (1995). Clustering of Shaker-type K^+ channels by interaction with a family of membrane-associated guanylate kinases. *Nature (London)* **378,** 85–88.
Konishi, T. (1994). Activity-dependent regulation of inwardly rectifying potassium currents in nonmyelinating Schwann cells in mice. *J. Physiol. (London)* **474,** 193–202.
Kornau, H.-C., Schenker, L. T., Kennedy, M. B., and Seeberg, P. H. (1995). Domain interaction between NMDA receptor subunits and the postsynaptic density protein PSD-95. *Science* **269,** 1737–1740.
Korotzer, A. R., and Cotman, C. W. (1992). Voltage-gated currents expressed by rat microglia in culture. *Glia* **6,** 81–88
Kouri, J., Noa, M., Diaz, B., and Niubo, E. (1980). Hyperpolarization of rat peritoneal macrophages phagocytosing latex particles. *Nature (London)* **283,** 868–869.
Kubo, Y., Miyashita, T., and Kubokawa, K. (1996). A weakly inward rectifying potassium channel of the salmon brain. Glutamate 179 in the second transmembrane domain is insufficient for strong rectification. *J. Biol. Chem.* **271,** 15729–15735.
Kuffler, S. W., Nicholls, J. G., and Orkand, R. K. (1966). Physiological properties of glial cells in the central nervous system of amphibia. *J. Neurophysiol.* **29,** 768–787.
Kusaka, S., Horio, Y., Fujita, A., Matsushita, K., Inanobe, A., Gotow, T., Uchiyama, Y., Tano, Y., and Kurachi, Y. (1998). Expression and polarized distribution of an inwardly rectifying potassium channel, K_{AB}-2/Kir4.1 (Kir1.2), in mammalian retinal pigment epithelial cells. Submitted for publication.
Kusaka, S., and Puro, D. G. (1997). Intracellular ATP activates inwardly rectifying K^+ channels in human and monkey retinal Müller (glial) cells. *J. Physiol. (London)* **500,** 593–604.

Leu, R. A., Shirin, M. M., Branton, D., and Chishti, A. H. (1994). Cloning and characterization of hdlg: The human homologue of the *Drosophila* discs large tumor suppressor binds to protein 4.1. *Proc. Natl. Acad. Sci. U.S.A.* **91,** 9818–9822.

McCloskey, M. A., and Cahalan, M. D. (1990). G-protein control of potassium channel activity in a mast cell line. *J. Gen. Physiol.* **95,** 205–227.

Mi, H., Deerinck, T. J., Ellisman, M. H., and Schwarz, T. L. (1995). Differential distribution of closely related potassium channels in rat Schwann cells. *J. Neurosci.* **15,** 3761–3774.

Mi, H., Deerinck, T. J., Jones, M., Ellisman, M. H., and Schwarz, T. L. (1996). Inwardly rectifying K^+ channels that may participate in K^+ buffering are localized in microvilli (also called nodal processes) of Schwann cells. *J. Neurosci.* **16,** 2421–2429.

Miller, R. H., Fulton, B. P., and Raff, M. C. (1989). A novel type of glial cell associated with nodes of Ranvier in rat optic nerve. *Eur. J. Neurosci.* **1,** 172–180.

Newman, E. A. (1987). Distribution of potassium conductance in mammalian Müller (glial) cells: A comparative study. *J. Neurosci.* **7,** 2423–2432.

Newman, E. A., Frambach, D. A., and Odette, L. L. (1984). Control of extracellular potassium levels by retinal glial cell K^+ siphoning. *Science* **225,** 1174–1175.

Nilius, B., and Reichenbach, A. (1988). Efficient K^+ buffering by mammalian retinal glial cells is due to cooperation of specialized ion channels. *Pfluegers Arch.* **411,** 654–660.

Oakley II, B., and Green, D. G. (1976). Correlation of light-induced changes in retinal extracellular potassium concentration with c-wave of the electroretinogram. *J. Neurophysiol.* **39,** 1117–1133

Orkand, R. K., Nicholls, J. G., and Kuffler S. W. (1966). Effect of nerve impulses on the membrane potential of glial cells in the central nervous system of amphibia. *J. Neurophysiol.* **29,** 788–806.

Perry, V. H., and Gordon, S. (1988). Macrophages and microglia in the nervous system. *Trends Neurosci.* **6,** 273–277.

Pfrieger, F. W., and Barres, B. A. (1995). What the fly's glia tell the fly's brain. *Cell* (*Cambridge, Mass.*) **83,** 671–674.

Philippi, M., Vyklicky, L., and Orkand, R. K. (1996). Potassium currents in cultured glia of the frog optic nerve. *Glia* **17,** 72–82.

Raff, M. C., Miller, R. H., and Noble, M. (1983). A glial progenitor cell that develops *in vitro* into an astrocyte or an oligodendrocyte depending on culture medium. *Nature* (*London*) **303,** 390–396.

Raff, M. C., Williams, B. P., and Miller, R. H. (1984). The *in vitro* differentiation of a bipotential glial progenitor cell. *EMBO J.* **3,** 1857–1864.

Schwartzkroin, P. A., and Prince, D. A. (1980). Changes in excitatory and inhibitory synaptic potentials leading to electrogenic activity. *Brain Res.* **183,** 61–76.

Schwartzkroin, P. A., and Wyler, A. R. (1980). Mechanisms underlying epileptiform burst discharge. *Ann. Neurol.* **7,** 95–107.

Tada, Y., Horio, Y., and Kurachi, Y. (1998). Inwardly rectifying K^+ channels in retinal Müller cells; Comparison with the K_{AB}-2/Kir4.1 channel expressed in HEK293T cells. *Jpn. J. Physiol.* **48,** 71–80.

Takumi, T., Ishii, T., Horio, Y., Morishige, K., Takahashi, N., Yamada, M., Yamashita, T., Kiyama, H., Sohmiya, K., Nakanishi, S., and Kurachi, Y. (1995). A novel ATP-dependent inward rectifier potassium channel expressed predominantly in glial cells. *J. Biol. Chem.* **270,** 16339–16346.

Tejedor, F. J., Bokhari, A., Rojero, O., Gorczyca M., Zhang, J., Kim, E., Sheng, M., and Budnik, V. (1997). Essential role for *dlg* in synaptic clustering of *Shaker* K^+ channels *in vivo*. *J. Neurosci.* **17,** 152–159.

Wilson, G. F., and Chiu, S. Y. (1990). Ion channels in axon and Schwann membrane at paranodes of mammalian myelinated fibers studied with patch-clamp. *J. Neurosci.* **10,** 3263–3274.

Index

E

F

G

H

I

J

K

L

P

ISBN 0-12-153346-8

90065

9 780121 533465